Geomagnetic Observations and Models

IAGA Special Sopron Book Series

Volume 5

Series Editor

The International Association of Geomagnetism and Aeronomy is one of the eight Associations of the International Union of Geodesy and Geophysics (IUGG).

IAGA's Mission
The overall purpose of IAGA is set out in the first statute of the Association:

- to promote studies of magnetism and aeronomy of the Earth and other bodies of the solar system, and of the interplanetary medium and its interaction with these bodies, where such studies have international interest;
- to encourage research in these subjects by individual countries, institutions or persons and to facilitate its international coordination;
- to provide an opportunity on an international basis for discussion and publication of the results of the researches; and
- to promote appropriate standardizations of observational programs, data acquisition systems, data analysis and publication.

Volumes in this series:

The Earth's Magnetic Interior
Edited by E. Petrovský, D. Ivers, T. Harinarayana and E. Herrero-Bervera

Aeronomy of the Earth's Atmosphere and Ionosphere
Edited by M.A. Abdu, D. Pancheva and A. Bhattacharyya

The Dynamic Magnetosphere
Edited by W. Liu and M. Fujimoto

The Sun, the Solar Wind, and the Heliosphere
Edited by M.P. Miralles and J. Sánchez Almeida

Geomagnetic Observations and Models
Edited by M. Mandea and M. Korte

For titles published in this series, go to
http://www.springer.com/series/8636

Geomagnetic Observations and Models

Editors

Mioara Mandea

Institut de Physique du Globe, Paris, France

Monika Korte

GFZ German Research Centre for Geosciences
Potsdam, Germany

Springer

planetearth
Earth Sciences for Society

IAGA

Editors
Prof. Dr. Mioara Mandea
Université Paris Diderot
Institut de Physique du Globe de Paris
rue Thomas Mann 5
75205 Paris, France
mioara@ipgp.fr

Dr. Monika Korte
GFZ German Research Centre for Geosciences
Telegrafenberg
14473 Potsdam, Germany
monika@gfz-potsdam.de

ISBN 978-90-481-9857-3 e-ISBN 978-90-481-9858-0
DOI 10.1007/978-90-481-9858-0
Springer Dordrecht Heidelberg London New York

Cover illustration: Models of magnetic fields produced by the core or the lithosphere, are shown at the top of the core or respectively Earth's surface. These models are obtained from geomagnetic high accuracy data, provided by ground (magnetic observatories and repeat stations), aeromagnetic and satellite measurements. Figure produced by Martin Rother (Helmholtz-Zentrum Potsdam, Deutsches GeoForschungsZentrum - GFZ).

Printed on acid-free paper

Springer is part of Springer Science+Business Media (www.springer.com)

Foreword by the Series Editor

The IAGA Executive Committee decided in 2008, at the invitation of Springer, to publish a series of books, which should present the status of the IAGA sciences at the time of the IAGA 2009 Scientific Assembly in Sopron, Hungary, the "IAGA Special Sopron Series". It consists of five books, one for each of the IAGA Divisions, which together cover the IAGA sciences:

Division I – Internal Magnetic Field
Division II – Aeronomic Phenomena
Division III – Magnetospheric Phenomena
Division IV – Solar Wind and Interplanetary Field
Division V – Geomagnetic Observatories, Surveys and Analyses.

The teams of Editors of the books contain members of the IAGA Executive Committee and of the leadership of the respective Division, with, for some of the books, one or a few additional leading scientists in the respective fields.

The IAGA Special Sopron Series of books are the first ever (or at least in many decades) with the ambition to present a full coverage of the present status of all the IAGA fields of the geophysical sciences. In order to achieve this goal each book contains "overview papers", which together summarize the knowledge of all parts of the respective field. In book no. 5, on geomagnetic observations, all papers are of that kind. These major review papers are, in the other four books, complemented with invited reviews of special questions presented in Sopron. Finally, in some of the books a few short "contributed" papers of special interest are included. Thus, we hope the books will be of interest to both those who want a relatively concise presentation of the status of the sciences and to those who seek the most recent achievements.

I want to express my thanks to the editors and authors who have prepared the content of the books and to Petra van Steenbergen at Springer for good cooperation.

Kiruna, Sweden
November 2010

Bengt Hultqvist

Contents

Introduction

The magnetic field is one of the oldest observable Earth properties. The first measured geomagnetic field component, declination, was linked to the use of compasses, having their origin in the human curiosity of the north-pointing characteristics of loadstone. Starting with the 16th century measuring the Earth's magnetic field components (firstly only declination, thereafter also inclination and finally the full field vector) has become more systematic and kept improving over time.

Observation and modeling are the prerequisites to describe the Earth's magnetic field and to understanding the processes generating it. It is the determination of Division V, with its three working groups, to support

(i) the global efforts of measuring the geomagnetic field and its changes by
(ii) encouraging common data standards and global data availability and
(iii) providing global magnetic field models of core field and lithosphere.

The scope of this book is defined by the activities in the three working groups and covers the three topics: geomagnetic observation, data and modeling.

A network of ground-based geomagnetic observatories forms the backbone of Earth's magnetic field observation since Alexander von Humboldt's time in the early 19th century. Today these continuously recording stations are an important complement to the new wealth of data obtained from satellites observing the Earth's magnetic field from low Earth orbits. In the first chapter, an overview over the present observatory network is given with special emphasis on recent efforts to improve the global coverage.

The era of satellites observing the full magnetic vector field relatively close to Earth started in 1979/1980, when the MAGSAT satellite was in orbit for 6 months. Longer term field monitoring of this kind, however, only started 20 years later, when the Danish Ørsted, the German CHAMP and the Argentinian SAC-C satellites were launched in close succession. Both Ørsted and CHAMP have continued to provide global magnetic field data in 2010. A follow-up mission consisting of a constellation of 3 satellites is in preparation by ESA under the name of Swarm. All aspects of magnetic satellite missions are discussed in Chapter 2.

Satellites provide a limited resolution of the field distribution at the Earth's surface due to their orbit altitude. Repeat station measurements carried out in regular time intervals on well-defined locations supplement observatory recordings, which are spatially sparse in large parts of the world. Aeromagnetic and marine measurements are dedicated to detailed mapping of the (regional) lithospheric magnetic field.

Chapters 3 and 4 provide information on the purposes and techniques of these kinds of observations, highlighting recent activities.

Chapter 5 gives an overview over commonly used instruments to measure the Earth's magnetic field and particularly outlines new developments to facilitate the recording of high-quality geomagnetic field data.

Data quality, documentation and availability are the key requirements for suitable interpretation to gain a better understanding of all the geomagnetic field contributions and their temporal variations. For satellite data, the proper calibration of vector measurements is a challenging task. Data processing, however, is generally done by the institution in charge of the mission, ensuring a fixed data format, a homogeneous quality level and an easy data availability. All other kinds of observations, however, including those by geomagnetic observatories, are carried out by very different institutions and agencies in individual countries. Common standards for data quality, common data formats, and global data availability are difficult issues. Chapters 6 and 7 describe the efforts by the geomagnetic community, IAGA working groups and organisations like the World Data Centers to improve quality and global availability of geomagnetic data and ensure a good documentation of them.

Magnetic measurements from any platform in fact give the sum of all field contributions from the sources internal and external to Earth. A straightforward separation of the individual contributions is impossible and many scientific studies deal with different aspects of this problem. Approximate descriptions of the strength of different external variations, however, are provided by geomagnetic indices. These are obtained by standardized rules as special data products, mostly from geomagnetic observatory recordings. The planetary geomagnetic activity index Kp probably is the most widely known, but several other indices more suitable for special applications exist. Chapter 8 provides a comprehensive overview of magnetic indices and their relevance.

Geomagnetic field models obtained by inverse techniques from the measured data are widely used tools for studies of secular variation and the underlying processes in the Earth's core. Moreover, they can provide declination predictions for any location on Earth for navigational purposes. The most commonly used global modeling technique is spherical harmonic analysis. This method and other techniques applied to obtain global or regional models of the magnetic core or lithospheric field are described in detail in Chapter 9.

The following chapter is dedicated to an important IAGA product, the International Geomagnetic Reference Field IGRF. This field model consists of standardized descriptions of the geomagnetic core field, is updated every 5 years with predictive secular variation for the next 5 years, is easily available and field predictions from this model can be obtained interactively from several websites.

The IGRF is extremely useful for all applications where a standard reference is needed, but for scientific applications requiring the highest possible accuracy or including descriptions of some external or lithospheric field contributions, more specialized models are developed. After a brief overview, three examples of most recently obtained versions of this kinds of core field models are discussed in Chapter 11. The next chapter provides an overview over important results and findings regarding the geodynamo process as obtained from recent core field models. More about this topic can be found in the book by Division I within this series. Finally, Chapter 13 deals with mapping and interpreting the lithospheric field, including a brief summary of the enormous efforts undertaken by several international groups that resulted in the first global and digital map of magnetic anomalies, WDMAM, in 2007. Several examples of geological and tectonic interpretations are presented.

Our thanks go to all authors who have contributed to make this volume a true community effort and a comprehensive overview over the current status of observing and modelling the geomagnetic field. We also would like to express our thanks to the following reviewers, whose comments improved the original contributions: A. Balogh, A. Chulliat, G. Duma, C. Finn, R. Haagmans, A. Jackson, D. Jault, V. Korepanov, P. Kotzé, F. Lowes, G. Plank, M. Rajaram, A. Thomson, K. Whaler.

Paris, France
Potsdam, Germany
June 2010

Mioara Mandea
Monika Korte

Contributors

Sobhana Alex Indian Institute of Geomagnetism, Plot 5, Sector 18, New Panvel, Navi Mumbai 410218, India, salex@iigs.iigm.res.in

David R. Barraclough Formerly with British Geological Survey (now retired) Edinburgh, Scotland, drbarraclough@hotmail.com

David A. Clark CSIRO Materials Science and Engineering, Lindfield 2070, NSW, Australia, david.clark@csiro.au

Angelo De Santis Istituto Nazionale di Geofisica e Vulcanologia (INGV), V. Vigno Murata 605, 00143 Rome, Italy, angelo.desantis@ingv.it

Jérôme Dyment Institut de Physique du Globe de Paris, 1, rue Jussieu, 75005 Paris, France, jdy@ipgp.fr

Christopher Finlay Institut für Geophysik, ETH Zürich, 8092 Zürich, Switzerland, cfinlay@erdw.ethz.ch

Danielle Fouassier Observatoire Magnétique, 45340 Chambon la Fôret, France, fouassie@ipgp.jussieu.fr

Mohamed Hamoudi Helmholtz Centre Potsdam, GFZ German Research Centre for Geosciences, Telegrafenberg, F 407, 14473, Potsdam, Germany, hamoudi@gfz-potsdam.de

Donald C. Herzog National Geophysical Data Center/CIRES E/GC 325 Broadway, Boulder, CO 80305, USA, don.herzog@noaa.gov

Ivan Hrvoic GEM Systems Inc., 135 Spy Court, Markham, ON, Canada L3R 5H6, Info@gemsys.ca

Anca Isac Surlari National Geomagnetic Observatory, 077130 Moara Vlasiei Ilfov, Romania, margoisac@yahoo.com

Toshihiko Iyemori World Data Center for Geomagnetism, Kyoto, Data Analysis Center for Geomagnetism and Space Magnetism, Graduate School of Science, Kyoto University, Oiwake-cho, Kitashirakawa, Sakyo-ku, Kyoto 606-8502, Japan, iyemori@kugi.kyoto-u.ac.jp

Evgeny P. Kharin World Data Center for Solar Terrestrial Physics, Molodezhnaya, 3, Moscow, 117296, Russia, kharin@wdcb.ru

Stavros Kotsiaros DTU Space, Technical University of Denmark, DK-2100 Copenhagen Ø, Denmark, skotsiaros@space.dtu.dk

Weijia Kuang Planetary Geodynamics Laboratory, NASA Goddard Space Flight Center, Greenbelt, MD, USA, Weijia.Kuang-1@nasa.gov

Vincent Lesur Helmholtz Centre, GFZ German Research Centre for Geosciences, 14473 Potsdam, Germany, lesur@gfz-potsdam.de

Hans-Joachim Linthe Helmholtz Centre, GFZ German Research Centre for Geosciences, Potsdam, Adolf-Schmidt- Observatorium Niemegk, D-14823 Niemegk, Germany, linthe@gfz-potsdam.de

Susan Macmillan British Geological Survey, Edinburgh EH9 3LA, UK, smac@bgs.ac.uk

Aurélie Marchaudon LPC2E (Laboratoire de Physique et Chimie de l'Environnement et de l'Espace), CNRS et Université d'Orléans, 45071 Orléans cedex 2, France, aurelie.marchaudon@cnrs-orleans.fr

Jürgen Matzka DTU Space, National Space Institute, Technical University of Denmark, 2100 Kobenhavn, Denmark, jrgm@space.dtu.dk

Susan McLean National Geophysical Data Center E/GC 325, Broadway, Boulder, Colorado, USA, Susan.McLean@noaa.gov

Michel Menvielle LATMOS-IPSL (Laboratoire Atmosphères, Milieux, Observations Spatiales), Université Versailles St-Quentin, CNRS/INSU, Univ. Paris Sud, Boîte 102, 4 place Jussieu, 75252 Paris Cedex 05, France, michel.menvielle@latmos.ipsl.fr

Lawrence R. Newitt Boreal Language and Science Services, 310 1171 Ambleside Drive, Ottawa, ON, Canada K2B 8E1, lnewitt@sympatico.ca

Masahito Nosé World Data Center for Geomagnetism, Kyoto, Data Analysis Center for Geomagnetism and Space Magnetism, Graduate School of Science, Kyoto University, Oiwake-cho, Kitashirakawa, Sakyo-Ku, Kyoto 606-8502, Japan, nose@kugi.kyoto-u.ac.jp

Nils Olsen DTU Space, Technical University of Denmark, DK-2100 Copenhagen Ø, Denmark, nio@space.dtu.dk

Michael E. Purucker Raytheon at Planetary Geodynamics Lab, Goddard Space Flight Center, Code 698, Greenbelt, MD 20771, USA, michael.e.purucker@nasa.gov

Yoann Quesnel CEREGE – Université Paul Cézanne Aix-Marseille III, 13545 Aix-en-Provence cedex 04, France, quesnel@cerege.fr

Jean L. Rasson Institut Royal Météorologique de Belgique, Centre de Physique du Globe, Dourbes, Viroinval, B-5670 Belgium, jr@oma.be

Sarah J. Reay British Geological Survey, Murchison House Edinburgh EH9 3LA, UK, sjr@bgs.ac.uk

Jan Reda Institute of Geophysics, Polish Academy of Sciences, Warsaw, Poland, jreda@igf.edu.pl

Jean-Jacques Schott Ecole et Observatoires des Sciences de la Terre, Université de Strasbourg, F-67084 Strasbourg, France, jj.schott@eost.u-strasbg.fr

Natalia A. Sergeyeva World Data Center for Solid Earth Physics, 117296, Moscow, Russia, nata@wdcb.ru

Andrew Tangborn Joint Center for Earth Systems Technology, University of Maryland, Baltimore County, Baltimore, MD 21250, USA, tangborn@umbc.edu

Erwan Thébault Institut de Physique du Globe de Paris, CNRS/INSU UMR7154, Université Paris Diderot, F-75005 Paris, France, ethebault@ipgp.fr

Alan W.P. Thomson British Geological Survey, Edinburgh EH9 3LA, UK, awpt@bgs.ac.uk

Hiroaki Toh Data Analysis Center for Geomagnetism and Space Magnetism, Graduate School of Science, Kyoto University, 6068502 Kyoto, Japan, toh@kugi.kyoto-u.ac.jp

Christopher W. Turbitt British Geological Survey, Edinburgh EH93LA, UK, cwtu@bgs.ac.uk

Dongmei Yang Institute of Geophysics, China Earthquake Administration, Haidan District, Beijing 100081, China, yangdm@cea-igp.ac.cn

Chapter 1

The Global Geomagnetic Observatory Network

Jean L. Rasson, Hiroaki Toh, and Dongmei Yang

Abstract The paper provides a snapshot of the global geomagnetic observatory network at the time of the Sopron IAGA Assembly in August 2009. The need for these observatories is explained and the evolution and outlook of the observing techniques are examined. We present three projects addressing the upgrade of existing observatories, the creation of new ones and the problems of observing in the oceanic regions.

1.1 The Network at the Time of the Sopron IAGA Assembly (August 2009)

1.1.1 Introduction

Magnetic observatories are something rather mysterious to the layman and even to many of our fellow scientists. As the human being lacks a sense for detecting and appreciating a magnetic field, the mere activity of "observing" it, sounds strange if not outright weird to the ordinary citizen. Yet the magnetic observatories have been around for more than 500 years and they were one of the first institutions involved in monitoring a "Global Change" on our planet.

Good magnetic observatories are needed more than ever for global modeling and navigation. Magnetic satellite missions, once said to be the death of ground based observations, are now demanding quality data from fixed observation points on the Earth. A good magnetic observatory is a place where precise, continuous long-term measurements of the geomagnetic field are made and from where definitive data are regularly published to the wider scientific community.

The art of observing the field has evolved over the years and today, as the 2009 Sopron IAGA assembly shows, many efforts are underway to improve the way we do it: new ways of observing the field are devised, instrumentation is continually improved and the efforts required to observe are lessened by easier observation procedures and improved instrument controls.

Magnetic observatories are important because they provide data for scientific research and practical applications. They have therefore a unique position in geomagnetism where their role impacts almost equally academia and commercial activity. However, experience shows that the "big money" rarely has benefitted the efforts to create, improve and upgrade observatories. Instead it is mainly the efforts of (a group of) motivated individuals, working within scientific institutions, with or without project monies, that manage to make things go forward. An essential component here is the availability of dedicated local staff and suitable estates and buildings for the (new) observatories.

Basically we see the observatory network as one big planetary facility and we continuously try to make it better by improving:

- The coverage: this means installing more observatories on the planet
- The distribution: by trying to cover the planet evenly with observatories

J.L. Rasson (✉)
Institut Royal Météorologique de Belgique, Centre de Physique du Globe, No 2 Rue du Centre de Physique, Dourbes, Viroinval, B-5670 Belgium
e-mail: jr@oma.be

M. Mandea, M. Korte (eds.), *Geomagnetic Observations and Models*, IAGA Special Sopron Book Series 5,
DOI 10.1007/978-90-481-9858-0_1, © Springer Science+Business Media B.V. 2011

- The observations proper: better instruments, staff, accuracy, lower instrument noise in time and amplitude, continuity of data
- The storage and forwarding of data: data centers, availability of data

Our paper will deal with the scientific contributions presented at the Sopron IAGA 2009 Assembly regarding this observatory network. But principally we will deal with the improvements as explained above. We will first review the INDIGO project, a collaboration between the British Geological Survey, Institut Royal Météorologique de Belgique and agencies in all the continents to improve existing observatories and help create new ones. Next, the China Earthquake Administration's (CEA) centralized effort to upgrade about 30 magnetic observatories will be presented as a case study. Finally the problem of covering the ocean regions on Earth with observatories is addressed in "Filling the Gaps".

1.1.2 Highlights from the Sopron IAGA Assembly in 2009

The Assembly presented a rich programme dealing with magnetic observatories in the world network. During the Assembly lectures, Dr. C. Reeves pointed out in "Geomagnetism and the Exploration of Global Geology" that "a programme lasting decades needs also to eliminate secular variation from its results so that adjacent survey blocks may be stitched together objectively. It is a singular success for IAGA that the International Geomagnetic Reference Field (IGRF) has been adopted almost universally by magnetic anomaly surveyors to achieve this". On another note, the Assembly lecture by Dr A Rodger "The Mesosphere as a Link in Sun-Climate Relationships" revealed to us the possibility to measure the magnetic field in the mesosphere (altitude 50–100 km) from an Earth based station, using a laser to inspect the sodium atoms there [Moussaoui et al., 2009]. Do we witness a new observation method in the making here?

During presentations and discussions, it turned out that the geomagnetic observatory data user community requested increasing data and products from magnetic observatories, leading to new tasks for them:

- Proposal for a new observatory data product: quasi-definitive data
- Towards a metadata standard for geomagnetic observatory data
- 1-second INTERMAGNET standard magnetometry

The "Quasi Definitive Data" is required in order to speed-up the delivery of good quality data to users unable or unwilling to wait until the beginning of the next year when observatories traditionally release their definitive data. With this new product, quasi definitive data should be available after a few months only. Data users also realize that the phasing out of yearbooks, with the advent of exclusive digital databases, leaves them in the dark about many events and changes occurring in geomagnetic observatories. Therefore a metadata standard would come in handy to show the way as to how to replace or complement the observatory yearbooks. A growing number of observatories collect data at a 1 Hz sampling rate, because there is a demand for higher time resolutions from the user community. INTERMAGNET is in the process of defining a standard for this 1 Hz variometric data, and manufacturers are now proposing instruments able to comply with its 0.01 s timing accuracy and 0.01 nT noise limit [Korepanov et al., 2007].

Also the widespread network and good availability of geomagnetic observatory data resulted in it being used as proxies in the following cases:

- Expected trends in the geomagnetic sq field due to secular changes in the earth's magnetic field allow monitoring of the long term changes in the ionosphere
- Ionospheric reflection of the magnetic activity described by the new index η: Ionospheric data are sparse in time and in space in opposition to the magnetic data
- Geomagnetic activity (Ap index) and polar surface air temperature variability

1.1.3 The Network

The network is presented in Fig. 1.1, with a distinction made for the observatories having attained INTERMAGNET certification (more details at www.intermagnet.org). As is to be expected, a non

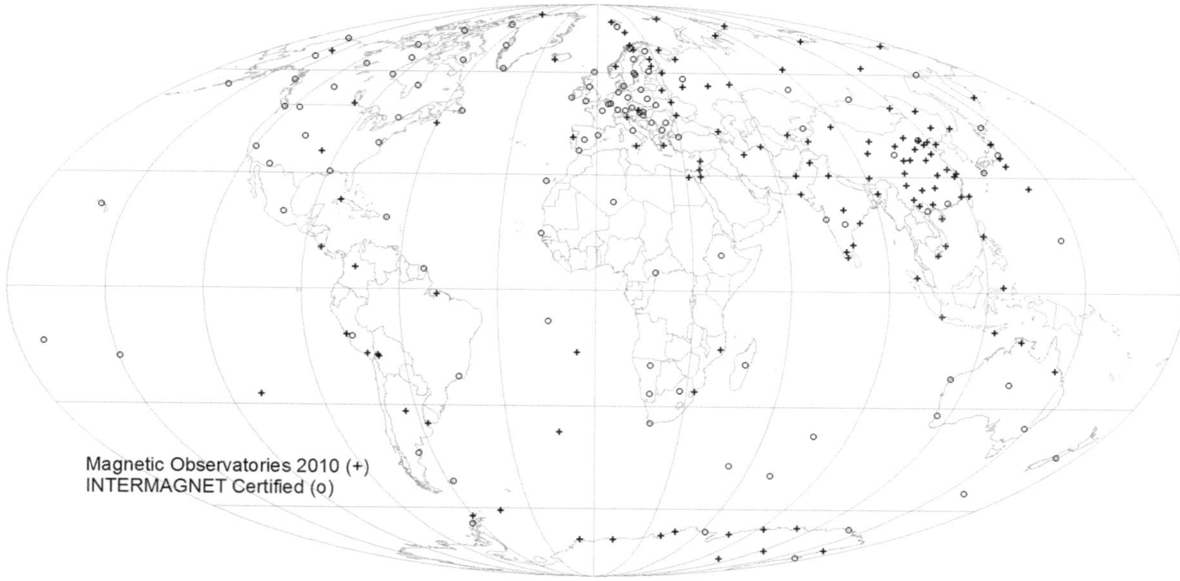

Fig. 1.1 The global network of active geomagnetic observatories as it exists beginning 2010 (Mollweide projection). The names and IAGA codes of these observatories can be found on http://www.meteo.be/IAGA_WG_V.1/ along with the station details like latitude and longitude, elevation and operation years

uniform distribution of the observatories on the globe is clearly seen.

There is a bias of more observatories on the emerged lands and also in the northern hemisphere. It is noteworthy to see that the Chinese network almost reaches the density of stations that we witness in Europe. Our Chapter 2 therefore looks into the centralized effort China is deploying to build its high quality observatories.

The interest for geomagnetic data from the South Atlantic anomaly is growing and following the renovation of the Vassouras and Trelew observatory in 1993 [Rasson et al., 1996], many programs targeted that region to install or upgrade observatories there, including the difficult to access St Helena and Tristan da Cunha islands [Korte et al., 2009].

1.1.4 INDIGO: Better Geomagnetic Observatories at the Right Place

The INDIGO project aims to provide the means for achieving quality observatories in a selected set of places on the Earth. Equipment, software, training and data processing, when missing on site, are given to colleagues worldwide so that they might improve or start their own geomagnetic observations.

Usually existing premises are used and/or reconfigured to provide an adequate hosting of the INDIGO equipment (Fig. 1.2). If necessary, local staff are trained in observatory operations and observing skills.

The installation of a successful observatory also depends on the motivation of the local observatory staff; with the help of the INDIGO project, most common problems can be overcome.

A state-of-the-art magnetic observatory is an expensive piece of infrastructure in equipment and in manpower, since automatic observations are not possible at the moment. Therefore the good ones tend to be clustered in richer countries. INDIGO tries to provide help where there are gaps in the observatory world map.

The INDIGO effort has therefore been directed towards Asia, Africa and Latin-America (Fig. 1.3).

The original objective of the INDIGO project was to make use of fifteen EDA fluxgate variometers donated to the British Geological Survey (BGS), coupled with a low power digitiser to make filtered one-minute digital daily recordings.

Fig. 1.2 Pictures of INDIGO installations. *Top*: Variometer house at Maputo, Moçambique. The newly erected variometer house in Pelabuhan Ratu, Indonesia. *Middle*: Arti observatory in Russia. *Bottom*: the new Kupang observatory in Timor-West, Indonesia

Use of the more precise and stable DMI variometer was later adopted for some installations. Absolute instruments like ZEISS, Ruska and Tavistock Difluxes as well as Geometrics and GEM proton magnetometers are used.

The digitizer is based on a 16 bits ADAM ADC module and a GPS receiver with all sampling controlled by a PIC16F877 microcontroller programmed in BASIC or C. It also controls the optional proton magnetometer.

The data logger is based on a JAVA platform. Usually a PC is used with two programs running in parallel: EDA2GDAS and GDASVIEW. The former communicates with the latter and with the digitizer, so that filtering and formatting to various file formats (including INTERMAGNETs imfv1.22) is performed along with graphical display (Fig. 1.4).

More recently it was realized that it would be possible to do away with the PC altogether, by connecting a USB memory stick directly to the digitizer through a custom stick writer. This allows us to log the data in the required format without use of a PC. This results in huge savings in terms of power and cost and would allow a modest battery to power the system for days. Monitoring of the data log can be done on an optional PC running a new piece of java soft: INDIGOwatch.

The relative success of the INDIGO project has made possible the availability of its data to the scientific community. Although the definitive data has yet to be processed, preliminary results are encouraging.

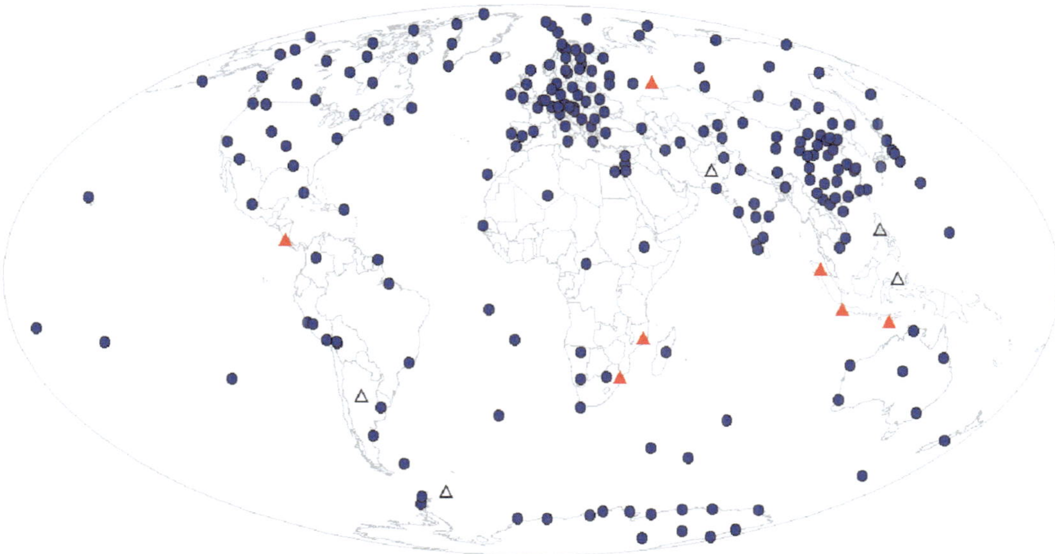

Fig. 1.3 Location of the INDIGO observatories: *Blue dots*: existing observatories, *red triangles*: INDIGO installations, *white triangles*: projected actions

Fig. 1.4 Typical INDIGO hardware: *Clockwise from top left*: The EDA triaxial fluxgate sensor in Nampula; The Tavistock DIflux operated in Karachi; Absolutes in Nampula with a Ruska DIflux and a Geometrics proton; DMI and GEM recording magnetometers in Kupang, Timor-W; EDA console and INDIGO logger in Maputo, Mozambique

Variometer data has been delivered to commercial companies, and in situ absolute measurements have allowed the computation of annual means where none were available previously (notably South-East Asia).

The data is stored on a BGS server and is currently accessed with a password. There is also a website centralising all present and past data on the different INDIGO observatories: instruments in use, serial numbers, scale values, preliminary baselines, monthly bulletins, site plans, pictures and history.

The goal of INDIGO, at this stage, is the creation or upgrade of its magnetic observatories with

INTERMAGNET certification. INDIGO observatories have not yet reached the stage where the certification criteria can be met, but some are very close. The main short term objective is training the observatory staff in preparing their definitive data.

Several new observatories are actively involved in becoming INDIGO's: Quetta (Pakistan), Muntinlupa (Philippines), Pilar and Orcadas-del-Sur (Argentina), Tondano (Indonesia).

We foresee more observatories becoming involved, particularly in Africa where difficult conditions are threatening the operation of some of the few existing observatories. Such observatories often benefit from lower cultural noise in comparison to established Western observatories; they are far more sparsely distributed than Western observatories, and are located close to regions of particular scientific interest such as the dip equator or South-Atlantic anomaly.

New INDIGO equipment is constantly being projected and developed: full solar panel electrical supplies, RF wireless transmission of data, MkIII data logger.

INDIGO is the brainchild of retired but very active colleague John C Riddick, C Turbitt and S Flower from BGS together with one of us (JLR) [Ameen et al., 2009].

1.2 Advances in a Newly Upgraded Network: The China Earthquake Administration (CEA) Effort

1.2.1 Observatories in China: Short History Up to Twentieth Century

There were four major periods for the development of the magnetic observatories in China's mainland.

- Before the International Geophysical Year (IGY)

  The first magnetic observatory in China was constructed in Beijing in 1870 by Russia. It ceased working in 1882. There were six other magnetic observatories with different periods of operation but all of them ceased working before 1944. One exception was Sheshan (SSH) Observatory which was constructed in 1874 by the French missionary and is still in operation now [Zongqi Chen, 1944].

- During the IGY

  During the IGY, construction of seven magnetic observatories including Beijing (BJI), Changchun (CNH), Wuhan (WHN), Guangzhou (GZH), Lanzhou (LZH), Lhasa (LSA) and Urumqi (WMQ) observatories was initiated. Together with SSH, these eight observatories became the backbone of the Chinese magnetic network.

- From 1966 to 1979

  The 1966 Xingtai Earthquakes triggered the beginning of the Chinese research on earthquake prediction. Various observation methods were utilized including magnetic observation. More than 200 magnetic stations were set up around the country in the following years. These stations were sponsored by different organizations and/or institutions with a variety of observational procedures and quality controls [Rongsheng Gao et al., 1999].

- After 1979

  In 1979, all of the magnetic observatories and stations were put under the administration of the SSB (the State Seismological Bureau, now the CEA). An organization (now the Geomagnetic Network of China) in the Institute of Geophysics, SSB (IGSSB, now IGCEA) was responsible for the technical support and data management of the magnetic network. The network was readjusted several times taking into account the distribution and the observational environment of the stations [Anlong Cheng et al., 1990]. Half of the stations ceased working. Meanwhile, instrumentation at 29 of the stations was improved to make them operate as observatories. Seven observatories out of these 29 joined the existing 8 older observatories to form the primary observatories. Data from these 15 observatories have been being archived in the World Data Center system ever since.

1.2.2 Planning the Major Upgrade: Towards INTERMAGNET Standards

In the late 1990s, the CEA began to consider the modernization of the geomagnetic network. Taking into account the importance of continuous reliable geomagnetic data, they started cautiously at the beginning. Twenty one sets of the Chinese GM-3 type triaxial fluxgate magnetometers were installed in twenty observatories and 33 sets of the Chinese CTM-DI type fluxgate theodolites were installed in 33 observatories in 2001. The tri-axial fluxgate magnetometers had been running simultaneously with the analog variometers at the observatories for several years to compare data from the old system and the new system in order to achieve and accumulate experience about the modern techniques in fluxgate magnetometer observation, data processing, data quality control and dissemination. After that, the CEA succeeded in finding financial support from the central government to upgrade the whole network towards INTERMAGNET standards. At the end of the year 2007, every observatory has been equipped with at least two tri-axial fluxgate magnetometers to make variation recordings, one fluxgate theodolite and one proton precession magnetometer to make absolute measurements. For the most important observatories, there is a Chinese GM-4 type tri-axial fluxgate magnetometer, a Danish suspended FGE magnetometer, a Chinese CTM-DI fluxgate theodolite, a Hungarian MINGEO DIM fluxgate theodolite, a Chinese proton magnetometer and also continuous recording of total field by the Canadian GSM-19F Overhauser effect magnetometer. The sampling rate for the tri-axial fluxgate magnetometers and Overhauser effect magnetometers is 1 Hz. A GPS signal is applied in the data acquisition system to give accurate time stamps.

1.2.2.1 Description of Present and Future Network

At present, there are 40 magnetic observatories in China's mainland (Fig. 1.5) sponsored by the CEA (China Earthquake Administration, formally China Seismological Bureau (CSB) and State Seismological Bureau (SSB)). All of them are equipped with at least

Fig. 1.5 Distribution of the magnetic observatories in China's mainland

two tri-axial fluxgate magnetometers, one fluxgate theodolite and one proton precession magnetometer. For the most important observatories, an additional fluxgate theodolite and Overhauser effect magnetometer are also deployed, the latter to continuously record the total field.

1.2.2.2 Criteria for Geographical Distribution

Taking into account the features and changes of the main magnetic field and the external magnetic field, the whole network was designed to be distributed with proportional spacing. The distance between adjacent observatories is expected to be around 1000 km for the primary observatories and around 500 km for the secondary ones. If possible, a backup observatory is maintained near the primary observatory who might suffer from anthropogenic disturbances in the near future so as to replace it in case that the anthropogenic disturbances become severe. For examples, the Jinghai observatory is to replace Beijing observatory and Chongming observatory to replace Sheshan observatory although Jinghai and Chongming also suffer from slight influences due to DC electric trains.

The relatively short distances among the observatories have advantages. For example, the detection of errors in data by using the method of interobservatory comparison is justifiable if the observatories are close together. It also helps us to understand some detailed changes in both the internal part and external part of the geomagnetic field.

1.2.2.3 Instrumentation

- Past

  Before the year 2001, all the observatories used analog variometers to make variation recording and the absolute instruments were quite different from one observatory to another. In late 2001, twenty observatories began to use fluxgate magnetometers for variation recording and all of the observatories began to use fluxgate theodolites and proton precession magnetometers for absolute measurements.
- Now

  With the upgrade of the network, by the end of the year 2007, every observatory has been equipped with at least two tri-axial fluxgate magnetometers to make variation recording, one fluxgate theodolite and one proton precession magnetometer to make absolute measurement. For the most important observatories, there is a Chinese GM-4 type tri-axial fluxgate magnetometer, a Danish suspended FGE magnetometer, a Chinese CTM-DI fluxgate theodolite, a Hungarian MINGEO DIM fluxgate magnetometer, a Chinese proton magnetometer and also continuous recording of total field by the Canadian GSM-19F Overhauser effect magnetometer. The sampling rate for the tri-axial fluxgate magnetometers and Overhauser effect magnetometers is 1 Hz. The GPS signal is applied in the data acquisition system to give accurate time stamps.
- Future

  It is in discussion that instruments with sampling rate $\geq$10 Hz and resolution $\leq$0.01 nT might be used in the network to enhance the application of magnetic observations in earthquake monitoring and prediction.

1.2.2.4 Buildings: Describing the Way the Buildings are Designed and How Many Per Observatory [China Earthquake Administration, 2004]

At each observatory, there is at least one absolute house (AH) (Fig. 1.6 and Fig. 1.7) and one variation room (VR) (Fig. 1.6) of different sizes. Usually there are six pillars in the AH (Fig. 1.8) and four pillars in the VR for the primary observatories, and four pillars in the AH and 2 pillars in the VR for the secondary ones. The minimum distance between adjacent pillars is three meters. In the absolute house, this distance will be larger because the pillars are not arranged in a line in order to make the azimuthal marks being visible at each pillar. In most cases the VRs are underground or semi-underground chambers to help keeping the temperature stable in the room.

For the observatories which are located in important regions and are expected to be a place for instrument testing or training courses, there are also laboratory buildings in different form.

For observatories also equipped with a proton vector magnetometer (PVM), there will be a hut for the PVM

Fig. 1.6 Overview of the Quanzhou magnetic observatory. 1: Absolute house, 2: Proton vector magnetometer hut, 3: Variation room, 4: Electric hut, 5: Instrument comparison hut and 6: Office

Fig. 1.7 The Quanzhou magnetic observatory absolute house

(Fig. 1.6). This hut should be 30 m away from the AH, VR and laboratory.

All the above facilities are made of magnetic free materials.

It is recommended that each observatory should set up two azimuthal marks. The main mark should be at least 150 m away from the AH to achieve a high accuracy in declination measurements. The backup mark can be a little bit nearer to the AH. It will take the place of the main mark in bad weather when the main mark is not visible. The two marks are used in the routine

Fig. 1.8 The Quanzhou magnetic observatory absolute house featuring the pillars for the absolute measurement of the geomagnetic vector

absolute measurements and the consistency between the observations is checked continuously.

1.2.2.5 Staff: Training Level, Quantity

Usually, besides the magnetic observation, there are other kinds of measurements at the magnetic observatories such as telluric electric field, magneto-telluric observation, crustal deformation, water level and water temperature etc. The observatory staffs are required to maintain all the instruments, process data and maintain logs to tell what they do and why they do it in data processing. In case that the instrument is not working properly, they should try to find out what is wrong. If they can not solve the problem, they should report to the administration department as soon as possible. So in the common cases, there are at least four staffs at an observatory. At least two of them should be good at absolute measurements.

There are three kinds of training courses provided to the observatory staff.

The first one is open to the whole network and is organized in a college sponsored by the CEA. The courses include fundamental theories related to observation and interpretation of data, such as geophysics, signal processing techniques, electronics and computer science. This kind of training is a kind of routine training and takes one and a half months.

The second one is also open to the whole network. It is organized by the Geomagnetic Network of China and focuses on some special topics related to the practical operation such as the techniques on absolute measurements, instrument calibration, data processing and log records. This kind of training lasts one week and aims at solving certain technical problems revealed in the annual reviews of the operation of the network. In cases that operation of an instrument is necessary for trainees, the training will be organized in small groups. For example, a training course was held to share experiences and exchange ideas after one year of running the new digital magnetic network in late May, 2009 with 104 participants from magnetic observatories and stations. Trainings on calibration of proton vector magnetometers were carried out simultaneously at Lanzhou, Wuhan and Jinghai observatories in October and November 2009.

More than seventy participants went to one of the three observatories in groups including around ten people.

The third one is open to regional network and is usually organized by one local earthquake administration (LEA) or jointly organized by several LEAs. This kind of training helps the participants to discuss in much more detail the problems encountered in work. Usually experienced observers from observatories will give lectures during the training course.

1.2.3 Modern Centralized Data Processing

1.2.3.1 A Centralized Approach

There are four levels of information nodes in the network: the observatories, the local earthquake administrations (LEAs), the China Earthquake Network Center (CENC) and the Geomagnetic Network of China (GNC). The Oracle Database is used at each node to archive data including metadata, observations, data processing logs, and data products. The technique of the Oracle advanced replication (materialized view) is used to transfer data from lower node to upper node. Except for GNC, the three nodes at observatories, LEAs and CENC archive all kind of observation data including geomagnetic, telluric, crustal deformation, water level and temperature data etc.

Each variometer has internet access and the observation data is transferred from the data acquisition systems (DAS) to the Oracle database automatically once per day. The observatory staff will check the database each morning. If the automatic procedure failed, they collect data manually from the DAS and transfer data to the database. After that, they will view the daily variation, remove artificial influence in the record and write in the log to tell which part of the data is modified, how and why the modifying is done and who is responsible for the work. All these data are transferred to the upper nodes once a day.

On Monday and Thursday afternoon every week, the staff will make absolute measurements at least twice with each set of the DI fluxgate theodolites. The baseline values and quasi-definitive data will be calculated on the following day and are archived in the Oracle database. These data will begin to be transferred automatically to the upper nodes in March 2010.

On Tuesday and Friday afternoon, the staffs are asked to view the intranet WEB site of GNC to see if there are still any bad data in the variation recordings from the GNC's point of view. This helps the staff to identify artificial influences or abnormal behavior of instruments that are seldom seen. If the staffs agree with the GNC, they can correct the data and update the log in their database and the updated data and log will be transferred to the upper nodes automatically. If they do not agree with the GNC, they can initiate a discussion via the forum on the intranet WEB site. After the discussion, the side who wrongly interprets the data will correct their mistake.

Besides the routine work on quality check of the observations in the network, the GNC is in charge of producing K indices, catalogues of magnetic storms and figures of spatial distributions of typical parameters of magnetic field such as annual changes of the main field, daily variation, daily range, amplitude of SCs and storms and so on. The catalogues and figures are published on the web pages.

1.2.3.2 Data Quality Assessment: How-to

Data quality assessment is done from five points of view (Fig. 1.9):

- Integrity (behavior of the instrument and power supply)
- Mistakes in data processing and log and how soon the mistakes are corrected (working quality of the local staffs)
- Noise level (background)
- Efficiency in absolute measurement (working quality of the local staffs)
- Reliability (observational environment)

We examine them individually:

- Integrity

  The integrity of each set of the variometers is calculated monthly and yearly and the reason for missing data is analyzed. These are retrieved from the log record in the database filled by the observatory staffs. The log includes the following records:

  - Date*
  - Observatory ID*

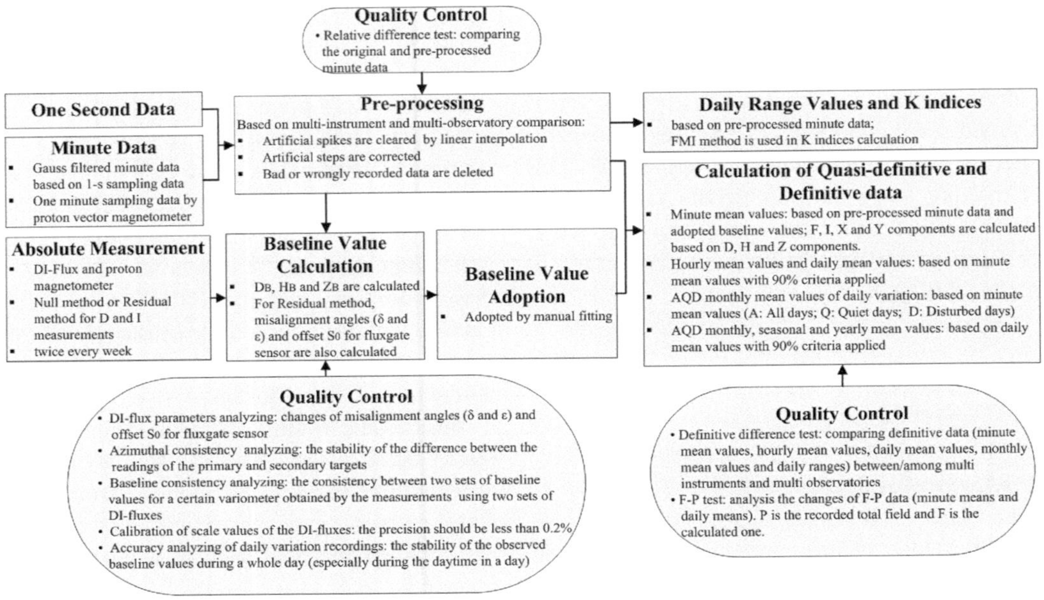

Fig. 1.9 The flowchart for the centralized quality control of geomagnetic observatory data

- Point ID (Instrument ID)*
- Component ID*
- Starting time for missing data*
- Ending time for missing data*
- Duration, in minutes, for missing data*
- Type of missing data (0 for originally missing; 1 for deleted by observatory staff)
- Event ID for missing data
- Event description for missing data
- Person responsible for editing the record
- Date and time for the filled record*
- Version of the software*

Records for items with * are written in the database automatically by the data processing software. Others are written by the observatory staff. There are standard expressions for the event IDs and event descriptions. They have been included in the data processing software. The observatory staff will select the correct expression in the list. In this way, when talking about the same event, the event records for different observatories are the same. That makes the statistical analyses quite easy and accurate.

- Mistakes in data processing and log and how soon the mistakes are corrected

The GNC staff checks the daily variations for the past three or four days every Tuesday morning and Friday morning. They stamp marks to each record of daily variation to tell the quality of the record. For example, mark "9" means the record is good; mark "4" means part of the recording is bad and should be modified; mark "2" means the record is OK but we cannot use it to calculate daily range because recording of daily extreme is missing; mark "3" means the recording is totally wrong or all data is missing so no further calculation can be made for this record. This procedure is called "first check".

As soon as the stamped marks are saved in the database during the quality check procedure, everyone who is interested in the results will be able to see the table shown on the intranet website of GNC. If they do not agree with the GNC, they can publish their opinion on the forum on the website. The GNC will check the notes on the forum every Thursday morning and Monday morning. If the GNC is wrong, they will correct the mark. Otherwise they will explain the reason why they think the record is not good. This procedure is called "inquiry and answer".

Every two weeks, the GNC will return back to the records two weeks ago to see if the bad records found in first check have been corrected. If the staff corrected the record, then mark “8” will be given to it. If the record has been modified but still not good, mark ‘5’ will be stamped to it. This procedure is called “second check”.

In the middle of a month, the GNC will return back to the last month to see if the bad records found in first check and second check have been corrected. Mark “7” will be given to the newly arrived correct data and mark “6” to the newly arrived but still not correct data. This procedure is called “monthly check”.

By the end of March, the annual check will be made for the last year. All the data record for last year will be copied to a backup table. Mark “D” will be given to newly received correct data and mark “W” to the newly received bad data.

In this way, we can understand the existing mistakes in data processing, how soon the mistakes are corrected, and whether mistakes of data processing are a problem to the whole network or only to some special observatories.

The quality of the log is checked only monthly and annually. Usually we tend to believe what the local staffs fill in the log but do not permit too much “reason unknown” or “other reasons” without detailed explanation. In case that event ID is obviously wrong, we will contact that observatory and discuss with the local staffs.

- Noise level

  It is important to understand the weakest magnetic signal that can be detected at an observatory. The following method is used in the magnetic network to evaluate the background noise level at observatories.

  Firstly, five extremely quiet 3 h intervals for each month are chosen. Initially, four observatories taken from the typical positions in the southern, northern, eastern and western China are selected to calculate the 3 h standard deviations of the first differences for H, D, Z and F components for the whole month. Combining all the standard deviations, five 3 h time periods are determined during which the magnetic field is quietest.

  Secondly, the first differences of the five 3 h recordings are calculated for each component and each observatory.

  Thirdly, maximum peak-peak values of the first differences, with the top 20% being ignored, are calculated for each component and each observatory.

  Fourthly, noise level is estimated by computing averages for the five maximum peak-peak values for each component and each observatory.

  Fifthly, the spatial distribution of the noise level is analyzed to find the general changes of noise level related to latitude and longitude. If the noise level for a certain observatory is quite different from the general levels for other observatories, this observatory will be asked to find the reason and solve the problem.

- Efficiency in absolute measurement

  Regularly, the observatory staffs are asked to make absolute measurements on Monday afternoon and Thursday afternoon every week. Each time, at least two measurements for each set of DI-flux’s should be made. If the two observed baseline values do not agree well with each other, then repeat measurements should be made. If the observed baseline values drift much from the latest ones three or four days ago, then the observers should try to find out if the observer is magnetically clean, if there is any magnetic pollution near the absolute house or pillars, if the drift coincides with the temperature change in the variation room or if the DI-flux is in good status. After they find the reason and solve the problem as much as they can, they need to repeat the absolute measurements and fill in the absolute log. This quality assessment is designed to determine the following: if the absolute measurements have been made frequently enough and the repeated measurements have been made when necessary and if the observed baseline values are reasonably accurate.

- Reliability

  The reliability is checked from two points of view. One is concerned with the accuracy of the daily variations. The other is the reliability of the definitive data.

  On the accuracy of the daily variations, we intend to understand how exactly the daily variations are recorded by the variometers. In addition to the above careful checking of the daily variations, every observatory is asked to make absolute measurements at a frequency of once per hour in the daytime once in a year. By calculating the peak-peak value and standard deviation and analyzing the temporal changes of the baseline values in this day, the performance of the magnetometer will be clear.

Analysis of the reliability of the definitive data starts with the midnight means of the definitive data.

Firstly, plots of groups of the midnight means and their residuals (subtracting the reference ones taken from a certain observatory) are viewed and distinctly bad data are given stamp "E".

Secondly, the logs for absolute measurement are checked. If the bad data came from artificial influences, then the stamp "E" is changed to stamp "W" which means the data is wrong and can not be improved. Otherwise, we go to the baseline values to see if the adopted baseline values are suitable or the observed baseline values are reasonably correct.

Thirdly, we make a linear fit to the data without stamps "W" and "E", calculate the annual linear drifts and plot the contour map of the annual drifts. From the contour map we will see if the linear change at a certain observatory agrees with the adjacent observatories. If not, data for that observatory will be checked in detail to try to find the reason.

Finally, we analyze the seasonal change and 27-day variation from the midnight data with linear trend being removed. This is something not exactly related to quality assessment but it can help us to understand more about the data and the importance to have continuous good data. For example, it was first thought that definitive data is commonly used to study the secular variation of the main field so it is not a serious problem for an observatory to have bad data for several days in a year. But when we studied the 27-day variation using the Chree superposed-epoch method, it was realized later that the lack of data in some days was a big problem. Realizing the importance and application of the data will help the observers to perfect their work and try their best to reduce the numbers of missing data and bad data.

1.3 Filling the Gaps: Sea-Bottom Observatories

1.3.1 Rationale

There are more than one hundred geomagnetic observatories that are presently active over the globe. However, a major problem of the world's ground geomagnetic observation network is its biased spatial distribution, viz., the observatories are mostly operated not in the oceans but on the continents, and in the northern hemisphere rather than in the southern hemisphere. This is mainly because the network has grown spontaneously without any agreed international guidelines for suitable observatory locations. It has lead to a concentration of high-quality geomagnetic observatories mostly in developed countries, where governments manage to understand the importance of monitoring our planet's magnetic field such that they are able to operate one or several observatories per nation for periods ranging from decades to more than 100 years. The most fatal outcome of this concentration is that the oceans, which occupy more than 70% of our planet's surface, are left almost unattended.

It, therefore, is desirable to extend the network to the seafloor in order to improve our knowledge of the spatial distribution of the geomagnetic field, which in the past has been very important for navigation purposes and now is still indispensable for monitoring the Earth's electromagnetic (EM) environments. The deep seafloor has not yet been made use of as an observing platform into the Earth's electrical structure by EM induction methods such as Magnetotellurics, which gives us a biased image of the true electrical Earth. This alone can constitute the rationale for establishing very expensive geomagnetic observatories at the seafloor.

It is also among the major and essential roles of the geomagnetic observation network on the ground, rather than mobile crafts in earth orbits, to monitor the smooth and gradual but significant change in the Earth's main field, i.e., the so-called geomagnetic secular variation. The phenomenon is of particular interest for geoscientists who study dynamics within the Earth's metallic core. Electromagnetically, the geomagnetic secular variation is one of the few direct observational constraints that are applicable to research on the geodynamo processes.

It has been more than three decades since observation of the geomagnetic field entered into a satellite era. Two successive and successful European Low Earth Orbit (LEO) satellites (e.g. Ørsted: [Neubert et al., 2001] and CHAMP: [Reigber et al., 2002]) have provided and are providing geomagnetic data of unprecedentedly good quality not to mention their quantity. This resulted in a better spatial resolution of the geomagnetic field models not only for the Earth's main field [IAGA, 2005] but for the lithospheric contribution to the geomagnetic field [Sabaka et al., 2004]. The satellite data enabled detection of the high-resolution geomagnetic secular variation as well [Holme and Olsen, 2006].

Advent of geomagnetic measurements from space has changed the role of the ground geomagnetic observation network. It is now possible to study the dynamics of the ionospheric current system more accurately using the simultaneous data recorded both on the Earth's surface and in space. It is also noteworthy that the dynamics of the ocean can be explored by data from seafloor geomagnetic observatories because the EM fields at the seafloor are known to be very sensitive to barotropic motions of the conducting seawater that couples with the Earth's main field [Sandord, 1971; Luther et al., 1991; Segawa and Toh, 1992].

1.3.2 Where to Deploy?

World Data Centre for Geomagnetism, Kyoto [WDC Kyoto, 2008] catalogued 120 land geomagnetic observatories that were active as of January 1, 2008. Although there still exist many other geomagnetic observatories not catalogued by that centre, the 120 land observatories are sufficient in number for monitoring the Earth's main field expanded by spherical harmonics up to degree 13, provided that they are distributed evenly all over the Earth's surface. Chave et al. [1995] reported that 92 observatories are enough to cover the entire globe by less than 2000-km spacing. They recommended, for a first step forward, eight seafloor geomagnetic observatories, i.e., two in the Atlantic, three in the western Pacific, and three in the southern hemisphere. Early simulation results [Barker and Baraclough, 1985; Langel et al., 1995] have shown that the maximum improvement of the biased spatial distribution can be achieved by putting a seafloor geomagnetic observatory in the southern Pacific Ocean. Those results again highlight the necessity of increased numbers of geomagnetic observatories not in the northern hemisphere but in the southern hemisphere.

Two Japanese observatories are now operating at the seafloor in the western Pacific since August, 2001 and June, 2006. They are respectively NWP on the Northwest Pacific Basin, and WPB on the West Philippine Basin. NWP is five years older than WPB. Figure 1.10 shows the locations of the two seafloor geomagnetic observatories in the western Pacific as well as the 120 existing geomagnetic

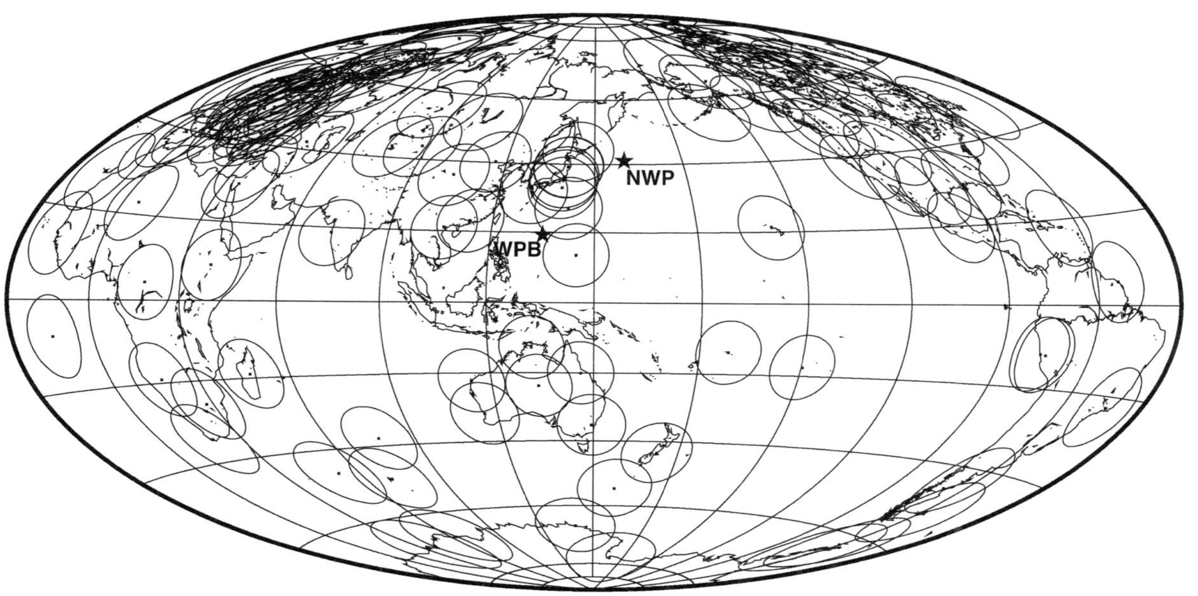

Fig. 1.10 The 120 presently active land geomagnetic observatories (small solid squares) catalogued in 2008 by World Data Centre for Geomagnetism, Kyoto with circles of a 1000-km radius surrounding each of them. Locations of the two Japanese seafloor observatories are also shown

observatories on land. Detailed location of the two seafloor observatories are provided in Fig. 1.11 and their location and timing are summarized in Table 1.1. Preferable locations of future geomagnetic observatories at the seafloor can be found outside the circles in Fig. 1.10 that surround each observatory with a radius of 1000 km. Note that the two presently active seafloor observatories in the western Pacific are neither in ideal locations nor at the recommended sites by Chave et al. [1995]. In determination of their locations, logistic requirements should be taken into consideration, since they are not cabled observatories and thus require frequent (once in at least a few years) servicing in order to maintain the observation site and retrieve data from those observatories. This means that they are still pilot seafloor observatories, although addition of these observatories yielded, for the first time in history, a regional geomagnetic reference field model over the western Pacific [Toh et al., 2007] based on the ground geomagnetic observation. Need for more observatories at the seafloor especially in the southern hemisphere has not yet been fulfilled.

1.3.3 Seafloor Environments

The ocean is the largest heat sink on earth and thus the deep seafloor is thermally very stable. Thermometers of the seafloor geomagnetic observatories showed very little temperature variations ranging from 0.5 to 1.5°C over 7 years at NWP and from 0.85 to 0.95°C over 2.5 years at WPB. The deep seafloor, therefore, is an ideal platform for long-term geomagnetic observation if thermal stability is the only requirement for the observation.

However, there are many other factors that impede our advance into the seafloor. Good conducting seawater makes the medium opaque for EM waves, which leaves us acoustic waves with very limited band-width as the only alternative for communication to uncabled seafloor observatories from the surface. In addition, seawater has so corrosive a nature that any long-term observation at the seafloor cannot be maintained without pertinent protection against incessant corrosion. Very high pressure at the seafloor, on the other hand, imposes critical limits on available volume for not only scientific equipment of seafloor observatories but also energy sources such as primary lithium cells. The Japanese seafloor observatories are both open ocean observatories and thus subject to extremely high pressure. It is almost 560 times as high as unit atmospheric pressure at NWP, while it is more than 570 times at WPB. Problems associated with high pressure at the seafloor may be relaxed at coastal observatories under shallow seas. However, advantages of open ocean observatories such as thermal stability and absence of man-made noises, in turn, will be lost for them. Vibration of seafloor instruments by strong benthic currents may get even worse due to enhanced tidal flows at the coast. Furthermore, coastal observatories in regions of western boundary currents will have to prepare a special countermeasure against them.

Presence of the conducting seawater affects the EM measurements by seafloor observatories as well. Incoming geomagnetic disturbances of external origin are strongly attenuated at the seafloor. This geophysical filtering by the conductive seawater cuts off short-period geomagnetic variations. This masks the uppermost part of the electrical Earth from exploration by natural source EM induction methods. The attenuation is stronger for horizontal geomagnetic components than for horizontal electric and vertical geomagnetic components, and dependent on electrical structures beneath the seafloor. This, in turn, can be exploited for delineating the electrical conductivity structure in the Earth if horizontal geomagnetic components are observed simultaneously on the sea surface and at the seafloor. This method (so-called 'Vertical Gradient Sounding' method; e.g., [Law and Greenhouse, 1981] is applicable for seafloor EM array studies when a permanent observatory or a temporary geomagnetic station on land is available in close proximity to the array. Another practical merit of the geophysical filtering is that it may reduce the required sampling rate of seafloor geomagnetic observatories. Presence of the geophysical filter makes one minute sampling even too redundant depending on the ocean depth of a specific seafloor observatory. This relaxes requirements for each seafloor observatory in terms of power supply, data storage and its transmission.

Because the oceanic crust normally consists of strongly magnetized igneous rocks such as basalts and gabbros, extensive marine magnetic survey is indispensable for determination of specific locations of each seafloor observatory. This is necessary to minimize directional errors of the measured

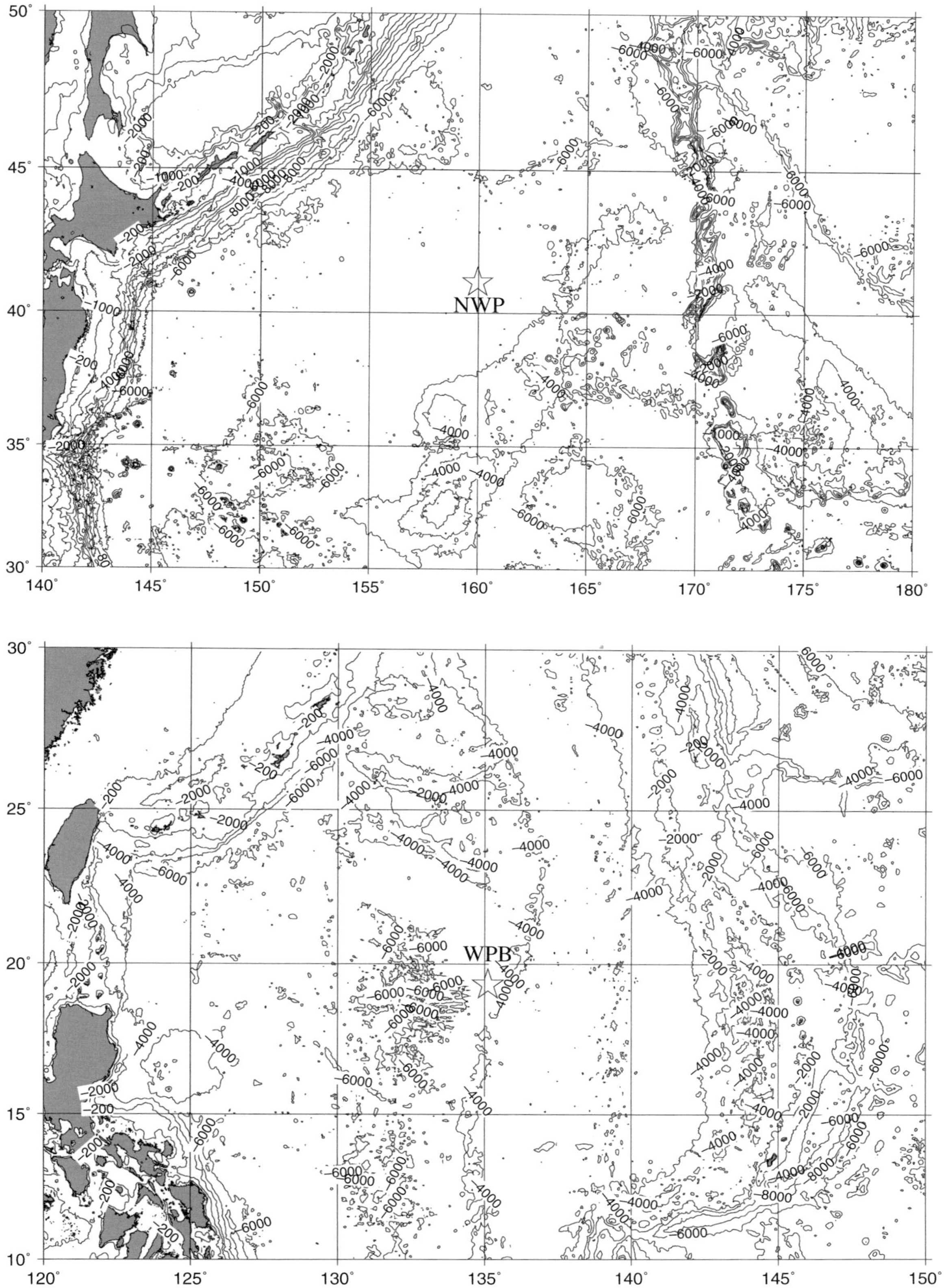

Fig. 1.11 Bathymetric contours in the vicinity of WPB and for NWP

Table 1.1 Summary of the presently operating seafloor geomagnetic observatories

Site Name	Latitude[a] [N]	Longitude[a] [E]	Depth[a] [m]	Since [UTC]
NWP	41°06′08″	159°57′47″	5580	August 1, 2001
WPB	19°19′18″	135°06′41″	5690	June 17, 2006

[a]Positions are based on the WGS-84 geodetic datum.

declination/inclination by the observatory as well as to yield precise station correction estimates for the observatory. If thickness of less magnetized sedimentary layers is large, the effect of the magnetic crust may be reduced. However, the pelagic sediments are normally less than 1 km thick for open ocean observatories. For instance, NWP sits on sediments as thin as 375 m above Magnetic Lineation M8 (~129 Ma); [Shipboard Scientific Party of ODP Leg 191, 2000] and WPB on 510-m thick sediments above Magnetic Lineation 21 (~49 Ma); [Salisbury et al., 2006]. Fortunately, the amplitudes of marine magnetic anomalies around the two seafloor observatories are moderate because they are both on the very old seafloor. It, however, may not necessarily be the case with all of the future seafloor geomagnetic observatories especially for those in the vicinity of mid-ocean ridges. Local magnetic anomalies there could be large enough to deflect not only declinations but also inclinations by more than 10 degrees.

1.3.4 Specific Solutions for Seafloor Geomagnetic Instrumentation

Multidisciplinary efforts aimed at constructing seafloor observatories resulted in success of a few pilot data acquisitions, e.g. Beranzoli et al. [2003] even in the hostile environment of the seafloor described in the previous section. In North America, an interdisciplinary ocean bottom observatory was built in the Monterey Bay, which was originally uncabled [Romanowicz et al., 2006], but efforts are still being made to make it a cabled coastal observatory. It should be noted here that these attempts have been mostly made and repeated in coastal regions such as the Mediterranean Sea rather than in open oceans. This is partly due to the fact that the coastal seafloor observatories can be serviced at better frequencies, and with easier operation for cable extension. However, the need for open ocean observatories is also keen. We, therefore, will focus on specific solutions for open ocean observatories in this section.

1.3.4.1 Underwater Housing of the Observatory

Figure 1.12 shows an outer view and a schematic of a presently operating seafloor geomagnetic observatory in the western Pacific [Toh et al., 2004]. To overcome the high pressure and the perpetual corrosion mentioned before, all sensors, electronics and lithium batteries are housed in pressure-tight glass spheres that are further mounted on a non-magnetic titanium frame.

Glass spheres are the most inexpensive solutions for buoyancy as well as deep-sea housing compared with other candidates such as metallic cylinders. Non-metallic housing is of particular merit for seafloor observatories that conduct EM measurements. It is favorable especially for geoelectric measurements to minimize electrochemical effects around the seafloor instruments not to mention the intensive corrosion of any metals by seawater. The glass spheres, however, need careful maintenance not only prior to but also after sea experiments since they are inferior to metallic housing in repetitive durability. The glass spheres sometimes require complete replacement after long-term deployment at sea.

Titanium is presently the best choice among the available metallic materials for use at the seafloor. It can be used for both frames and housing. It is heavier than aluminum but lighter than stainless steel in terms of density and provides the best corrosion protection compared with the two kinds of metal commonly used at sea. It, however, is desirable to avoid use of metals as much as possible in order to minimize unexpected biases in absolute geomagnetic measurements (see the next sub-section as well). Fibre reinforced plastic is another candidate material for future geomagnetic observatories at the seafloor provided that it is shock-resistant enough to withstand deployments and recovery by research vessels even in rough sea conditions.

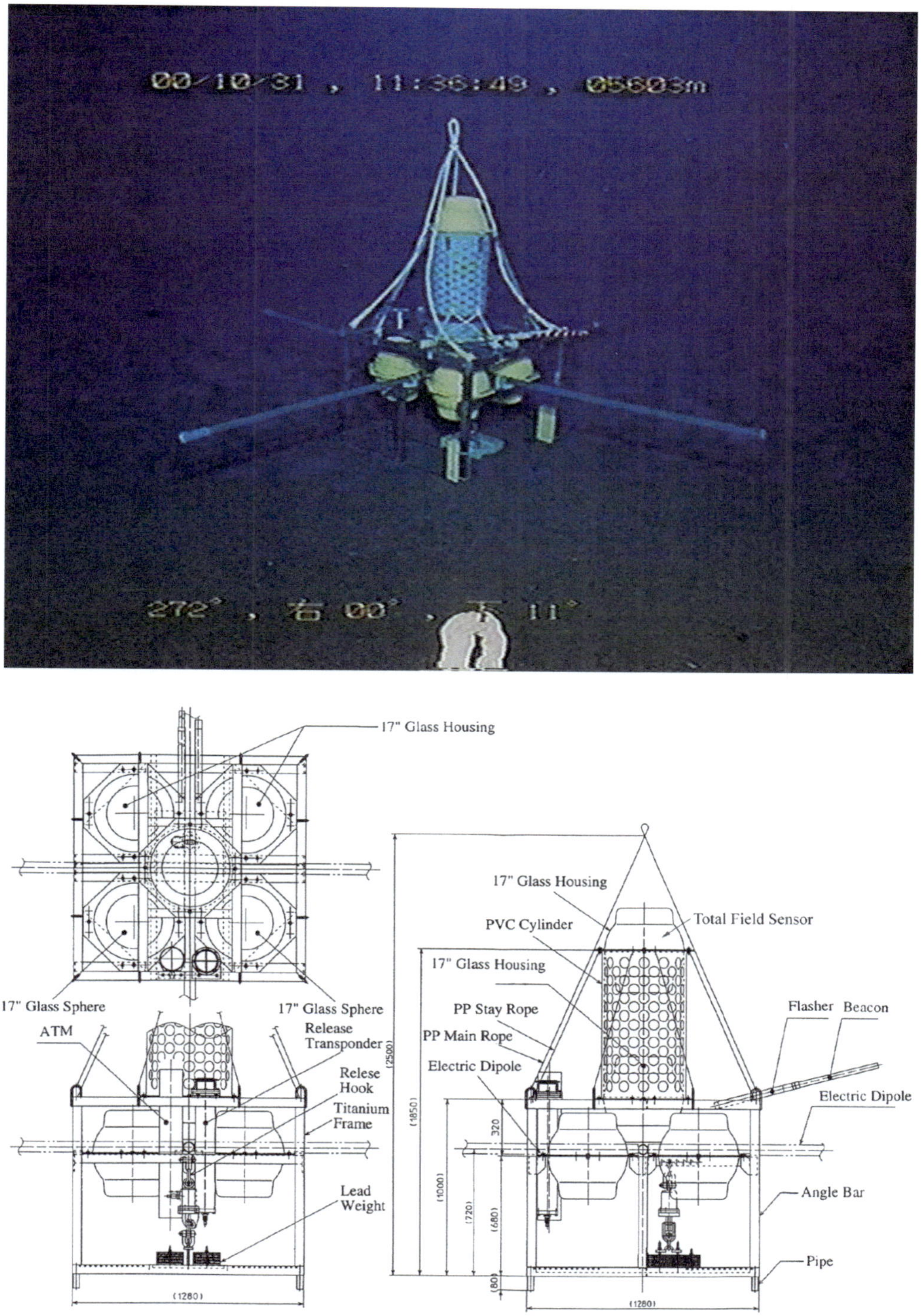

Fig. 1.12 *(Top)* The seafloor geomagnetic observatory operating at NWP *(Bottom)* The plan of the observatory

1.3.4.2 Total Field Measurement

Overhauser-type [Overhauser, 1953] proton precession magnetometers were adopted as sensors for the absolute total field intensity measurements mainly because of their very low power consumption rate. The other merit of the Overhauser sensor, i.e., capability of continuous measurements at very high sampling rates, is not of particular interest in the case of seafloor geomagnetic observatories.

The Overhauser sensors had been tested by several sea experiments prior to real deployments of the seafloor observatories in the western Pacific, and were proved to have absolute precision of better than 0.2 nT [Toh and Hamano, 1997]. Instrumental biases were also examined using the geomagnetic total force collected during the instruments' travels to/from the seafloor [Toh et al., 1998]. They are known to be less than 10 nT now.

As shown in Fig. 1.12 (bottom), the total field sensor is contained in the top glass sphere. The glass sphere is mounted on top of the non-magnetic titanium frame so as to make the height of the scalar sensor as high as approximately 2 m from the seafloor. This is to circumvent the strong magnetic gradient around the seafloor. Care should also be given for prevention of instruments' vibration that increases with instruments' height. 50 kg weight in water was found sufficient for that purpose.

1.3.4.3 Component Measurements

Carefully pre-calibrated fluxgate-type vector magnetometers are now being used for component measurements by the seafloor geomagnetic observatories. The calibration included various aspects of the system response such as scale factors for both canceling and sensing coils of the fluxgates, temperature coefficients for each axis of the magnetic sensors, alignment measurements between the attitude measuring frame and the measuring frame for EM components, and so on. Refer to Toh et al. [2006] for details of calibration.

Among all the calibration items, it was crucial for the component measurements to reveal the true system response of tilt meters. It was found that noise level of our original tilt meters was sensitive to tilt biases. Namely, the tilt sensors were very quiet when the actual tilt was almost nil, while they tended to become noisy if the real tilt got larger values. A time-consuming survey on this issue revealed that the original detectors of the tilt meters amplified noises as proportional to the absolute angles of the actual tilts. They, therefore, were replaced by new detectors that return flat noise level of ~3 arc seconds against tilt angle variations.

It was not until the tilt detectors were replaced that observation of vector geomagnetic secular variations was made possible even at the seafloor. Because seafloor instruments normally change their attitudes from their initial position given by free-fall installation from the sea surface as they settle firmly at the seafloor, it is essential to monitor precisely the attitude change in order to realize detection of the geomagnetic secular variation in the component measurements. Figure 1.13 shows a sample plot of 3-component geomagnetic data at the seafloor before tilt correction. The measured tilt data itself is shown in Fig. 1.14 as well as the tilt-corrected geomagnetic data in Fig. 1.15.

1.3.4.4 Orientation at the Seafloor

Attitude determination of the vector magnetometer of a seafloor geomagnetic observatory is the very key to the success of the observatory as described in the previous section. Without it, it would be almost impossible to distinguish the real geomagnetic secular variation from the change in the baselines of each geomagnetic axis, even in the ideal case of negligible instrumental drifts. Physical mechanisms that keep the vector magnetic sensors stay in a level surface (e.g., two-axis gimbals) are not recommended here, because the leveling precision of such equipment cannot be made better than several tens of arc seconds. By this low precision, magnetic errors arising from the attitude ambiguity sum up to more than 5 nT. It is very difficult to force the sensors automatically rotate back to the horizontal reference using the natural gravity field. It is rather easier just to measure the instruments' tilts as precisely as possible, and then correct for them. If the precision of the tilt meters can be made as good as a few arc seconds, the baseline ambiguity becomes smaller than one nanotesla, which is well below the INTERMAGNET requirement [Kerridge, 2001].

The remaining issue in attitude determination of a seafloor geomagnetic observatory is its orientation. To know the instruments' azimuth from the true north

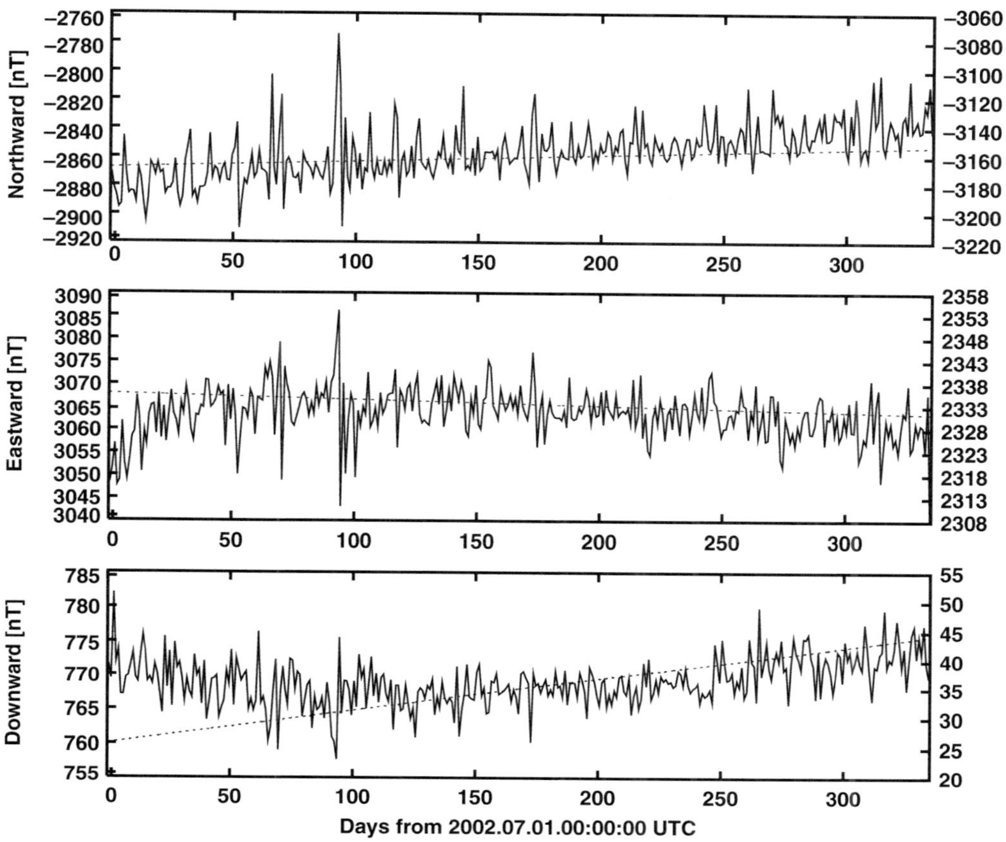

Fig. 1.13 One-year daily means of tilt-uncorrected geomagnetic three components at NWP. *Thin dotted lines* are predicted vector secular variations by a global geomagnetic field model [Olsen, 2002]

is another big issue with regard to the seafloor geomagnetic observatory. There are a few means that can be solutions of the absolute orientation determination at the seafloor such as acoustic array measurements. Here we report usage of gyrocompasses adopted in the seafloor geomagnetic observatories in the western Pacific.

Small fibre optical gyros (FOGs) were selected for our seafloor observatories and contained in one of the pressure-tight glass spheres with their power supply in order to determine the observatories' orientation with respect to the geographical north. The small FOGs were developed and manufactured originally for autonomous underground vehicles that bore through the Earth digging pilot tunnels. Their precision is around 0.2° in rms.

The FOGs, however, generate magnetic noises and consume large amounts of energy when they operate. They, therefore, are normally scheduled to be switched on at the seafloor once every several months. The intermittent operations save considerable amount of energy so that the power supply issue is settled by enclosing lithium primary cells with a capacity of 150 Ah per FOG. The magnetic noises which they produce affect the vector magnetometers more than the scalar Overhauser magnetometers, since the fluxgate magnetometers are mounted at the same level on the titanium frame (see Fig. 1.12 (bottom)) as the FOG. Horizontal spacing between the FOG and the fluxgate magnetometer is approximately 1 m. However, the seafloor geomagnetic data showed that the noise level at the time of the FOG operations increases up to 1 nT at most. Hence, the vector geomagnetic measurements are now kept running even during the FOG operations. Misalignment between the attitude and EM measuring frames is determined accurately

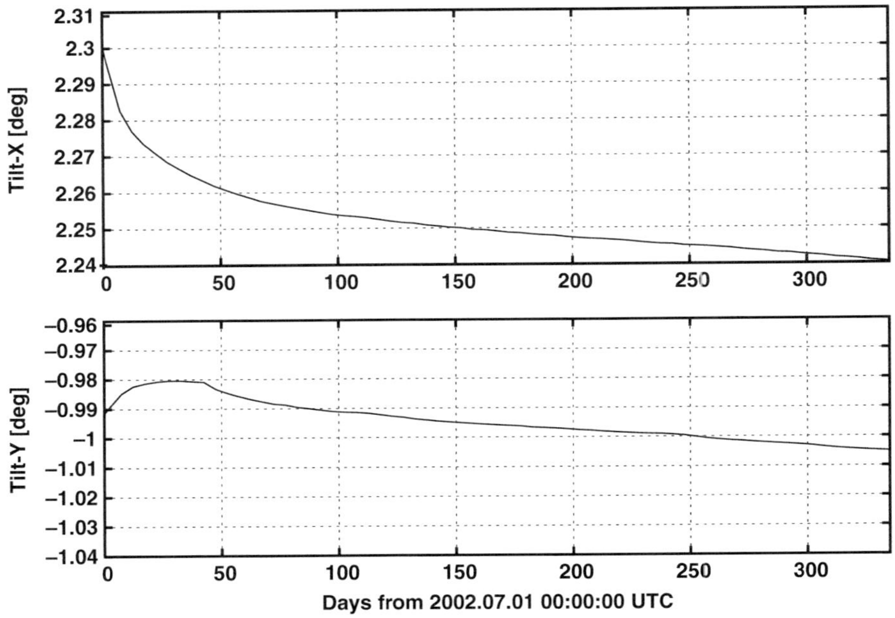

Fig. 1.14 Simultaneous tilt variations with the raw vector geomagnetic time-series in Fig. 1.13

using laser beams at the time of instrument assemblage on-board. The misalignment was usually less than 0.2° and used in orientation correction procedures after data retrieval. Total duration of the intermittent FOG operations is normally set to be more than two days at the seafloor. Because the FOG returns azimuthal data by one-minute interval when operated, standard errors of the seafloor geomagnetic observatories' orientation become as small as around 10 arc seconds.

1.3.5 Shortcomings Still Preventing the Full Absolute Accuracy: How to Eliminate Them?

The major challenge still remaining in the present seafloor geomagnetic observatories is how to measure the absolute direction at the seafloor. Other problems peculiar to the seafloor environments such as the presence of marine magnetic anomalies have been almost sorted out to date by experience accumulated during more than eight years on the deep seafloor of the Northwest Pacific Basin.

Strong marine magnetic anomalies especially around mid-ocean ridges still need considerable attention because they are associated with large station corrections for the seafloor observatory. Apart from very strong magnetic anomalies, details of the magnetic anomalies can be surveyed precisely enough to be corrected. The survey can be made by a combination of deep-towed scalar proton and vector fluxgate magnetometers conducted prior to actual installation of a geomagnetic observatory at a specific site on the seafloor.

Unnecessary instrumental motions by strong benthic currents and/or motionally induced EM fields by the so-called oceanic dynamo effect are the issues also peculiar to the seafloor environments. However, the observed EM time-series at NWP as long as six years has shown that those effects may be negligible at least for open ocean observatories. Gradual change in tilts is more important than the vibrations of the EM instruments by ocean bottom currents. The tilt change, which is large particularly at the beginning of seafloor installation, can be monitored precisely by available tilt meters as described earlier.

Although offsets of each magnetic sensor of a vector magnetometer can be measured as precisely as desired by prior calibrations on land, the magnetic bias of a seafloor geomagnetic observatory itself is usually difficult to know beforehand. As for the Japanese observatories, they were estimated by collecting vector

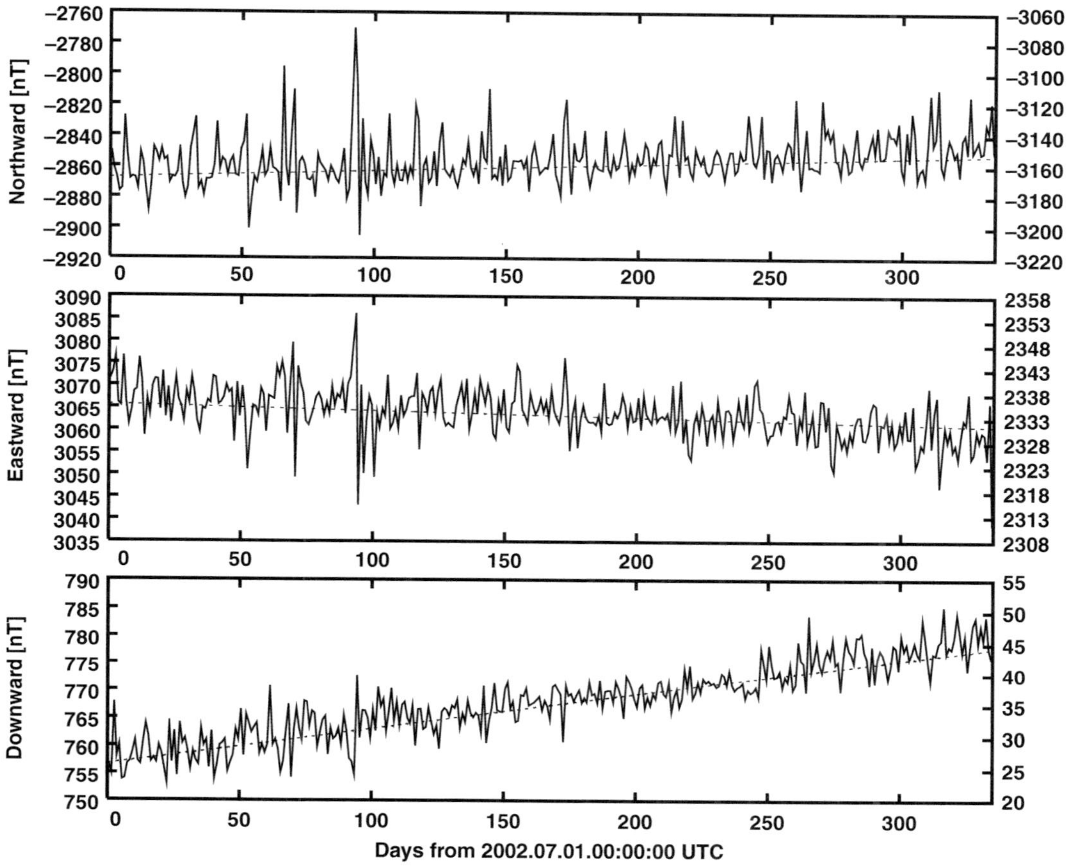

Fig. 1.15 Same as Fig. 1.13 but for the corrected vector geomagnetic variations

geomagnetic data during the travels of the seafloor instruments to/from the seafloor. The biases were proved to be smaller than 10 nT provided that sufficient time for aging from the initial instrument construction was given to each instrument. Instrumental drift of a variograph is another source of possible errors, since fluxgate sensors are known to have non-negligible drift rates. However, comparison of the observed absolute geomagnetic total force by the Overhauser magnetometer and synthetic total force using actually observed 3-component geomagnetic field by a fluxgate sensor [Toh et al., 2006] has shown that the variograph drift is small enough to be negligible. It may be mainly because of the extremely stable temperature condition on the seafloor.

Determination of the absolute directions of the seafloor geomagnetic observatories still needs further investigation. The precision of the determination is now solely dependent on that of the gyrocompasses, which cannot be switched on continuously. Temporal variations of the observatories' azimuths, therefore, cannot be monitored at this moment. Because it is power requirements rather than the magnetic noise generation by the gyrocompasses that prevents frequent azimuth determinations, it is desirable to have cabled seafloor geomagnetic observatories in the future. Because the cabled observatories are free from the power supply issue, at least monthly operation of the gyrocompasses will be enabled. The cable link is also preferable for the sake of the absolute geomagnetic total force measurements by Overhauser magnetometers. One problem of the present absolute scalar geomagnetic measurements is how to ensure pertinent tuning for the absolute magnetometer. It is difficult for the Overhauser sensor to go back into a suitable range automatically once it completely runs out of measurable range. If the sensor is connected by a cable to the shore, it can be monitored continuously and tuned manually or even automatically again. The cable link will also save the present labor paid in replacement of the seafloor instruments by remotely operated vehicles (submersibles) in order to keep the positions of the old and new instruments within 10 m.

1.3.6 The Future of Seafloor Observatories: Where Do We Head from Here?

Lack of international consensus on where to deploy the next geomagnetic observatory has long been a major obstacle for extending the ground geomagnetic observation network into the sea. It is necessary to reach an agreement with the seafloor community in the near future so as to establish effective international collaboration to install the very first international geomagnetic observatory at the seafloor. This is especially because an open ocean seafloor geomagnetic observatory is too expensive to be maintained for, say, as long as 10 years simply by one nation. International cooperation is the essential factor here. As for the venue, the southern hemisphere is preferred much more than the northern hemisphere that is over-crowded by continental geomagnetic observatories as mentioned earlier. It is recommended to select the venue from the following four marine regions: the Southeastern Pacific, the Eastern Equatorial Pacific, the Middle Indian Ocean and the South Atlantic Ocean. Starting with the four open ocean observatories is a good compromise leading to the full installation of eight seafloor geomagnetic observatories recommended by Chave et al. [1995] in the future.

Cable links to the open ocean observatories may be time-consuming, costly and logistically difficult tasks even if they are well-planned in advance. It is rather important at this stage to accumulate experience over cabled observatories using data from cabled coastal observatories such as the one in Monterey Bay.

One possible shortcut to have a cabled seafloor observatory quickly is to make use of existing seafloor observation networks that are already linked to the shore. One candidate is the DART system which is deployed mainly over the Pacific and maintained by NOAA in order to monitor hazardous tsunamis [cf. Borrero et al., 2009]. Since seafloor geomagnetic observatories are sensitive enough to detect motionally induced magnetic signals generated by tsunami propagation, there may be chances to urge research institutions in charge of such seafloor observation networks to integrate geomagnetic observatories into their systems. At any rate, it would be impossible to study the ionospheric current system in detail even in the era of satellite observation of our planet's magnetic field without seafloor geomagnetic observatories.

References

Ameen A, Ashfaque M, Borodin P, Brenes J, Daudi E, Efendi N, Flower S, Hidayat M, Husni M, Kampine M, Kusonski O, Langa A, Monge I, Mucussete A, Murtaza M, Nhatsave A, Oca Santika K, Rasson J, Riddick J, Suharyadi D, Turbitt C. and Yusuf M, (2009) Indigo: Better Geomagnetic Observatories where we Need Them, IAGA Sopron Poster session 502-MON-P1700-0162

Anlong C, Jinping Z, Yufen G (1990) The adjustment and optimization of geomagnetic fundamental network in China (in Chinese). Seismol Geomagnetic Observ Res 11(6):42–49

Barker FS, Barraclough DR (1985) The effects of the non-uniform distribution of magnetic observatory data on secular variation models. Phys Earth Planet Inter 37:65–73

Beranzoli L et al (2003) Mission results from the first GEOSTAR observatory (Adriatic Sea, 1998). Earth Planet Space 55:361–373

Borrero JC et al (2009) The tsunami of 2007 September 12, Bengkulu province, Sumatra, Indonesia: post-tsunami field survey and numerical modelling. Geophys J Int 178(1): 180–194. doi:10.1111/j.1365-246X. 2008.04058.x

Chave AD et al (1995) Report of a workshop on technical approaches to construction of a seafloor geomagnetic observatory, Woods Hole Oceanogr. Inst Tech Rep WHOI-95-12, 43 pp. http://darchive.mblwhoilibrary.org:8080/handle/1912/514?show=full

China Earthquake Administration (2004) Specification for the construction of seismic station: geomagnetic station (in Chinese). Seismological Publishing House

Holme R, Olsen N (2006) Core surface flow modeling from high-resolution secular variation, Geophys. J Int 166: 518–528

International Association of geomagnetism and aeronomy (IAGA) division V, working Group VMOD: geomagnetic field modeling 2005. The 10th-Generation International Geomagnetic Reference Field. Geophys J Int 161:561–565

Kerridge, D, (2001) INTERMAGNET: Worldwide near-real-time geomagnetic observatory data. Proceedings of Workshop on Space Weather, ESTEC, http://www.intermagnet.org/publications/IM_ESTEC.pdf

Korepanov V, Klymovych Ye, Kuznetsov O, Pristay A, Marusenkov A, Rasson JL, (2007) New INTERMAGNET Fluxgate Magnetometer. Publs Inst Geophys Pol Acad Sc C-99(398):291–298

Korte M, Mandea M, Linthe HJ, Hemshorn A, Kotzé P, Ricaldi E (2009) New Geomagnetic Field Observations In the South Atlantic Anomaly Region. Ann Geoph 52:65–81
Langel RA, Baldwin RT, Green AW (1995) Toward an improved distribution of magnetic observatories for modeling of the main geomagnetic field and its temporal change. J Geomag Geoelectr 47:475–508
Law LK, Greenhouse JP (1981) Geomagnetic variation sounding of the asthenosphere beneath the Juan de Fuca Ridge. J Geophys Res 86(B2):967–978. doi:10.1029/JB086iB02p00967
Luther DS, Filloux JH, Chave AD (1991) Low-frequency, motionally induced electromagnetic fields in the ocean: 2. Electric field and Eulerian current comparison. J Geophys Res 96(C7):12797–12814
Moussaoui N, Holzlöhner R, Hackenberg W, Bonaccini Calia D (2009) Dependence of sodium laser guide star photon return on the geomagnetic field. A A 501(2):793–799
Neubert T et al (2001) Ørsted satellite captures high-precision geomagnetic field data. Eos Trans Am Geophys Union 82:81–88
Olsen N (2002) A model of the geomagnetic field and its secular variation for epoch 2000 estimated from Ørsted data. Geophys J Int 149:454–462
Overhauser AW (1953) Polarization of nuclei in metals. Phys Rev 92:411–415
Rasson J, Giannibelli JC, Pelliciuoli AO (1996) A new digital magnetic observatory in Trelew Patagonia. Rom J Geophys 17:37–42
Reigber C, Luhr H, Schwintzer P (2002) CHAMP mission status. Adv Space Res 30:129–134
Romanowicz B et al (2006) The Monterey Bay broadband ocean bottom seismic observatory. Ann Geophys 49:607–623
Rongsheng G, Qizheng S (1999) Geomagnetic observation in China (in Chinese). Seismol Geomagnetic Observ Res 20(5): 47–53
Sabaka TJ, Olsen N, Purucker ME (2004) Extending comprehensive models of the Earth's magnetic field with Ørsted and CHAMP data. Geophys J Int 159:521–547
Salisbury MH et al (2006) 2. Leg 195 Synthesis: Site 1201—A geological and geophysical section in the West Philippine Basin from the 660-km discontinuity to the mudline. Proceedings of Ocean Drilling Program, Scientific Reports 195:27
Sanford TB (1971) Motionally induced electric and magnetic fields in the sea. J Geophys Res 76:3476–3492
Segawa J, Toh H (1992) Detecting fluid circulation by electric field variations at the Nankai trough. Earth Planet Sci Lett 109:469–469
Shipboard Scientific Party of ODP Leg 191 (2000) Northwest Pacific seismic observatory and hammer drill tests, Proceedings of Ocean Drilling Program, Initial Reports 191
Toh H, Hamano Y (1997) The first realtime measurement of seafloor geomagnetic total force—Ocean Hemisphere Project Network. J Japan Soc Mar Surv Tech 9: 1–13
Toh H, Hamano Y, Goto T (1998) A new seafloor electromagnetic station with an Overhauser magnetometer, a magnetotelluric variograph and an acoustic telemetry modem. Earth Planets Space 50:895–903
Toh H et al (2004) Geomagnetic observatory operates at the seafloor in the northwest Pacific Ocean. Eos Trans AGU 85:467–473
Toh H, Hamano Y Ichiki M (2006) Long-term seafloor geomagnetic station in the northwest Pacific: A possible candidate for a seafloor geomagnetic observatory. Earth Planet Space 58:697–705
Toh H, Kanezaki H, Ichiki M (2007) A regional model of the geomagnetic field over the Pacific Ocean for epoch 2002. Geophys Res Lett 34
World Data Centre for Geomagnetism, Kyoto (2008) Data catalogue, No. 28, 182 pp.
Zongqi C (1944) Review of geomagnetic observation in China (in Chinese). Acad Trans 1:99–126

Chapter 2

Magnetic Satellite Missions and Data

Nils Olsen and Stavros Kotsiaros

Abstract Although the first satellite observations of the Earth's magnetic field were already taken more than 50 years ago, continuous geomagnetic measurements from space are only available since 1999. The unprecedented time-space coverage of this recent data set opened revolutionary new possibilities for exploring the Earth's magnetic field from space.

In this chapter we discuss characteristics of satellites measuring the geomagnetic field and report on past, present and upcoming magnetic satellite missions. We conclude with some basics about space magnetic gradiometry as a possible path for future exploration of Earth's magnetic field with satellites.

2.1 Introduction

Exploring the Earth's magnetic field from space began about 50 years ago with the launch of the *Sputnik 3* satellite in 1958. However, data for global field modeling were first obtained by the *POGO* satellite series that measured the magnetic field intensity between 1965 and 1971. The first high-precision vector measurements were taken by the *Magsat* satellite in 1979–80. More recently, the launch of the satellites Ørsted (Denmark, February 1999), CHAMP (Germany, July 2000) and SAC-C (Argentina/US/Denmark, November 2000) opened revolutionary new possibilities for exploring the Earth's magnetic field from space. In the near future, the *Swarm* satellite constellation mission, comprising of three satellites to be launched in 2012, will provide even better opportunities.

A few papers have recently been published on magnetic satellite missions for modeling the geomagnetic field (e.g., Hulot et al. 2007; Olsen et al. 2010). In the present chapter we therefore only describe very briefly the various satellite missions and concentrate on information about data availability and orbit characteristics that has not been published elsewhere. In addition, we provide an outlook on the principles of magnetic space gradiometry as a possible future way to go for exploring the Earth's magnetic field from space.

2.1.1 Basic Equations

Magnetic field investigations of the Earth's core and crust are typically done in the quasi-static (or pre-Maxwell) approximation, which requires that the time scales in consideration are longer ($\gg$1 s) compared to the time required for light to pass the length scale of interest (less than a few thousand km). In this approximation displacement currents can be neglected, and the magnetic field $\mathbf{B}$ is given by

$$\nabla \times \mathbf{B} = \mu_0 \mathbf{J} \tag{2.1}$$

where $\mu_0 = 4\pi 10^{-7}$ Vs(Am)$^{-1}$ is vacuum permeability, and current density

$$\mathbf{J} = \mathbf{J}_e + \mathbf{J}_m \tag{2.2}$$

(expressed in units of A/m^2) is the sum of free charge current density $\mathbf{J}_e$ and the equivalent current density, $\mathbf{J}_m = \nabla \times \mathbf{M}$, due to material of magnetization $\mathbf{M}$ (units of A/m). The sources of magnetic fields

N. Olsen (✉)
DTU Space, Technical University of Denmark, Juliane Maries Vej 30, DK-2100 Copenhagen Ø, Denmark
e-mail: nio@space.dtu.dk

M. Mandea, M. Korte (eds.), *Geomagnetic Observations and Models*, IAGA Special Sopron Book Series 5,
DOI 10.1007/978-90-481-9858-0_2, © Springer Science+Business Media B.V. 2011

are therefore electric currents (for instance in the Earth's core, the ionosphere or the magnetosphere) and/or magnetized material (for instance in the Earth's crust). Outside its sources (i.e., in regions with vanishing current density, $\mathbf{J} = 0$), the magnetic field, $\mathbf{B} = -\nabla V$, is a Laplacian potential field and can be derived from a scalar magnetic potential V. This condition is fulfilled for magnetic measurements taken in the non-conducting atmosphere at or close to the Earth's surface. In that case the magnetic field has similar properties as the gravity field and the same methods for studying both fields might be used (Blakely 1995). However, the similarity is only true under certain assumptions and a closer look at the differences and similarities of $\mathbf{g}$ and $\mathbf{B}$ is helpful.

2.1.2 Magnetic vs. Gravity Field

The source of the magnetic field $\mathbf{B}$ are electric currents, while the source of the gravity field is mass density ρ_m (units of kg m^{-3}). The governing equations connecting sources and fields are different for the magnetic, resp. gravity, field. Also the constraining equations are rather different, since the magnetic field is always solenoidal ($\nabla \cdot \mathbf{B} = 0$) while the gravity field is always irrotational ($\nabla \times \mathbf{g} = 0$). Collecting governing and constraining equations yields

$$\nabla \times \mathbf{B} = \mu_0 \mathbf{J}, \qquad \nabla \cdot \mathbf{B} = 0 \tag{2.3a}$$

$$\nabla \cdot \mathbf{g} = -4\pi G \rho_m, \qquad \nabla \times \mathbf{g} = 0 \tag{2.3b}$$

where $G = 6.6743 \times 10^{-11}$ m^3 kg^{-1}s^{-2} is the gravitational constant.

The different constraining equations ($\nabla \cdot \mathbf{B} = 0$, resp. $\nabla \times \mathbf{g} = 0$) lead to different general representations for the two fields. In the case of the magnetic field, a toroidal-poloidal decomposition (Backus 1986, Sabaka et al. 2010) is possible, which allows to represent $\mathbf{B}$ in terms of the two scalar functions Φ and Ψ:

$$\mathbf{B} = \nabla \times \hat{\mathbf{r}}\Phi + \nabla \times \nabla \times \hat{\mathbf{r}}\Psi \tag{2.4}$$

where $\hat{\mathbf{r}}$ is the unit vector in radial direction. In contrast, the gravity field can always be represented by one scalar potential U:

$$\mathbf{g} = -\nabla U \tag{2.5}$$

where U is connected to the sources of the gravity field through Poisson's equation

$$\nabla^2 U = 4\pi G \rho_m. \tag{2.6}$$

Note that this does not necessarily mean that $\mathbf{g}$ is a Laplacian potential field.

Outside there respective sources, which means in vacuum in the case of the gravity field ($\rho_m = 0$), and in a non-magnetized insulator (like the atmosphere) in the case of the magnetic field (resulting in $\mathbf{J} = 0$), Eq. 2.3 lead to similar and symmetric equations for the magnetic and gravity field:

$$\nabla \cdot \mathbf{B} = 0 \quad \nabla \times \mathbf{B} = 0 \tag{2.7a}$$

$$\nabla \cdot \mathbf{g} = 0 \quad \nabla \times \mathbf{g} = 0, \tag{2.7b}$$

which means that both are Laplacian potential fields

$$\mathbf{B} = -\nabla V \quad \nabla^2 V = 0 \tag{2.8a}$$

$$\mathbf{g} = -\nabla U \quad \nabla^2 U = 0. \tag{2.8b}$$

This opens the possibility to use similar methods for the investigation of $\mathbf{B}$ and $\mathbf{g}$ (e.g., Blakely 1995).

In particular, it allows for expanding the magnetic potential V globally in series of spherical harmonics:

$$V = V^{\mathrm{int}} + V^{\mathrm{ext}}$$
$$= a \sum_{n=1}^{N_{\mathrm{int}}} \sum_{m=0}^{n} \left(g_n^m \cos m\phi + h_n^m \sin m\phi\right) \left(\tfrac{a}{r}\right)^{n+1} P_n^m(\cos\theta) \tag{2.9a}$$

$$+ a \sum_{n=1}^{N_{\mathrm{ext}}} \sum_{m=0}^{n} \left(q_n^m \cos m\phi + s_n^m \sin m\phi\right) \left(\frac{r}{a}\right)^n P_n^m(\cos\theta) \tag{2.9b}$$

(Chapman and Bartels 1940; Langel 1987), where $a = 6371.2$ km is a reference radius, (r, θ, ϕ) are spherical coordinates, P_n^m are the associated Schmidt semi-normalized Legendre functions, N_{int} is the maximum degree and order of the expansion coefficients (Gauss coefficients) g_n^m, h_n^m describing internal sources, and N_{ext} is that of the coefficients q_n^m, s_n^m describing external sources. Analysis of horizontal and vertical magnetic field components allows to determine g_n^m, h_n^m and q_n^m, s_n^m, thereby enabling a separation of internal and external sources.

The corresponding internal potential recovered from satellite data may include some signal from ionospheric sources below the satellite which, however, can be minimized through data selection (by selecting night-time data when ionospheric sources are weakest). In any case it is important to recognize that satellites move through an electric plasma (the ionosphere), and the existence of electric currents at satellite altitude does, in principle, not allow to describe the observed field as the gradient of a Laplacian potential. The magnetic field produced by the in-situ currents (which results in a toroidal magnetic field) is, however, (mathematical) orthogonal to the Laplacian potential field caused by sources in the Earth's interior and the magnetosphere (which are poloidal fields). Provided that the data are properly sampled in space and time, the existence of a toroidal field has therefore only marginal impact on the determination of the potential field.

2.2 Characteristics of Magnetic Satellite Data

Compared to magnetic measurements obtained at ground by geomagnetic observatories or repeat station, data from magnetic satellite are different in several aspects: Firstly, it is not possible to decide whether an observed magnetic field variation is due to a temporal or spatial change, since the satellite moves (with a velocity of about 8 km s^{-1} for an altitude of about 400 km). Secondly, satellites map the entire Earth (apart from the *polar gap*, a region around the geographic poles that is left unsampled if orbit inclination is $\neq 90°$). Thirdly, the observations are taken over different regions with the same magnetometer, which minimizes spurious effects due to different instrumentation. And finally, a spatially low-pass-filtered map of the magnetic field is obtained, since measurements taken from an altitude of, say, 400 km corresponds roughly to averaging over an area of this dimension. As a consequence, the effect of local magnetic heterogeneities is reduced.

2.2.1 Orbit, Time and Position

In general a satellite moves around Earth along an elliptical orbit, but for many of the satellites used for geomagnetic field modeling the orbit ellipticity is small. As sketched in the left panel of Fig. 2.1, orbit inclination i is the angle between the orbit plane and the equatorial plane. A perfectly polar orbit implies $i = 90°$, but for practical reasons most satellite orbits have inclinations that are different from 90°. This results in "polar gaps", which are regions around the geographic poles that are left unsampled. As an example, the right part of Fig. 2.1 shows the ground track of one day (January 2, 2001) of Ørsted satellite data.

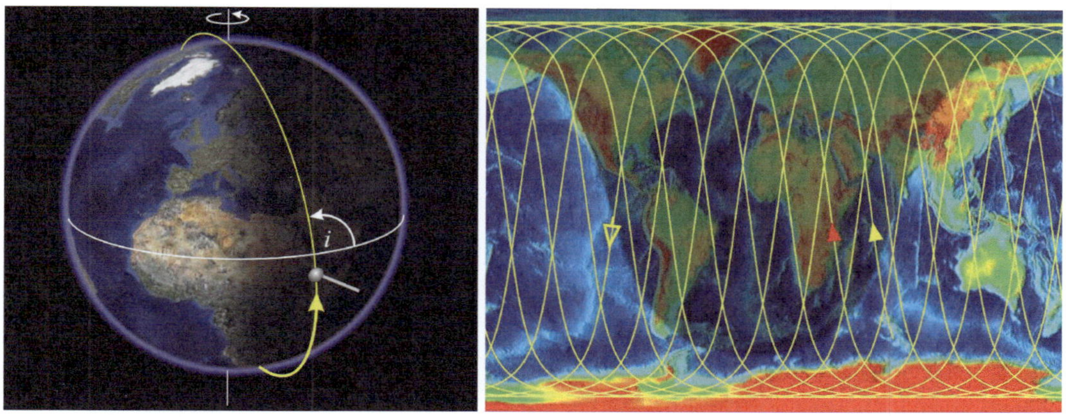

Fig. 2.1 *Left*: The path of a satellite at inclination i in orbit around the Earth. *Right*: Ground track of 24 h of the Ørsted satellite on January 2, 2001 (*yellow curve*). The satellite starts at $-57°$ N, 72° E at 00 UT, moves northward on the morning side of the Earth, and crosses the Equator at 58° E (*yellow arrow*). After crossing the polar cap it moves southward on the evening side and crosses the equator at 226° E (*yellow open arrow*) 50 min after the first equator crossing. The next Equator crossing (after additional 50 min) is at 33° E (*red arrow*), 24° westward of the first crossing 100 minutes earlier, while moving again northward

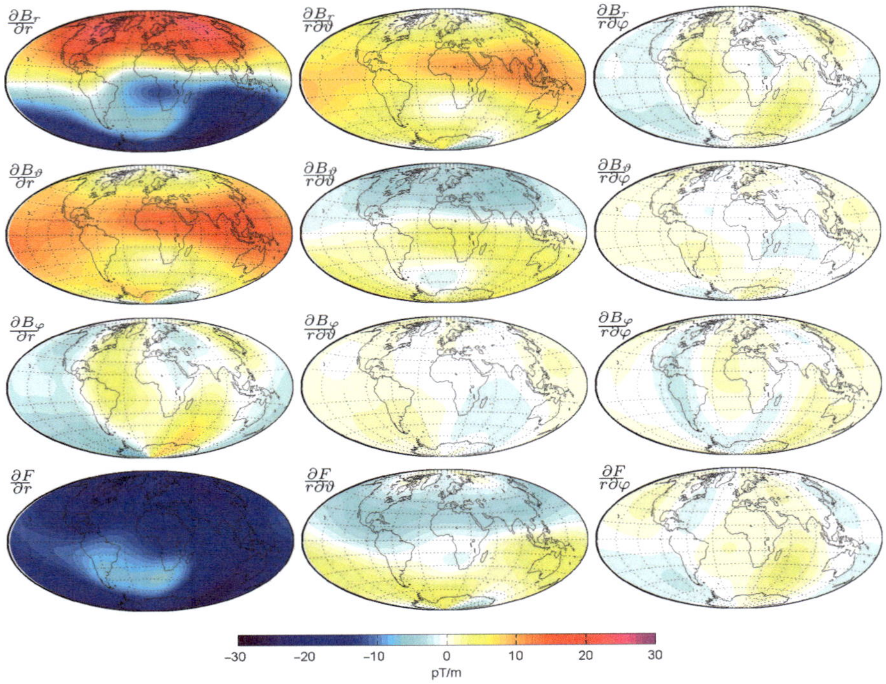

Fig. 2.2 Magnetic field change in radial (*left*), North-South (*middle*) and East-West direction (*right*) for the magnetic elements B_r, B_ϑ, B_ϕ and $F = |\mathbf{B}|$ at 400 km altitude

An error in the satellite position $\delta\mathbf{r}$ transforms directly into a magnetic field error $\delta\mathbf{B}$. Figure 2.2 shows the contribution of a position error in radial (left column), North-South (middle column), resp. East-West direction (right column) to the magnetic field error. A position error in radial direction has by far the largest impact, with maximum errors of 28, 18, 8 and 28 pT/m for B_r, B_θ, B_ϕ resp. $F = |\mathbf{B}|$. Least influence has a position error in East-West direction (which means horizontal cross-track for a polar orbiting satellite).

The precise determination of time and position was a major technical challenge befor GPS became available, and wrong position is the largest contribution to the magnetic error budget of previous satellite missions like POGO. However, GPS has dramatically improved the situation, resulting in a position accuracy better than a couple of meters (in many cases much better) and a timing error of less than a few ms. Recall that a timing error of 5 ms corresponds to an along-track position error of 40 m (for a satellite at 400 km altitude moving with 8 (km s^{-1}), which transforms to a magnetic field error of less than 0.5 nT. For present satellites like Ørsted and CHAMP the contribution of position and timing errors to the overall magnetic error budget is well below 1 nT.

2.2.2 Calibration and Alignment of Satellite Data

The strength of the magnetic field can be measured in space absolutely with scalar magnetometers. Examples of such absolute instruments are proton precession magnetometers (in particular the Overhauser instrument), Alkali metal vapor magnetometers, and Helium magnetometers. In contrast to scalar instruments, vector magnetometers are non-absolute instruments; their output has to be calibrated and aligned. Reviews of magnetometers for space applications are given by Acuña (2002) and Primdahl (1998).

Data Calibration

The conversion of the raw vector magnetometer readings into scaled magnetic field components (in units of nT) in the orthogonal coordinate system of the magnetometer, is called *calibration*. Most vector magnetometers used for space applications are fluxgate instruments. They are calibrated in-orbit by comparing the instrument readings with the magnetic field intensity F_{OVH} measured simultaneously with an absolute scalar magnetometer, which in case of the Ørsted and CHAMP missions is an Overhauser instrument.

In following we will assume that the Vector Fluxgate Magnetometer (VFM) is a linear instrument, an assumption that has been proven to be valid for the Ørsted and CHAMP instruments. In that case the instrument output $\mathbf{E} = (E_1, E_2, E_3)^T$ (in engineering units, eu) is connected to the applied magnetic field $\mathbf{B}_{\mathrm{VFM}} = (B_1, B_2, B_3)^T$ (in the orthogonal magnetometer coordinate system) according to

$$\mathbf{E} = \underline{\underline{\mathbf{S}}} \cdot \underline{\underline{\mathbf{P}}} \cdot \mathbf{B}_{\mathrm{VFM}} + \mathbf{b} \tag{2.10}$$

where

$$\mathbf{b} = \begin{pmatrix} b_1 \\ b_2 \\ b_3 \end{pmatrix} \tag{2.11a}$$

is the offset vector (in eu),

$$\underline{\underline{\mathbf{S}}} = \begin{pmatrix} S_1 & 0 & 0 \\ 0 & S_2 & 0 \\ 0 & 0 & S_3 \end{pmatrix} \tag{2.11b}$$

is the (diagonal) matrix of sensitivities (in eu/nT), and

$$\underline{\underline{\mathbf{P}}} = \begin{pmatrix} 1 & 0 & 0 \\ -\sin u_1 & \cos u_1 & 0 \\ \sin u_2 & \sin u_3 & \sqrt{(1 - \sin^2 u_2 - \sin^2 u_3)} \end{pmatrix} \tag{2.11c}$$

is a matrix which transforms a vector from the orthogonal magnetic axes coordinate system to the non-orthogonal magnetic sensor axes coordinate system. The nine parameters $b_i, S_i, u_i,\ i = 1, \ldots, 3$ (some of which may depend on temperature, cf. Olsen et al. (2003)) completely describe the linear VFM magnetometer.

These parameters are estimated by means of a linearized least-squares approach, minimizing the mean squared difference between $F_{\mathrm{VFM}} = |\mathbf{B}_{\mathrm{VFM}}|$ and the field intensity F_{OVH} measured with the absolute scalar magnetometer. Details of this in-flight calibration of vector magnetometers is given by Olsen et al. (2003).

Data Alignment

Merging the calibrated vector data with attitude data and transforming them to vector components $\mathbf{B}_{\mathrm{ECEF}} = (B_r, B_\theta, B_\phi)^T$ (i.e., the upward, southward, and eastward components of the magnetic field) in an *Earth-Centered-Earth-Fixed (ECEF)* coordinate system requires one additional calibration step, called data *alignment*. For this it is necessary to precisely determine the rotation (Euler angles) between the star imager and the vector magnetometer. This requires models of the star constellation and of the ambient magnetic field. The former model is known with high precision (e.g., Hipparcos catalog). The limiting factor for the alignment is the accuracy of the ambient magnetic field *to be known at the time and position of each data point.*

It is important to recognize that alignment of spaceborne magnetometers is rather different from alignment of observatory magnetometers, mainly due to the fact that the satellite moves. Alignment of ground magnetometers is possible without any field model, by turning the magnetometer by respectively 180° around the three magnetometer axes and taking additional measurements. Proper combination of the magnetometer readings taken during this procedure, which is performed at the same position (no spatial change of the ambient magnetic field) and almost instantaneously (within a few minutes, to minimize the influence of temporal changes of the ambient field; remaining temporal field changes are corrected for by subtracting the field changes monitored by a nearby variometer) removes the ambient magnetic field, and hence no magnetic field model is required.

Such a procedure is, however, not possible for satellite magnetometers in-orbit, due to movement of the satellite, and therefore satellite magnetometer alignment requires a model of the ambient magnetic field.

Let $\underline{\underline{\mathbf{R}}}_3$ be the matrix which rotates the magnetic field $\mathbf{B}_{\mathrm{ECEF}}$ from the *ECEF* system to the magnetic field $\mathbf{B}_{\mathrm{ICRF}} = \underline{\underline{\mathbf{R}}}_3 \cdot \mathbf{B}_{\mathrm{ECEF}}$ in the *International Celestial Reference Frame (ICRF)*; $\underline{\underline{\mathbf{R}}}_3$ is derived from satellite position and time (Seeber 2004). Next, $\underline{\underline{\mathbf{R}}}_2$ is a matrix which rotates the magnetic field $\mathbf{B}_{\mathrm{SIM}} = \underline{\underline{\mathbf{R}}}_2 \cdot \mathbf{B}_{\mathrm{ICRF}}$ from the *ICRF* frame to the star imager (*SIM*) frame; this matrix is constructed from the attitude data measured by the star imager. Finally, $\underline{\underline{\mathbf{R}}}_1$ is the matrix which rotates from the *SIM* coordinate system to the orthogonal magnetometer (*VFM*) coordinate system;

this rotation is described by the three Euler angles (α, β, γ) that have to be determined.

These Euler angles are estimated in orbit by assuming that the magnetic field vector in the *ECEF* system, $\mathbf{B}_{\mathrm{ECEF}} = -\mathrm{grad}\ V$, can be described by means of a Laplacian potential, cf. Eq. (2.9). In that case the relationship between the magnetic vector in the magnetometer coordinate system, $\mathbf{B}_{\mathrm{VFM}}$, and the magnetic potential V in the *ECEF* coordinate system is given by

$$\mathbf{B}_{\mathrm{VFM}} = \underline{\underline{\mathbf{R}}}_1 \cdot \underline{\underline{\mathbf{R}}}_2 \cdot \underline{\underline{\mathbf{R}}}_3 \cdot \mathbf{B}_{\mathrm{ECEF}} = -\underline{\underline{\mathbf{R}}}_1 \cdot \underline{\underline{\mathbf{R}}}_2 \cdot \underline{\underline{\mathbf{R}}}_3 \cdot \mathrm{grad}\ V \qquad (2.12)$$

and the Euler angles describing $\underline{\underline{\mathbf{R}}}_3$ can be determined either using an a-priori magnetic field model (i.e. a given potential V) or by co-estimating V together with the Euler angles.

The assumption that the measured magnetic field can be described by a Laplacian potential field is not strictly fulfilled in case of satellite data, because of field-aligned currents crossing the satellite orbit. Their magnetic field contribution can not be described by a Laplacian potential, and may modify the Euler angles *at a given time instant*, but when averaging over a sufficiently long period, and assuming that the Euler angles are time-independent, it is assumed that this effect averages out.

More details on satellite magnetometer alignment can be found in Olsen et al. (2003, 2006).

2.3 A parade of Magnetic Satellite Missions

Ground based measurements were the only data source for exploring the geomagnetic field before the first space-borne measurements were taken by the *Sputnik 3* satellite in 1958 (Dolginov et al. 1962). The first global magnetic satellite data have been obtained by the *POGO* satellites (data from the earlier satellites *Cosmos 26* and *Cosmos 49* in 1964 were of much poorer quality).

Here we concentrate on satellite missions measuring the high-precision data that are necessary for modeling the geomagnetic field. In addition to those, there are quite a few satellites that provide magnetic field data for investigations of ionospheric and magnetospheric currents. However, accuracy of the data obtained by most of these satellites does not allow to use them for field modeling, and therefore these satellites are not described here. There are, however, a few exceptions: data taken by the DE-1 and DE-2 satellites have for example been used for field modeling after careful re-calibration of the data since these satellites flew during a period without high-precision satellites in orbit.

Table 2.1 lists key parameters of satellites that have been used for modeling the geomagnetic field. In following we will briefly describe these satellites, some of which are shown in Fig. 2.3. The data of most of these satellites are available at www.space.dtu.dk/English/Research/Scientific_data_and_models/Magnetic_Satellites.aspx

2.3.1 POGO (OGO-2, OGO-4, OGO-6)

The *Polar Orbiting Geophysical Observatories (POGO)* were the first satellites which globally measured the Earth's magnetic field. They were equipped with optically pumped rubidium vapor absolute magnetometers. The POGO series consists of six satellites, but only three of them flew at sufficient low altitudes to be of interest for field modeling: OGO-2 measured the field between October 1965 and September 1967; OGO-4 between July 1967 and January 1969, with a few weeks of data overlap with OGO-2; and OGO-6 operated between June 1969

Table 2.1 High precision magnetic satellite missions

Satellite	Operation	Inclination	Altitude	Data
OGO-2	Oct 1965–Sep 1967	87°	410–1510 km	scalar only
OGO-4	Jul 1967–Jan 1969	86°	410–910 km	scalar only
OGO-6	Jun 1969–Jun 1971	82°	400–1100 km	scalar only
Magsat	Nov 1979–May 1980	97°	350–550 km	scalar and vector
Ørsted	Feb 1999–	97°	650–850 km	scalar and vector
CHAMP	Jul 2000–	87°	310–450 km	scalar and vector
SAC-C/ Ørsted-2	Jan 2001–Dec 2004	97°	698–705 km	scalar only
Swarm	2012 – 2016	88°/87°	530/ $<$ 450 km	scalar and vector

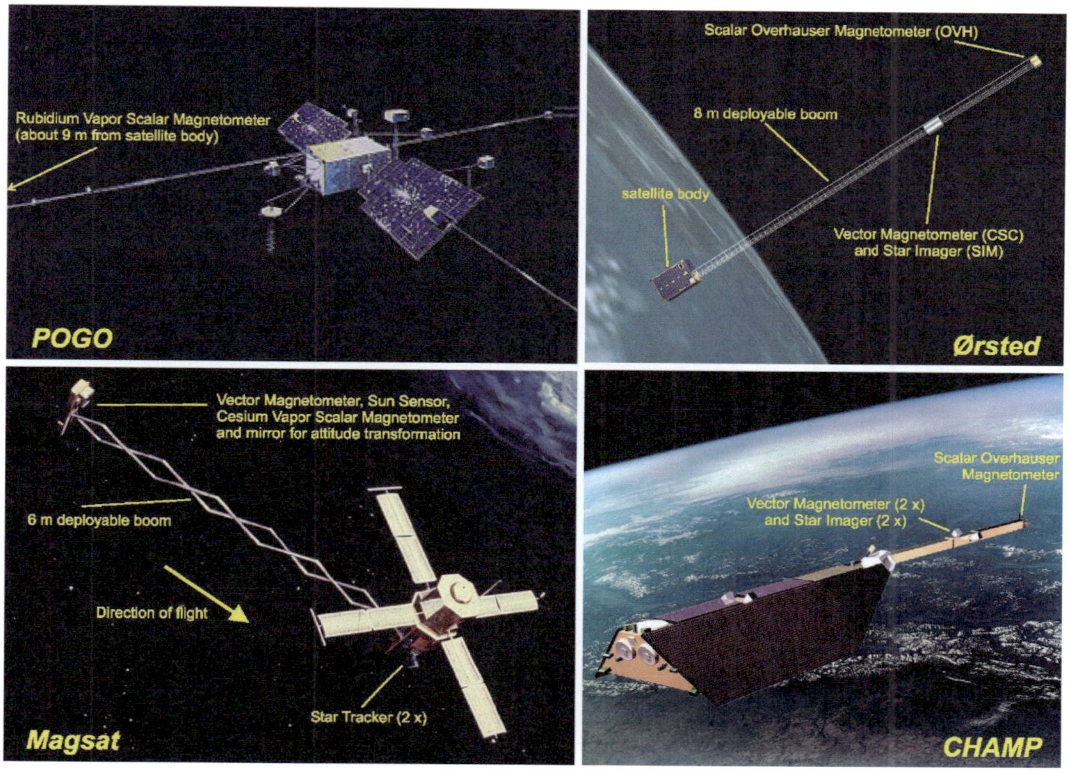

Fig. 2.3 Artist's view of the POGO, Magsat, Ørsted and CHAMP satellites

and June 1971. The top panel of Fig. 2.4 shows the altitude of orbit perigee, resp. apogee (which is the lowest, resp. highest, altitude of the orbit), and mean orbit altitude for OGO-2, -4 and -6; its middle panel presents local time of the ascending node (which is the equator crossing of the south-going part of the orbit). Intrinsic measurement error of all three satellites is believed to be below 1 nT, but contribution due to position uncertainty results in an effective magnetic error of up to 7 nT. Data availability is shown in the lower panel of Fig. 2.4. See Cain (2007) for more information on the POGO satellites.

2.3.2 Magsat

The US satellite *Magsat* (Oct 1979 to June 1980) made the first precise, globally distributed vector measurements of the magnetic field. As shown in the top part of Fig. 2.5, the mean orbit altitude decreased from about 450 km in November 1979 to about 350 km at the end of the mission. The satellite was in a near-polar orbit with an inclination of 97°. Such an inclination results in an orbit that is fixed relatively to the sun (fixed local time); in the case of Magsat the local time of the ascending node was 17:40 (±20 min). The satellite carried a cesium 133 vapor optically pumped scalar magnetometer and a triaxial fluxgate magnetometer measuring the vector field at 16 Hz with a resolution of ±0.5 nT. Attitude was measured using two star-trackers on the spacecraft; transformation of attitude determined by these star trackers to the vector magnetometer at the tip of the boom was done using a complicated optical system. Attitude errors limit the vector data accuracy to about 4 nT rms. Data availability is shown in the bottom panel of Fig. 2.5. See Purucker (2007) for more information on the satellite.

2.3.3 Ørsted

The Danish *Ørsted* satellite is the first satellite mission after Magsat for high-precision mapping of the Earth's magnetic field. It was launched on 23rd February 1999 into a near polar orbit. Being the first satellite of the *International Decade of Geopotential Research*, the

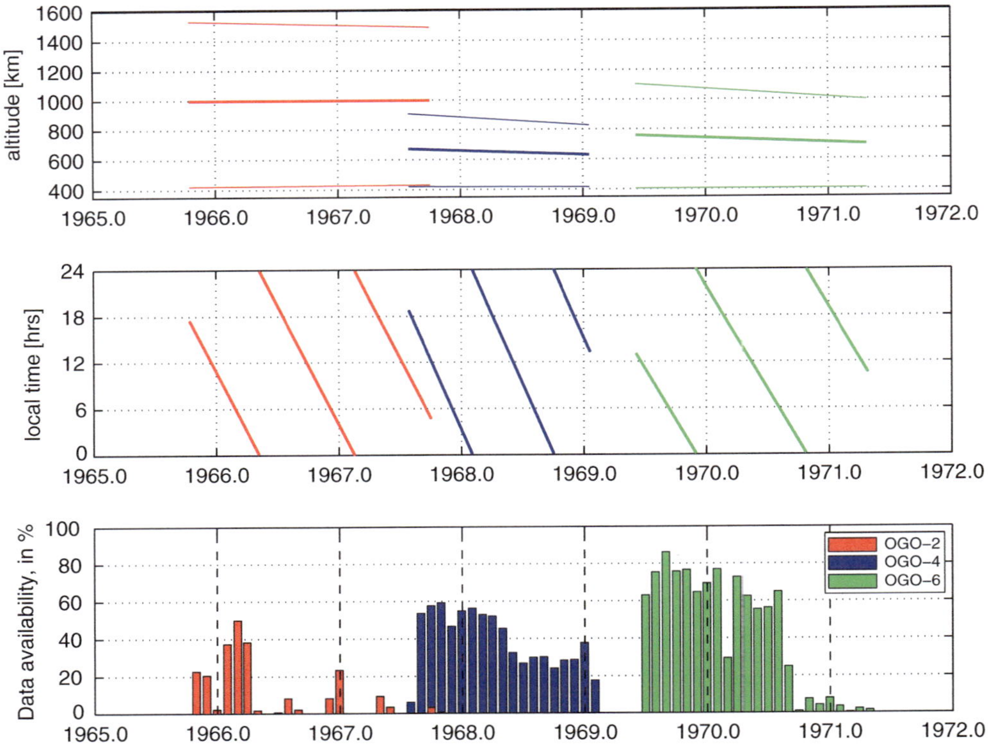

Fig. 2.4 Top: altitude of orbit perigee, resp. apogee (*thin lines*), and of mean altitude (*thick lines*) for OGO-2, -4 and -6. *Middle*: Local time of the ascending node (i.e., equator crossing of the south-going track). *Bottom*: Data availability

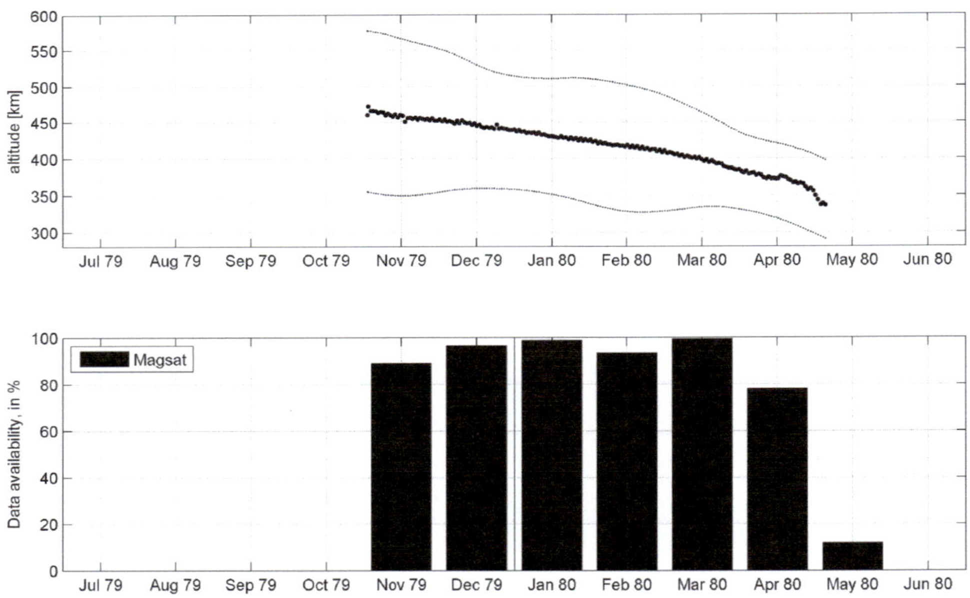

Fig. 2.5 *Top*: altitude of orbit perigee, resp. apogee (*thin lines*), and of mean altitude (*thick lines*) for the Magsat satellite. *Bottom*: Magsat satellite data availability

satellite and its instrumentation has been a model for other present and forthcoming missions like CHAMP and *Swarm*.

Ørsted is flying in a gravity gradient stabilized configuration; attitude maneuvers are performed using magnetic torquers. The orbit has an inclination of 96.5°, a period of 100.0 minutes, a perigee at 650 km and an apogee at 860 km. The orbit plane is slowly drifting, the local time of the equator crossing decreases by 0.91 min day^{-1}, starting from an initial local time of 02:26 on 23rd February 1999 for the south-going track, as shown in the upper panel of Fig. 2.6.

The satellite is equipped with an 8 m long deployable boom carrying the magnetic field instruments. A proton precession Overhauser magnetometer (OVH), measuring the magnetic field intensity with a sampling rate of 1 Hz and an accuracy better than 0.5 nT, is mounted at the tip of the boom. At a distance of 6 m from the satellite body is the optical bench with the CSC (Compact Spherical Coil) fluxgate vector magnetometer mounted closely together with the Star Imager (SIM). The CSC samples the magnetic field at 100 hz (burst mode, at polar latitudes) or 25 Hz (normal mode) with a resolution better than 0.1 nT. It is calibrated using the field intensity measured by the OVH (cf. Section 2.2.2). After calibration, the agreement between the two magnetometers is better than 0.33 nT rms. Due to attitude errors, the accuracy of the vector components (B_r, B_θ, B_ϕ) is limited to 2 to 8 nT (4 nT rms), depending on component.

The lower part of Fig. 2.6 shows the availability of Ørsted scalar data (MAG-F, black) and of vector data (MAG-L, blue). Since attitude data (red) are essential for providing vector data, drop outs of the SIM (for instance due to thermal problems, or due to blinding of the instrument by the Sun, Moon, or Earth) limit the availability of Ørsted vector data. The satellite is fully illuminated by the Sun for a few months roughly every 2 to 2.5 years (shaded areas in the Figure); this happened for the first time from July to November 2000. Since the satellite is not designed for this situation (nominal lifetime of 14 months ended in April 2000), thermal problems results in decreased data availability during these periods, as shown in the bottom panel of the Figure. As of March 2010, after more than 11 years in space, the satellite is still healthy and provides high-precision magnetic data—since 2005 only of the field intensity. See Neubert et al. (2001) and Olsen (2007) for more information on the satellite.

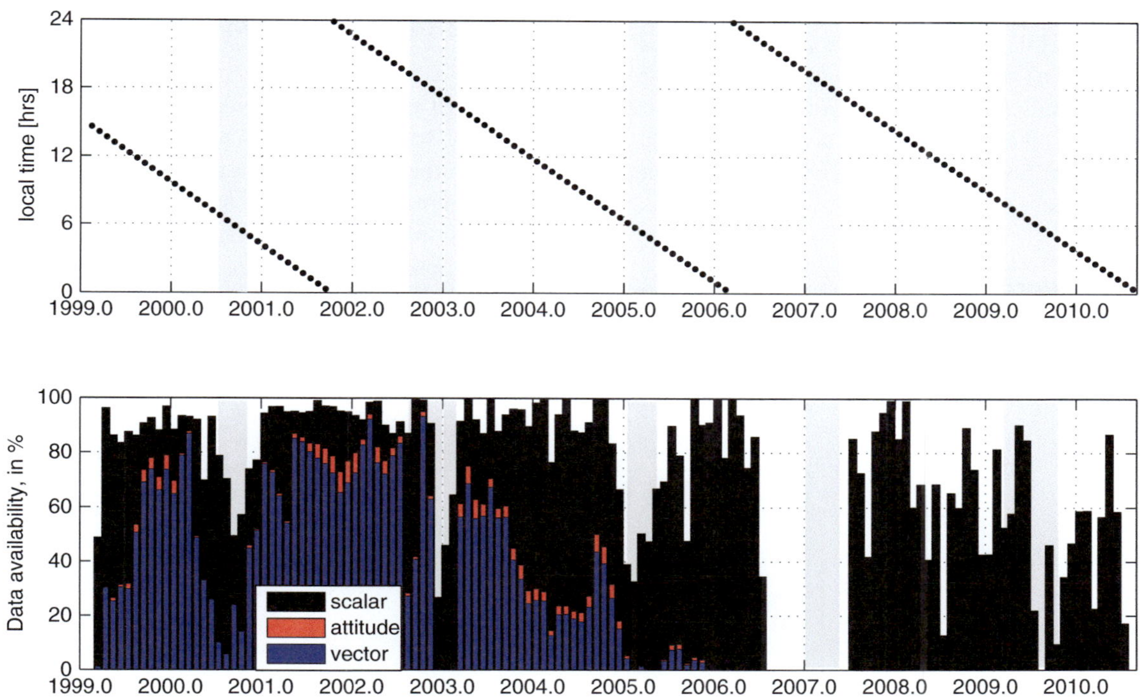

Fig. 2.6 *Top*: Local time evolution of the Ørsted orbit ascending node. *Bottom*: Data availability as of September 2010

2.3.4 SAC-C/Ørsted-2

A copy of the Ørsted boom and payload (but with a Scalar Helium Magnetometer instead of the Overhauser magnetometer) was launched in November 2000 on-board the Argentinean satellite SAC-C. This satellite is in a sun-synchronous orbit (local time of the ascending node is 22:00 ± 10 minutes) at about 700 ± 25 km altitude. Due to a broken connection in a coaxial cable, no high-precision attitude data (and hence no reliable vector data) are available. Figure 2.7 shows the availability of the SAC-C scalar data (MAG-F).

2.3.5 CHAMP

The German CHAMP satellite was launched on July 15, 2000 into a near polar (inclination 87.3°) orbit with an initial altitude of 454 km. Atmospheric re-entry happened on 19th September 2010. The satellite advanced one hour in local time within eleven days (see Fig. 2.8). Instrumentation is very similar to that of Ørsted; however, attitude is obtained by combining measurements taken by two star imager heads, to minimize attitude error anisotropy. Accuracy of the scalar measurements is similar to that of Ørsted (better than 0.5 nT), while that of the vector components is better than 2 nT when attitude is measured by both star imager heads (which happens for more than 60% of the time), otherwise the same accuracy as for Ørsted is achieved. See Maus (2007) for more information on the satellite.

2.3.6 *Swarm*

The *Swarm* constellation mission was selected by the European Space Agency (ESA) in 2004. Scheduled for launch in 2012, the mission comprises a constellation of three identical satellites, with two spacecraft flying side-by-side at lower altitude (450 km initial altitude) separated in longitude by 1.4° (corresponding to about 150 km at the equator). This configuration allows for an instantaneous estimation of the East-West gradient of the magnetic field. The third satellite will fly at higher altitude (530 km) and at a different local time compared to the lower satellite pair. The local time difference between the orbits of the higher satellite and of the lower pair increases from 0 h at launch to 6 h after 2–3 years, allowing for better determination of the space-time structure of large-scale magnetospheric fields. Each of the three *Swarm* satellites takes high-precision and high-resolution measurements of the strength, direction and variation of the magnetic field, complemented by precise navigation, accelerometer and electric field measurements. In combination they provide the necessary observations that are required to separate and model the various sources of the geomagnetic field. Fig. 2.9 shows the satellites and their instrumentation. More details on the mission can be found in Friis-Christensen et al. (2006).

2.4 The Years After *Swarm*

Swarm will provide a mapping of the Earth's magnetic field which is superior to the one that is possible with the present missions Ørsted, CHAMP and SAC-C. However, no follow-up mission is presently planned to measure the geomagnetic field after 2015. Looking toward the years after *Swarm*, there is a wide spectrum of directions that exploration of the geomagnetic field from space could take. The end points of this spectrum may be described by the following two scenarios.

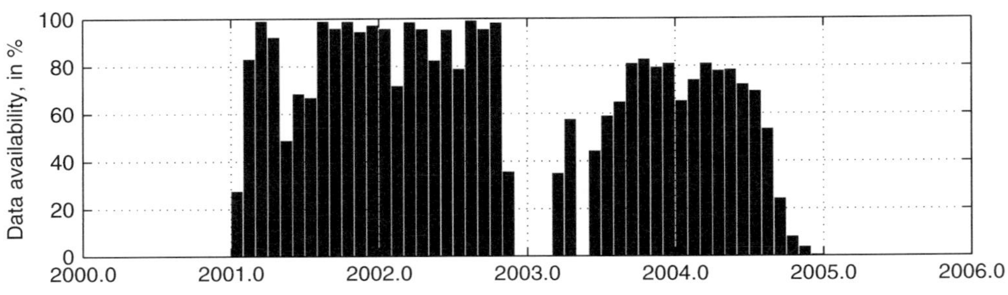

Fig. 2.7 SAC-C magnetic scalar data availability

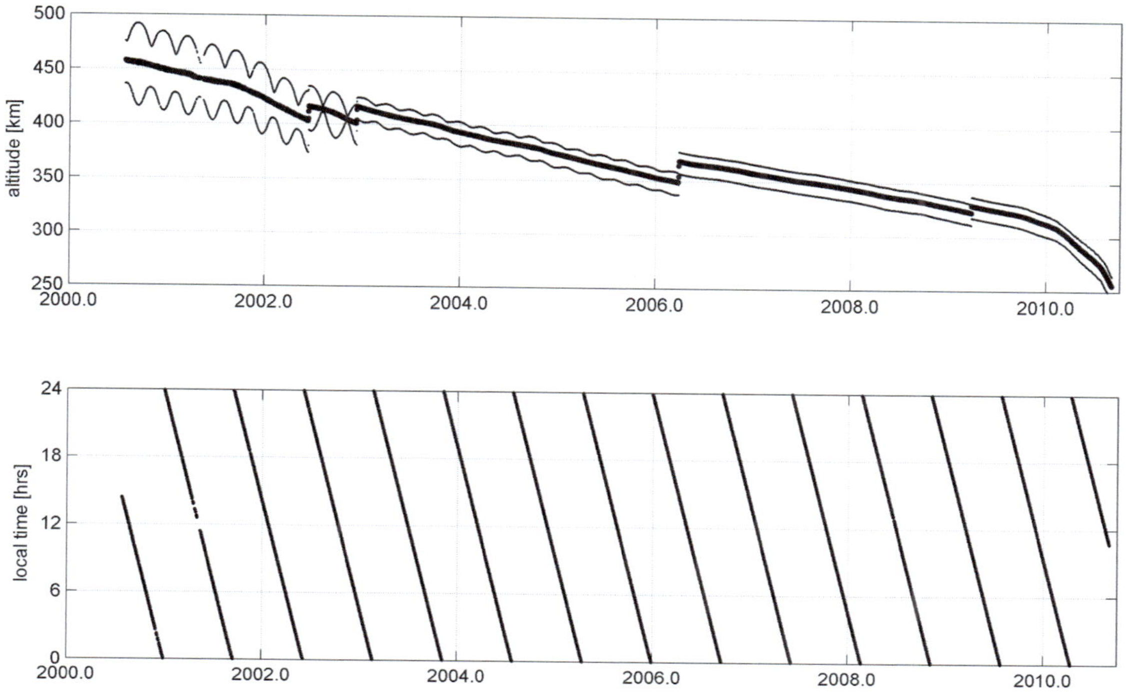

Fig. 2.8 *Top*: altitude of orbit perigee, resp. apogee (*thin lines*) and of mean altitude (*thick lines*) for the CHAMP satellite. *Bottom*: Local time evolution of the ascending node of the CHAMP orbit

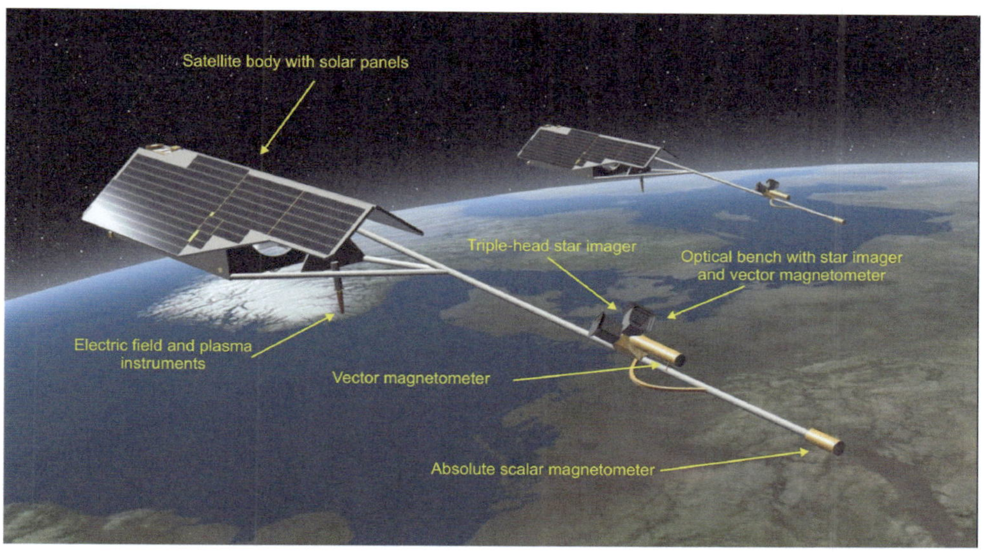

Fig. 2.9 Two (out of three) *Swarm* satellites

A fleet of satellites for monitoring core field evolution

The combination of data taken by the missions Ørsted, SAC-C, CHAMP and *Swarm* will result in more than 15 years of (almost) continuous data for studying variations of the geomagnetic field. But also after 2016 a continuous magnetic field monitoring from space is needed, both for fundamental science and technical applications: Geomagnetic reference models, like the IGRF (the de-facto industry standard of the geomagnetic field, used for instance for navigation, exploration and drilling) or the WMM (the standard model of NATO and the military), have to be updated regularly, and a true global coverage of magnetic field observations, which is a prerequisite for obtaining good field models, can only be done from space. Monitoring of the geomagnetic field evolution from space is possible with relatively simple and cheap satellites dedicated for taking high-precision magnetic field measurements. Although vector data, as obtained for instance by the Ørsted satellite, are optimal for this task, measuring the magnetic field intensity (scalar field) alone would also be highly beneficial from a scientific point of view, and technically much simpler and cheaper. A "fleet" of scalar-only satellites measuring the field intensity at various altitudes and local times is therefore a very attractive option for the years after *Swarm*. The scientific benefit of such a mission should be investigated in detail.

Magnetic Space Gradiometry

The two lower-altitude side-by-side flying *Swarm* satellites will measure for the first time the East-West gradient of the magnetic field, which contains valuable information on North-South oriented features of crustal magnetization. A similar resolution of East-West oriented structures is, however, not possible with *Swarm*, since this requires measurement of the North-South (or, alternatively, of the radial) magnetic field gradient.

It is helpful in this context to look at the analogy between missions for exploring Earth's magnetic and gravity fields. Ørsted and CHAMP are satellites for measuring the field only (**B** in case of Ørsted, and **B** and **g** in case of CHAMP), but not its gradient, while GRACE and *Swarm* measure one component of the gradient (in case of GRACE a quantity related to the North-South gradient of the gravity field, and in case of *Swarm* the East-West gradient of the magnetic field). ESA's GOCE mission, launched in March 2009, measures the full gravity gradient tensor. A magnetic equivalent to GOCE would be a natural next step for exploring the geomagnetic field from space after *Swarm*. Such a mission allows for an isotropic high-resolution mapping of crustal magnetization and for an in-situ determination of electric current density in space.

Measuring the full magnetic field gradient tensor in space is a major technical challenge, and may not be easy to obtain in near future. However, determination of some elements of the gradient tensor is technically simpler; cf. the East-West gradient estimation with *Swarm*. Measuring the radial gradient is for instance possible using tethered missions or by a satellite constellation flying in "cartwheel" configuration (e.g., Wiese et al. 2009). Regardless of the technical difficulties in measuring the full magnetic gradient tensor in space we discuss in the following section the additional information that is provided by the magnetic gradient tensor.

2.5 Outlook: Space Magnetic Gradiometry

In three-dimensional space the spatial variation of a vector like the magnetic field **B** can be linearly approximated by the derivatives of the field components in the three different spatial directions. This approximation defines the gradient tensor, consisting of $3 \times 3 = 9$ spatial derivatives and forming a second rank tensor.

Each element of the magnetic gradient tensor represents a directional filter and emphasizes certain magnetic field structures. This allows for enhancement of certain features of the field, or suppression of specific undesirable contributions. The magnetic gradient tensor is therefore a powerful tool for detecting hidden magnetic field structures. Magnetic gradiometry is used for regional studies based on near-surface data, see e.g., Pedersen and Rasmussen (1990); Schmidt and Clark (2006) and references therein for more details. However, application to satellite data of global coverage is somehow different from the gradiometry used in exploration geophysics, because of the existence of electric currents at satellite altitude and the necessity to use spherical rather than Cartesian coordinates.

In this section we discuss some properties of the magnetic gradient tensor, making different assumptions. We start from the most general case of a solenoidal field ($\nabla \cdot \mathbf{B} = 0$). This condition reduces the number of independent elements of the tensor from 9 to 8. Each solenoidal field can be decomposed into poloidal and toroidal parts according to the Mie representation of vector fields, cf. Eq. (2.4). It is possible to construct the gradient tensor for each of these two parts of $\mathbf{B}$. In current-free regions ($0 = \mu_0 \mathbf{J} = \nabla \times \mathbf{B}$) the gradient tensor is symmetric, which reduces the number of independent elements further from 8 to five. In that case $\mathbf{B}$ is a Laplacian potential field, the toroidal part vanishes, and the remaining poloidal part can be decomposed into internal and external field parts.

2.5.1 General Case: B as a Solenoid Vector Field

The magnetic gradient tensor is conveniently derived by means of tensor calculus. More specifically, the gradient tensor elements can be constructed from the covariant derivative $B_{p;q}$ which is a generalization of the directional derivative. The covariant derivative of the magnetic field $\mathbf{B}$ is given by

$$B_{p;q} = \frac{\partial B^p}{\partial y^q} + \Gamma^p_{kq} B^k$$

where B^p denotes the component of the magnetic field vector $\mathbf{B}$, y^q is the direction of the differentiation, and summation has to be taken over all three values of k. Γ^p_{kq} are the Christoffel symbols, a set of coefficients specifying the differentiation for a given specific coordinate system. For more details on the covariant derivative see Talpaert (2002).

In spherical coordinates (r, θ, ϕ) we have $B^p = (B_r, B_\theta, B_\phi)$ and the only non-zero Christoffel symbols are

$$\begin{aligned}
\Gamma^\theta_{r\theta} &= \Gamma^\phi_{r\phi} = \tfrac{1}{r} \\
\Gamma^r_{\theta\theta} &= \Gamma^r_{\phi\phi} = -\tfrac{1}{r} \\
\Gamma^\phi_{\theta\phi} &= \tfrac{\cot\theta}{r} \\
\Gamma^\theta_{\phi\phi} &= -\tfrac{\cot\theta}{r}.
\end{aligned}$$

Using the differential directions

$$\frac{\partial}{\partial y^r} = \frac{\partial}{\partial r}, \frac{\partial}{\partial y^\theta} = \frac{1}{r}\frac{\partial}{\partial \theta}, \frac{\partial}{\partial y^\phi} = \frac{1}{r\sin\theta}\frac{\partial}{\partial \phi},$$

the magnetic gradient tensor in spherical coordinates follows as

$$\nabla\mathbf{B} = \begin{pmatrix} \frac{\partial B_r}{\partial r} & \frac{1}{r}\frac{\partial B_r}{\partial \theta} - \frac{1}{r}B_\theta & \frac{1}{r\sin\theta}\frac{\partial B_r}{\partial \phi} - \frac{1}{r}B_\phi \\ \frac{\partial B_\theta}{\partial r} & \frac{1}{r}\frac{\partial B_\theta}{\partial \theta} + \frac{1}{r}B_r & \frac{1}{r\sin\theta}\frac{\partial B_\theta}{\partial \phi} - \frac{\cot\theta}{r}B_\phi \\ \frac{\partial B_\phi}{\partial r} & \frac{1}{r}\frac{\partial B_\phi}{\partial \theta} & \frac{1}{r\sin\theta}\frac{\partial B_\phi}{\partial \phi} + \frac{1}{r}B_r + \frac{\cot\theta}{r}B_\theta \end{pmatrix}. \tag{2.13}$$

The trace of the tensor corresponds to the divergence of the field, and because of $\nabla \cdot \mathbf{B} = 0$ the trace of the magnetic gradient tensor is always zero,

$$\nabla \cdot \mathbf{B} = 0 \Leftrightarrow tr\,(\nabla\mathbf{B}) = 0, \tag{2.14}$$

which reduces the number of independent tensor elements from 9 to 8.

A powerful aspect of the gradient tensor is that it provides information on the in-situ electrical current density $\mathbf{J} = (J_r, J_\theta, J_\phi)^T$ by taking the difference of the gradient tensor and its transpose:

$$(\nabla\mathbf{B}) - (\nabla\mathbf{B})^T = \mu_0 \begin{pmatrix} 0 & -J_\phi & +J_\theta \\ +J_\phi & 0 & -J_r \\ -J_\theta & +J_r & 0 \end{pmatrix}. \tag{2.15}$$

In a current-free region ($\mathbf{J} = 0$) the gradient tensor is symmetric since Eq. (2.15) leads to $(\nabla\mathbf{B}) = (\nabla\mathbf{B})^T$. This condition reduces the number of independent tensor elements to 5.

Note that in the spherical coordinate frame all tensor elements except those in the first column contain, in addition to the field derivatives along the direction θ or ϕ, also contributions from the field components, cf. Eq. (2.13). They thus represent a mixture of contributions from the field and its spatial derivative. Since the former is dominated by the large-scale main field (i.e., contributions described by spherical harmonic degrees up to, say, $n = 13$) this contribution can in good approximation be neglected when only looking for small-scale features, and Eq. (2.13) reduces to

$$
\nabla \mathbf{B} \approx \begin{pmatrix} \frac{\partial B_r}{\partial r} & \frac{1}{r}\frac{\partial B_r}{\partial \theta} & \frac{1}{r\sin\theta}\frac{\partial B_r}{\partial \phi} \\ \frac{\partial B_\theta}{\partial r} & \frac{1}{r}\frac{\partial B_\theta}{\partial \theta} & \frac{1}{r\sin\theta}\frac{\partial B_\theta}{\partial \phi} \\ \frac{\partial B_\phi}{\partial r} & \frac{1}{r}\frac{\partial B_\phi}{\partial \theta} & \frac{1}{r\sin\theta}\frac{\partial B_\phi}{\partial \phi} \end{pmatrix}. \tag{2.16}
$$

In that approximation the tensor elements contain only contributions from the field derivatives along each direction.

2.5.2 Toroidal-Poloidal Decomposition

As stated in Section 2.1.2, the magnetic field can be written uniquely as the sum of a toroidal and a poloidal part (cf. Backus 1986; Sabaka et al. 2010). The toroidal and poloidal fields are each associated with a scalar function from which the fields can be derived via appropriate curl operations, see Eq. (2.4). In spherical coordinates this decomposition is given by

$$
\mathbf{B} = \underbrace{\begin{pmatrix} 0 \\ \frac{1}{\sin\theta}\frac{\partial}{\partial\phi}\Phi \\ -\frac{\partial}{\partial\theta}\Phi \end{pmatrix}}_{\text{toroidal part}} + \underbrace{\begin{pmatrix} -\Delta_s(r\Psi) \\ \frac{1}{r}\frac{\partial}{\partial\theta}(r\Psi)' \\ \frac{1}{r\sin\theta}\frac{\partial}{\partial\phi}(r\Psi)' \end{pmatrix}}_{\text{poloidal part}} \tag{2.17}
$$

with $(r\Psi)' = \frac{d(r\Psi)}{dr}$ and $\Delta_s = \frac{1}{r^2\sin\theta}\frac{\partial}{\partial\theta}\left(\sin\theta\frac{\partial}{\partial\theta}\right) + \frac{1}{r^2\sin^2\theta}\frac{\partial^2}{\partial\phi^2}$ as the horizontal part of the Laplacian.

According to Eq. (2.13) the gradient tensor of the toroidal, resp. poloidal, part of the magnetic field follows as

$$
\nabla \mathbf{B}_{\text{tor}} = \begin{pmatrix} 0 & -\frac{1}{r\sin\theta}\frac{\partial\Phi}{\partial\phi} & \frac{1}{r}\frac{\partial\Phi}{\partial\theta} \\ \frac{1}{\sin\theta}\frac{\partial^2\Phi}{\partial r\partial\phi} & \frac{\partial}{r\partial\theta}\left(\frac{1}{\sin\theta}\frac{\partial\Phi}{\partial\phi}\right) & \frac{\cos\theta}{r\sin\theta}\frac{\partial\Phi}{\partial\theta} + \frac{1}{r\sin^2\theta}\frac{\partial^2\Phi}{\partial\phi^2} \\ -\frac{\partial^2\Phi}{\partial r\partial\theta} & -\frac{1}{r}\frac{\partial^2\Phi}{\partial\theta^2} & -\frac{\partial}{r\partial\theta}\left(\frac{1}{\sin\theta}\frac{\partial\Phi}{\partial\phi}\right) \end{pmatrix} \tag{2.18a}
$$

$$
\nabla \mathbf{B}_{\text{pol}} = \begin{pmatrix} -\frac{\partial}{\partial r}\Delta_s(r\Psi) & -\frac{1}{r}\frac{\partial}{\partial\theta}\left(\Delta_s(r\Psi) + \frac{1}{r}(r\Psi)'\right) & -\frac{1}{r\sin\theta}\frac{\partial}{\partial\phi}\left(\Delta_s(r\Psi) + \frac{1}{r}(r\Psi)'\right) \\ \frac{\partial}{\partial r}\left(\frac{\partial(r\Psi)'}{r\partial\theta}\right) & -\frac{1}{r}\Delta_s(r\Psi) + \frac{1}{r^2}\frac{\partial^2}{\partial\theta^2}(r\Psi)' & \frac{\partial}{\partial\theta}\left(\frac{1}{r^2\sin\theta}\frac{\partial(r\Psi)'}{\partial\phi}\right) \\ \frac{1}{\sin\theta}\frac{\partial}{\partial r}\left(\frac{\partial(r\Psi)'}{r\partial\phi}\right) & \frac{\partial}{\partial\theta}\left(\frac{1}{r^2\sin\theta}\frac{\partial(r\Psi)'}{\partial\phi}\right) & -\frac{1}{r}\Delta_s(r\Psi) + \frac{1}{r^2\sin\theta}\left(\frac{1}{\sin\theta}\frac{\partial^2}{\partial\phi^2} + \cos\theta\frac{\partial}{\partial\theta}\right)(r\Psi)' \end{pmatrix}. \tag{2.18b}
$$

The general constraining Eq. (2.14) holds for both the toroidal and the poloidal field gradient tensor: the trace of each tensor is zero, which confirms that both the toroidal and the poloidal parts are solenoidal fields. However, in general these tensors are not symmetric; they are only symmetric in the case of vanishing current density (i.e., $\nabla \times \mathbf{B} = \mu_0 \mathbf{J} = 0$).

2.5.3 Laplacian Potential Approximation

Let us now assume that the magnetic measurements are taken in a source-free region (absence of currents, $\mathbf{J} = 0$). In that case the magnetic field, $\mathbf{B} = -\nabla\, \mathbf{V}$ is a Laplacian potential field. The potential V can be separated into an internal and an external part, $V = V^{\text{int}} + V^{\text{ext}}$, each of which may be expanded in series of spherical harmonics, see Eq. (2.9). The total magnetic field $\mathbf{B}$ follows as

$$
\mathbf{B} = \underbrace{\begin{pmatrix} -\frac{\partial V^{\text{int}}}{\partial r} \\ -\frac{1}{r}\frac{\partial V^{\text{int}}}{\partial\theta} \\ -\frac{1}{r\sin\theta}\frac{\partial V^{\text{int}}}{\partial\phi} \end{pmatrix}}_{\mathbf{B}^{\text{int}}} + \underbrace{\begin{pmatrix} -\frac{\partial V^{\text{ext}}}{\partial r} \\ -\frac{1}{r}\frac{\partial V^{\text{ext}}}{\partial\theta} \\ -\frac{1}{r\sin\theta}\frac{\partial V^{\text{ext}}}{\partial\phi} \end{pmatrix}}_{\mathbf{B}^{\text{ext}}}. \tag{2.19}
$$

According to Eq. (2.13) the gradient tensor is then given by:

$$
\nabla \mathbf{B} = \begin{pmatrix} -\frac{\partial^2 V}{\partial r^2} & -\frac{\partial}{\partial r}\left(\frac{1}{r}\frac{\partial V}{\partial\theta}\right) & -\frac{\partial}{\partial r}\left(\frac{1}{r\sin\theta}\frac{\partial V}{\partial\phi}\right) \\ -\frac{\partial}{\partial r}\left(\frac{1}{r}\frac{\partial V}{\partial\theta}\right) & -\frac{1}{r}\frac{\partial V}{\partial r} - \frac{1}{r^2}\frac{\partial^2 V}{\partial\theta^2} & -\frac{\partial}{\partial\theta}\left(\frac{1}{r^2\sin\theta}\frac{\partial V}{\partial\phi}\right) \\ -\frac{\partial}{\partial r}\left(\frac{1}{r\sin\theta}\frac{\partial V}{\partial\phi}\right) & -\frac{\partial}{\partial\theta}\left(\frac{1}{r^2\sin\theta}\frac{\partial V}{\partial\phi}\right) & -\frac{1}{r}\frac{\partial V}{\partial r} - \frac{\cot\theta}{r^2}\frac{\partial V}{\partial\theta} - \frac{1}{r^2\sin^2\theta}\frac{\partial^2 V}{\partial\phi^2} \end{pmatrix} \tag{2.20}
$$

where $V = V^{\text{int}}$ or $V = V^{\text{ext}}$ respectively. Of course the general constraining Eq. (2.14) holds also for this case and results in the Laplace equation, $\nabla^2 V = 0$, for the potential V. In addition, the gradient tensor is now symmetric, in accordance to what was stated before: in case of vanishing currents (source-free region) only five tensor elements have to be specified in order to fully determine the tensor.

Each tensor element emphasizes magnetic field structures of a particular orientation. In terms of a spherical harmonic expansion this corresponds to a sensitivity to a certain degree n and order m. Writing the spherical harmonic expansion of the internal magnetic field potential, Eq. (2.9a), in complex form yields

$$V^{\text{int}} = a \sum_{n=1}^{N_{\text{int}}} \sum_{m=0}^{n} \left(\frac{a}{r}\right)^{n+1} \gamma_n^m P_n^m e^{im\phi}$$

with $\gamma_n^m = g_n^m - ih_n^m$ from which the magnetic field components follow as

$$\mathbf{B} = \begin{pmatrix} B_r \\ B_\theta \\ B_\phi \end{pmatrix} = \begin{pmatrix} -\frac{\partial V}{\partial r} \\ -\frac{1}{r}\frac{\partial V}{\partial \theta} \\ -\frac{1}{r\sin\theta}\frac{\partial V}{\partial \phi} \end{pmatrix}$$

$$= \begin{pmatrix} (n+1)P_n^m \\ -\frac{dP_n^m}{d\theta} \\ -\frac{im}{\sin\theta}P_n^m \end{pmatrix} \left(\frac{a}{r}\right)^{n+2} \gamma_n^m e^{im\phi}.$$

Compared to the radial dependence $\propto (a/r)^{n+2}$ of the field, the gradient tensor has radial dependence $\propto (a/r)^{n+3}$. In addition to this different radial dependency, the expansion coefficients of tensor elements describing the radial derivative (i.e., the left column of the tensor, Eq. (2.20) using the approximation leading to Eq. (2.16)) are those of the field components multiplied by $(n+2)$, which emphasizes magnetic structures described by high *degree* spherical harmonics. Likewise, coefficients of tensor elements related to the East-West derivative (right column of the tensor) are multiplied by $im/\sin\theta$, which emphasizes the high *order* terms. Note that it is not possible to determine the zonal terms ($m = 0$) from the East-West gradient alone. This, on the other hand, means that these tensor elements are not disturbed by contributions from the magnetospheric ring-current, which is one of the largest sources of "noise" for geomagnetic field modeling.

2.5.4 Magnetic Field Gradient Tensor Visualization

Figure 2.10 shows the magnetic field components $\mathbf{B} = (B_r, B_\theta, B_\phi)^T$ of the small-scale part of the crustal field (spherical harmonic degrees $n = 16 - 60$) at 300 km altitude, according to model MF6 by Maus et al. (2008). The corresponding magnetic gradient tensor $\nabla\mathbf{B}$ is shown in Fig. 2.11. Since the crustal field has a "flat" power spectrum at long wavelengths, a sharp cut-off at a certain spherical harmonic degree (e.g., $n = 15$) in the wavenumber domain (corresponding to setting all coefficients with degree $n < 16$ to zero)

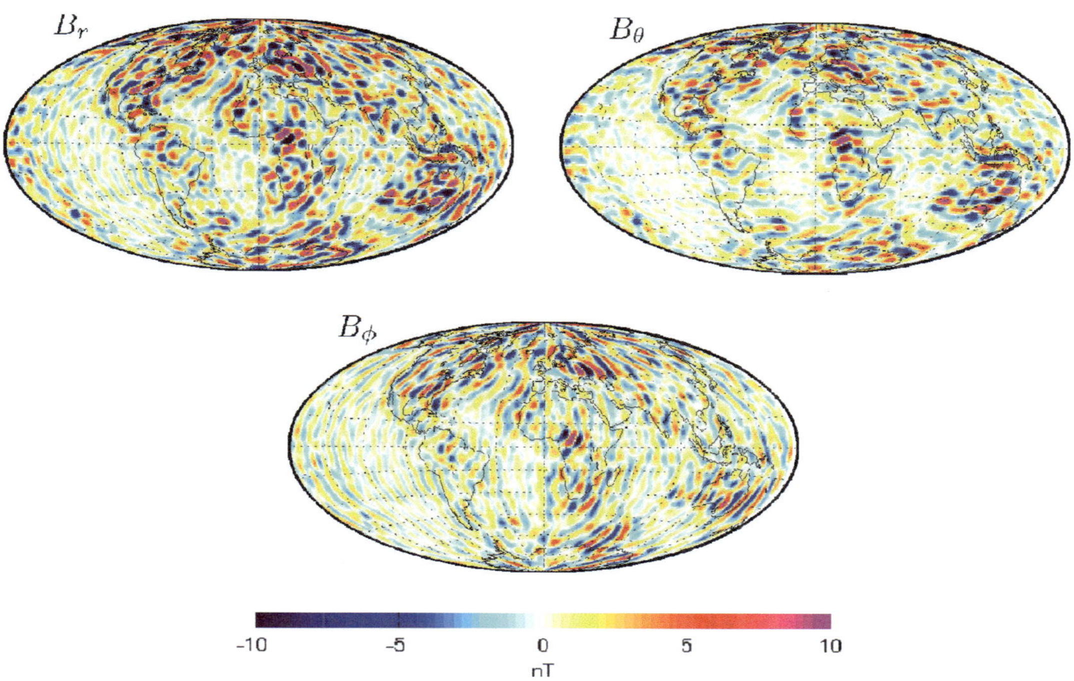

Fig. 2.10 Magnetic field components (spherical harmonic degrees $n > 15$) at 300 km altitude due to the crustal magnetization

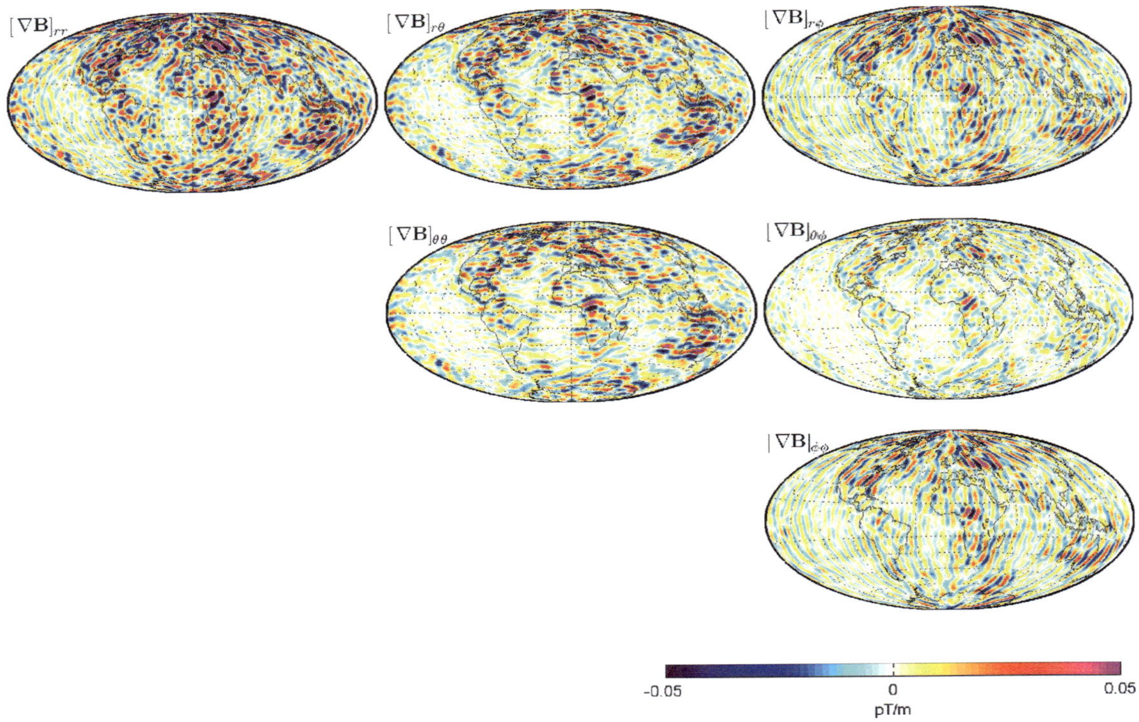

Fig. 2.11 Magnetic field gradient tensor elements (spherical harmonic degrees $n > 15$) at 300 km altitude due to the crustal magnetization

leads to ringing in the space domain. Therefore, the crustal field has been filtered using a high-pass Hann filter.

Each tensor element acts as a directional filter, emphasizing structures in particular orientations and thus depicting certain features of the field. For the Cartesian case Schmidt and Clark (2006) describe some properties of the magnetic gradient tensor that also hold for the spherical frame, as confirmed by Fig. 2.10. In following $[\nabla\mathbf{B}]_{pq}$ denotes the tensor element corresponding to the derivative of the pth magnetic field component in direction q.

- $[\nabla\mathbf{B}]_{rr}$ outlines steep boundaries.This tensor element gives a clear visualization of the magnetic anomaly delineations associated with the North-Atlantic mid-oceanic ridge between Europe and North America.
- $[\nabla\mathbf{B}]_{\theta\theta}$ and $[\nabla\mathbf{B}]_{r\theta}$ outline East-West boundaries.
- $[\nabla\mathbf{B}]_{\phi\phi}$ and $[\nabla\mathbf{B}]_{r\phi}$ delineate North-South boundaries. Both elements provide information on the behavior of the crustal field in the East-West direction, and combination of the two elements gives information about horizontal, resp. vertical, magnetization direction. For example, the delineated anomalies in the ocean bottoms mainly due to plate tectonics are amplified.
- $[\nabla\mathbf{B}]_{\theta\phi}$ outlines body corners, i.e., it sharpens the boundaries of crustal magnetic anomalies. Namely, this tensor element depicts the strongest anomalies. As an example, the Bangui anomaly is clearly outlined. This anomaly has been studied by Ravat et al. (2002) with gradient methods.

Comparing the gradient tensor elements with the magnetic field components shows that small scale anomalies are amplified and better resolved by the tensor elements. Therefore, the tensor not only enhances certain features of the crustal field, but also captures a more detailed signature of the field.

As mentioned earlier, measuring the magnetic gradient tensor from space is challenging. In case of the *Swarm* constellation mission (cf. Section 2.3.6) the East-West magnetic field difference is measured by the lower pair of satellites, which allows for an estimation of the East-West gradient part corresponding to the right column of the gradient tensor: The difference of the magnetic field vector measured by two satellites flying simultaneously with a longitudinal separation $\Delta\phi$ is $\Delta\mathbf{B} = \mathbf{B}(r,\theta,\phi) - \mathbf{B}(r,\theta,\phi + \Delta\phi) = -\mathrm{Re}\{\nabla\ \Delta V\}$, where ΔV is a spherical harmonic expansion with coefficients

$$\Delta\gamma_n^m = \gamma_n^m \left(1 - \mathrm{e}^{im\Delta\phi}\right). \qquad (2.21)$$

Hence by analyzing the difference of the magnetic field measured by the two satellites the Gauss coefficients γ_n^m of the internal potential are multiplied with some filter factors with filter gain $\mid\left(1 - \mathrm{e}^{im\Delta\phi}\right)\mid = \sqrt{2(1 - \cos m\Delta\phi)}$. For small values of $\Delta\phi$ this quantity becomes $\mid\left(1 - \mathrm{e}^{im\Delta\phi}\right)\mid \approx m\Delta\phi$, which approximates the East-West gradient. Fig. 2.12 shows the filter gain for three different values of longitudinal separation, $\Delta\phi$, of the satellites. *Swarm* aims at a determination of the lithospheric field up to spherical harmonics of degree and order 150 corresponding to a spatial scale of 270 km. As seen from the Figure, the optimal longitudinal separation of the satellites used to estimate the East-West gradient is $\Delta\phi \approx 1.4°$ for that case, which indeed is the value chosen for *Swarm*.

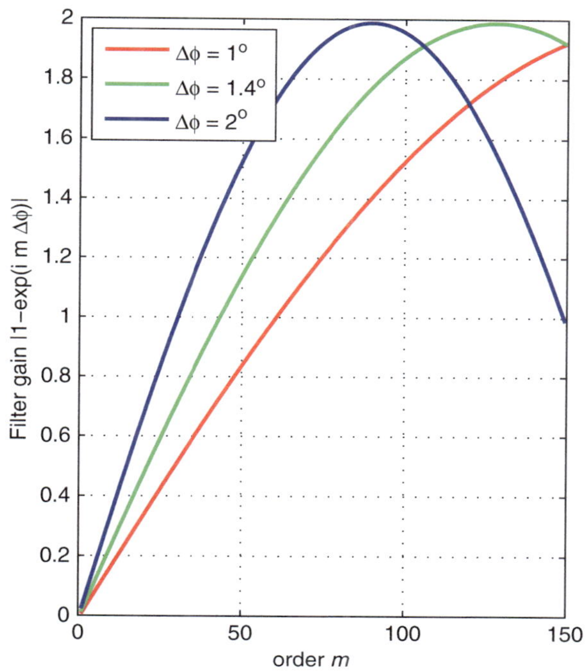

Fig. 2.12 Relative sensitivity of the East-West magnetic difference versus spherical harmonic order *m*, for three different longitude separations, $\Delta\phi$, of the spacecraft

The advantage of including information on the East-West gradient for improved determination of the crustal field has been demonstrated in a full 4.5 years mission simulation which was carried out as part of the preparation of *Swarm* (see Sabaka and Olsen (2006); Olsen et al. (2007); Tøffner-Clausen et al. (2010) for details). In that study, synthetic magnetic signals were generated for all relevant contributions to Earth's magnetic field (core and lithospheric fields, fields due to currents in the ionosphere and magnetosphere, due to their secondary, induced, currents in the oceans, lithosphere and mantle, and fields due to currents coupling the ionosphere and magnetosphere) and the *Comprehensive Inversion* scheme (Sabaka and Olsen 2006) has been used to recover the various field contributions. The study confirms the great advantage of including the part of the magnetic gradient tensor describing the East-West derivative for determination of the high-order crustal fields. Taking explicit advantage of this information reduces the error of the crustal field roughly by a factor of two (Tøffner-Clausen et al. 2010).

2.6 Summary

Although the first satellite observations of the Earth's magnetic field were already taken more than 50 years ago, continuous geomagnetic measurements from space are only available for about one decade. Since 1999 the satellites Ørsted, CHAMP and SAC-C provide a unique data set of high-precision magnetic field observations, and the unprecedented time-space coverage of these data opens revolutionary new possibilities for exploring the Earth's magnetic field from space. In the near future, the *Swarm* satellite constellation mission, comprising of three satellites to be launched in 2012, will provide even better measurements of the geomagnetic field. Data from *Swarm* will allow for a determination of the small-scale structure of the lithospheric field down to length-scales of 270 km. In addition, they will extend the continuous monitoring of core field changes that begun with the launch of Ørsted in 1999 by at least additional 4.5 years, i.e., until end of 2016.

References

M. Acuña (2002) Space-based magnetometers. Rev Sci Instrum 73:3717

Backus G (1986) Poloidal and toroidal fields in geomagnetic field modeling. Rev Geophys 24:75–109

Blakely RJ (1995) Potential theory in gravity and magnetic applications. Cambridge Press, Cambridge

Cain JC (2007) POGO (OGO-2, -4 and -6 spacecraft), In: Gubbins D, Herrero-Bervera E (eds) Encyclopedia of geomagnetism and paleomagnetism. 828–829 Springer, Heidelberg

Chapman S, Bartels J (1940) Geomagnetism, vol. I+II, Clarendon Press, Oxford, London

Dolginov S, Zhuzgov LN, Pushkov NV, Tyurmina LO, Fryazinov IV (1962) Some results of measurements of the constant geomagnetic field above the USSR from the third artificial Earth satellite. Geomagnetism Aeronomy 2:877–889

Friis-Christensen E, Lühr H, Hulot G (2006) Swarm: a constellation to study the Earth's magnetic field. Earth Planets Space 58:351–358

Hulot G, Sabaka TJ, Olsen N (2007) The present field, In: Kono M (ed) Treatise on Geophysics, vol 5. 33–75 Elsevier, Amsterdam

Langel RA (1987) The main field, In: Jacobs JA (ed) Geomagnetism, vol 1. Academic, London, pp 249–512

Maus S (2007) CHAMP magnetic mission. In: Gubbins D, Herrero-Bervera E (eds) Encyclopedia of geomagnetism and paleomagnetism. 59–60 Springer, Heidelberg

Maus S, Yin F, Lühr H, Manoj C, Rother M, Rauberg J, Michaelis I, Stolle C, Müller R (2008) Resolution of direction of oceanic magnetic lineations by the sixth-generation lithospheric magnetic field model from CHAMP satellite magnetic measurements. Geochem Geophys Geosyst 9(7):07021. doi:10.1029/2008GC001949

Neubert T, Mandea M, Hulot G, von Frese R, Primdahl F, Joergensen JL, Friis E-Christensen, Stauning P, Olsen N, Risbo T (2001) Ørsted satellite captures high-precision Geomagnetic field data. EOS Trans. AGU 82(7):81–88

Olsen N, Ørsted (2007) In: Gubbins D, Herrero-Bervera E (eds) Encyclopedia of geomagnetism and paleomagnetism. Pages 743–746 Springer, Heidelberg

Olsen N, Sabaka TJ, Gaya-Pique L (2007) Study of an improved comprehensive magnetic field inversion analysis for Swarm, DNSC Scientific Report 1/2007, Danish National Space Center, Copenhagen

Olsen N, Lühr H, Sabaka TJ, Mandea M, Rother M, Tøffner-Clausen L, Choi S (2006) CHAOS—a model of Earth's magnetic field derived from CHAMP, Ørsted, and SAC-C magnetic satellite data. Geophys J Int 166:67–75. doi:10.1111/j.1365-246X.2006.02959.x

Olsen N, Hulot G, Sabaka TJ (2010) Sources of the geomagnetic field and the modern data that enable their investigation, In: Freeden W, Nashed Z, Sonar T (eds) Handbook of geomathematics. in press Springer, Heidelberg

Olsen N, Tøffner-Clausen L, Sabaka TJ, Brauer P, Merayo JMG, Jørgensen JL, Léger JM, Nielsen OV, Primdahl F, Risbo T (2003) Calibration of the Ørsted vector magnetometer. Earth Planets Space 55:11–18

Pedersen L, Rasmussen T (1990) The gradient tensor of potential field anomalies: Some implications on data collection and data processing of maps. Geophysics 55(12):1558–1566

Primdahl F (1998) Scalar magnetometers for space applications. Geophys Monogr 103:85–99

Purucker ME (2007) Magsat. In Gubbins D, Herrero-Bervera E (eds) Encyclopedia of geomagnetism and paleomagnetism. 673–674 Springer, Heidelberg

Ravat D, Wang B, Wildermuth E, Taylor PT (2002) Gradients in the interpretation of satellite-altitude magnetic data: an example from central Africa. J Geodyn 33(1–2):131–142. doi:10.1016/S0264-3707(01)00059-X

Sabaka TJ, Olsen N (2006) Enhancing comprehensive inversions using the Swarm constellation. Earth Planets Space 58: 371–395

Sabaka TJ, Hulot G, Olsen N (2010) Mathematical properties relevant to geomagnetic field modelling. In: Freeden W, Nashed Z, Sonar T (eds) Handbook of geomathematics. in press Springer, Heidelberg

Schmidt P, Clark D (2006) The magnetic gradient tensor: its properties and uses in source characterization. Leading Edge 25(1):75–78. doi:10.1190/1.2164759

Seeber G (2004) Satellite geodesy, Walter de Gruyter, Berlin, New York

Talpaert Y (2002) Tensor analysis and continuum mechanics. Kluwer, Dordrecht

Tøffner-Clausen L, Sabaka TJ, Olsen N (2010) End-To-End Mission Simulation Study (E2E+), In: Proceedings of the Second International Swarm Science Meeting, ESA, Noordwijk, NL

Wiese D, Folkner W, Nerem R (2009) Alternative mission architectures for a gravity recovery satellite mission. J Geodesy 83(6):569–581. doi:10.1007/s00190-008-0274-1

Chapter 3

Repeat Station Activities

David R. Barraclough and Angelo De Santis

Abstract A repeat station is a site whose position is accurately known and where accurate measurements of the geomagnetic field vector are made at regular intervals in order to provide information about the secular variation of the geomagnetic field. In this chapter we begin by giving a brief history of the development of repeat station networks. We then describe the instruments used to make measurements at a repeat station. These include fixing the position of the station, finding the direction of true north and measuring the components of the geomagnetic field. Emphasis is given to techniques and instruments that are in current use. We next discuss the procedures that are used to reduce the measurements to a usable form and consider the uses to which the reduced data are put. Finally, we discuss the continued importance of such data in the present era of satellite geomagnetic surveys.

3.1 Introduction

We begin with a reminder: a repeat station is a site where accurate measurements of the geomagnetic field vector are made at regular intervals of, typically, two to five years. The position of a repeat station must be known very accurately and must be recorded in detail so that repeat measurements are always made at exactly the same location as earlier observations. As an aid to this, repeat stations are often marked in some way, for example by means of a non-magnetic pillar or a buried tile.

A possible physical explanation of the necessity for frequent repeat station occupations is that given for the need to update global geomagnetic field models such as the International Geomagnetic Reference Field (IGRF): from analyses of the secular variation derived from geomagnetic observatory time series, Barraclough and De Santis (1997) and De Santis et al. (2002) found evidence for chaotic behaviour of the geomagnetic field. This means that the field cannot be extrapolated for more than 5–6 years beyond the epoch of the latest reliable geomagnetic measurements.

This chapter gives a brief history of the development of repeat station networks; describes the instruments used to make measurements at a repeat station and the procedures that are used to reduce the data to a usable form; considers the uses to which repeat station data are put and discusses the continued importance of such data in the era of satellite geomagnetic surveys.

An earlier IAGA publication (Newitt et al. 1996) gives detailed guidance about setting up repeat stations, the instruments needed and the treatment of measurements made.

3.2 History of Repeat Stations

Observations of the geomagnetic field began with measurements of the declination. Because of their importance for navigation they were usually made at sea ports. Later, when magnetic phenomena aroused the interest of early savants, observations were also made

A. De Santis (✉)
Istituto Nazionale di Geofisica e Vulcanologia (INGV),
V. Vigno Murata 605, 00143 Rome, Italy
e-mail: angelo.desantis@ingv.it

M. Mandea, M. Korte (eds.), *Geomagnetic Observations and Models*, IAGA Special Sopron Book Series 5,
DOI 10.1007/978-90-481-9858-0_3,

in other cities. Before long it was realised that other elements of the geomagnetic field were of interest and these were also measured.

After the discovery of secular variation by Henry Gellibrand (1635) (or by Edmund Gunter, see Malin and Bullard 1981) it was realised that observations needed to be repeated at intervals to keep knowledge of the field up-to-date. Early accuracy requirements were not very severe and the interval between measurements tended to be quite long. Exact reoccupation of measurement sites was thus not too important, as exemplified in the data collections just cited. Examples of such collections include those made at London (Malin and Bullard 1981), Paris (Alexandrescu et al. 1996), Rome (Cafarella et al. 1993) and Edinburgh (Barraclough 1995). When demands on data accuracy increased, permanent geomagnetic observatories began to be established, once again often in or near to large towns or cities.

As interest in geomagnetic phenomena widened it became important to be able to describe how the field varied over extensive regions, for example over the territory of a particular country. The results were almost always expressed as contour charts of one or more of the geomagnetic elements and many early charts were based on rather heterogeneous collections of data derived from several sources rather than from a planned network of measurements. The earliest survey using a more or less regular network of points was that made in 1640 by Fathers Borri and Martini in Italy (e.g., Cafarella et al. 1992; De Santis and Dominici 2006). Measurements of declination were made at 21 sites and a simple declination map was drawn but was later lost (see also Malin 1987).

William Whiston (1721), between 1719 and 1720, measured the dip angle at 33 points in southern England and produced a (very idealised) contour map. Just over a century later James Dunlop (1830) made measurements of the horizontal intensity at 35 points in Scotland and northern England; unfortunately they were only relative values, expressed in terms of a value of 1.0 at Edinburgh (Gauss (1833) had yet to publish his method for measuring magnetic field in absolute units) so their usefulness is very limited.

Between 1834 and 1838 Edward Sabine and his co-workers Robert Were Fox, Humphrey Lloyd, John Phillips and James Clark Ross made an extensive magnetic survey of the British Isles (Sabine 1839, 1870). The project had been suggested at the third meeting of the British Association for the Advancement of Science in 1833 and was part of a growing interest in geomagnetic studies in the UK that Cawood (1979) has termed the "magnetic crusade". Measurements of declination, inclination, horizontal intensity and total intensity were made at 203 stations. The intensity results were all converted to total intensity values and were originally (Sabine 1839) presented as relative to a value of unity in London. They were later converted to absolute units (Sabine 1870). This survey, in the words of Sabine (1862), "deserves to be remembered as having been the first complete work of its kind planned and executed in any country as a national work, coextensive with the limits of the state or country, and embracing the three magnetic elements". In this same report Sabine also pointed out that such surveys are able "by their repetition at stated intervals to supply the best kind of data for the gradual elucidation of the laws and source of the *secular change* [Sabine's italics] in the distribution of the earth's magnetism". In furtherance of this, Sabine and his colleagues made observations of the same three magnetic elements as before at 105 stations in England, Wales and Scotland. Not all the later measurements were made at stations of the earlier network. In fact, data from only 29 locations were used to extract secular variation information and two of these were magnetic observatories, at Kew and Dublin.

To quote Sabine (1862) yet again, the example of the earlier of the two British surveys "was speedily followed by the execution of similar undertakings in several parts of the globe; more particularly in the Austrian and Bavarian dominions, and in detached portions of the British Colonial Possessions, viz. in North America and India" (Kreil 1845, Lamont 1854, Lefroy 1883, Schlagintweit et al. 1861). By the end of the nineteenth century magnetic surveys had been made, at least once, in most of Western Europe, the USA, the Dutch East Indies and Japan—an extension to other countries of the magnetic crusade.

These surveys all fall into the category that Newitt et al. (1996) describe as ground surveys rather than repeat station surveys. The accuracy of the measurements was relatively low, there was little or no attempt to remove the effects of external magnetic fields, the station positions were not described in sufficient detail to enable exact reoccupation and the stations themselves were not marked by pillars, tiles or similar means. In Italy the first modern three-component

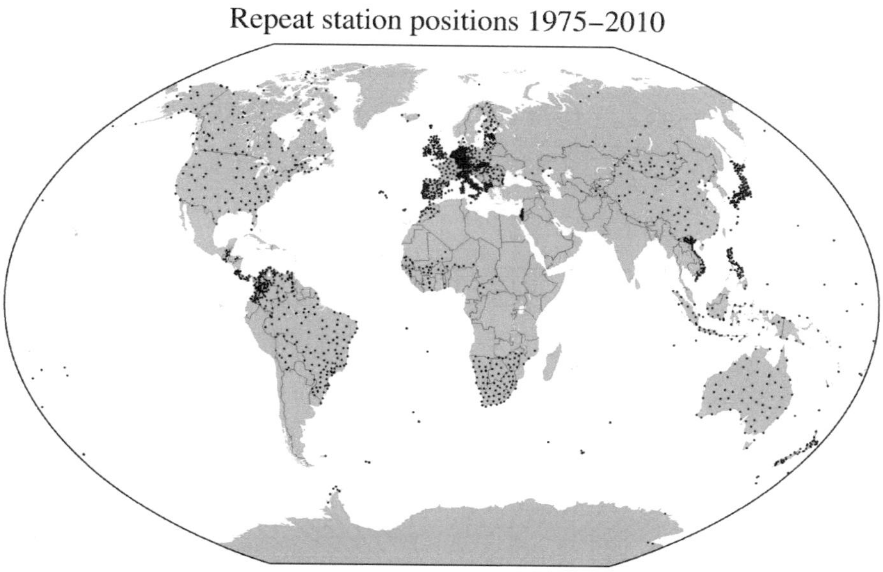

Fig. 3.1 Positions of all repeat stations that have been occupied at least twice since 1975 and have submitted their data to the World Data Centres

repeat station survey was undertaken by Chistoni and Palazzo in 1891–1892 (e.g., De Santis and Dominici 2006) and in the UK it was not until the work of Walker (1919) that a network of genuine repeat stations was established.

Meanwhile, a truly global network of repeat stations was being planned and established by the Department of Terrestrial Magnetism of the Carnegie Institution of Washington (CIW) under its first Director L A Bauer. As well as networks established by land surveying parties, including stations in Australia, Canada, South Africa and South America, repeat stations were set up by the surveying ships *Galilee* and *Carnegie* at most of their ports of call and on many mid-oceanic islands (see Good 2007).

Since the beginning of the twentieth century many countries have established and have continued to reoccupy repeat station networks, in several cases building on the CIW work just mentioned. Figure 3.1 is a map, kindly provided by Dr Susan Macmillan of the British Geological Survey, showing the distribution of repeat stations that are currently used for modelling the main geomagnetic field and its secular variation. Most countries in Western Europe are covered, as are Canada, the USA, Australia and New Zealand. Northern South America, East and South Africa, China, Indo-China, Japan and Indonesia all have extensive networks.

3.3 Instruments and Procedures

3.3.1 *Establishing the Position of a Station*

A repeat station is usually located in a remote place, where man-made contamination is low or negligible. Where it is possible, to improve the global distribution of land-based data, a repeat station is placed on an island (e.g., the repeat station on Capri Island, Italy; Fig. 3.2).

The positions of early survey stations were determined by astronomical means: the position of the Sun or a star was measured using a sextant or similar instrument for latitude and a chronometer to measure the difference between local and standard time for longitude. More recently, in well-surveyed parts of the world, positions can be determined with sufficient accuracy from large-scale maps. Nowadays the Global Positioning System (GPS) provides an even simpler method of determining station positions.

In many cases, for reasons of security, there is no surface indication of the presence of a repeat station, its position being marked typically by a buried tile. To enable such a station to be found readily on subsequent reoccupations it is usual to select a

Fig. 3.2 View from Capri Island repeat station (Italy). Courtesy of G. Dominici

set of well-spaced and prominent features that will act as reference objects. The station description then includes bearings or alignments using these objects which enable the buried marker to be found.

3.3.2 Establishing the Direction of True North

To determine a value of declination it is necessary to know the direction of true north as well as that of the horizontal component of the geomagnetic field. The classical method for finding the former is an astronomical one involving observations of the positions of the Sun or a star at accurately known times. Rather laborious calculations and the use of an almanac or other astronomical tables or, nowadays, the use of appropriate computer software then give the desired bearing. The Sun or star must, of course, be visible and this can be a problem in many regions. One solution is to measure accurately and record the bearing of one or more prominent features, such as the reference objects described in the previous section. This is not a complete solution, however, as the object or objects selected may be destroyed or may move slightly between visits.

Newitt et al. (1996) give details of how to make the necessary astronomical observations and of the calculations involved. They also include listings of two Fortran programs for inputting sets of Sun observations and for performing the computations to determine the azimuth of a reference mark.

Nowadays the direction of true north is commonly found using a gyroscopic device that can be attached to the theodolite that is used to support the instruments that measure the direction of the geomagnetic field. This device is also known as a gyro-theodolite or a north-seeking gyroscope. The gyroscope rotates at very high speed, typically 22000 rpm, about a horizontal axis and is pulled out of its initial spin-plane by the Earth's rotation. The spin-axis oscillates about the meridian plane until it finally settles pointing true north. Newitt et al. (1996) give details of three methods for using a gyro-theodolite to find the direction of true north.

An observation with the gyro-theodolite takes about half an hour. The instrument must be protected from the weather and this is usually achieved by placing the equipment in a tent. Since the theodolite must not be moved between the gyro measurements and those that measure the geomagnetic field the tent must not contain any magnetic materials. This means using either a specially made tent or one that has been carefully modified by the removal of all ferrous material. The gyro is magnetic and it must therefore be removed from the theodolite before making the magnetic measurements, taking great care not to move the theodolite. Here it should be noted that the theodolite must be non-magnetic.

3.3.3 Measuring the Magnetic Field

In this section we give a brief survey of instruments that have been and are being used in repeat station surveys. Fuller treatments of the subject of geomagnetic instrumentation are given by Wienert (1970), Forbes (1987) and Korepanov (2006).

During the nineteenth and early twentieth centuries the instruments used to measure the geomagnetic field elements at observatories and survey stations, including repeat stations, used as sensors either suspended or pivoted magnets. Declination was initially measured using sophisticated versions of the magnetic compass, dip measurements used a dip needle and the intensity of the field was measured by Gauss's method or a development of it. Instruments used at observatories and in the field were very similar except that the latter were designed to be rather more portable.

Towards the end of the nineteenth century the unifilar magnetometer was developed. This enabled astronomical observations, determination of the declination and measurement of the horizontal intensity by Gauss's method, or Lamont's variation of it, to be made with the same instrument. One of the most widely used versions of this instrument was the Kew pattern magnetometer.

Dip circles are inherently inferior to compasses and instruments using suspended magnets because of the difficulty in designing pivots for the dip needle that allow free movement and that do not wear easily. They were replaced quite early by earth-inductors which use a coil of many turns of wire that can be rotated rapidly about an axis lying along a diameter of the coil. If the axis is not parallel to the direction of the geomagnetic field an alternating voltage is induced in the coil. To measure dip the direction of the rotation axis is adjusted until no signal is detected in a galvanometer connected across the coil.

By the 1920s instruments using electrical methods were coming into use. Examples of these, both of which used Helmholtz coils to produce a region of uniform field at their centre in which was suspended a small magnetic needle, were the Schuster-Smith coil (Smith 1923) and the Dye coil (1928). The former was used for measuring the horizontal intensity, the latter the vertical component. Observations could be made much more quickly and easily with these instruments than with the Kew pattern magnetometer or similar instruments.

Like their predecessors these electrical magnetometers were heavy and bulky. Two much more portable instruments that were much used in surveying work as well as at magnetic observatories were the quartz horizontal-force magnetometer (QHM) (La Cour 1936) and the magnetometric zero balance (balance magnétometrique zéro, BMZ) (La Cour 1942). The former was used for measuring both declination and horizontal intensity, the latter the vertical component. Neither was an absolute instrument and both needed regular calibration.

Fluxgate magnetometers were developed during World War II, initially for submarine detection. A fluxgate magnetometer measures the component of the geomagnetic field along the axis of the sensor, which is a rod or ring of high-permeability material with a non-linear relationship between the applied field and the magnetic induction. The core is surrounded by two coils of wire through one of which an alternating electrical current is passed. This drives the core through an alternating cycle of magnetic saturation, i.e., magnetised—unmagnetised—inversely magnetised—unmagnetised—magnetised. This constantly changing field induces an electrical current in the second coil, and this output current is measured by a detector. In the presence of an ambient magnetic field the core will be more easily saturated in alignment with that field and less easily saturated in opposition to it. The resulting biassed sinusoidal excitation creates a distorted AC signal in the second coil. Detection of the even harmonics in this signal provides a DC output that is proportional to the field being measured.

Three fluxgate elements arranged orthogonally constitute a vector magnetometer and such instruments are often used as variometers at magnetic observatories and in field surveys, including repeat station surveys where there are no nearby observatories that can be used for the reduction of the observations (see Section 3.3.4).

A draw-back to the use of fluxgate magnetometers to measure the field along the sensor axis with high accuracy is their sensitivity to ambient temperature. This is no longer a problem when a fluxgate sensor is used as a null detector and this is exploited in the fluxgate theodolite (also known as a DI fluxgate, declination-inclination magnetometer or DIM). In this instrument a single-axis fluxgate sensor is mounted parallel to the axis of the telescope of the non-magnetic theodolite. Rotating the assembly about a vertical axis and finding the position where the sensor gives zero output gives the direction—read off the horizontal circle of the theodolite—perpendicular to the horizontal component of the geomagnetic field. With the telescope and fluxgate positioned in the plane of the magnetic meridian (at right angles to the direction just determined) rotation about the horizontal axis enables a null position to be found which is in the direction perpendicular to the total geomagnetic field vector. From these two directions, and knowledge of the direction of true north, values of declination and inclination can be derived. In each of these determinations a total of four measurements of the null direction are made, inverting the telescope and rotating the telescope/fluxgate assembly, so as to eliminate errors due to misalignment of the fluxgate axis with respect to that of the telescope and to remanent magnetisation of the instrument. Details of these measurement procedures are given by Newitt et al. (1996). The fluxgate theodolite is used in many modern repeat station networks for the measurement of declination and inclination (see also Korepanov 2006).

For the measurement of the total field intensity, the instrument of choice is the proton precession magnetometer. This was developed in the 1950s and consists of a container full of a hydrogen-rich liquid such as water or paraffin surrounded by a coil of wire. A direct current is passed through the coil, producing a strong field in the liquid. The protons in the liquid are aligned with this field rather than with the ambient geomagnetic field. When the current is switched off they gradually realign themselves with the geomagnetic field and in so doing precess about it. The precession frequency f is given by the expression

$$f = \gamma_p' F/2\pi,$$

where γ_p' is the proton gyromagnetic ratio at low field strengths for a spherical sample of water at 25°C; the value adopted by IAGA is 2.675153362×10^8 $T^{-1}s^{-1}$ (Mohr et al. 2008). An observation of total field intensity in absolute measure thus reduces to a measurement of frequency, which can be performed with high accuracy, and a knowledge of γ_p', which is known to high precision. Conventional proton precession magnetometers are based on this principle. Proton precession magnetometers, based on the Overhauser effect (Overhauser 1953) have several advantages over the conventional type and are becoming widely used. They can be cycled more rapidly than conventional types, they are more sensitive, they use less power and their polarising field is weaker, meaning that they can be sited closer to other sensors than traditional proton precession magnetometers.

Proton precession magnetometers are used to measure total intensity at the repeat station site, usually by recording the field continuously at a nearby position whilst the declination and inclination observations are being made. The site difference between the main site and the site of the recording proton precession magnetometer is determined by preliminary measurements made at both sites simultaneously. A proton precession magnetometer is also used to survey the area around the repeat station site, both when the site is being selected initially and at each successive reoccupation. Measurements are made on a regular grid usually oriented north-south and east-west with a spacing of a metre. The aim is to detect any significant departures from smooth field gradients. Any such departures in the initial survey would lead, if possible, to the selection of a site with smoother gradients. Such departures detected during a reoccupation would suggest man-made contamination. If the sources cannot be found and removed a new, nearby site has to be found.

3.3.4 Data Reduction

Measurements made at a repeat station include contributions from the main geomagnetic field that originates in the core, from sources in the Earth's crust and from sources above the Earth's surface. For secular variation modelling—the most important use to which repeat station data are put—only the first of these is required. The crustal contribution can be assumed to be constant over the time scales important for secular variation and is therefore removed when differences between measurements made at different times are computed. The effects of external fields, originating in the ionosphere and magnetosphere, and the effects of currents induced in the crust by these external fields, must be removed as completely as possible. This is the aim of the data reduction procedures.

In mid-latitude regions where there are nearby magnetic observatories the simplest and most economical method of data reduction is to use data from these observatories. It is important to make sure that measurements at the observatory are characteristic of the field and its variations over a wide enough region that includes the repeat station under consideration.

An advantage of the ready availability of internet access is that nowadays it is possible to follow in real time the behaviour of geomagnetic field components from the websites of many institutions that operate magnetic observatories. In this way, one can be made aware of the beginning of a period of disturbed magnetic activity and can then postpone repeat station measurements until the field returns to quiet conditions.

It is usual to reduce mid-latitude repeat station measurements to the value that would be measured during a quiet night-time interval since external magnetic fields have their smallest values during the night hours at these latitudes. Alternatively, the repeat station values are sometimes reduced to an annual mean value since the effects of most external sources are removed by taking means over a year. In either case the assumption is made that the effects of external sources are the same at the observatory and at the repeat station. It is also assumed that the secular variation is the same over the time interval between the epoch of observation at the repeat station and that to which the reduction is made. This implies that

$$E_S(t) - \boldsymbol{E_S} = E_O(t) - \boldsymbol{E_O}, \quad (3.1)$$

where $E_S(t)$ is the observed value of element E at the repeat station at epoch t; $\boldsymbol{E_S}$ is the corresponding value at the station reduced to either a quiet night-time value or to an annual mean; $E_O(t)$ is the value of the element E at the observatory at epoch t; and $\boldsymbol{E_O}$ is either the value of the element E at the observatory for a quiet night-time interval or an annual mean value of E at the observatory. (Note that the use of bold-face symbols does not imply that the corresponding variables are vectors; it is simply a useful device to differentiate measured and reduced or mean values.) Therefore

$$\boldsymbol{E_S} = E_S(t) + C \quad (3.2)$$

where

$$C = \boldsymbol{E_O} - E_O(t).$$

If the repeat station is bracketed in latitude by two nearby observatories and has a similar longitude to the observatories, data from the two observatories can be used with appropriate interpolation. This is often the case for a country such as the UK and Italy which are well endowed with observatories and are relatively long and thin in a north-south direction. The correction factor C in Eq. (3.2) then becomes

$$C = \frac{\Delta\varphi_2 \left(\boldsymbol{E_{01}} - E_{01}(t)\right) + \Delta\varphi_1 \left(\boldsymbol{E_{02}} - E_{02}(t)\right)}{\varphi_2 - \varphi_1}$$

where φ_1 is the latitude of the observatory north of the repeat station; φ_2 is the latitude of the observatory to the south; $\Delta\phi_1 = \phi_1 - \phi_S$; $\Delta\phi_2 = \phi_S - \phi_2$; ϕ_S is the latitude of the repeat station and the subscripts 01 and 02 indicate that the values of the element E refer to the northern and southern observatory, respectively. Gaps in the observatory minute value data can lead to problems when using this method of data reduction. Several different strategies have been proposed for dealing with these difficulties (e.g., Mandea 2002; Schott and Linthe 2007; Herzog 2009; Marsal and Curto 2009; Newitt 2009; Love 2009). None of them appears to be completely effective when the missing data constitute more than about 10% of the total during more than moderately disturbed magnetic periods.

If there are no observatories sufficiently near to the repeat station survey area or if the morphology of the external field variations is complicated (as, for example, in the auroral zones), the recommended procedure is to use a variometer (nowadays usually using fluxgate sensors) to record the geomagnetic field variations continuously at a point near to the repeat stations. The aim is to make the repeat station plus variometer as similar to a standard magnetic observatory as possible. The variometer is therefore operated for several days during which absolute observations are made at frequent intervals at the repeat station so as to provide baseline values as at an observatory.

A quiet night-time interval is chosen from the variometer record and the repeat station measurements are reduced to this value. Newitt et al. (1996) and Korte and Fredow (2001) give further details of this procedure. They also discuss sources of error in this procedure and in that using nearby observatory data.

Another recently proposed method of data reduction uses a comprehensive field model such as CM4 (Sabaka et al. 2004) which takes into account lithospheric and external (ionospheric and magnetospheric) fields. These contributions can then, in principle, be removed from the repeat station measurements. Of course, this technique depends on the quality of the global model removed and in particular its accuracy over the area and time interval concerned (Matzka et al. 2009).

3.4 Uses of Repeat Station Data

Repeat stations provide an important source of vectorial data for main field and secular variation modelling. Their data contribute to global modelling of the geomagnetic field, and are essential for producing regional field models and charts, the former being important for scientific studies and the latter for navigational purposes. In the present satellite era, more and more global and regional models are based on satellite magnetic data with the addition of ground data, i.e., observatory and repeat station data, to stabilise the inversion (noise at satellite altitude is amplified when downward continued to the ground) and to take into account the true vertical gradient of the field. While spherical harmonic models are used for global representations (e.g., Sabaka et al. 2004), polynomial or spherical cap harmonic models are used for modelling restricted regions of the Earth. Spherical cap harmonic modelling is used either in its original version (Haines 1985) or in its more recent revision (Thébault et al. 2006). Although the repeat station data are imperfect and noisy, their inclusion in regional modelling greatly reduces the ambiguity in the vector components at different altitudes. They are also important for upward and downward continuation purposes (Korte and Thébault 2007).

When using repeat station data for secular variation modelling it is important to estimate the overall error budget due to all the steps in the measurement and reduction procedures. Ideally, accuracies should be comparable to those normally achieved at the best magnetic observatories. i.e., better than 1 nT for magnetic components and better than $0.1'$ for angular elements (Newitt et al. 1996). However, the various steps contribute errors that add together statistically producing a greater overall error: for instance, there will certainly be an error due to the measurement operations themselves and also to possible changes in environmental conditions (temperature changes being the most critical), together with errors coming from possible crustal and/or external contaminations, from imperfect reduction, and other known or unknown factors. Thus, more realistic errors are of about 5 nT in the components, and about 0.5–1 min of arc in inclination and declination (Newitt et al. 1996). This should be taken into account when modelling, usually by weighting repeat station data differently from observatory annual (or monthly) means.

Of special interest are repeat stations placed near or at airports, where magnetic declination is measured at special calibration pads for aircraft compass certification and checks (Loubser and Newitt 2009). These provide important information for aircraft navigation (Rasson and Delipetrov 2006). This kind of measurement is being requested more and more often by airport authorities, sometimes with a frequency of once per year and thus provides some of the most up to-date information for regional field modelling.

There are other indirect but still important uses of repeat station data: for instance, of particular interest is their possible use for estimating the Koenigsberger ratio of lithospheric rocks, i.e., for discriminating between remanent and induced lithospheric magnetisation (Hulot et al. 2009; Shanahan and Macmillan 2009). The former is practically constant on geological time scales while the latter tends to be proportional to the geomagnetic field. A comparison between repeat station intensity measurements made since 1900 and those made since 1960 shows a decrease that could be ascribed to the corresponding main field decay over the past century (Shanahan and Macmillan 2009).

Special mention should be made of a particular type of measurement that, although not actually repeat station measurements, show some similar aspects: finding how the North (or South) magnetic pole moves with time (e.g., Newitt et al., 2009). Here the same position is not reoccupied but the movement of a place with particular magnetic characteristics, i.e., where the magnetic inclination is $\pm 90°$ is determined. These direct measurements should not be necessary at epochs with accurate geomagnetic field models. However, they provide independent verification of the model and enable the velocity at which the pole moves to be calculated.

This velocity may be related to jerks (Mandea and Dormy 2003), one of the fastest (and most intriguing) features of the recent secular variation.

3.5 State of the Art of Repeat Station Activities

Because of the increased need for ground measurements in connection with recent satellite missions (e.g., Matzka et al. 2010), the time between repeat station measurements has been getting smaller. In the past this interval was between 5 and 10 years but it is now between 2 and 5 years. IAGA Working Group V-MOD deals with several aspects of repeat station activities. According to its website these are: (a) to maintain a catalogue of regional and global magnetic surveys, models and charts; (b) to promote and set standards for magnetic repeat station surveys and reporting; (c) to define operating procedures and classification standards; (d) to encourage agencies to submit repeat data in appropriate formats to World Data Centres (WDCs); (e) to maintain a catalogue of national repeat station network descriptions; (f) to promote international interest in surveying, modelling and analysis of the international geomagnetic field, both globally and on a regional scale.

A regional magnetic survey questionnaire was circulated in 2000 to which 49 countries responded out of 82 contacted. The results confirmed the uneven distribution in space and time of repeat station data. From Fig. 3.1 it is evident that there are significant gaps over large parts of the Earth's surface: for example Mexico in North America, Chile and Argentina in South America, significant parts of Asia and Africa. Fortunately, regarding the last-named continent, some recent international efforts have improved the situation in Southern Africa (Korte et al. 2007). Even in Europe, where there appears to be a large number of data, some national repeat station networks are no longer active over their complete extent. For instance, Russia contributes only three stations in the most recent WDC data set.

To produce better regional and global models we need to have measurements in as many areas as possible, especially in remote regions, such as polar areas, the deep seafloor (e.g., Vitale et al. 2009) and on ocean islands (e.g., Matzka et al. 2009). Particular efforts have been made in this direction: a recent special issue of Annals of Geophysics collects some examples of measurements in remote regions (De Santis et al. 2009).

Another important aspect in repeat station activities is their close relationship with operating observatories. The absence of a nearby observatory together with the possibility of missing data can cause severe problems for repeat station surveys. IAGA encourage both the continuing operation of existing observatories and the establishment of new ones where this is possible.

At a European level, the European network of repeat stations (MagNetE) has had a central role in the last 10 years. It has organised four workshops so far (at Niemegk in 2003, Warsaw in 2005, Bucharest in 2007, Helsinki in 2009) and the next will be held in Rome in 2011. This recently established network among the European institutions dedicated to repeat station surveys has improved agreement between the institutions concerning intervals between reoccupations and measurement techniques. An up-to-date review of the situation of repeat station activities in Europe has been recently presented at the recent MagNetE workshop in Helsinki (Duma 2009). One of the next objectives of this network will be the preparation and realisation of a European magnetic declination map centred at some recent epoch: the latest discussions in the MagNetE community have proposed 2006.5. Any chosen map or model, although relating to a particular past epoch, should also provide predictive information, although the problem of extrapolation into the future is not a simple one and better solutions must be sought. One recently proposed alternative to the commonly used polynomial extrapolation reconstructs the temporal measurements in an ideal phase space where fitting and extrapolating techniques can be better performed (e.g., De Santis and Tozzi 2006). This technique is based on the nonlinear behaviour of the geomagnetic field in time (e.g., Barraclough and De Santis 1997).

Another aspect which is important is to look at the repeat station network in terms of efficiency: a large number of stations is not as important as the fact that they can be reoccupied more frequently, especially in those areas where secular variation is more rapid or significantly different from the rest of the country or of the continent (e.g., South Atlantic, South Africa, Australia). When resources are not available to reoccupy all the stations more frequently, a good compromise is to chose a subset of them, the so-called "super" or "class A" repeat stations (e.g., McEwin

1993) which are reoccupied more frequently, e.g., once every year or two years. Repeat stations of this type, which are rather similar to observatories, are better able to follow the secular change on time scales of a year or so.

3.6 Conclusions

The geomagnetic field is a fundamental property of our planet; it changes with space and time in a complex fashion. Most of it originates in turbulent motions in the outer metallic fluid core of the Earth at around 3000 km depth, with time scales from years to millennia. On the Earth's surface and above the potential from which it can be derived is usually represented by spherical harmonics as solutions of Laplace's equation in spherical coordinates. Very accurate observations of the field are made at geomagnetic observatories. However these are sparse and do not cover the Earth's surface as uniformly as needed. In order to follow the evolution of the field in time and space with sufficient accuracy it is necessary to complement the observatory measurements with other kinds of data, such as repeat station observations. The latter have high enough accuracy and can be made in less time than those at the observatories. Repeat station measurements lead to an increase in the spatial detail of secular variation studies, improving both regional and global geomagnetic field modelling in space and time. An optimum scheme for the reduction of repeat station measurements is to use a combination of the different techniques discussed above. In particular, a variometer installation, recording for one or more days whilst the repeat station observations are being made, and a nearby observatory are highly recommended. Use of a comprehensive model such as CM4 or another more recent global model that takes into account the external field contributions should also be considered. By studying the behaviour of magnetic indices such as K and Dst whilst measurements are being made we can also assess the quality of the final reduced repeat station component values.

References

Alexandrescu M, Courtillot V, Le Mouël J-L (1996) Geomagnetic field direction in Paris since the mid-sixteenth century. Phys Earth Planet Inter 98:321–360

Barraclough DR (1995) Observations of the Earth's magnetic field made in Edinburgh from 1670 to the present day. Trans R Soc Edinb Earth Sci 85:239–252

Barraclough DR, De Santis A (1997) Some possible evidence for a chaotic geomagnetic field from observational data. Phys Earth Planet Inter 99:207–220

Cafarella L, De Santis A, Meloni A (1992) Secular variation from historical geomagnetic field measurements. Phys Earth Planet Inter 73:206–221

Cafarella L, De Santis A, Meloni A (1993) Il catalogo geomagnetico storico italiano. Istituto Nazionale di Geofisica, Rome

Cawood J (1979) The magnetic crusade: science and politics in early Victorian Britain. Isis 70:493–518

De Santis A, Dominici G (2006) Magnetic repeat station network in Italy and magnetic measurements at heliports and airports. In: Rasson, Delipetrov (eds) (2006), pp 259–270

De Santis A, Tozzi R (2006) Nonlinear techniques for short term prediction of the geomagnetic field and its secular variation. In: Rasson, Delipetrov (eds) (2006), pp 281–289

De Santis A, Barraclough DR, Tozzi R (2002) Nonlinear variability of the recent geomagnetic field. Fractals 10: 297–303

De Santis A, Arora B, McCreadie H, (eds) (2009) Geomagnetic measurements in remote regions. Ann Geophys 52(1):special issue

Duma G (2009) Magnetic repeat station measurements in Europe. MagNetE-Report, Update 2009-1, pp 25

Dunlop J (1830) An account of observations made in Scotland on the distribution of the magnetic intensity. Trans R Soc Edinb 12:1–65

Dye DW (1928) A magnetometer for the measurement of the Earth's vertical intensity in c.g.s. measure. Proc R Soc Lond A117:434–458

Forbes AJ (1987) General instrumentation. In: Jacobs JA (ed) Geomagnetism, vol 1. Academic, London and Orlando, Florida

Gauss CF (1833) Intensitas vis magneticae terrestris ad mensuram absolutam revocata. Dieterich, Göttingen

Gellibrand H (1635) A discourse mathematical on the variation of the magnetical needle. William Jones, London

Good GA (2007) Carnegie Institution of Washington, Department of Terrestrial Magnetism. In: Gubbins D, Herrero-Bervera E (eds) Encyclopedia of geomagnetism and paleomagnetism. Springer, Dordrecht

Haines GV (1985) Spherical cap harmonic analysis. J Geophys Res 90:2583–2591

Herzog DC (2009) The effects of missing data on mean hourly values. Proc. XIIIth IAGA Workshop. In: Love JJ (ed) Golden CO, Open File Report 2009–1226, pp 116–126

Hulot G, Olsen N, Thébault E, Hemant K (2009) Crustal concealing of small-scale core-field secular variation. Geophys J Int 177:361–366

Korepanov V (2006) Geomagnetic instrumentation for repeat station survey. In: Rasson, Delipetrov (eds) (2006), pp 145–166

Korte M, Fredow M (2001) Magnetic repeat station survey of Germany 1999/2000. Sci Tech Rep STR01/04, Geoforschungszentrum, Potsdam

Korte M, Thébault E (2007) Geomagnetic repeat station crustal biases and vectorial anomaly maps for Germany. Geophys J Int 170:81–92

Korte M, Mandea M, Kotzé P, Nahayo E, Pretorius B (2007) Improved observations at the southern African geomagnetic repeat station network. S Afr J Geol 110:175–186

Kreil K (1845) Magnetische und geographische Orstbestimmungen in Böhmen in den Jahren 1843–1845. Abh. K. Böhm. Ges. Wiss. Series 5 vol 4

La Cour D (1936) Le quartz-magnétomètre QHM. Commun. Magn Dan Meteorol Inst 15

La Cour D (1942) The magnetometric zero balance (BMZ). Commun Magn Dan Meteorol Inst 19

Lamont J (1854) Magnetische Karten von Deutschland und Bayern, nach den neuen Bayerischen und Österreichischen Messungen, unter Benützung einiger älterer Bestimmungen. München

Lefroy JH (1883) Diary of a magnetic survey of a portion of the Dominion of Canada chiefly in the North-Western territories executed in the years 1842–1844. London

Loubser L, Newitt L (2009) Guide for calibrating a compass swing base, International Association of Geomagnetism and Aeronomy, Hermanus, pp 35

Love JJ (2009) Missing data and the accuracy of magnetic-observatory hour means. Ann Geophys 27:3601–3610

McEwin AJ (1993) The repeat station network and estimation of secular variation in Australian region. Explor Geophys 24:87–88

Malin SRC (1987) Historical introduction to geomagnetism. In: Jacobs J (ed) Geomagnetism, vol 1. Academic, London and Orlando, pp 1–49

Malin SRC, Bullard E (1981) The direction of the Earth's magnetic field at London, 1570–1975. Philos Trans R Soc Lond A299:357–423

Mandea M (2002) 60, 59, 58, ... How many minutes for a reliable hourly mean? Proceedings of Xth IAGA Workshop, Hermanus, pp 112–120

Mandea M, Dormy M (2003) Asymmetric behaviour of the magnetic dip poles. Earth Planets Space 55:153–157

Marsal S, Curto JJ (2009) A new approach to the hourly mean computation problem when dealing with missing data. Earth Planets Space 61:945–956

Matzka J, Olsen N, Fox Maule C, Pedersen LW, Berarducci AM, Macmillan S (2009) Geomagnetic observations on Tristan da Cunha, South Atlantic Ocean. Ann Geophys 52:97–105

Matzka J, Chulliat A, Mandea M, Finlay C, Qamili E (2010) Direct measurements: from ground to satellites. Space Sci Rev In press

Mohr PJ, Taylor BN, Newell DB (2008) CODATA recommended values of the fundamental physical constants: 2006. J Phys Chem Ref Data 37:1187

Newitt LR (2009) The effects of missing data on the computation of hourly mean values and ranges. Proceedings of XIIIth IAGA Workshop. In: Love JJ (ed.), Golden CO, Open File Report 2009–1226, pp 194–201

Newitt LR, Barton CE, Bitterly J (1996) Guide for magnetic repeat station surveys. International Association of Geomagnetism and Aeronomy, Boulder, Colorado

Newitt LR, Chulliat A, Orgeval J-J (2009) Location of the North Magnetic Pole in April 2007. Earth Planets Space 61: 703–710

Overhauser AW (1953) Polarisation of nuclei in metals. Phys Rev 92:411–415

Rasson JL, T Delipetrov (eds) (2006) Geomagnetics for aeronautical safety: a case study in and around the Balkans. Nato Advanced Study Science Series, Springer, Dordrecht

Sabaka TJ, Olsen N, Purucker ME (2004) Extending comprehensive models of the Earth's magnetic field with Ørsted and CHAMP data. Geophys J Int 159:521–547

Sabine E (1839) Report on the magnetic isoclinal and isodynamic lines in the British Islands. Eighth Rep Br Assoc Adv Sci 49–196

Sabine E (1862) Report on the repetition of the magnetic survey of England, made at the request of the General Committee of the British Association. Rep Br Assoc Adv Sci 1861: 250–279

Sabine E (1870) Contributions to terrestrial magnetism No. XII. The magnetic survey of the British Islands, reduced to the epoch 1842.5. Philos Trans R Soc Lond 160: 265–275

Schlagintweit H de, Schlagintweit A de, Schlagintweit R de (1861) Results of a scientific mission to India and High Asia undertaken between the years MDCCCLIV, and MDCCCLVIII, vol I. by order of the Court of Directors of the Honourable East India Company, Leipzig and London

Schott J, Linthe HJ (2007) The hourly mean computation problem revisited. Publ Inst Geophys Pol Acad Sci. C-99:398

Shanahan TJG, Macmillan S (2009) Status of Edinburgh WDC global survey data. Presentation at June 2009 MagNetE Workshop, Helsinki

Smith FE (1923) On an electromagnetic method for the measurement of the horizontal intensity of the Earth's magnetic field. Philos Trans R Soc Lond A223:175–200

Thébault E, Schott JJ, Mandea M (2006) Revised spherical cap harmonic analysis (R-SCHA): validation and properties. J Geophys Res 111:B01102. doi:10.1029/2005JB003836

Vitale S, De Santis A, Di Mauro D, Cafarella L, Palangio P, Beranzoli L, Favali P (2009) GEOSTAR deep seafloor missions: magnetic data analysis. Ann Geophy 52: 57–64

Walker GW (1919) The magnetic re-survey of the British Isles for the epoch January 1, 1915. Philos Trans R Soc Lond A219:1–135

Whiston W (1721) The Longitude and Latitude Found by the Inclinatory or Dipping-needle; Wherein the Laws of Magnetism are also discover'd. To which is prefix'd, An Historical Preface; and to which is subjoin'd, Mr. Robert Norman's New Attractive, or Account of the first Invention of the Dipping Needle. J Senex and W Taylor, London

Wienert KA (1970) Notes on geomagnetic observatory and survey practice. UNESCO, Brussels

Chapter 4

Aeromagnetic and Marine Measurements

Mohamed Hamoudi, Yoann Quesnel, Jérôme Dyment, and Vincent Lesur

Abstract Modern magnetic measurements have been acquired since the 1940s over land and the 1950s over oceans. Such measurements are collected using magnetometer sensors rigidly fixed to the airframe or towed in a bird for airborne or in a fish in marine surveys using a cable long enough to avoid the ship/airplane magnetic effect. Positioning problems have been considerably reduced by the Global Positioning System (GPS). Considerable progress has been made in geomagnetic instrumentation increasing the accuracy from ~10 nT or better in the 1960s to ~0.1 nT or more nowadays. Scalar magnetometers, less sensitive to orientation problems than the fluxgate vector instruments, are the most commnonly used for total-field intensity measurement. Optical pumping alkali vapor magnetometers with high sampling rate and high sensitivity are generally used aboard airframes whereas proton precession magnetometers (including Overhauser) are favored at sea. Scalar magnetic anomalies are calculated by subtraction of global core field models like the International Geomagnetic Reference Field (IGRF) after subtraction of an external magnetic field estimate using magnetic observatories or temporary magnetic stations. The external field correction using an auxiliary station is often not possible in marine measurements. However comprehensive models such as CM4 can be used to provide adequate core and external magnetic fields, particularly for almost all early magnetic measurements which were not corrected for the external field. In the case of airborne measurements such global models help to define a reference level for global mapping of the anomaly field. The current marine dataset adequately covers most of the Northern Hemisphere oceanic areas while major gaps are observed in the southern Indian and Pacific oceans. Airborne measurements cover all the world, except oceanic areas and large part of Antarctica. Data are however often not available when owned by private companies. The data released are mainly owned by governmental agencies. The derived airborne/marine magnetic anomaly maps combined with long-wavelength satellite maps help scientists to better understand the structure and the evolution of the lithosphere at local, regional and global scales. Marine magnetic observations are also made at depth, near the seafloor, in order to access shorter wavelengths of the magnetic field for high resolution studies. Airborne High Resolution Anomaly Maps (HRAM) are also nowadays the new trends pushing towards the generalisation of the Unmanned Aerial Vehicles (UAV) or Autonomous Underwater Vehicles (AUV) or Remotely Operated Vehicles (ROV) magnetic surveys.

4.1 General Introduction

It has been known for some two thousand years that pieces of magnetized rocks attract (or repel) each other. However Gilbert's statement in the very beginning of the 17th century – that the Earth behaves itself as a great magnet – is a milestone in Earth's magnetism. Latter on, by the mid-19th century, it has been realized that magnetic instruments (e.g., magnetic theodolite), normally operated for measuring the Earth's magnetic field variations, might be employed to discover magnetic ore bodies (Telford et al. 1990).

M. Hamoudi (✉)
Helmholtz Centre Potsdam, GFZ German Research Centre for Geosciences, Telegrafenberg, F 407, 14473, Potsdam, Germany
e-mail: hamoudi@gfz-postdam.de

M. Mandea, M. Korte (eds.), *Geomagnetic Observations and Models*, IAGA Special Sopron Book Series 5,
DOI 10.1007/978-90-481-9858-0_4,

Advances in building magnetic instrumentations have been very rapid since World War II with the first Magnetic Airborne Detector (MAD) designed for submarines and mines detection. From the early 1950s, not only geological national agencies but also oil and mining companies have shown a great interest for aeromagnetic and marine surveys.

It is worth to mention that the marine magnetic data acquired during the 1950s and early 1960s have played an important role in uncovering plate tectonics, starting a revolution in geosciences. As a consequence, the systematic acquisition of marine magnetic data helped to decipher the age of ocean crusts and carry out paleogeographic reconstructions. This has led to a first order picture of the Earth's lithosphere evolution for the last 200 millions years. Moreover, advancement in magnetic instrumentations as well as positioning systems have allowed, on one hand to achieve the required precision for the global satellite mapping, and on the other hand to derive high-resolution magnetic mapping at low altitude both at sea and on land. Geological mapping has directly benefitted from the available airborne and marine magnetic surveys. However, more efforts are needed for a full coverage of the Earth's surface with magnetic survey data.

This manuscript is intended to give a non exhaustive review of the progress in aeromagnetic and marine magnetic surveys over more than half of century. Let us note that the paper is built around two distinct parts: the first one is devoted to aeromagnetic surveys and the second to marine magnetics. Even if the progress and evolution of both fields are closely related, they are described separately in order to better emphasize their specificities.

4.2 Introduction to Aeromagnetics

The main or global magnetic field of the Earth is generated in the conducting fluid outer core by geodynamo processes (Braginski and Roberts 1995). This field is much stronger than the field generated in the Earth's lithosphere. During magnetic surveys, numerous sources of the magnetic field contribute to the measured signal, but the main target is the field generated in the magnetized rocks.

Potential field exploration methods, like gravity and magnetism, are considered as passive methods (Heiland 1929, 1940) because the measured signal is the response permanently generated by physical property contrasts in the rocks. These gravity or magnetic responses may be measured remotely, without having direct access to the rocks. The physical contrasts considered are either density contrasts for gravity, or remnant and induced magnetization contrasts for magnetism. We have known since Newton that all rocks contribute to the observed gravity field, but only magnetized rocks generate a magnetic field. Geological formations may be very strongly or very weakly magnetized, depending on their magnetite (or any iron or sulphide oxide) content (Heiland 1940). Unlike density, magnetization is strongly temperature dependent (Kitte 2005) and exists only if this temperature does not exceed a certain temperature-threshold called the Curie temperature. This Curie temperature varies between 300°C and 1200°C for iron sulphides or oxides (Frost and Shive 1986), and is approximately 580°C for magnetite at atmospheric pressure (Blakely 1988). Above the Curie temperature, spontaneous magnetization vanishes, and minerals exhibit paramagnetic susceptibility that has a small effect compared to magnetization (Kittel 2005). Therefore rocks are essentially non-magnetic at temperatures greater than the Curie temperature of the most important magnetic mineral in the rocks. Taking into account a normal geothermal gradient, sources of the measured magnetic signal are then restricted to 30–40 km depth, except in old cratonic areas where the Curie depth at which the temperature reaches the Curie temperature – may be greater (Hamoudi et al. 1998).

The success of the magnetic method and its wide spread use are due to the numerous discoveries of iron ore deposits since the mid-19th century, all around the world, in USA, Canada, (Heiland 1929, 1940), in Russia (Logachev 1947), and even before in Sweden in northern Europe (Sundberg and Lundberg 1932, Heiland 1940). The surveys were initially ground based. The depth of an ore body, assuming it can be represented as a line of poles, may be determined using the vertical gradient of the vertical field component by differencing the field at different height levels. Experiments were first proposed at the end of 19th century to measure the field at different levels in a mine and at different depths of a shaft (Heiland 1940). For ore bodies with a large depth extent, measuring the field at the Earth's surface and on a platform a few meters above was proposed. However Eve and Keys

(1933) rapidly reached the limit of the method and it became apparent that because platforms failed to give sufficient changes in the distance to the source, geophysicists were going to have to "take to the air" (Heiland 1940). Captive balloons have been used above the Kiruna ore body (Sundberg and Lundberg 1932, Heiland 1940). It was soon established that measurements in airplanes by an automatic recording device had many benefits among them great speed of survey, applicability to inaccessible areas, and "direct depth determination" (Heiland 1935, 1940, Logachev 1947).

Airborne geophysical methods have grown since their inception in the 1930s. Submarine and mine detection during World War II gave an impetus to improvements in apparatus and methods for aeromagnetic surveys (Wyckoff 1948). Since then, continuous improvements in instrumentation and positioning systems have made the aeromagnetic survey a powerful tool in multi-scale exploration. Nowadays magnetometry, spectrometry or radiometry, electromagnetic and gravity surveys are concurrently, or separately, conducted onboard moving platforms. These methods were developed beside other important geophysical methods for mineral and oil exploration. In the following, only airborne magnetic surveys will be discussed in detail. For a long time, the most distinguishing features of the aeromagnetic method, in comparison with other geophysical prospecting methods, were its low cost and its data acquisition speed (Heiland 1940; Reford and Sumner 1964) especially when compared to seismic campaigns in oil exploration. The availability of the Global Positioning System (GPS) by the early 1990s, particularly in its differential form, together with the very high sensitivity and accuracy of the magnetometers, dramatically reduced the error budget in aeromagnetic surveys. Subtle magnetic variations can now be resolved (Millegan 2005; Nabighian et al. 2005) and high-resolution aeromagnetic surveys (HRAM) are industry standard. These achievements have pushed toward "lower and lower" altitude and "higher and higher" resolution. Safety is now a crucial issue, pilots and geophysicists already having paid a heavy price with 21 crashes and 48 fatalities between 1977 and 2001 (Urquhart 2003). Advances in miniaturized electronics, GPS technology, and sensors (magnetometers, video cameras) coupled with sophisticated guidance, navigation and control systems enable the development of small Unmanned Aerial Vehicles (UAVs) for survey missions operating for extended periods of time over large geographical areas (Lum et al. 2005, Lum 2009). However, to bridge the gap between long wavelengths, say larger than 600 *km*, resolved by near-Earth orbiting satellite measurements and short wavelengths of less than about 200 km, resolved by aeromagnetic data, measurements of the magnetic field aboard stratospheric balloon flying at 30−40 km altitude prove to be also useful (Cohen et al. 1986; Achache et al. 1991; Tsvetkov et al. 1995; Nazarova et al. 2005; Tohyama et al. 2007). Most aeromagnetic data processing procedures are now fairly standard even though some minor differences still exist in leveling and gridding. There is not yet any standard format for the magnetic data as in seismic industry with the SEG-format (Paterson and Reeves 1985). Important efforts in archiving data for future use, particularly raw unfiltered data, have still to be made either by national agencies or by private companies. The best examples of problems that may arise from non-standard data archiving are given by the compilation of the 29 available aeromagnetic datasets used for the World Digital Magnetic Anomaly Map (WDMAM) project (Korhonen et al. 2007). This compiled magnetic anomaly map, containing all available wavelengths, is very useful for geological and tectonic mapping of the crust. However, the quality of each dataset covering a specific region has been hard to estimate as very few compilations have complete metadata information (Hamoudi et al. 2007). When available, metadata show compilations to be in different coordinate systems and projections. All compilations resulted from the stitching together of smaller surveys carried out at various altitudes in which the individual panels were, or were not, upward continued to a common altitude. Often, this information is provided but in general the mean altitude, or the mean terrain clearance with respect to mean sea level, is not systematically known. Panels inside each individual compilation were derived for different epochs using for the reference field either local polynomials or global models. In most cases, it was difficult to find out which model had been used to reduce the data. Because the quality of these global models is continuously improving, keeping track of the reference field used to derive the anomaly field is fundamental. Despite continuous technological developments in surveying techniques, progress in geophysical and geological interpretation of potential field data, especially the magnetic field, are slow. It is only recently that the relationships between magnetite and geologic processes have been deeply studied (Reynolds and Schlinger 1990; Reynolds et al. 1990; Frost Shive 1986; Grant 1985a, b).

In the following sections, we present some general aspects of aeromagnetic measurements. The progress of magnetic instrumentation, aircrafts and aeronautical techniques from the 1930s pioneering era to nowadays are described in the first part. The second part concerns data acquisition from the first step of survey design to the final one of mapping the crustal magnetic field. Geological and geophysical interpretation of the aeromagnetic data is beyond the scope of the present paper.

4.3 History of Aeromagnetics

The magnetic exploration method is one of the oldest geophysical methods. It is directly tied to knowledge of terrestrial magnetism. This method was applied as early as in the mid 17th century for the location of ore bodies (Heiland 1940) and especially iron-bearing formations (Heiland 1929). The attraction of compass needles to these latter formations led to extensive use of magnetic compass as a prospecting tool in many countries, among them, Sweden, Finland, Russia, and the USA during the 19th and beginning of 20th centuries (Heiland 1929; Nabighian et al. 2005). Adolph Schmidt developed the first terrain suitable device for measurement of magnetic anomalies of geological structures in the 1920s (Heiland 1929). These magnetometers were also based on a magnetic needle system. They were used for relative measurements of Z and H (vertical and horizontal resp.) magnetic components with an uncertainty of ± 2 nT. They were used in mineral as well as oil exploration. The Earth inductor inclinator magnetometer also called the Earth inductor (Heiland 1940), was the first instrument not based on needles and which could measure both the inclination and various components of the Earth's magnetic field by the voltage induced in rotating coil (Heiland 1940; Nabighian et al. 2005).

The first recorded attempt to measure the magnetic field onboard an airframe seems to be that of Edelmann who designed in 1910 a vertical balance to be used in a balloon (Heiland 1935). Hans Lundberg, using a captive balloon above Kiruna's ore body (Sweden) in 1921, realised the first airborne magnetic measurements (Eve 1932). Pioneering aircraft surveys in 1936 and 1937 were also reported by Logachev in the former Soviet Union. The magnetometer he developed and used was an induction coil designed for measurement of the vertical component of the Earth's magnetic field. The 1936 flight test at 1000 m and 300 m altitude along a 60 km length line using an open-cockpit aircraft was above a weak ($-230, +1430$ nT) but known magnetic anomaly. The measurements obtained along 3 flight lines were compared to ground data using Schmidt's balance. They showed clearly the same anomaly although with a shift in the maximum location between the three lines. Logachev attributed this shift to errors in orientation and the divergence in values of the anomaly was ascribed to instrumental errors. His first magnetometer had about 1000 nT accuracy and 72 kg weight. The unit used at that time was the gamma (1gamma = 1 nT = 10^{-9} T). The second airborne experiment took place in 1937. It was conducted over a strong – of the order of 30000 gammas – magnetic anomaly. Six flight lines, 30 km long, were realised with an altitude of either 200 m or 300 m (Logachev 1947) depending on the weather conditions. The main result was to prove the feasibility of magnetic surveys from an airplane. The second version of his magnetometer was only 30 kg weight and a better accuracy and accuracy of about 100 nT. The third and probably most important aeromagnetic survey reported at that time was above one of the largest ore deposits in the world – The Kursk ore body – and highest related magnetic field anomaly (Logachev 1947). The main goal of the experiment was to determine whether the depths of the upper and lower limits of the Kursk ferrous quartzites could be computed from aeromagnetic data (Logachev 1947). A total of 22 traverses (lines) were flown at an altitude between 500 and 1600 m, among them four lines laid out approximately at right angles to the strike of the geological structures. Heiland (1935) also described such experiments using an Earth's magnetic inductor. He also reported advantageous aspects of magnetic surveys from the air (Heiland 1935).

World War II certainly favoured technology developments, starting from the early 1940s. The main goal was then submarine and mine detection. Victor Vacquier with the Gulf Research & Development Company investigated in 1940 and 1941 the properties of iron-cored devices as a sensitive element of a magnetometer. He then helped developing the magnetometer (Vacquier 1946; Wyckoff 1948; Reford and Sumner 1964, Hanna 1990). The instrument, also known as a fluxgate sensor, was suitable for airborne magnetic prospecting and could measure very weak fields of about 1 nT. This instrument was the base of the MAD (Marine Airborne Detector) heavily used for military purposes. The comprehensive history of the development of the airborne magnetometer based on fluxgate sensors can be found in Muffly (1946), Reford and Sumner (1964) and in Hanna (1990). Gulf research & Development Company in 1946 made modifications to the magnetometer for geophysical exploration. The USGS (United States Geological Survey) was also involved in airborne magnetometer developments in late 1942 (Hanna 1990). Aeroservice Corporation made a successful test flight in April 1944. Three traverses were flown at different altitudes along a line over an area in Pennsylvania (USA) where the USGS had previously made a ground survey (Hanna 1990). Different tests were also conducted in various environments (wood, swamp land) with single engine aircraft.

The need for more powerful aircraft soon became apparent during these test-flights and the use of cooperatively USGS-US Navy twin-engine aeroplanes allowed more extensive oil prospecting aeromagnetic surveys prior to the security classification restrictions being lifted in 1946 (Hanna 1990). A large number of manuscripts announcing the arrival of aeromagnetic techniques were then published. The first offshore aeromagnetic survey was conducted in 1946 over the coastal gulf of Mexico by Balsley (1946). Also more than 16,000 line kilometres of magnetic data were collected in 1944 over the northernmost part of Alaska (Hildenbrand and Raines 1990). Composite magnetic anomaly maps of the conterminous US were published in 1982 (Sexton et al. 1982). The first aeromagnetic anomaly map of the former Soviet Union and its adjacent areas was published in 1979 (Zonenshain et al. 1991). For this, an instrument with an accuracy of 2 nT labelled AM-13 (Reford and Sumner 1964; Zonenshain et al. 1991) was developed. It was based on a saturable core. In 1947 the Canadian Federal government initiated systematic national aeromagnetic surveys as an aid to both geological mapping and mineral prospecting. The Geological Survey of Canada was using a modified war surplus two-axis fluxgate magnetometer (AN/ASQ-1) acquired from the U.S. Navy. The aeromagnetic map sheet was published in 1949 at various scales (Hood 1990; 2007). More than 9,500 aeromagnetic anomaly maps of Canada and adjacent areas have been published between 1949 and 1990. This amounts to more than 9,650,000 line kilometres (Hood 1990; 2007). The first national aeromagnetic map of Canada was published in 1967 (Nabighian et al. 2005). Starting from 1969, during 22 years, the Canadian International Development Agency (CIDA) funded surveys that were carried out in more than ten countries in Africa among them Botswana, Burkina Faso, and Zimbabwe. CIDA funded also the survey in Brazil in South America, and in Pakistan (Hood 2007). The aeromagnetic method was adopted by Australia. There, the first aeromagnetic survey was flown in 1947 (Doyle 1987; Horsfall 1997). Systematic national airborne geophysical surveys by the Bureau of Mineral Resources in Australia took place in 1951 (Tarlowski et al. 1992; Horsfall 1997). More than 4,000,000 line kilometres were flown with a reconnaissance survey altitude of 150 m above ground at line spacings between 1.5 and 3.2 km. The first aeromagnetic anomaly map of Australia was published in 1976 (Tarlowski et al. 1992, 1996). The fourth edition has recently been released (Milligan and Franklin 2004) using more than 10,000,000 line kilometres. Many countries among them Finland, the former Soviet Union, and South Africa (Hildenbrand T.G. and Raines 1990; Hood 2007) have also established cost effective national airborne geophysics programs. The systematic national aeromagnetic survey of Finland started in 1951 (Korhonen 2005, Nabighian et al. 2005). The high altitude survey, around 150 m, was completed in 1972 and a new one was then started, at low altitude in the range 30–40 m and with 200 m line spacing. The first Finnish national aeromagnetic anomaly map was published in 1980 (Nabighian et al. 2005, Airo 2002). The aeromagnetic anomaly map of the Fennoscandian shield was released later on (Korhonen et al. 1999).

The history of aeromagnetic methods and their evolution is closely related to the technology evolution. Indeed, the technology has progressively evolved and the capabilities of the initial electronic equipment, at the beginning rudimentary by today's standards, developed all the while. The fluxgate magnetometer was widely used in aeromagnetic surveys until mid-1960s. Its major disadvantage for the airborne applications is that it must be oriented. It has been supplanted by proton precession magnetometers that were introduced in the mid-1950s (Germain-Jones 1957; Nabighian et al. 2005, Hood 2007). The proton precession magnetometer is a scalar magnetometer and hence does not require any precise orientation (Reford 1980; Hood 2007). Furthermore it is very easy to operate and maintain. Its main limitation comes from its discontinuous operating mode, related to the proton's polarization (Nabighian et al. 2005). The Overhauser variant of the proton precession magnetometer uses Radio Frequency (RF) excitation that allows continuous oscillations and thereby alleviates the sampling rate problem (Nabighian et al. 2005). It is also widely used for marine surveys. Notice that the fluxgate and Overhauser magnetometers are also commonly used onboard Earth orbiting magnetic satellites like the Danish Oersted satellite (Nielsen et al. 1995), the German flight mission CHAMP (Reigber et al. 2002) and planetary missions like the Lunar (Binder 1998; Hood et al. 2001) or the Martian (Acuña et al. 1999) missions. In 1957, almost at the same time as proton precession magnetometers became available, optically pumped alkali vapour magnetometers were introduced. The first instrument was used in airborne surveys in 1962 (Reford 1980; Jensen 1965). Three types of instruments were developed by different companies and are based on different alkali gases: Rubidium or Cesium, Potassium or Helium (Jensen 1965). Today, these optically pumped alkali vapour magnetometers are the most often used instruments in magnetic surveys for airborne, shipborne or ground exploration (Nabighian et al. 2005). It is worth mentioning that these magnetometers have excellent sensitivity, nowadays of the order of $1\text{pT}(\sqrt{\text{Hz}})^1(1\text{pT} = 10^{-3}\ \text{nT} = 10^{-12}\ \text{T})$, and a very high sampling rate e.g., 10 Hz is common. Although the gradiometer mode was experimented with in the 1960s (Hood 2007), it was only in the early seventies that measuring the horizontal and vertical magnetic gradients was recognized as very important for enhancing near surface magnetic sources and for reducing noise level (Paterson and Reeves 1985). The advantage of a measured vertical gradient over a calculated one has long been debated (Grant 1972; Doll et al. 2006). The former is the difference in magnetic intensity between two sensors and the latter is derived using gridded total field maps using Fourier analysis or other method (Grant 1970). The first experiments using two sensors in which one sensor is fixed and the second being towed in a bird some tens of meters below started in the 1960s for petroleum exploration (Paterson and Reeves 1985). Even though the measurements were made with high accuracy using an optically pumped Rubidium sensor (Slack et al. 1967), the results were not convincingly superior to the computed gradient. A system for measuring the vertical gradient using rubidium-vapour sensors rigidly mounted in a twin boom was soon adopted by Geological Survey of Canada (Hood 2007). The separation between the sensors was 1.83 m (6 ft). This is not very useful for petroleum exploration but well adapted for mineral surveys (Hood et al. 1976). The gradiometers have been improved both for vertical, with separation of 0.5 or 1 m, and horizontal, with separation of 1 or 1.7 m, measurements and have been used in many high-resolution applications (Doll et al. 2006). The next generation of magnetometers that will quantitatively enhance the accuracy of mapping, are cryogenic magnetometers based on the electrical superconducting property of conducting material in low temperature liquid-helium (Zimmerman and Campbell 1975, Stolz et al. 2006). The acronym of this magnetometer is SQUID, standing for Superconducting Quantum Interference Device. Until recently the main limitation on the extensive use of SQUID magnetometer for airborne purposes came from the constraint related to liquid Helium maintenance (Stolz, 2006). A prototype was designed in 1997 and a portable version was operated as a full tensor gradiometer (three components of the gradient in each direction of the Cartesian coordinate system of the field) in 2003. The very sensitive sensor (a few femto Tesla) is towed from a helicopter and is suspended on a long cable to eliminate noise from the aircraft. This system was commissioned in 2008 and airborne geophysical surveys are being conducted by private companies (Exploration Trends & Development in 2008).

The onboard recording data system is an important part of the airborne geophysical survey. Analogue recording was a limiting factor in data acquisition and hence in the quality of surveys. In multi-channel data acquisition, the digital data recording associated with high storage capacity and the ability to verify and store unlimited quantity of aeronautical and geophysical parameters during the flight surveys, greatly improved the quality of surveys. The high sampling rates of magnetic readings and the accuracy of measurement are now easily handled and high-resolution mapping is achieved (Paterson and Reeves 1985).

Aeronautical evolution together with positioning and attitude system improvements significantly contributed to the progresses made in aeromagnetic survey accuracy, especially over oceanic areas where the positioning issue is crucial. Indeed, the snapshot photographic technique developed in 1952 by Jensen over land had an accuracy of 50 m. The flight path recovery is derived by extrapolation between points. Loran-C (Decca system) Radio positioning was then used over land and offshore, the accuracy achieved offshore was ≈500 m (Hofmann-Wellenhof et al. 2003; Urquhart 2003). The Doppler radar that provides a positioning accuracy around 5 and 10 m along tracks later replaced this system. Then the Inertial Navigation System was introduced, giving 5–10 m relative lateral position accuracy along track. The Mini-Ranger radio systems when used allowed 2 m accuracy over 75 km range (direct line of sight). The best positioning system since the 1980s is the satellite based GPS. The first use of GPS for detailed offshore aeromagnetic survey was made in 1985 (Hood 2007). The position accuracy of an aircraft with a single receiver is of the order of 20 m in the horizontal plane and much larger for the height (2–3 times horizontal error). The differential GPS mode (dGPS) allows much higher accuracy, of the order of a fraction of a centimetre in the carrier phase dGPS. Laser altimeters, now currently used in detailed or high-resolution surveys, provide centimetre precision altitude (Exploration trends & Development in 2008). This very high accuracy allows safe flights in drape mode. Unmanned Aerial Vehicles (UAV) and Remotely Operated Vehicles (ROV) are even safer. The miniaturization technology for magnetic sensors and electronic systems (data acquisition, compensation, data transmission, etc.), automated flight control system and GPS navigation have enabled the design and development of drones of short range and long range cruising (Miles et al. 2008). The first UAV survey was operated in 2004 with high endurance, more than 10 hours at a speed of 75 km h^{-1} (Anderson and Pita 2005). Recent UAV, with 3-m wingspan and 18 kg mass, are more powerful with an endurance of 15 h and can travel at 100 km per hour. These UAV can be operated from sites near survey areas or from marine vessels. The main limitation for most of the UAV is their control by Line of Sight Communications. A remote operator near the region being flown is thus still required. New capabilities include autonomous tridimensional flight paths, long-range continuous satellite telemetry of geophysical data and flight parameters, radar altimeter and cooperative aeromagnetic surveying using teams of UAV controlled from a single ground station (Exploration trends & Development, 2005; Lum et al. 2005).

4.4 Data Acquisition and Reduction

4.4.1 Instrumentation

The geomagnetic field being a vector, magnetometers can be divided into two categories that differ both in terms of functionality and principle of operation. Vector magnetometers measure the magnetic induction value in a specific direction in 3-dimensional space whereas scalar magnetometers measure only the magnitude of the field regardless of its direction. Vector magnetometers are often mainly used as variometers, particularly in geomagnetic observatories, whereas scalar magnetometers are generally used as absolute instruments. In survey applications, one of the earliest instruments used was the Swedish mining compass developed in the mid-nineteenth century (Nabighian et al. 2005). This device resulting from the modification of the mariner's compass is based on a light suspended needle. It measures the inclination I and the declination D of the field. The revolution in geomagnetic surveys, at least for airborne magnetometers, came with the advent and development of the earth inductor (Logachev 1947; Heiland 1953; Reford and Sumner 1964). Various components of the magnetic field could then be measured from the electric voltage induced in the rotating coil. Nowadays magnetometers are not based on magnetic needles but use quantum mechanics properties of the atoms and nuclei for scalar magnetometers (Telford et al. 1990) and on ring-core saturation of a high magnetic permeability alloy for vector magnetometers (Muffly 1946; Vacquier 1946; Wyckoff 1948). The picoTesla precision and sensitivity reached nowadays are unprecedented.

4.4.2 Fluxgate Magnetometers

The airborne fluxgate magnetometer was originally designed and developed in 1941 by Victor Vacquier (Wyckoff 1948). It was built for use from low-flying aircraft as a detection device for submarines during World War II. After modification of the airborne instrument, it was also a first ship-towed instrument for marine magnetic surveys. It became apparent that the device had possibilities for studying geologic features. Many airborne magnetic surveys were carried out using fluxgate detectors between 1945 and 1985.

A fluxgate magnetometer consists of two identical soft magnetic cores. Special low noise core material, usually μ-metal or Permalloy (Vacquier 1946), with high magnetic permeability and low energy requirements for saturation, are used to obtain very sensitive fluxgates with a low level of noise. These cores are wound with primary and secondary coils and are mounted in a parallel configuration with the windings in opposition. An alternating current (AC) of frequency f (50 to 1000 Hz) is passed through the primary coils, generating a large, artificial, and varying magnetic field in each coil. This field drives periodically the cores into saturation. This coil configuration produces induced magnetic fields in the two cores that have the same strengths but opposite orientations at any given time during the current cycle. If the cores are in an external magnetic field, such as the Earth's Magnetic field, the component of the external field parallel to the artificial field reinforces it in one of the cores. It is anti-parallel in the other core, reducing the artificial field. As the current and the artificial field strength increase, saturation will therefore be reached at different times in the two cores. When the electric current decreases, the two cores fall below saturation at different times. These differences are sufficient to induce a measurable voltage in a secondary detection coil at a frequency $2f$. The detected signal is proportional to the strength of the magnetic field in the direction of the cores. This type of magnetometer has an accuracy of about 0.5 nT to 1 nT but has a wide dynamic range. A modern version of this type of magnetometer includes three-axis fluxgate magnetometers designed for vector measurements. They are also suitable for magnetic compensation in planes. It should be mentioned that fluxgate devices have been intensively used in near Earth orbiting geomagnetic satellites since Magsat (Acuña et al. 1978; Langel et al. 1982), Oersted (Nielsen et al. 1995, 1997) and CHAMP (Reigber et al. 2002, 1999) but they are supplanted by scalar magnetometers for airborne applications (Paterson and Reeves 1985).

4.4.3 Nuclear Precession Magnetometers

Nuclear precession magnetometers polarize the atomic nuclei of a substance contained in a bottle by applying an electric current in the coil circling it. These nuclei starts precessing when the current is switched off. As the behavior of the nuclei returns to normal, the frequency of precession called the Larmor frequency of the nuclei is measured. It can be correlated to magnetic induction strength. Let us briefly review some of the common scalar nuclear precession magnetometers:

- Proton Precession magnetometers
- Overhauser Effect magnetometers
- Optical Pumping Alkali Vapor Magnetometers

4.4.3.1 Proton Precession Magnetometers

A proton precession magnetometer was developed by Varian Associates in the mid-1950s (Reford 1980) and very rapidly became the most popular magnetometer for all type of surveys (Reford 1980). It uses hydrogen as precessing atoms. Liquids such as water, kerosene and methanol can also be used because they all offer very high proton densities (hydrogen nuclei). A standard proton precession magnetometer uses a high intensity artificial DC around the sensor to generate a strong static magnetic field to polarize the protons. The polarizing DC current is then switched off which causes the protons in the liquid to precess around the ambient Earth's field as a top rotates and precesses around the Earth's gravity field. The Larmor frequency of the precession is proportional to the ambient magnetic field strength and the proportionality factor is called the nuclear gyro-magnetic ratio. This ratio depends only on fundamental constants and therefore proton precession magnetometers are absolute instruments. A simple coil can detect the precession signal of the protons. The signal lasts for 1–2 s. The power required to polarize the protons may be significant (Telford et al. 1990). Nevertheless, the standard proton precession magnetometer is by far the cheapest portable magnetometer. Its main advantages are its operating simplicity without the need for orientation of the sensor and a high accuracy (0.1 to 1 nT). For airborne applications its main limitations are related to its low sampling rate and limited dynamic range (Ripka 1996).

4.4.3.2 Overhauser Effect Magnetometers

Overhauser effect magnetometers (Overhauser 1953; Dobrin and Savit 1988) are based on the principle of nuclear magnetic resonance. They have been developed from the proton precession principle. An Overhauser magnetometer uses radio-frequency power to excite the electrons of a special chemical dissolved in the hydrogen-rich liquid. The electrons pass on their excited state to the hydrogen nuclei, altering their spin state populations, and polarizing the liquid, just like in a standard proton magnetometer but to a greater extent and with a much lower power requirement (Nabighian et al. 2005). Actually, the total magnetization vector of the hydrogen liquid is larger in an Overhauser magnetometer than in a proton precession magnetometer. This allows sensitivity to be improved. Since the liquid can be polarized while the signal is being measured, Overhauser magnetometers have a much higher speed of cycling than standard proton precession magnetometers. Overhauser magnetometers are efficient magnetometers available with high precision (~1 pT) and high sampling rate (10 samples per second) suitable for Earth's field measurements. However it should be noted that Overhauser as well as free proton precession sensors have signal to noise ratios (S/N) that are dependent upon the field strength conditions (Geometrics, Technical Report TR-120, 2000). In areas where the geomagnetic field is weak, in the south Atlantic for instance, their S/N deteriorates. The Overhauser magnetometer is commonly used onboard near-Earth geomagnetic satellite like Oersted (Nielsen et al. 1995) and CHAMP (Reigber et al. 2002).

4.4.3.3 Optical Pumping Alkali Vapor Magnetometers

The principle of optical pumping of electrons of a gas or a vapor was first described by Kastler (1954), then by Hawkins (1955) and by Dehmelt (1957). The concept of optical pumping is based on energy transition (or pumping) by circularly polarized optical-frequency radiation of electrons from one of two closely spaced energy levels to a third higher level, from which they fall back to both of the initial ground levels. The use of Zeeman transitions in the alkali metals for magnetometry was first suggested by Bell and Bell and Bloom (1957) using Sodium (and Potassium) vapor to detect the resonance. They suggested also the use of Rubidium or Cesium vapor. Potassium is also used in some magnetometers (Pulz et al. 1999). Recently Leger et al. (2009) described an absolute magnetometer based on ^{4}Helium to be used aboard the three satellites of the future near-Earth geomagnetic SWARM mission. We know from quantum physics that the electron can only take on a limited number of orientations with respect to the ambient magnetic field vector. Each of these orientations will have a slightly different energy level. This electron energy differentiation in the presence of an external magnetic field is called Zeeman splitting. The differences in energy from one Zeeman level to the next are proportional to the strength of the ambient field. It is these energy differences between the Zeeman levels that are measured to determine the Earth's magnetic field strength. For an ambient field of $\sim 50,000$ nT, the splitting energy will correspond to a frequency in the range of a few hundred kHz (Nabighian et al. 2005; Smith 1997; Parsons and Wiatr 1962). The frequency of resonance used was 700 kHz for the first alkali vapor magnetometers (Parsons and Wiatr 1962). Nowadays, the frequency of the oscillating signal varies between 70 kHz and 350 kHz whereas the free proton precession and Overhauser magnetometers use an oscillating signal of 0.9 kHz and 4.5 kHz respectively (Parsons and Wiatr 1962) (Geometrics, Technical Report TR-120, 2000). It is then clear that the higher frequencies of the optical pumping magnetometers as compared to precession magnetometers provide better response and reproduction of the magnetic field signal. Alkali vapor instruments have excellent sensitivity, better than $0.01\ \mathrm{nT}(\sqrt{\mathrm{Hz}})^{-1}$, and high sampling rate – values as high as 10 Hz – are commonly used in magnetic surveys (Nabighian et al. 2005). The comprehensive theoretical and technical descriptions of the alkali vapor optical pumping magnetometer may be found in Parsons and Wiatr (1962). A good discussion may also be found in Telford et al. (1990).

4.5 Survey Design

Aeromagnetic surveys are undertaken at the early stage of petroleum exploration before any other geophysical method (Reford and Sumner 1964; Reford 1980; Paterson and Reeves 1985; Nabighian et al.

2005). The aim was to determine the depth of the basement crystalline rocks underlying sedimentary basins. Sedimentary formations are assumed non-magnetic as their magnetic signal is very weak, below the resolution and accuracy of the measurements. The estimation of the basin thickness is thus indirectly derived. However, the steady improvement of magnetometer sensitivity, the high resolution and accuracy achieved by the magnetometers, and the very high accuracy (sub-metric) of the positioning systems allowed by dGPS (Parkinson and Enge 1996) make it possible nowadays to outline weakly magnetized layers. According to Paterson and Reeves (1985), very small variations in magnetite concentration inducing anomalies as low as 0.1 nT can be correlated with diagenetic processes in hydrocarbon accumulations, and some hydrocarbon-related structures can now be detected in weakly magnetic sedimentary rocks (Grauch and Millegan 1998). The discovery of many structural oil traps by this method within the Sichuan Basin in China is the most typical example (Zhana 1994). However regional and detailed aeromagnetic surveys continue to be primary mineral exploration tools. These surveys allow variations in the concentration of various magnetic minerals, primarily magnetite, to be mapped. Aeromagnetic methods are therefore indirect exploration methods as magnetite is only a "marking" element. The goal in aeromagnetic surveys is the search for mineralization such as iron-oxide-copper-gold deposits as well as skarns and massive sulfides or heavy mineral sands (Nabighian et al. 2005). One of the main applications is the recognition and delineation of structural or stratigraphic environments favorable for mineral deposits of various types such as carbonatites, kimberlites (as host rock for diamonds), porphyritic intrusions, faulting and hydrothermal alterations (Keating 1995; Allek and Hamoudi 2008; McCafferty and Gosen 2009). High resolution aeromagnetic surveys are therefore very powerful tools for general geologic mapping (Reynolds et al. 1990, Bournas 2001). Depending upon the geological problem to be addressed, its framework, and all the related economical, scientific and technical constraints, aeromagnetic surveys are flown with a wide variety of geometric and metrological characteristics. The geometrical characteristics are mainly the flight lines and control lines (Tie lines) spacing, and the terrain clearance, or barometric height, or height above mean sea level. The metrological characteristics are related to accuracy and resolution of the magnetometers, the sampling rates, and the positioning system. All these points, together with the recording of the magnetic field temporal variations at a base station, are fundamental in order to achieve high quality final mapping. More details are given in "Data acquisition Section 4.6".

4.5.1 Flight Direction and Line Spacing

During aeromagnetic survey design, the flight path direction is selected mostly on the basis of the geological strike. For general reconnaissance mapping purposes the flight lines are usually oriented along cardinal directions, north-south or east-west (Cordell et al. 1990; Horsfall 1997). In the case of more specific surveys related to mineral exploration, it is then preferred to orient the flight lines in the direction perpendicular to the geological strike to maximize the magnetic signature. Control lines (Tie lines) are flown perpendicular to flight lines to provide a method of eliminating temporal variation of the magnetic field using pairs of values recorded at the intersections. This process is called leveling. It will be further described in Section 4.8.3. As a rule of thumb the tie lines spacing is in general 10 times the profile lines spacing. In polar regions a rate of 5 to 1 is often adopted (Bozzo et al. 1994). Cordell et al. (1990) recommended this rate for United States of America Midcontinent aeromagnetic surveys. However, for some petroleum exploration the ratio may be as low as 3 to 1 (Horsfall 1997) or even 2 to 1 (Reeves 2005). In the past, line spacing of 3000 m was generally adopted for surveys over sedimentary basins. These kinds of surveys are now flown with 500 m line spacing (Cady 1990). Flight-line spacing was limited in the past to 1500 m over crystalline areas whereas they are now flown at 400–500 m or even 200 m. Surveys dedicated to mineral exploration are usually flown at 200 m line-spacing, sometimes as close as 50 m line-spacing for very high resolution exploration surveys (Horsfall 2). Flight-line spacing is generally determined by average depth to crystalline basement (Reid 1980; Cordell et al. 1990), by the degree of detail required in final mapping (Horsfall, 1997) and the size of the target to detect (Horsfall 1997; Reeves 2005). The financial resources available for the survey are also crucial in this choice. Reid (1980) showed that in order to avoid aliasing in the short wavelength of the signal, neither the flight-line spacing, nor inline sampling rate should exceed twice the average target depth in total-intensity surveys. These results were obtained using spectral analysis and the power spectrum expectation relationship of Spector and Grant (1970). Equivalent analysis for the gradiometer surveys led Reid (1980) to recommend flight-line spacing equal to the average depth to crystalline basement. The sampling rate of the modern magnetometers is very high, up to 1 kHz. Generally, the speed of the aircrafts used is not a limitation to achieve the expected resolution of the mapping. Indeed, the aircraft speed used during the surveys varies between 220 and 280 kmh^{-1} – typically 250 kmh^{-1} – corresponding to about 69 ms^{-1}. Using a modern magnetometer with a sampling rate of 10 Hz, the along line spacing is then around 7 m (Horsfall 1997; Reeves 2005). The achieved sensitivity of the Alkali Vapor Optically Pumped Magnetometers is of the order of $0.01\,\text{nT}(\sqrt{\text{Hz}})^{-1}$. To benefit from such a high accuracy and achieve high resolution field mapping, it becomes crucial to be able to remove the signal due to other sources of noise, like aircraft magnetic interference (Hardwick 1984a, b).

4.5.2 Survey Flight Height

Let us recall that the magnetic field decreases as the inverse cube of the distance from the magnetic source to the sensor, at least for elementary dipolar sources. Therefore to detect small variations in the magnetic field, surveys must be flown close to the ground. The magnetic sources may be covered by non-magnetic material and the ground clearance is the distance between the sensor and the Earth's surface. Regional mineral/petroleum surveys were usually flown in the 1970s at a constant ground clearance of 150 m (Horsfall 1997; Bournas 2003; Allek and Hamoudi 2008). Table 4.1 shows the line spacing and corresponding height used for recent aeromagnetic surveys

The main limitation on the survey height is related to flight safety. Aircraft performance is the main factor in maintaining ground clearance. In areas with highly varying topography, or rugged terrain, fixed-wing aircraft may not be suitable for surveys. Whenever possible financially, helicopter are by far the best platform to use in rugged terrain to keep a small ground clearance. Regional geological purpose surveys might be conducted at higher altitudes ranging from 500 to 1000 m with appropriate line spacing, but then, only broad features will be outlined (Reid 1980).

Table 4.1 Survey lines spacing and corresponding height values (From Horsfall 1997)

Line spacing (m)	400	200	100
Height (m)	100	80	60

4.6 Data Acquisition

Airborne magnetic survey quality has benefited from technological developments and miniaturizing devices. The amount and variety of data collected during an airborne survey is so large, due to the fast sampling rate achieved, that a computer is necessary for acquisition and storage. Analog magnetic data are digitized and stored (either in a data logger or in the computer). The navigation data necessary for flight-path and data recovery for geophysical field mapping are also stored. The flight paths are nowadays recorded on a color video camera (Fig. 4.1). The flight-path tracking cameras and aerial photographs used in pre-GPS and radio era were an essential component of coordinate and flight-path recovery (Le Mouël 1969; Luis 1996; Horsfall 1997). Indeed they were time synchronized to geophysical data via an onboard timer. The accuracy achieved with such tracking systems ranges between 50 m and 1 km for horizontal coordinates. With the help of radio navigation or inertial devices the vertical accuracy was improved to about 30 m (Le Mouël 1969). For low altitude mineral/petroleum exploration, radar altimeter led to accuracy of the order of ± 10 m in the mid-seventies (Allek and Hamoudi 2008), whereas ± 1 m accuracy can easily be reached with modern devices (Reeves 2005). The altimeter data are used to validate each crossover during tie-line leveling of the magnetic data.

With GPS navigation there is no need for video recording in flight-path recovery. It is mainly used for a posteriori checking of the accuracy of the navigation (Horsfall 1997) and in the case of special "cultural" signatures appearing in the geophysical data. GPS not only provides a very accurate positioning system especially in its differential technique (dGPS) (Parkinson and Enge 1996) but also very accurate time reference. This time is synchronized with geophysical data and recorded on the data acquisition system.

Fig. 4.1 The onboard aircraft data acquisition system mounted on a rack for airborne geophysics (Courtesy of Geometrics)

Moreover, the synchronization with the base station where the geomagnetic field is continuously recorded is essential for removal of the daily diurnal field variation from the total-field recorded onboard the aircraft. The data acquisition system, as presented in the Fig. 4.1, incorporates a monitor where outputs from the real-time geophysical and navigation instruments are displayed and a monitor associated with a color video camera. The present data acquisition equipment is almost self autonomous and need only be programmed before take-off so that surveys are often flown with only the pilot onboard. This policy allows longer survey flights (Reeves 2005). The actual accuracy and resolution really achieved in an airborne survey is dramatically reliant on the aircraft navigation system used. The constant ground clearance normally specified for the survey requires altitude measurement. Survey aircrafts are then fitted with radar altimeters beside the classical barometric altimeter. The data of these altimeters are also recorded by the data acquisition system. When combined with the aircraft GPS height, the radar altimeter allows the 3D flight path and the Digital Elevation Models (DEM) of the area being flown to be derived. For the very high-resolution surveys, a laser altimeter is also added onboard. Measurements undertaken in draped mode have considerably improved and they have become almost a standard in both heliborne and fixed-wing magnetic surveys. Pre-computed heights along the flight line in the new versions of navigation software let the pilot follow the draping more effectively and safely than the previous intensive computer's CPU time versions where the position along the profile was computed in real-time from grids (Exploration Trend & Development in 2008).

4.6.1 Magnetic Compensation of Aircraft

Two configurations are possible for the sensors of an airborne magnetic survey. In the first, and classical one, the magnetic sensor is located in a bird and towed as far as possible below the aircraft to reduce its magnetic effect. In the second configuration, the magnetic sensor is fixed to the aircraft either in a tail stinger (Fig. 4.2) or

Fig. 4.2 Magnetometer sensor in a stinger (*top*) on a tail of a Piper Navajo aircraft and (*bottom*) on a Bell helicopters (Courtesy of Novatem)

on the wingtip in the horizontal gradiometer configuration. A fixed installation of a total field magnetometer sensor on an aircraft is much more desirable than the towed bird configuration first for safety reasons and second because the bird configuration is not error or noise-free. The fixed configuration usually shows the best signal-to-noise ratio provided that all the magnetic disturbing effects of the aircraft are removed or compensated (Horsfall 1997; Reeves 2005).

Let us assume that the total field $B(P, t)$ measured at time t and point $P\,(x, y, z)$ by an airborne magnetometer in the fixed-sensor configuration may be modeled as the sum of three components (Williams 1993):

$$B\,(P, t) = B_i\,(x, y, z, t) + B_e\,(x, y, z, t) + B_{\text{dist}}\,(\theta_1, \theta_2, \theta_3) \tag{4.1}$$

where B_i is a function of space and time and represents the intensity of the Earth's magnetic field at a point P. The three angles $\theta_1, \theta_2, \theta_3$ are the plane heading, roll and pitch respectively.

This is the quantity of interest in the survey while both the remaining terms may be considered as disturbances or interferences. The function B_e varies with time and represents the diurnal variation (or transient external magnetic field). It varies significantly during the survey flight but is considered uniform in a limited area around the base station where the magnetic field is recorded simultaneously. The third function B_{dist} is the disturbance field generated by the aircraft. This disturbance field is a function of the attitude of the aircraft. Among all the disturbances the most significant are:

(1) Its remnant magnetization – i.e., permanent magnetic effects B_{perm}
(2) Its induced magnetization generated by the Earth's magnetic B_{ind}
(3) Eddy currents caused by the electrical conductor moving through the Earth's magnetic field and their magnetic effects B_{eddy}.
These effects are not easy to compensate and the solution is to move the sensors away from these sources.
(4) Magnetic effects of electric currents from the instruments, generators and avionics. Shielding and grounding the electric cables reduce these effects. The first three magnetic interference sources should be minimized in order to produce reliable magnetic data that can be related to geological features. This minimization is called "magnetic compensation of the aircraft". Two approaches have been proposed to mitigate these disturbances (Hardwick 1984a, Williams 1993). The first approach called "passive magnetic compensation" uses permanent magnets at various places (Geometrics, MA-TR15 technical Report). This method is however a trial-and-error method. It is time-consuming and moreover it does not compensate for motion of the aircraft (Reeves 2005). The second approach proposed by Leliak (1961) is referred to as "active" and uses a compensator. The system was originally designed for use with military magnetic detection systems. Leliak (1961) proposed building an analytical model of the disturbances (Williams 1993; Gopal et al. 2008; Pang and Lin 2009). Let us assume that the disturbance field may be written as:

$$B_{\text{dist}} = B_{\text{perm}} + B_{\text{ind}} + B_{\text{eddy}} \tag{4.2}$$

With

$$\begin{aligned}
B_{\text{perm}} &= a_1 \cos X + a_2 \cos Y + a_3 \cos Z \\
B_{\text{ind}} &= a_4 B_t + a_5 B_t \cos X \cos Y + a_6 B_t \cos X \cos Z \\
&\quad + a_7 B_t \cos^2 Y + a_8 B_t \cos Y \cos Z \\
&\quad + a_9 B_t \cos^2 X \\
B_{\text{eddy}} &= a_{10} B_t \cos x \dot{\cos} X + a_{11} B_t \cos X \dot{\cos} Y \\
&\quad + a_{12} B_t \cos X \dot{\cos} Z + a_{13} B_t \cos Y \dot{\cos} X \\
&\quad + a_{14} B_t \cos Y \dot{\cos} Y + a_{15} B_t \cos Y \dot{\cos} Z \\
&\quad + a_{16} B_t \cos Z \dot{\cos} X + a_{17} B_t \cos Z \dot{\cos} Y \\
&\quad + a_{18} B_t \cos Z \dot{\cos} Z
\end{aligned} \tag{4.3}$$

where $\cos X$, $\cos Y$ and $\cos Z$ are the direction cosines of the Earth's magnetic field along the longitudinal, transverse and vertically down instantaneous major axes of the aircraft respectively while $\dot{\cos} X$, $\dot{\cos} Y$ and $\dot{\cos} Z$ represent their first time derivatives. These direction cosines are defined as:

$$\begin{aligned}
\cos X &= \tfrac{T}{B_t} \\
\cos Y &= \tfrac{L}{B_t} \\
\cos Z &= \tfrac{V}{B_t}
\end{aligned} \tag{4.4}$$

where T, L, and V are the components of the total field B_i along traverse, longitudinal and vertical axes of the aircraft respectively. T is positive to port, L is positive forward and V is positive downward (see Fig. 4.3). The

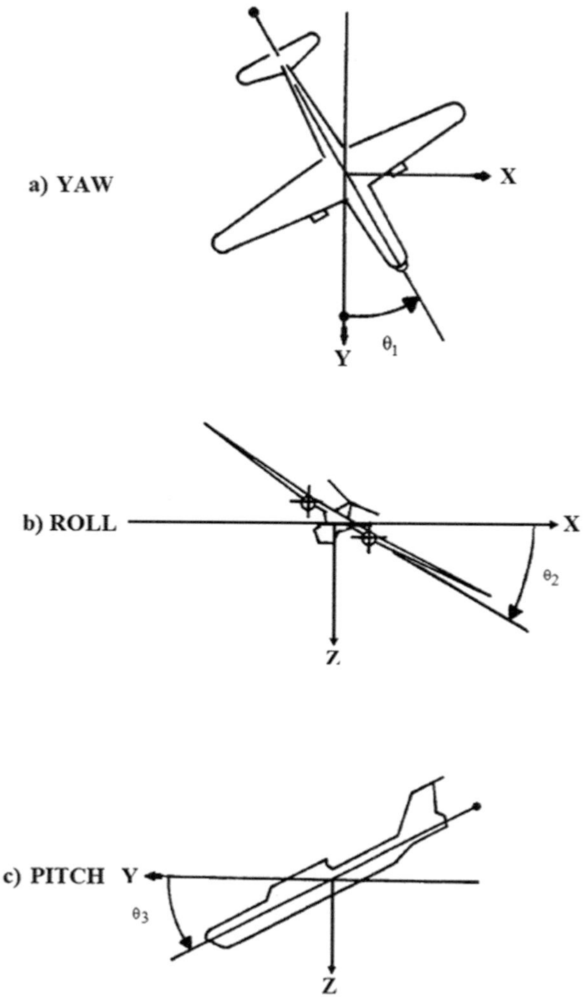

Fig. 4.3 The attitude Yaw, Roll and Pitch angles of the moving platform, in the transverse, longitudinal and vertical axes. Modified from Rice (1993)

determination of the 18 coefficients of Eq. (4.3) gives the magnetic field compensation. Hardwick (1984a) suggests adding a DC term for full compensation. Different numerical methods (ridge regression, least-squares, neural network, FIR model...) have been developed. Commercial compensators (Fig. 4.1) are based on such algorithms. Vector measurements are however necessary to solve the problem. A fluxgate vector magnetometer is added to the payload and must be rigidly mounted in a magnetically quiet location of the aircraft, far from the engines. In some configurations, the fluxgate is mounted in the middle section of the tail stinger. An active compensator is then composed of:

- Three-component fluxgate vector magnetometer.
- Multi-channel data acquisition and signal processor circuitry, to record signals from the scalar (Cesium, protons or Helium type) and vector magnetometers, GPS differential receiver board, Analog processor board.
- A main microcomputer with software, real-time clock, digital output.

Magnetic compensation for aircraft and heading effects is done usually in real-time. Raw magnetic values are also stored for later use if necessary. Active magnetic compensation begins with a calibration phase where all the magnetic interference values are determined in the absence of local magnetic anomalies. This undertaken at high altitude, generally between 1000 and 4000 m (Williams 1993; Reeves 2005) to minimize the influence of any local magnetic anomalies on the data following a specific geometry. To define the response of the aircraft in the Earth's magnetic field during the maneuvers, one has to derive a set of coefficients using magnetic data from scalar magnetometer and the attitude data from the fluxgate. Typical compensation maneuvers consist of a series of pitches, rolls, and yaws on four orthogonal headings (Hardwick 1984a; Reeves 2005) with 30 to 35 degree bank turns between each heading. This calibration procedure takes about 6 minutes of flying time (Hardwick 1984a; Reeves 2005). The standard amplitudes for aircraft attitude parameters are Pitch of $\pm 5°$, roll of $\pm 10°$ and yaw of $\pm 5°$. Each individual maneuver lasts about $30s$. The angles are relatively small which allow to use approximation of their trigonometric functions, i.e. $\sin\theta_i \approx \theta_i$ and $\cos\theta_i \approx 1 - \theta_i^2/2$. The compensation maneuvers are flown each time a new compensation is required, for instance if the magnetic field characteristics over a survey area are new. The effectiveness of the compensation is usually evaluated by the "Figure of Merit" (FOM) (Hardwick 1984a; Reeves 2005). The FOM is defined as the absolute sum of the total intensity anomaly measurements, along the four cardinal directions and compensated for the plane signal. In the seventies, a FOM of 12 nT was typical for regional surveys (Reeves 2005) whereas it has decreased nowadays from below 1 nT down to 0.3 nT after compensation (Horsfall 1997; Ferris et al. 2002).

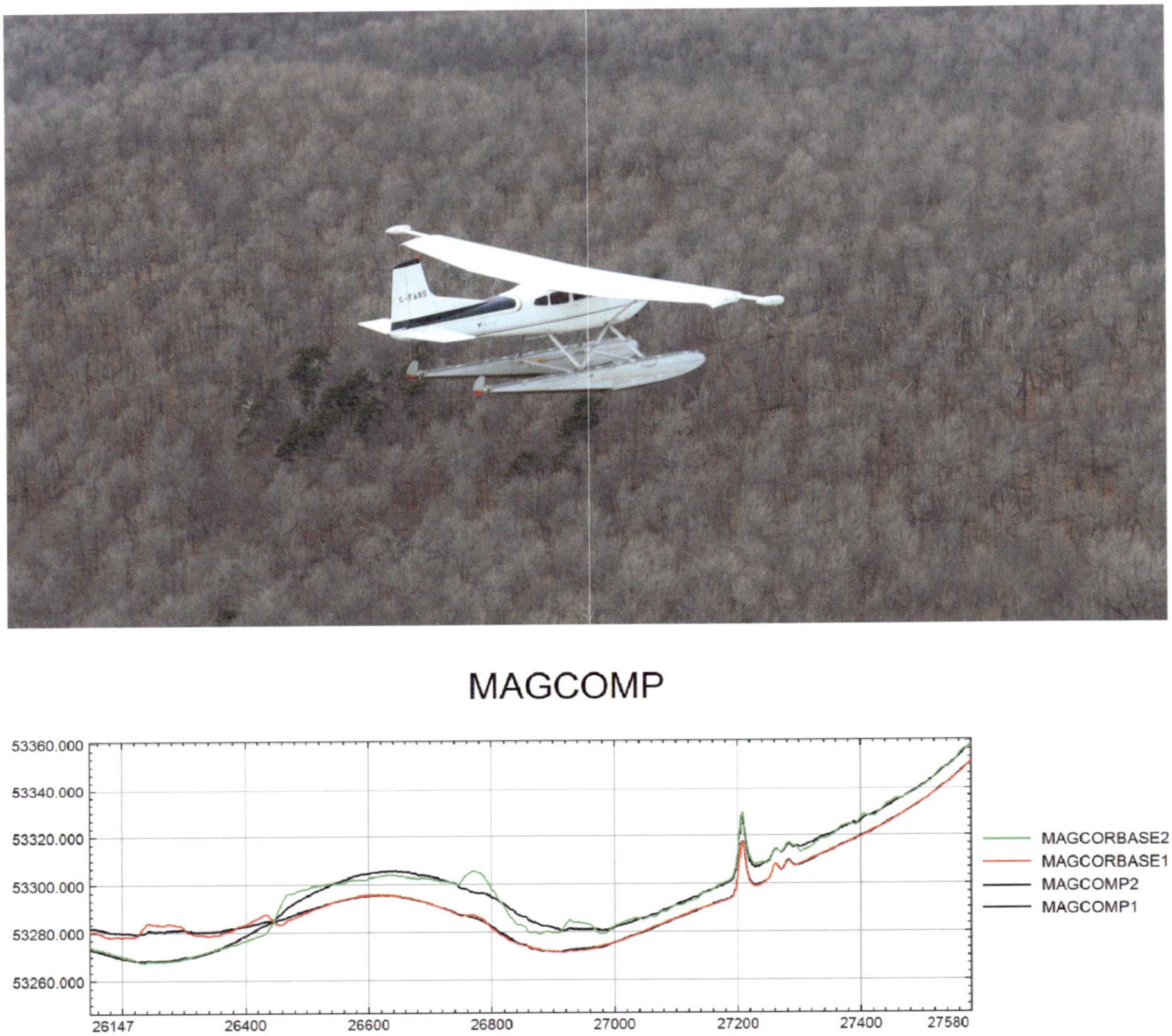

Fig. 4.4 Compensation effect of the interference generated by the aircraft on the two sensors located on wing-tip pods of aircraft (*top*). Red and green curves are uncompensated raw data and black curves are compensated data (*bottom*)(Courtesy of Novatem)

The magnetic component generated by the heading error of the aircraft is reduced to less than 1 nT. The calibration response is then stored in the memory of the compensator and subtracted from the incoming data during the survey operation (Horsfall 1997; Reeves 2005). Figure 4.4 illustrates the compensation effect on the data recorded by the two high-resolution magnetic sensors located in the wing-tip pods for measurements of the horizontal gradient.

4.7 Data Checking and Reduction

Digital recording and processing are nowadays commonly used in airborne surveys. The traditional relationship between those who are collecting and compiling data, and those who are using and interpreting these data has slightly changed. It is very easy to digitally handle the huge amount of collected data during the survey but great care has to be taken when checking data to avoid introducing false anomalies (Reford 1980). There is usually a two-step data verification, the first is the in-flight checking and the second after the flight.

4.7.1 In-Flight Data Checking

The data from the magnetometer(s), altimeter(s) and navigational system are displayed either on the monitor or on the graphic printer outputs when available onboard the aircraft and should reveal any major in-flight problem (Horsfall 1997).

4.7.2 Post-Flight Checking

At the end of each day, the data recorded onboard are verified and preliminary analyses are undertaken.

(a) Statistical analysis of each line flown and potential problem detection are among the first analyses performed.
(b) Detection and isolation of spikes and spurious recording (Fig. 4.5) are important for the data quality. This detection is usually based on the fourth-difference operator according to the following equation:

$$\Delta Q_i = Q_{i-2} - 4Q_{i-1} + 6Q_i - 4Q_{i+1} + Q_{i+2} \quad (4.5)$$

where Q is any measured quantity onboard the aircraft (uncompensated or compensated magnetic field data, Radar altimeter data or Barometric altimeter data) or at the base-station. The datum is considered valid if the operator returns a result less than a fixed threshold. The appropriate choice of threshold value is empirically determined.
(c) Calibration test line to ensure that equipment is operating within tolerances.
(d) Detection of any high frequency magnetic anomalies generated by any "cultural" anthropogenic noise like pipelines or railways (Horsfall 1997; Reeves 2005) is done. The video recording during flight-line data acquisition may be of great help to identify the perturbation sources.
(e) Checking the compliance of the flight path with survey specifications is necessary.
(f) The base station is checked to ensure the diurnal variation stays within the survey specifications.

It should be emphasized that partial or total re-flying will occur if one or more of the following conditions holds:

- The magnetic diurnal variation exceeds the survey specifications
- The aircraft's speed derived from the GPS navigation system is abnormal

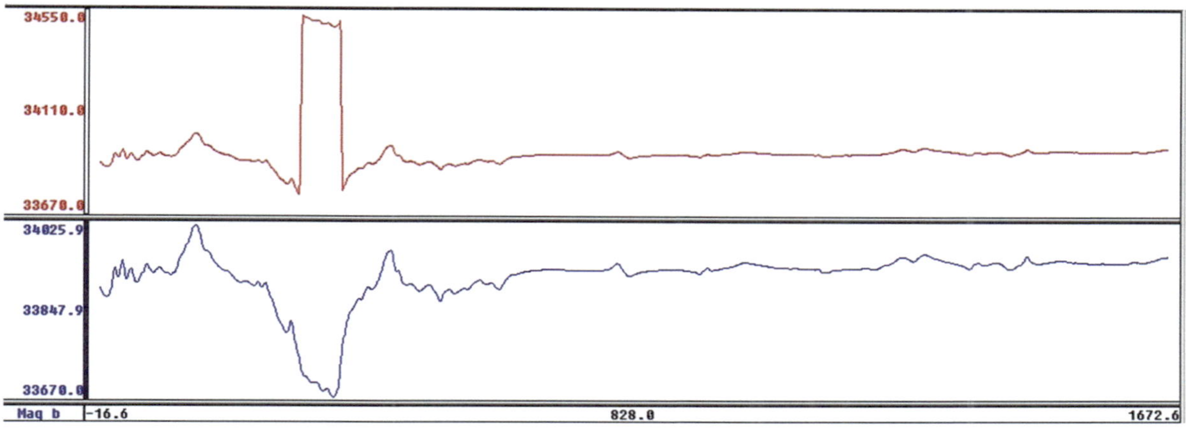

Fig. 4.5 Detection of spurious data along a line (*top*) in post-flight checking and correction (*bottom*). Units are nT for total field and kilometers for distance along flight line

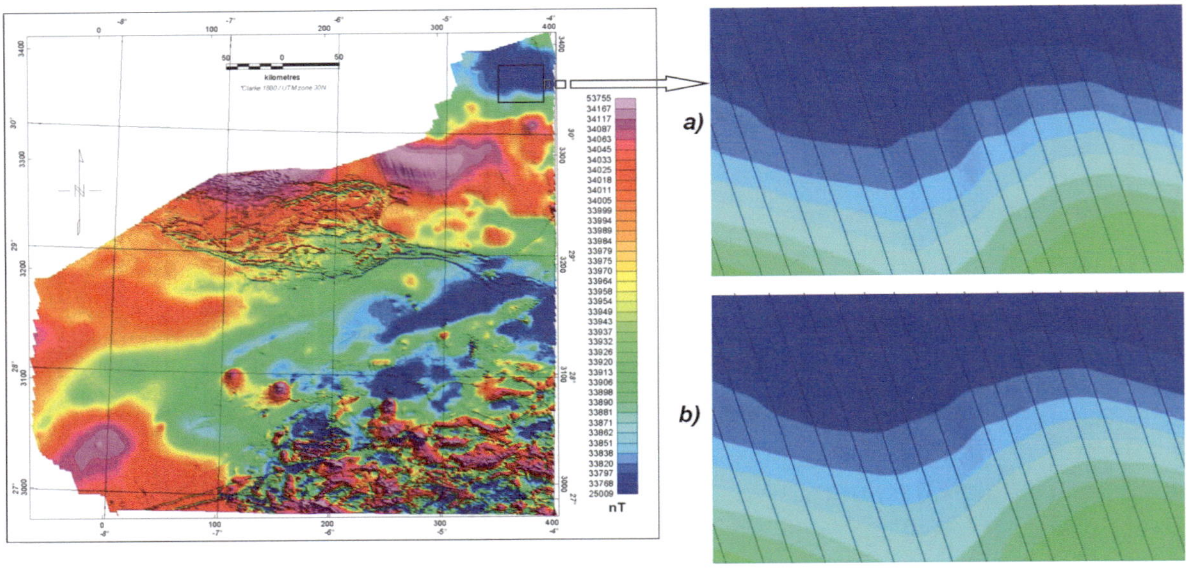

Fig. 4.6 lag effect correction of the magnetic field in the towed bird sensor configuration. (**a**) Raw data. (**b**) Corrected data (after Allek (2005))

- Final differentially corrected flight path deviates front the intended flight more than the survey specifications.
- The final differentially corrected altitude deviates from the flight altitude specifications.

In addition:

- Magnetic data channels contain multiple spikes.
- GPS data shall include at least four satellites for accurate navigation and flight path recovery.
- There should be no significant gaps in any of the digital data, including GPS and magnetic data.

After checking the continuity and integrity of the data, correcting the on-board recorded data (flight path, time, and geophysical data) and importing base station data, a database is created. The data are posted to the database on a flight by flight basis. The final steps in the daily processing of the data after validation are:

- Producing diurnally corrected airborne reading. For this, the base level value of the base station magnetic data has to be estimated. It is then subtracted from the digital diurnal data and the resultant values are added to the time synchronized digital onboard magnetic data.
- Merging of the geophysical data and navigational (aeronautical) data (geographic location, time)

The procedures described above are daily duties throughout of the survey. Data should be validated by the technical certifying authority in charge of the project (Reeves 2005) before ending the survey. The global database containing all the relevant information related to the survey is then created. In the case of towed bird magnetic sensor configuration, magnetic data should be corrected for the lag effect that is responsible for the "zigzag" shape perturbation of the anomaly field (Fig. 4.6). Further processing is however needed before gridding and mapping the magnetic anomaly field.

4.8 Data Processing

4.8.1 Magnetic Anomaly Field Determination

Let us recall that the "vector magnetic anomaly field" is the magnetic induction generated by the rock magnetization (or susceptibility) heterogeneities of the Earth or planetary crust. In the case of the Moon, this quantity has been directly measured on some sites by the Apollo Astronauts. This is because the Moon, unlike the Earth, does not have a global internal field (Ness 1971). In the case of the Earth, the geomagnetic field is clearly dominated by the core's contribution, which represents almost 99% of the amplitude of the internal signal. The total intensity of the internal magnetic field (i.e., magnitude of the combination of the core and crustal fields) has then to be measured very accurately in order to be able to recover the crustal field. Let us consider an orthonormal cartesian coordinate system (O, x, y, z) where O is the origin of the system, and the axes Ox, Oy and Oz are respectively directed toward the geographic North, the East and Downward. The crustal sources lie in the lower half-space ($z > 0$). We assume also that the survey area – call it the domain D – belonging to the upper half space is of limited extent so that the planar approximation holds. If the survey area is too large for this approximation to be valid, the survey area can be divided into small pieces to fulfill this requirement. At any point $P(x, y, z)$ of D, the instantaneous total magnetic field measured at time t may be expressed as:

$$\mathbf{B_t}(P,t) = \mathbf{B_N}(P,t) + \mathbf{B_a}(P,t) + \mathbf{B_e}(P,t) \qquad (4.6)$$

where: $\mathbf{B_N}$ is the core field or normal field, also called the main field, $\mathbf{B_a}$ is the crustal field or anomaly field and $\mathbf{B_e}$ is the transient external field. It is worth recalling that the amplitude of the main field varies from roughly 20000 nT to 65000 nT from the equator to the pole respectively. Its modeled spatial wavelengths vary from 2500 km to 40000 km. Its time variation, called secular variation, has to be taken into account in two cases: (1) if panels of adjacent surveys based on data collected and processed at different epochs have to be merged. (2) if the survey time span exceeds a year. The external field varies from some few 10^{-3} nT to some 10^{3} nT during magnetic storms and from 10^{-3} s characteristic time scale to 22 years for the solar cycle (Cohen and Lintz 1974; Courtillot et al. 1977). During magnetically disturbed days, acquisition should stopped. Typically, a day is disturbed if the diurnal activity is greater than 5 nT over a chord of 5 min in length. Sometimes values of 2 nT over 30 s are used. The most important external magnetic field contribution that is necessarily recorded during ground or airborne surveys is the diurnal variation.

4.8.2 Temporal Reductions/Corrections

When there is no permanent geomagnetic observatory available in their vicinity, airborne and land surveys generally include a base station magnetometer that continuously samples the magnetic field time variations during the data acquisition flight period. Usually, the fixed station is operated in the centre of the surveyed area. It is still a matter of debate on how many base stations are needed for large surveys in order to adequately represent the highly varying diurnal magnetic field. The problem was first pointed out by Whitham and Loomer (1957) and Whitham and Niblett (1961) (see Nabighian et al. (2005)). This problem is even more difficult to handle in the case of marine measurements or airborne surveys over oceanic areas (Luis 1996; Luis and Miranda 2008). When it is not necessary to recover the total field, it is then simpler to use a gradiometer technique rather than a single sensor magnetometer. In this multi-sensor configuration (Fig. 4.4), the common features – i.e., normal field and external time varying field – are removed by calculating the differences between the signals recorded at different instruments. With the significant improvements in aeromagnetic survey instrumentation (resolution of magnetometers less than 0.1 nT and high precision positioning systems with an accuracy of less than a meter) and processing, the assumption of uniform temporal magnetic variations is only partially justified (Reeves 1993). Clearly, in some specific areas like those under the influence of the Equatorial Electrojet (EEJ), the non-uniformity of the temporal variations should be taken into account for data correction (Rigoti et al. 2000). The uncorrected effect of the EEJ, after subtraction of the base station data was reported by Rigoti et al. (2000) to amounts to 70 nT over a distance of 250 km (or 0.28 nT km^{-1}). This gradient may be as large as 1 nT km^{-1} (Rigoti et al. 2000). Close to auroral zones, or areas with high electric conductivity, where induction effects may be important (see for example Milligan et al. (1993)), the temporal variations may be considered uniform only for very short distances, not exceeding 50 km from the base station.

Let us assume that transient external magnetic field variations are zero-mean when averaged over a long time interval, at least a year. Then, the values of corrected for time variations may be derived using the data collected at the base station and at the nearest magnetic observatory. This procedure has been described by Le Mouël (1969) and used for example for the Azores Island aeromagnetic survey by Luis et al. (1994). Let us briefly recall the relationship to be used for correcting the data for these effects. First, let us denote O for observatory, S for Base-station and P for any given point in space where the total-field is measured. We note $\bar{B}_t^{\mathrm{an}}(O)$ and $\bar{B}_t^{\mathrm{sur}}(O)$ the annual mean and survey time-interval mean of the total-field at a given observatory O close to the domain D. The $\bar{B}_t^{\mathrm{an}}(S)$ and $\bar{B}_t^{\mathrm{sur}}(S)$ are the corresponding means at the base-station S. $\bar{B}_t^{\mathrm{an}}(P)$ is the annual mean at any measurement point P along the flight lines. Following Le Mouël (1969) two simplifying assumptions are necessary for time variation corrections. The first one assumes that the transient variations are the same at the base-station and at the measurement point P of D, which gives:

$$B_t(P,t) - \bar{B}_t^{\mathrm{an}}(P) = B_t(S,t) - \bar{B}_t^{\mathrm{an}}(S) \tag{4.7}$$

The problem is then to compute $\bar{B}_t^{an}(S)$ when the duration of the survey is less than a year. This done by assuming that:

$$\bar{B}_t^{\mathrm{sur}}(S) - \bar{B}_t^{\mathrm{an}}(S) = \bar{B}_t^{\mathrm{sur}}(O) - \bar{B}_t^{\mathrm{an}}(O) \tag{4.8}$$

i.e., by assuming that the differences of the mean between the base-station S and the closest observatory O, due to the time difference, are the same. The combination of (4.7) and (4.8) leads to:

$$\begin{aligned}\bar{B}_t^{\mathrm{an}}(P) =& (B_t(P,t) - B_t(S,t)) + (\bar{B}_t^{\mathrm{sur}}(S) - \bar{B}_t^{\mathrm{sur}}(O)) \\ &+ \bar{B}_t^{\mathrm{an}}(O)\end{aligned} \tag{4.9}$$

Equation (4.9) is convenient to derive to a common epoch the static total field at any points P of the domain D. In order to use the Eq. (4.9), it is necessary to have either an observatory or a repeat station nearby (Le Mouël 1969; Chiappini et al. 2000; Supper et al. 2004) to estimate the mean field and the secular variation values. In the case where no observatories or repeat stations are available, different approaches have been proposed. As an example, in the case of the aeromagnetic survey of the Azores Islands, Luis (1996) suggests approximating the mean observatory values $\bar{B}_t^{\mathrm{an}}(O)$ and $\bar{B}_t^{\mathrm{sur}}(O)$ by their corresponding values derived from the International Geomagnetic Reference Field (IGRF) models (for details about the IGRF and other reference field, refer to section 4.9). Then the general equation reduces to:

$$\begin{aligned}\bar{B}_t^{\mathrm{an}}(P) =& (B_t(P,t) - B_t(S,t)) \\ &+ (\bar{B}_t^{\mathrm{an}}(S) - \bar{B}_t^{\mathrm{sur}}(S))_{IGRF} + \bar{B}_t^{\mathrm{sur}}(S)\end{aligned} \tag{4.10}$$

The quantity defined in (4.11) below, is a good approximation of the "secular variation" of the main magnetic field over the area of interest.

$$\delta B(S) = (\bar{B}_t^{\mathrm{an}}(S) - \bar{B}_t^{\mathrm{sur}}(S))_{IGRF}/\Delta t \tag{4.11}$$

where Δt is time interval of the survey. The temporal corrections to apply to sampled data in order to derive the total field are given by:

$$\bar{B}_t^{\mathrm{an}}(P) = (B_t(P,t) - B_t(S,t)) + \bar{B}_t^{\mathrm{sur}}(S) + \delta B(S).\Delta t \tag{4.12}$$

The Eq. (4.9) described above is valid in a general framework. It can be used for surveys of limited geographic extension and/or flown over short time span. However it is used in a simplified form for helicopter-borne surveys and for surveys performed within a radius of approximately 50 km (Paterson and Reeves 1985) and up to 100 km (Whitham and Niblett 1961; Le Mouël 1969) of the base-station. Such surveys are typically those for oil and mineral exploration. Indeed, in these cases the time-corrected field is simply defined by:

$$B_t(P) = (B_t(P,t) - B_t(S,t)) \tag{4.13}$$

where measurements along the flight-lines and at the base-station are time-synchronized. Assuming that the total-field magnetic anomaly distribution is time-invariant, the values obtained at the intersections between flight-lines and control-lines should be almost the same. Any significant difference is then attributed to uncorrected temporal variation.

4.8.3 Magnetic Leveling

As previously described, aeromagnetic surveys are flown according to a designed and planned network (Reeves 1993) of flight-lines (L) and almost orthogonal control-lines also called Tie-lines (T). The Tie-line spacing is generally greater than the one for flight-lines. As a rule of thumb a rate of 10 to 1 is usually used while 5 to 1 is adopted in high latitude regions (Bozzo et al. 1994). In areas where geologic features lack a dominant strike, a rate of 1 to 1 has been used (Nabighian et al. 2005). This network provides a mean to assess the quality of temporal data reduction. The differences at the intersecting points of the network should be close to zero if the coordinates of the points are accurately determined in each direction (Paterson and Reeves 1985; Reeves 1993; Nabighian et al. 2005) as is generally the case for modern positioning systems like dGPS. Because the L-T differences at the intersection points are usually not negligible (Fig. 4.7), different empirical strategies have been developed to minimize the closure errors.

The process introduced to minimize these errors is called magnetic leveling. It was originally developed as an alternative to the use of base station data reduction (Whitham and Niblett 1961; Reford and Sumner 1964; Foster et al. 1970; Mittal 1984). The most common procedure is probably the two step method. The first step is a linear first order correction. A constant correction is calculated, based on the statistical mean of the closure errors or determined by least-squares minimization and distributed equally to each data point along the lines. In the second step a low order polynomial correction is adjusted to reduce the mis-ties below a specified minimum, usually 0.01 nT (Reeves 1993; Bozzo et al. 1994, Nabighian et al. 2005). Some algorithms consider tie-lines as fixed and adjust only the flight-lines. In the pre-GPS era, the L-T leveling errors were characterized by high amplitude values up to 20 nT with zero average. The achieved differences are now commonly of a few nT with an average over the length of a line of the order of 3 nT for a small extent or helicopter-borne survey (Reeves 1993). Once the leveling is complete, the total field may be gridded (Bhattacharyya 1971; Briggs 1974; Hansen 1993) and contoured using any available technique of digital enhancement provided that the data distribution is dense over the surveyed area (Fig. 4.8).

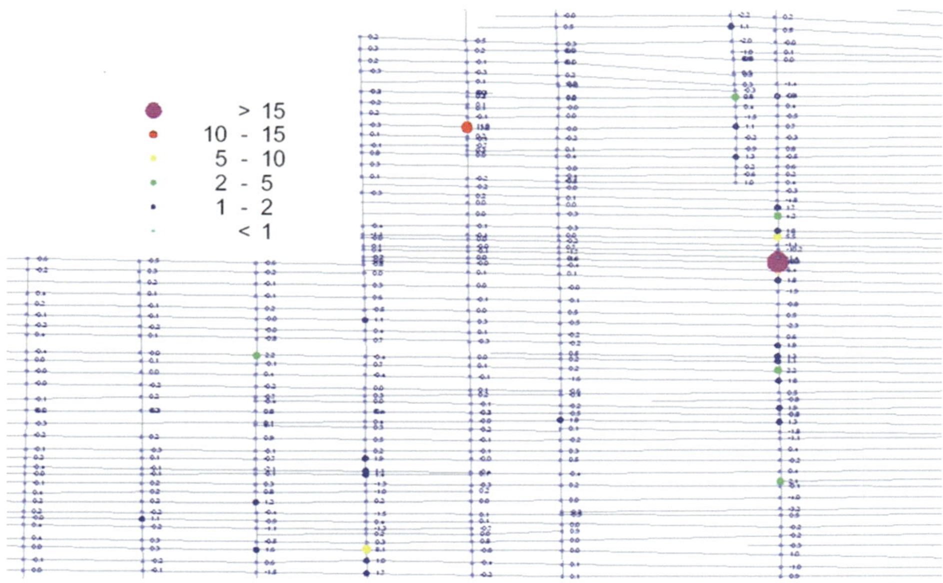

Fig. 4.7 Example of Tie-line cross-differences from an aeromagnetic survey over the Hoggar shield (Algeria). Radii of the colored circles are proportional to the difference in nT of the field

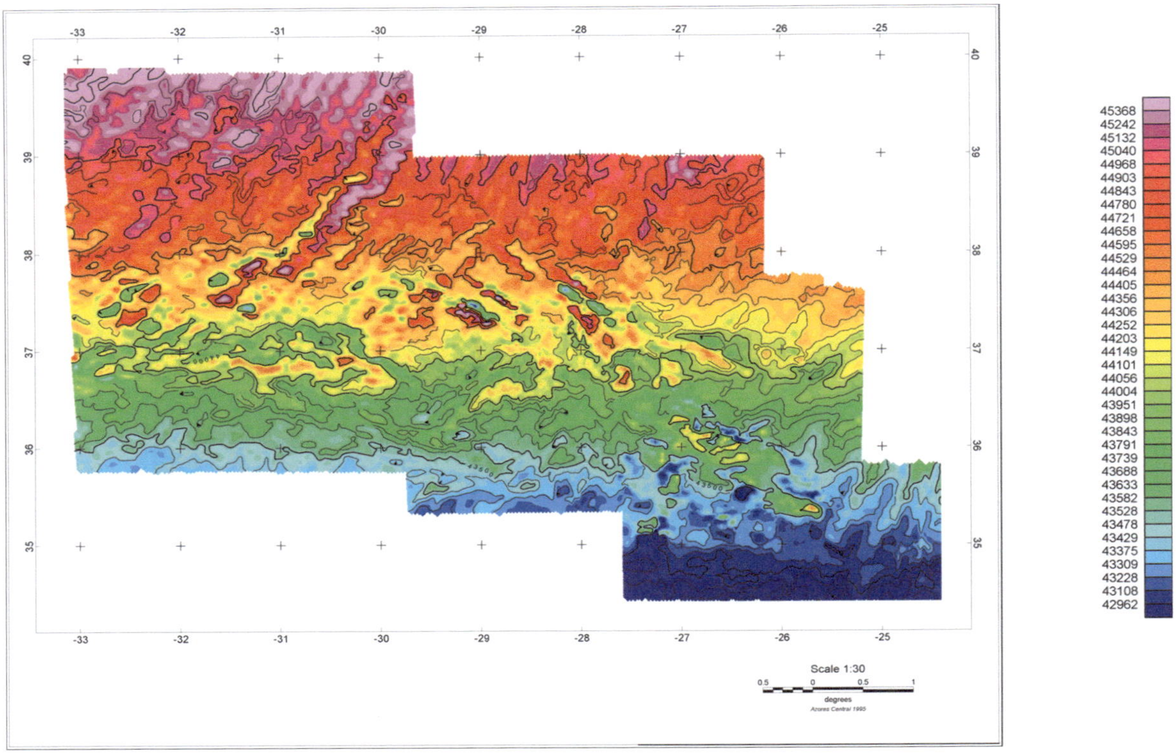

Fig. 4.8 Total field over the Azores Islands derived from aeromagnetic data (Luis and Miranda 2008). Color scale and contour lines are in nT

4.9 Lithospheric Field Mapping – Reference Field Correction

Depending on the time scale, the lithospheric or crustal anomaly field $\mathbf{B_a}$ is considered as a static field. Its characteristic wavelength ranges in from 10^{-5} km to 10^3 km. Its amplitude varies from few nT to some 10^3 nT at the Earth's surface. In some peculiar places, it reached 10^5nT and sometimes even larger than the main field (Heiland 1940; Logachev 1947). We may however assume that in general $|\mathbf{B_a}| \ll |\mathbf{B_N}|$ where $\mathbf{B_N}$ is the core field. It is also assumed that the orientation of the core field is almost uniform in the domain D. If this is not true, then the survey area can be divided into pieces small enough for the assumption to hold. Generally, for anomaly field mapping, only static fields are considered. The time duration of aeromagnetic surveying is usually short. In exceptional cases, they last long enough such that the data should be corrected for the secular variation of the main field (Luis 1996; Luis and Miranda 2008). If the external field is removed the Eq. (4.6) becomes:

$$\mathbf{B_t}(P) = \mathbf{B_N}(P) + \mathbf{B_a}(P) \tag{4.14}$$

Let us denote by $\mathbf{p}$ the unit vector in the core field direction, i.e.,:

$$\mathbf{p} = \frac{\mathbf{B_N}}{|\mathbf{B_N}|} \tag{4.15}$$

For airborne and marine magnetic surveys for which the magnitude of the field is usually measured, the total-field anomaly is defined as:

$$\Delta B = |\mathbf{B_t}(P)| - |\mathbf{B_N}(\mathbf{P})| \tag{4.16}$$

Bearing in mind that $\Delta B(P) \neq |\mathbf{B_a}|$, and under the assumptions given above, it is easy to show that:

$$\Delta B(P) \approx \mathbf{B_a}(P) \cdot \mathbf{p} \tag{4.17}$$

which is the projection of the field $\mathbf{B_a}$ onto $\mathbf{B_N}$. If ϕ is the angle between the two vectors, the error in the approximation (4.17) is proportional to $B_N \sin^2 \phi$ The total field B_t derived ultimately using the leveling process contains contributions from sources of deep origin – i.e., in the core – and contributions from sources of shallow origin – i.e., in the crust. For geodynamic studies or for oil/mineral exploration purposes we are mainly interested by the crustal magnetic field. It is then very important to try to accurately characterize the normal field B_N in order to derive the anomaly field B_a or more precisely its approximation everywhere over the area of interest. Nowadays, the most widely used reference fields are the International Geomagnetic Reference Field (IGRF) or its "definitive" version (DGRF) (Barton 1997), the BGS Global Magnetic Model (BGGM) in the oil industry, and other global models such as CM4 (Sabaka et al. 2002, Sabaka et al. 2004). An IGRF-like model is a mathematical expansion of the Earth's main magnetic field using spherical harmonics basis functions (Chapman and Bartels 1940) up to a given wavelength. Before the Magsat era (Langel 1982), the accuracy of such models was of the order of 100–200 nT at the Earth's surface. Starting from 1980 with Magsat scalar and vector data the accuracy achieved was of the order of 20 nT at the Earth's surface. The Danish initiative Oersted and the German CHAMP geomagnetic satellites orbiting the Earth since 1999 and 2000 respectively, make it possible to achieve global field models with an unprecedentedly high accuracy of 10 nT (Olsen 2002; Olsen et al. 2009; Lesur et al. 2008; Lesur et al. 2009; Maus et al. 2005, 2009).

The IGRF/DGRF models describe not only the static part of the geomagnetic field up to degree and order 13 but also its secular variation up to degree and order 8. These models are updated every 5 years. Following the IAGA-Division V announcement for global field models, the present IGRF model, with an extrapolation valid for the 2010–2015 time interval, is the 11th generation (Finlay et al. 2010). The Gauss coefficients of the IGRF/DGRF models are available from year 1900 through 2010 (Barton 1997; Macmillan et al. 2003; Macmillan and Maus 2005; Finlay et al. 2010). The DGRF models are very useful for gridding or assembling adjacent aeromagnetic surveys flown at different epochs (Hemant et al. 2007; Hamoudi et al. 2007; Maus et al. 2007). They allow the earliest surveys reduced with inaccurate old versions of the IGRF models and to be used by correcting them for a new common reference field epoch. Most industrial potential field softwares include the IGRF as the main field model. For many of the earliest surveys, even in the early 1970's, an arbitrary and often unspecified constant was subtracted from the measured data before contouring the residual field. It often happens that the original data are no longer available. In that case, the derived grids may not easily be incorporated in any compilation such as the World Digital Magnetic Anomaly Map (WDMAM) (Hamoudi et al. 2007, Maus et al. 2007). To correct and merge inconsistent or discontinuous grids, accounting for the secular variation of the field for different epochs, the comprehensive model CM4 (Sabaka et al. 2002, Sabaka et al. 2004) is probably more efficient than the IGRF models (Hamoudi et al. 2007). For surveys of limited geographic extent, derivation of a local polynomial expression for the normal field is certainly a better approach than global modeling one to improve the definition, resolution and the accuracy of the anomaly field (Le Mouël 1969; Luis 1996; Chiappini et al. 2000, Supper et al. 2004). Second or third order polynomials better constrain the spatial gradients of the normal field than IGRF does and accurately represent the long wavelength components of this field. The analytical expressions may be derived either using (x, y) Cartesian coordinates or longitude (λ) and latitude (ϕ) geographic coordinates. In the former case the general expression for the normal field is then given by:

$$B_N(x, y) = \sum_{i,j} a_{ij} \Delta x^i \Delta y^j \tag{4.18}$$

The indices (i, j) give the degree of the polynomial expansion, generally of maximum order less than or equal to three. The coefficients are calculated from the measured data by least-squares. The necessary condition to use such a method is that the anomaly field has zero-mean over the area of interest (Le Mouël 1969) and that there are no magnetic sources outside the survey area. As an example of using longitude and latitude geographic coordinates, Chiappini et al. (2000) used a second order polynomial for the magnetic anomaly map over Italy and surrounding marine areas of the form:

$$B_N(\phi, \lambda) = a_{00} + a_{10}\Delta\phi + a_{01}\Delta\lambda + a_{11}\Delta\phi\Delta\lambda + a_{20}\Delta\phi^2 + a_{02}\Delta\lambda^2 \tag{4.19}$$

Table 4.2 Numerical expression of the 2nd order polynomial over Italy and surrounding marine areas (Chiappini et al. 2000)

Coefficients	Values	Unit
$a_0 0$	45386.500	nT
$a_1 0$	342.10	nT degree^{-1}
$a_0 1$	69.034	nT degree^{-1}
$a_1 1$	−1.868	nT degree^{-2}
$a_2 0$	−4.438	nT degree^{-2}
$a_0 2$	1.457	nT degree^{-2}

where: $\Delta\phi = \phi - \phi_0$ and $\Delta\lambda = \lambda - \lambda_0$ with ($\phi_0 = 42°N, \lambda_0 = 12°E$) being the latitude-longitude of the central point of the survey area . The coefficients $a_{ij}(i = 1, 2, j = 1, 2)$ are listed in Table 4.2.

As an example of Cartesian expression of the normal field, the coefficients of the second order polynomial expression based on the UTM26 projection system derived for the aeromagnetic survey of the Azores islands by Luis (1996), using Equation (4.18)

Table 4.3 Numerical expression of the 2nd order polynomial over Azores Islands (Luis 1996)

Coefficients	Values	Unit
$a_0 0$	44184.0	nT
$a_1 0$	1.087	nT km^{-1}
$a_0 1$	4.215	nT km^{-1}
$a_1 1$	-0.66710^{-3}	nT km^{-2}
$a_2 0$	-0.11410^{-3}	nT km^{-2}
$a_0 2$	-0.74310^{-3}	nT km^{-2}

to degree 2 in both i and j are given in Table 4.3. In this case

$$\Delta x = x - x_0 \tag{4.20}$$

and

$$\Delta y = y - y_0 \tag{4.21}$$

with ($x_0 = 420, y_0 = 4250$) are the UTM coordinates of the central point of the survey expressed in kilometers.

Figure 4.9 shows the differences in the magnetic field between the IGRF90 model and the local second order polynomial approximation over the Azores Islands. These differences range between −100 and −20 nT inside the survey areas, the magnitude of the global IGRF90 derived field is smaller than the magnitude of field derived using local polynomial expression. We can also see clearly that the map is mainly dominated by the long wavelength of the IGRF field and that its gradient is poorly constrained.

Figure 4.10 presents the anomaly of the total field calculated using (4.12) as an example. Even if the field measurements are very accurate, say with less than 1 nT Root-Mean-Square (RMS) noise, the anomaly field accuracy and its precision is often dependent on the positioning system used. With old positioning systems, 10–20 nT accuracy was commonly reached

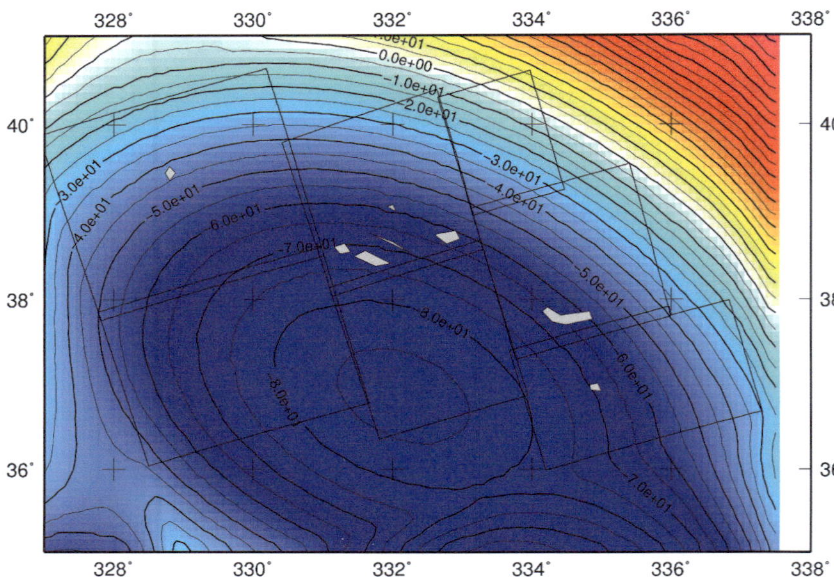

Fig. 4.9 Differences between IGRF90 and polynomial total fields over the Azores Islands (*grey polygons*). The seven aeromagnetic panels surveyed are shown by black rectangles. (Luis 1996)

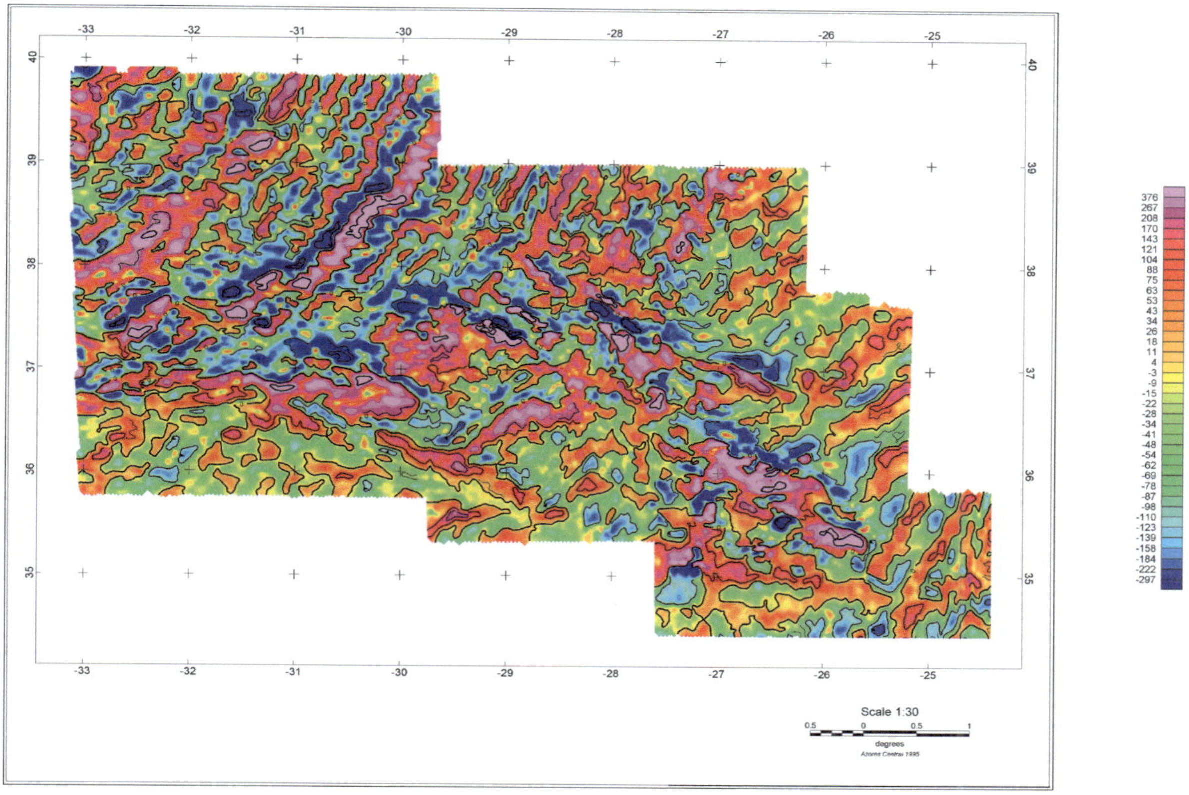

Fig. 4.10 Total field anomaly over the Azores islands derived from the aeromagnetic data. Color-scale and contour lines are in nT. (Data courtesy of J. Luis)

for the final maps. Nowadays, GPS (or dGPS), very high instrumental resolution of the order of a picoTesla and high frequency sampling rates up to ~1 kHz measurements are standard (Nabighian et al. 2005). This significantly reduces the noise affecting the anomaly field, allowing mapping of the magnetic heterogeneities with unprecedented high precision. Such surveys prove to be useful even in a sedimentary context where the magnetic signal is very weak.

4.10 Further Processing: Micro-leveling

Image processing and data enhancement of the aeromagnetic anomaly maps shed light on leveling errors still contaminating the reduced data (Minty 1991; Paterson and Reeves 1985). Once contoured, the anomaly field may appear as fully leveled. However, graphic shading representation of the field shows not only small-scale geologic features as expected but also short wavelength low amplitude oscillations oriented along the flight lines. Such organized noise has been called "corrugations" and its removal is called de-corrugation (Paterson and Reeves 1985) or micro-leveling (Minty 1991). Fig. 4.11a below shows the aeromagnetic anomaly field data above the Tindouf basin in southwest Algeria (Allek 2005). The flight-line azimuth is N160°. We can easily see high frequency noise oriented along the flight-lines. Fig. 4.11b presents the same anomaly field after micro-leveling.

Micro-leveling remains an empirical filtering process. Its principle – it is purely numerical with no underlying physics – is described by Minty (1991). It may be applied to any measured quantity, not just potential field data. An example would be radiometric data. As with classical tie-line leveling, many algorithms have been developed and are used for micro-leveling. However, great care should be taken when filtering the noise to preserve geologic features with the same properties (i.e., main direction and spectral content) as the noise (Fig. 4.12).

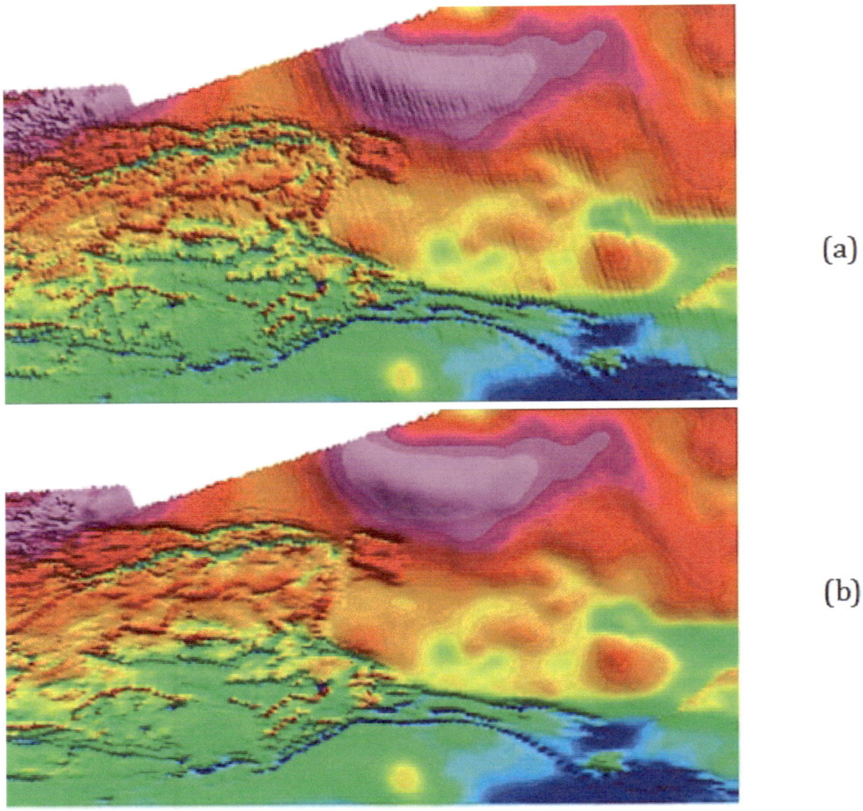

Fig. 4.11 Aeromagnetic anomaly field in the southwest of Algeria. Azimuth of flight-lines is N160°. (**a**) Before micro-leveling. (**b**) After micro-leveling (Allek 2005)

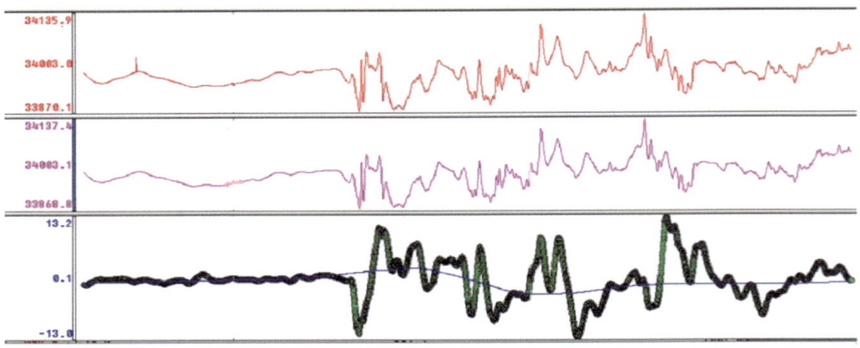

Fig. 4.12 Micro-leveling process. (*Top*) Raw aeromagnetic profile (*red*). (*Middle*) Micro-leveled profile (*pink*). (*Bottom*): High frequency error (*green*) and its smooth version (*blue*)

4.11 Interpolating, Contouring and Gridding

The anomaly of the total field, when all the errors have been corrected, can be interpolated and gridded. Various methods have been developed since the end of the sixties for automated contouring (Bhattacharyya 1969, 1971, O'Connell et al. 2005). The most popular and easy to use technique is probably the minimum curvature interpolation algorithm (Briggs 1974). Many sophisticated algorithms have been developed using kriging (Hansen 1993), fractal approaches (Keating 1993) or wavelets (Ridsdill-Smith and Dentith 1999). All these methods are proposed to alleviate the aliasing problem that may occur because the density of data is always so much greater along the flight-line direction than across flight lines. To cope specifically with this problem of different data density along and across flight lines, bi-directional gridding was developped (O'Connell et al. 2005; Reford 2006).

The final gridded data are then ready to be plotted at scales, ranging from at least 1 : 250,000 to less than 1 : 5,000, and/or further processed in the space or spectral domains for geologic interpretation.

4.12 Conclusions for Aeromagnetics

Almost a century has passed between the first attempt in 1910 to measure the geomagnetic field from the air in a captive balloon and nowadays using unmanned aircraft vehicles. Many millions of line kilometres have been flown by governmental agencies, companies, and academic institutions throughout the world. Many kinds of magnetometers have been used: Earth's inductor, fluxgate, proton precession, Overhauser, optically pumped alkali vapour and SQUID magnetometers. Scalar and vector measurements have been collected onboard fixed-wing aircrafts, helicopters and very recently with unmanned vehicles. At a lesser extent stratospheric balloons have also been used for geomagnetic field measurements. The accuracy of measurements evolved over a very wide range from mT in 1930s to some fT nowadays. The positioning uncertainties improved by several orders of magnitude for horizontal distances, from hundreds of meters with tracking by cameras and video recovery systems to less than a centimetre in carrier-phase dGPS. The high-resolution aeromagnetic method is not only useful in mineral and oil exploration but also for cultural research of ancient archaeological sites and military purposes such as unexploded ordnance (UXO), and submarine detection. For safety reasons the actual tendency for "lower and lower altitude" is a strong argument in favour of UAV development. Future directions in UAV research would be towards stretching the boundary of autonomous operation through an efficient trajectory generation and mission planning. The development of small inexpensive UAV will allow a flexible and robust distributed sensor network to replace limited manned flights or large UAV that concentrate expensive sensor and communication systems in a single agent with a large team of operators. Two kinds of UAV are foreseen: stratospheric high altitudes UAV for regional surveys and low altitudes high resolution UAV. They will contribute by better describing the broad spectrum of lithospheric field magnetic anomalies. Regional airborne and shipborne surveys cover a significant part of the Earth's surface. However, large parts remain still unsurveyed. Despite the large disparities between surveys, the compilation of huge amounts of released data – of the order of 5×10^{12} data points – collected over many decades has allowed the derivation the first global anomaly map at the Earth's surface within the framework of the World Digital Magnetic Anomaly Map project (Korhonen et al. 2007). In aeromagnetic surveying, in the same way that gradiometer data have been shown to be superior to single sensor data, it is expected that acquiring vector data will give more information about geologic structures and their physical properties than can be obtained using scalar measurements only. Efforts have also to be made to improve the geological interpretation of the magnetic anomaly field with respect to the petrology of rocks. These studies will have a substantial overlap with current initiatives that address the fields from rock and mineral physics to lithosphere and deep continental drilling.

4.13 Introduction to Marine Magnetics

About 70% of the Earth's surface is covered by water. Aeromagnetic surveys can only help to study the regional magnetic signal of the lithosphere over the oceans close to continents (e.g., Blakely et al. 1973, Malahoff 1982) or in remote oceanic areas with long-range high-altitude surveys like Project Magnet flights, whereas satellites also fly over the oceans but provide low resolution measurements. Therefore marine magnetic observations, defined here as magnetic measurements along ship tracks or from underwater autonomous vehicles, are the only way to study the magnetic signal over the oceans and seas at local and regional scales. This magnetic signal is due to the induced and remanent magnetization carried by the oceanic crust and uppermost lithosphere. For instance, when newly-formed crust cools at mid-oceanic ridges, it acquires a thermoremanent magnetization which 'freezes' the ambient magnetic field; the uneven sequence of geomagnetic field reversals recorded by the oceanic crust represents the best geophysical witness of lithospheric plate motions (Vine and Matthews 1963). This shows how crucial are the marine magnetic observations. Forward and inverse modeling approaches have been applied to retrieve the magnetic properties of the Earth's oceanic lithosphere (e.g., Parker and Huestis 1974; Schouten and Denham 1979; Pariso et al. 1996; Langel and Hinze 1998; Sichler and Hékinian 2002; Purucker and Whaler 2007). Apart from their obvious interests for marine geophysics and geology, other applications are nautical archaeology (e.g., Boyce et al. 2004; Van Den Bossche et al. 2004) as well as ocean engineering (i.e., pipeline or undersea cable detection). The few magnetic field observations for the latter two topics are not considered in this review study.

Acquiring magnetic measurements onboard a ship is not a straightforward task. First, compared to the planes used in aeromagnetics, oceanographic vessels are slow, implying less regional mapping capacity. Second, the magnetization of ships is usually very high, a problem solved by towing the magnetometer (at least) several hundred meters astern (and in some cases below) the ship – with less control on the sensor position and attitude. Third, the survey areas are usually quite remote from any magnetic observatory, making it difficult to estimate and subtract external field contributions from the Total-Field (TF) observations.

Magnetic measurements for scientific purposes really started in the 1950s. Indeed the submarine or mine detection during the Second World War and the Cold War triggered technological developments which considerably increased the accuracy of magnetometers. By 2010, all oceans had been covered by marine magnetic measurements, with gaps in the Southern Hemisphere. Such observations are made in most marine geophysical surveys with interests in the oceanic crust.

In the following sections, we present some general aspects of marine magnetic measurements. The first part concerns the global history of standard (scalar) observations from a statistical point of view and their main applications. The second part focuses on the typical sources of error when acquiring these data, and shows how to improve the quality of scalar marine datasets. The last part deals with peculiar instruments allowing vector and/or deep sea measurements and the corresponding processing techniques.

4.14 History of Marine Magnetics

4.14.1 The First Attempts

The first magnetic measurements at sea may have been made by a Chinese sailor with a compass onboard a ship about 2000 years ago. However, without any written reference to such an hypothetical event, we should attribute the first record of magnetic measurements at sea, in this case declination determinations, to Portuguese navigators. Merrill and MeElhinny (1983) mention that, in 1538–1541, João de Castro used a compass like a sun-dial with a magnetic needle to determine the azimuth of the sun at equal altitudes before and after noon. The half difference of these azimuths measured clockwise and anticlockwise respectively was the magnetic declination. He performed about 43 declination measurements when he commanded a ship that sailed to India and in the Red Sea. About a century and half later, in 1702, many similar observations led to the first declination chart of the whole Earth, published by Edmond Halley. Over two centuries more were needed to develop magnetic field theory (Gauss) and the first portable magnetometers.

Allan (1969) reports that a non-magnetic research ship named 'Carnegie' sailed between 1909 and 1929 and made magnetic measurements along widely spaced tracks in the Atlantic, Pacific and Indian oceans. During the Second World War, magnetometry at sea was used to detect submarines and mines (Germain-Jones 1957). The fluxgate magnetometer, originally developed as an airborne instrument for the detection of submarines, was converted for marine research at Lamont Geological Observatory (Allan 1969). The first measurements made with such a magnetometer towed behind a ship were reported by Heezen et al. (1953). Subsequently, the fluxgate magnetometer was largely superseded by the proton magnetometer, because the latter gives an absolute measurement of the field. Packard and Varian (1954) first developed this instrument, which was later adapted for land use by Waters and Phillips (1956) and modified for towing behind a ship by Hill (1959). Finally, in the late 1950s, the Scripps Institute of Oceanography and the United States Coast and Geodetic Survey made a detailed magnetic survey over a large area off the west coast of the United States (Mason 1958; Mason and Raff 1961; Raff and Mason 1961; Vacquier et al. 1961), opening the way for many marine magnetic surveys worldwide.

4.14.2 Evolution of the Global Dataset

Once the proton precession magnetometer became the standard instrument to measure the magnetic field over marine areas, oil and gas companies – who already used magnetic land prospection to help detect reservoirs – deemed marine magnetic surveys a complementary technique to reflection seismic. Although only a few public reports of marine magnetic prospection for oil and gas exploration are available, such exploration helped to spread the use of magnetometers at sea.

Figure 4.13 shows the evolution of the annual number of marine magnetic surveys over the world's oceans. These values are mainly extracted from the databases of Quesnel et al. (2009) and GEOphysical DAta System (GEODAS). The reader must be aware that many magnetic surveys carried out by private companies or by scientific institutes that did not share information on their data were not taken into account. The main trends should remain similar if these missing data were added. The values should be updated for the years since 2002: cruises in 2003–2010 will probably be released to the databases after 2010.

The histogram highlights how the number of cruises increased during the 1960s and 1970s, with a peak in 1972. Following Vine and Matthews (1963), these years mark the recognition of Plate Tectonics as the new paradigm for Earth Sciences, leading to an unprecedented effort of new marine data collection to validate the concept and derive first-order models of present and past global plate kinematics. Magnetic measurements were made routinely during most cruises and transits. Remarkably, the steady increase in number of surveys – hence in budgets allocated to these surveys – breaks in 1973, the year of a major international oil crisis.

Since the end of the 1970s, the annual amount of marine magnetic surveys has decreased regularly (Fig. 4.13), except for a small rebound in the late 1980s. Although many regional and local problems remain unsolved, plate kinematics is seen as understood at the first order, and the reduction of

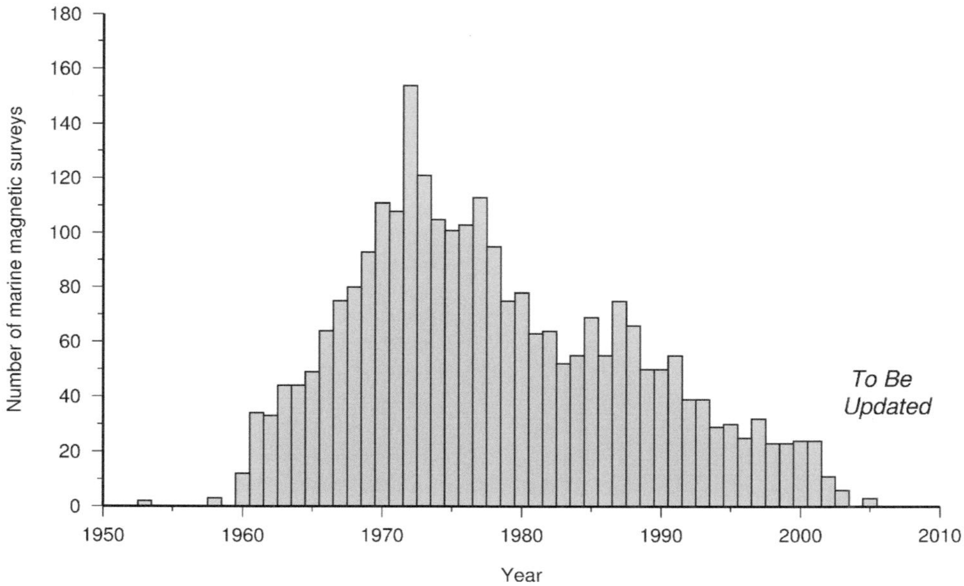

Fig. 4.13 Annual frequency of marine magnetic surveys since 1950. Most of these campaigns are stored at the GEODAS database. For years after 2002 the number of stored cruises is not fully updated

budgets led to the acquisition of magnetic measurements as a secondary consideration. Nowadays, only ~20 scientific cruises acquire magnetic measurements each year. Until recently, these cruises were noticeably supported by international scientific programs such as the International Ocean Drilling Program (IODP) – formerly the Deep-Sea Drilling Program (DSDP) and Ocean Drilling Program (ODP). Unfortunately, IODP recently decided to stop the systematic acquisition of marine magnetic measurements during their transits for budgetary reasons. Furthermore, the enforcement of Exclusive Economic Zones (EEZ) 200 nm (nautical miles) away from the coastal states and their future extension up to 300 nm under the UNCLOS (United Nation Convention for the Law Of the Sea) adds the difficulty of obtaining official permission to acquire data in these EEZ through the diplomatic channels, with 6 months notice.

Figure 4.14 represents the same data as Fig. 4.13 split into the Pacific, Atlantic and Indian oceans (left), or the Northern and Southern Hemispheres (right). Again, numbers for the recent years are probably underestimated. The Indian Ocean always had fewer cruises than the other oceans, partly because of its reduced size, and partly because of its remote location from the United States (US), Japan and Europe. The former Soviet Union collected numerous cruises over the Carlsberg Ridge (see, e.g., Merkouriev and DeMets 2006, and references therein), but these data are not considered in this study. In contrast, the northern Pacific Ocean was extensively investigated by US and Japanese research vessels. The Southern Hemisphere

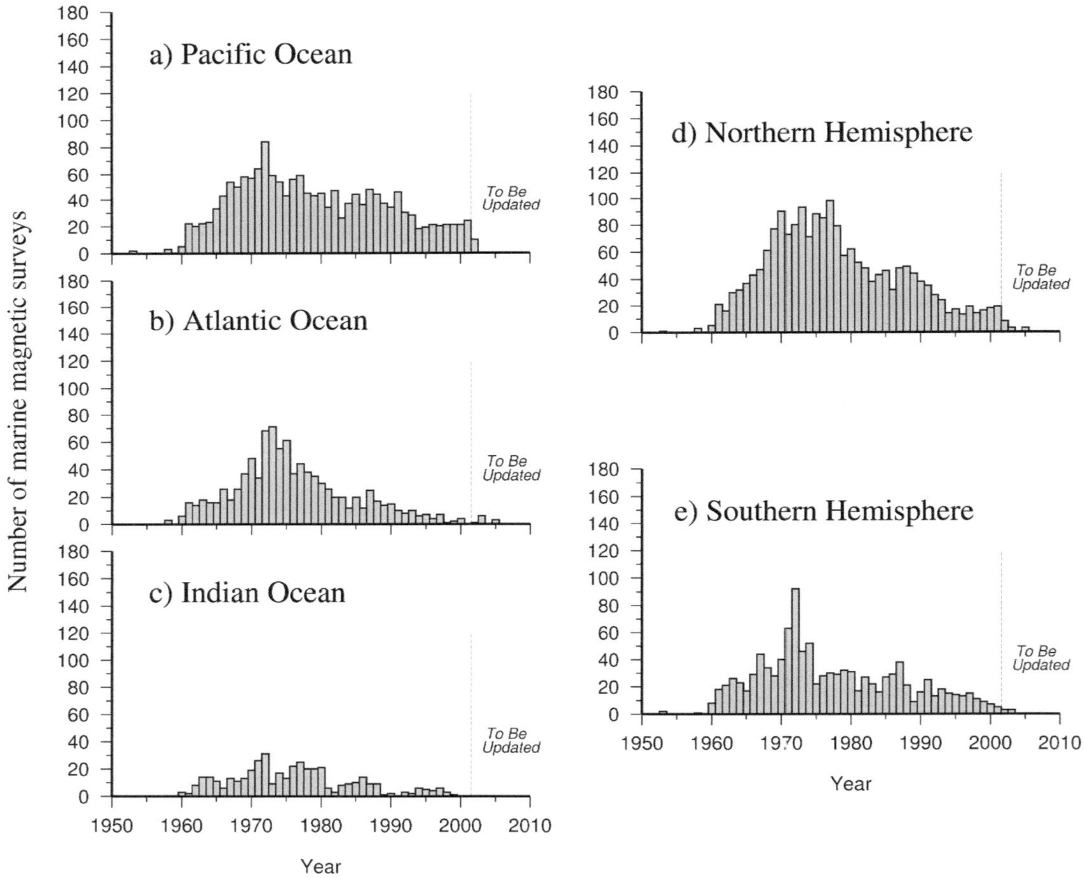

Fig. 4.14 Same as for Fig. 4.13, but only for surveys in the (**a**) Pacific Ocean, (**b**) Atlantic Ocean, (**c**) Indian Ocean, (**d**) Northern Hemisphere and (**e**) Southern Hemisphere. Note that these histograms could be biased by surveys belonging to two (or more) parts, but this should not greatly affect the main tendancies

was always much less explored by marine magnetic surveys than the Northern, except in 1972 when both hemispheres reached the same level.

Figure 4.15 shows the spatial evolution of the marine magnetic coverage from the first cruises before 1960 to the present. Vessels towing a magnetometer had already reached the mid-Pacific and mid-Indian oceans by 1960. A big transition occured in the 1960s (as Fig. 4.13 has already shown), when only the South Indian and South Atlantic oceans remained poorly covered by magnetics. Between 1970 and 1980, these gaps were partially filled. Since 1980, marine magnetic data coverage has not changed much, except the highest concentration in the Pacific Ocean near Antarctica. Areas close to the continents exhibit a lot of marine magnetic measurements. The final panel of Fig. 4.15 reveals a remaining dichotomy between the well-covered northern parts of the Pacific, Indian and Atlantic oceans versus their southern counterparts. It is obvious that further marine magnetic acquisition is needed for Antarctica and sub-Antarctic waters as well as for the Arctic Ocean (even if aeromagnetic data, not included in Fig. 4.15, exist in these areas). The dataset used to build this map (see end of Section 4.14.3) will be complemented by additional analog data (to be digitized) and some other cruises unavailable to Quesnel et al. (2009) to prepare an updated version of the marine magnetic dataset to be included in the next World Digital Magnetic Anomaly Map (WMAM; Korhonen et al. 2007, and T. Ishihara, pers. comm.).

4.14.3 Storage and Accessibility

A substantial fraction of the world marine magnetic observations from 1953 to present are available in digital format from the GEODAS database[1] hosted by the National Geophysical Data Center (NGDC). Some data are also stored by GEODAS in analog format as scanned documents. They appear mostly as handwritten charts where exact values of measurements plus time and space positioning are difficult to read. Some digital data were digitized from reports, a transfer resulting in additionnal errors.

Among the numerous research institutions which carried out marine magnetic surveys stored in the GEODAS database, the United States takes the lead with the Scripps Institute of Oceanography (over 550 cruises), the Lamont-Doherty Earth Observatory (over 540 cruises), the US Navy (about 130 cruises), the United States Geological Survey (USGS, about 120 cruises), the Woods Hole Oceanographic Institute (about 110 cruises), as well as universities like the University of Hawaii (about 130 cruises). Many other marine magnetic cruises were provided by Japanese institutions like the Japan Hydrographic and Oceanographic Department (JHOD; over 200 cruises) or the Geological Survey of Japan (about 40 cruises). France (about 180 cruises with about 50% from Ifremer), New Zealand (about 100 cruises), the United Kingdom (about 90 cruises), Australia (about 70) and South Africa (about 20 cruises) also contributed marine magnetic observations to the data base.

A few other databases storing marine magnetics exist. Some include cruises stored at NGDC, some not. Such databases belong to national and international research institutes, sometimes to specific laboratories. Free access to the data is usually straightforward for *bona fide* scientists for research purposes. Table 4.4 gives a non-exhaustive list of geophysical databases where marine magnetic observations are available. This table, and particularly internet URLs, are valid in 2010 and may change in the future.

Additionally, Germany performed numerous surveys (over 100; U. Barckhausen, pers. comm.) and contributed to world marine magnetic coverage. Similarly, the former Soviet Union (and later Russia) collected a large amount of data through systematic regional surveys undertaken, for example, in the North Atlantic and the Northwestern Indian oceans, amounting to about 2–3 millions of kilometers (S. Merkouriev, pers. comm.). Some of these data have been used by Verhoef et al. (1996), Merkouriev and DeMets (2006) and Merkouriev and DeMets (2008).

4.14.4 Scientific Objectives

Apart from oil and gas prospection (for which magnetics plays only a secondary role), the main application of marine magnetics is the study of the Earth's oceanic

[1]http://www.ngdc.noaa.gov/mgg/gdas/

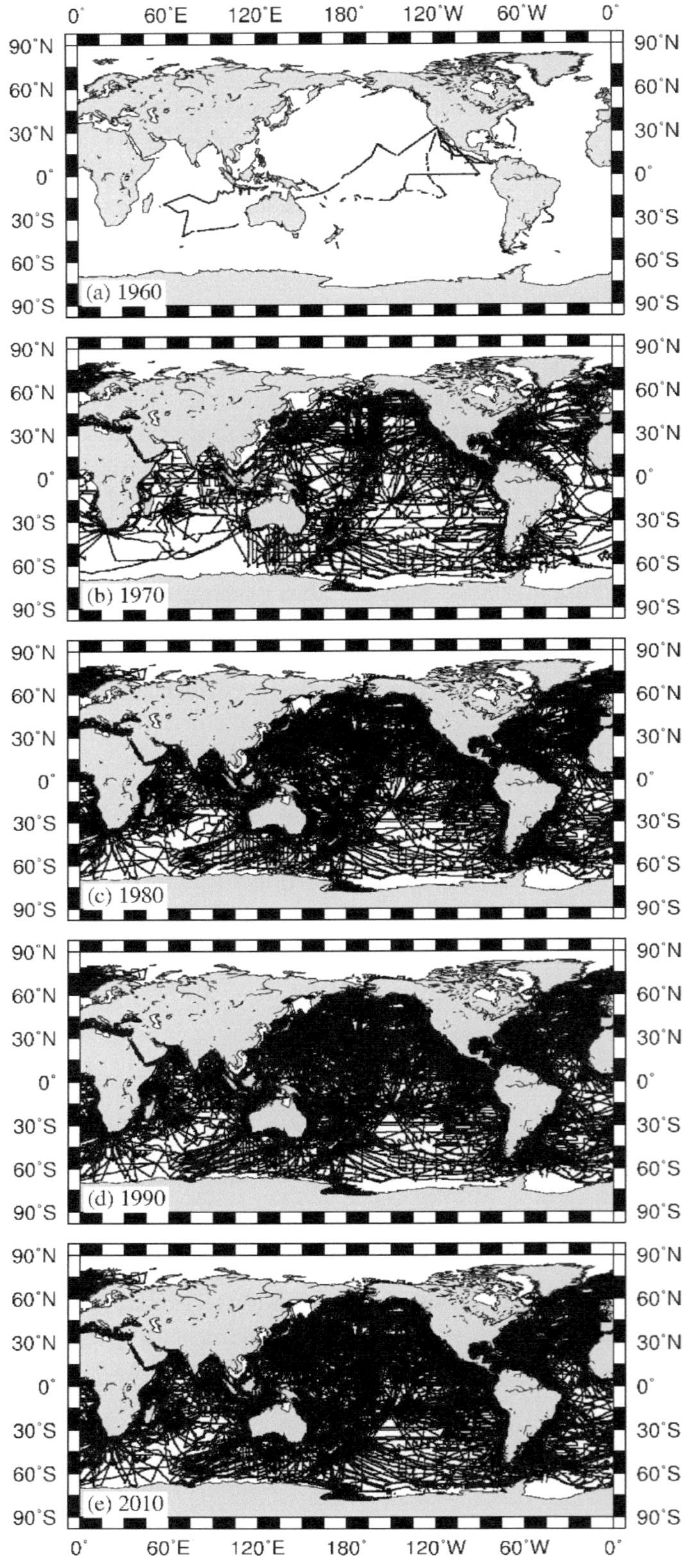

Fig. 4.15 Global marine magnetic survey coverage in (**a**) 1960, (**b**) 1970, (**c**) 1980, (**d**) 1990 and (**e**) 2010

Table 4.4 Databases with marine magnetic observations

Name/Acronym[a]	Institute[b]	URL (in 2010)
GEODAS	NGDC	http://www.ngdc.noaa.gov/mgg/gdas/
SISMER	IFREMER	http://www.ifremer.fr/sismer/
BAS	BAS/NERC	http://www.antarctica.ac.uk/bas_research/data/
JODC	JHOD JCG	http://www.jodc.go.jp/NEW_JDOSS_HP/MGD77_info_e.html
JAMSTEC	JAMSTEC	http://www.jamstec.go.jp/dataportal/
SeaDOG	NOC/NERC	http://www.noc.soton.ac.uk/cgibin/seadog/

[a]Acronyms are: GEODAS, GEOphysical DAta System; BAS, British Antarctic Survey; JODC, Japan Oceanographic Data Center; JAMSTEC, Japan Agency for Marine-Earth Science and Technology; SeaDOG, Sea Deep Ocean Geophysical data.

[b]NGDC, National Geophysical Data Center; IFREMER, Institut Francais de Recherche pour l'Exploitation de la Mer; JHOD, Japan Hydrographic and Oceanographic Department; JCG, Japan Coast Guard; NOC, National Oceanography Center; NERC, Natural Environment Research Council.

crust and uppermost mantle through its magnetization at all scales in time and space. Their contribution is essential to constrain the structure, age and evolution of ocean basins, from ridge to subduction. Other applications include constraints on the magnetic structure and properties of passive and active margins, mid-oceanic ridges, transform faults, subduction zones, seamounts, and fracture zones. Such constraints have implications for the geologic processes which affect or affected such areas. At smaller scales, archaeological prospection sometimes requires magnetic measurements to detect submerged constructions or sunken vessels.

4.15 Sources of Error, Evolution and Correction for Scalar Sea-Surface Measurements

This section describes the different problems associated with marine magnetic observations, from the acquisition to the storage, and the possible method of minimizing the resulting errors on the data. Further details can be found in Jones (1999) and Quesnel et al. (2009).

4.15.1 Magnetic Observation Accuracy

Magnetometers and sampling rates evolved since the first measurements. Here we show how this evolution reduced the systematic errors associated with marine magnetic data acquisition.

4.15.1.1 Definitions

Some common terms concerning magnetometers used at sea need to be defined. Most of the following definitions are well-described in Hrvoic (2007).

The **resolution** of a magnetometer corresponds to the minimum variation of the magnetic signal (in nT) that the measurement device (not the sensors) can detect. Conversely, the **sensitivity** reflects the minimum signal variation that the whole instrument can detect. It depends on the sensor noise level and is often represented in units of $(\mathrm{nT}(\sqrt{\mathrm{Hz}})^{-1}$ since the sensor frequency bandwith will also influence this noise.

The **drift** denotes a small variation of the magnetometer output with time and eventually temperature without any real change of the ambient magnetic field external to the instrument. It mainly concerns the sensor itself, even if the electronics of the measurement device can also be affected by temperature changes. To determine the drift, one must calculate the noise spectrum in the frequency domain: if this spectrum is flat, then no drift will occur with time. The **heading error** corresponds to the small change of the magnetometer output related to a change of the magnetic field direction with respect to the sensor. It can be due to ferromagnetic electronics close to the sensor. Finally, the range of heading directions for which the sensor cannot acquire any measurements is called the **dead zone**.

The aim is to reduce the last three parameters, and the resulting total error is expressed as the **absolute accuracy** of the magnetometer.

4.15.1.2 Fluxgate Magnetometers

Fluxgate magnetometers (three sensors oriented at right angles) were used for early surveys. At this time, these instruments had an estimated accuracy of several nT (Bullard and Mason 1961). Errors were amplified by the low sampling rate, recorded (and sometimes manually handled) every 5–10 min (Quesnel et al. 2009) and sometimes by the small distance between the instrument and the ship. Both effects have to be taken into account when using these old surveys. Proton precession magnetometers were soon preferred, since the three fluxgate sensors have to be very accurately oriented with respect to each other (possible orthogonality errors) and since such vector sensors have a significant drift with time and temperature, therefore requiring calibration.

Nowadays, fluxgate magnetometers offer better than 0.1 nT $(\sqrt{Hz})^{-1}$ sensitivity, for about 0.01 nT of resolution. Their final accuracy depends on the gyro tables on which they are mounted (Nabighian et al. 2005). The use of fluxgate magnetometers at sea is presented in Section 4.16.

4.15.1.3 Proton Precession Magnetometers

Hill (1959) suggested using nuclear spin (later called proton precession) magnetometers onboard ships. These instruments have the advantage of having no drift and therefore not requiring frequent calibrations. At this time, an absolute error of several nanoTeslas was usual, whereas the accuracy of modern proton magnetometers reaches 0.1 nT (Sapunov et al. 2001). More typical values would be 0.1 nT at 0.2 Hz for portable instruments (Nabighian et al. 2005).

The sampling rate in early surveys was generally a measurement every 30 s at a ship's speed of 10 knots (Allan 1969), adequately suited to a proton-precession magnetometer cycling every 10 s at most. Because the proton precession signal cannot be sampled during the polarization in the sensor, this sampling rate could not be increased. Another limitation was that the polarization requires a lot of energy, transported to the instrument through a thick armoured coaxial cable. The measured signal is transported back to the ship in analog form through the same cable and is very sensitive to any electric noise generated by various devices on the ship. Furthermore, the proton precession magnetometers do not prevent erroneous measurements from rotations or small motions of the sensor head during acquisition ('dead zone'; see Section 4.15.1.1).

4.15.1.4 Optically-Pumped or Alkali-Vapor Sensors

Since they provide excellent sensitivity (less than 0.01 nT) and very high sampling rates (more than 10 Hz) for a light and compact instrument (Nabighian et al. 2005), alkali vapor magnetometers are suitable to achieve high quality magnetic observations at sea. However, the fragility of the glass envelope and an intrinsic heading error limits their use.

4.15.1.5 Overhauser Effect Sensors

The Overhauser magnetometer is a variation of the proton precession instrument, which it has superseded for the last 20 years (Hrvoic 2007). This instrument is now widely used for marine surveys. It requires lower power than standard proton precession magnetometers, provides a dramatically higher signal to noise ratio, and avoids shipboard noise sources and data 'line loss' associated with the transmission of weak analog voltages usually met with proton precession sensors. Since the sensor can be polarized in tandem with precession signal measurement (because of different frequency bandwidths), faster sampling rates are also possible (Hrvoic 2007). Typically, the field can be sampled at 5 Hz with a resolution of 0.01 nT to 0.001 nT for a sensitivity of 0.015 nT at 1 Hz (Anderson et al. 1999). It also delivers very high absolute accuracy (0.2 nT), eliminating drift, heading error, and orientation problems.

4.15.2 Ship Noise

Due to their composition and engines, ships typically devoted to scientific surveys are magnetic. To reduce their magnetic effect, the magnetometer is towed at large distance from the ship, a method initiated in the 1960s (Bullard and Mason 1961; Laughton et al. 1960). Care must therefore be exercised with data prior to 1960. During GEODAS dataset analysis, Quesnel et al. (2009) nevertheless discovered small shifts of magnetic field mean level within many post-1960 cruises which were probably due to heading effects of the ship noise on the measurements. Indeed the ship's magnetism varies with the direction of the cruise (and so the orientation of the ship with regard to the sensor). It also evolves as the ship keeps a constant heading for some time, resulting in the acquisition of a viscous magnetization component. From theoretical work and experiments carried out at sea, Bullard and Mason (1961) estimated the effect of the ship on different headings in order to reduce the associated magnetic data. For instance, at the location of their experiment they found that a North-South survey will amplify this effect, whereas it is less than 1 nT at a distance of two times the ship length astern. At their time 1 nT was an acceptable error, but later surveys reached negligible values by towing the instrument at a greater distance. Nowadays, a cable 200 m-long or more is commonly used.

4.15.3 Position of the Ship

A very precise positioning measurement is needed for marine magnetic observations since the estimates of external and core magnetic field values (to be subtracted from TF measurements) vary spatially. The quality of positioning mainly depends on the date of the survey. Quesnel et al. (2009) also found obvious positioning errors in the GEODAS dataset such as cruises apparently located on land. Only comparison between adjacent and overlapping surveys can reveal the effect of such errors.

Accurate navigation in the open ocean was difficult in the past (Allan 1969). Field gradients of a few hundred nanoTeslas per km are not uncommon and the difficulty of matching up linear features from one area to another can be hazardous if ordinary 'dead-reckoning' navigation is used. Moored buoys fitted with radar reflectors were used to provide a reasonable relative accuracy in limited areas. The absolute position of a ship was believed known to within 100 m (Heirtzler 1964). The use of long-range radio navigation systems such as LORAN C or DECCA has improved the accuracy of magnetic surveys, where available (Allan 1969). At the end of the 1960s, the DOPPLER satellite navigation system, which combined a fixed accuracy of about 100 meters with world-wide coverage, brought greater precision to survey work (Talwani et al. 1966).

In the early 1990s, the Global Positioning System (GPS) appeared. At sea the error on position was initially less than 100 m (degraded mode), and was further reduced to less than 20 m in 2000. Therefore the towing distance must be determined as accurately as possible to properly differentiate the location of the ship and the measurement.

4.15.4 Date and Time of the Measurement

Errors in the acquisition time of marine magnetic measurements may affect the estimation of the external and core field (see Sections 4.15.6 and 4.15.7). However, even in the 1960s, the precision of clocks was acceptable to properly estimate these parameters at low and moderate sampling rates. Furthermore, higher sampling rates were later accompanied by higher precision of the time determination due to improvements in clock technology. Therefore such errors do not affect the quality of the computed magnetic anomaly values.

4.15.5 Transcription Errors

A valid measurement can be badly recorded. For early surveys (before magnetic tapes and, later, digital recording), manual data handling led to numerous erroneous values. Common errors are swaps of two digits of the total-field or resulting anomaly values (Quesnel et al. 2009). Therefore, one can not distinguish between an instrumental error and a transcription error. Along a track, such errors appear as spiky or shifted, isolated or grouped values that cannot be explained by commonly known sources of error.

Since such errors have very different amplitudes (10–100,000 nT) in the signal along track, it is difficult to assess their influence on the quality of a cruise. Quesnel et al. (2009) manually erased or corrected such erroneous data and/or applied filters to the noisy signal along-track. Finally, and after other kind of corrections, they were able to reduce the Root Mean Square (RMS) crossover differences (i.e., the difference between measurements at the intersection of two ship tracks) of their global dataset from 180 nT to 82 nT. Then, they adjusted the long-wavelength signal of each track and used a specific line-leveling method to reduce inconsistencies between different surveys. The resulting RMS of crossover differences of their dataset was 36 nT, which improved the coherency of magnetic maps at sea whatever the anomaly wavelength (Quesnel et al. 2009). We would like to point out here that although the principle is similar, the aeromagnetic leveling procedure is much more efficient because the flight-lines and tie-lines are orthogonal and contemporaneous.

Most of the marine magnetic data are stored in data bases as raw total-field measurements and the associated anomaly values. The estimates of external and core magnetic field values used to derive the anomaly values are usually not stored (Quesnel et al. 2009). In the next two sections, we consider the errors generated by the calculation of total-field anomaly values, whatever the quality of the raw total-field measurements. Moreover, if the date, time or location is erroneous, then the estimates of external and core magnetic field values will be inadequate, resulting in a poor magnetic anomaly value.

4.15.6 Estimation of the External Magnetic Field

For early cruises, reference stations such as the nearest magnetic observatory were sometimes used to reduce the external magnetic field effects, whatever the distance from this observatory. It resulted in a very poor estimation of the external magnetic field, especially when the vessel sailed in remote oceanic areas. Some attempts were made by Laughton et al. (1960) to use the mean of noon measurements as the absolute external field contribution on all measurement of a survey. They also minimized the diurnal variation by performing measurements at night. Finally, they corrected their measurements by about 5 to 15 nT to remove the external field.

Despite this exception, almost all marine magnetic data stored in the data base up to the 1990s are not corrected from the external field. A recently derived method to correct such data is to use Comprehensive Models such as CM4 (Sabaka et al. 2004) to estimate the external field at every time and location for the last fifty years (Ravat et al. 2003).

4.15.7 Estimation of the Core Magnetic Field

The International Geomagnetic Reference Field (IGRF) models of the survey period are commonly used to remove the core field from the total-field measurements at sea. They consist of spherical harmonic coefficients that predict the main field and its secular variation over 5-year intervals (Macmillan and Maus 2005). These models are regularly revised to generate a Definitive Geomagnetic Reference Field (DGRF). Therefore magnetic anomaly data of adjacent surveys carried out at different times may have different values resulting from an imprecise estimation of the core field.

Again, the Comprehensive Models can be used to subtract the core field from the initial total-field measurements (Ravat et al. 2003; Quesnel et al 2009). For recent epochs (after 2002) for which no Comprehensive Model is available yet, other geomagnetic field models such as CHAOS or GRIMM have to be used (Olsen et al. 2006; Lesur et al. 2008).

4.15.8 Summary of Marine Magnetic Observation Errors

In Table 4.5, we summarize the different errors (expressed in nT) associated with marine magnetic observations and their evolution over the last 50 years, allowing the reader to be aware of the quality of the data.

4.16 Unusual Instruments and Processing Approaches

In this section, we first describe how vector marine magnetic observations became possible over the last thirty years. The second part is devoted to deep water measurements.

Table 4.5 Evolution of errors associated with marine magnetic data

Type	Old data	Recent data	Solutions
Sensor[a]	0.1–1 nT	0.001–0.01 nT	Overhauser effect sensors
Ship noise	>1 nT	negligible	Large towing distance
Ship position[b]	1–100 nT	<1 nT	Radio, Doppler and later GPS
Date and time[c]	negl.	negl.	Manual/visual check of datasets
Transcription[d]	1–10000 nT	negl.	Digital recording, check of datasets
External field estimation[e]	1–100 nT	~1 nT	Mag. obs. data, CM4 or other models
Core field estimation[f]	10–100 nT	~1 nT	Mag. obs. data, CM4 or other models
Total error	0.1–10000 nT	0.001-1 nT	Cleaning and leveling of datasets[g]

[a]depending on the type of sensor, but we can consider the proton precession system as the most widely used for magnetometers at sea; for fluxgate sensors, see Section 4.16.1.
[b]depending on the ambient magnetic anomaly gradient, and difficult to quantify for recent data since GPS should provide very precise positioning.
[c]time, and sometimes date, is missing in few trackline datasets that we should not consider except if we retrieve this information. A small error in acquisition time should not greatly affect the resulting magnetic anomaly (Quesnel et al. 2009).
[d]often swap of one digit in a total-field value transcription.
[e]most of the marine magnetic data were not corrected for external field variations until recently; Mag. Obs., Magnetic Observatory; CM4, Comprehensive Model 4 of Sabaka et al. (2004).
[f]depending on the first IGRF models for early surveys.
[g]see Quesnel et al. (2009).

4.16.1 Vector Marine Magnetic Observations

Nowadays, fluxgate magnetometers offer better than 0.1 nT$(\sqrt{\mathrm{Hz}})^{-1}$ sensitivity, for about 0.01 nT of resolution. With such performance, it becomes possible to envisage the acquisition of vector magnetic measurements, i.e., the three components of the magnetic field, at sea. Such data would not substitute for absolute scalar measurements made with Overhauser magnetometers towed astern the ship. However, they may usefully complement these data. Indeed, the scalar magnetic anomaly of N-S trending structures near the Equator is almost zero, whereas the components of the vector anomalies still show some significant signal (Gee and Cande 2002; Engels et al. 2008). Furthermore, the three-component magnetic anomaly of elongated (2D) structures has the interesting property of having similar vertical and horizontal components, phase-shifted by $\frac{\pi}{2}$ (Isezaki 1986). Using this property, it is possible to estimate whether an anomaly is associated with a 2D or a 3D causative source, i.e., if the anomaly is a standard Vine and Matthews (1963) anomaly – an isochron of seafloor spreading, or a more complex structure such as a seamount or some kind of tectonic complexity. Furthermore, assuming that the anomaly is caused by an elongated (2D) body, it is possible to determine its orientation: it is the horizontal direction orthogonal to the vector anomaly, i.e., the direction along which the anomalous field is null. This ability to determine structural directions may be of importance in the case of single profiles or widely-spaced survey lines (such as those required for standard swath bathymetry), over sedimentary areas.

A major requirement to obtain accurate vector magnetic measurements is the knowledge of the sensor attitude, for instance by coupling the magnetic sensor to an inertial motion sensor. The final accuracy depends on that of the attitude sensor (Nabighian et al. 2005). Two types of instruments have been successfully tested: a towed vector magnetometer, in which both fluxgate magnetometer and inertial attitude sensors have been combined in a single 'fish' (Gee and Cande 2002; Engels et al. 2008) and Shipboard Three-Component Magnetometers (STCM), in which a three-component fluxgate magnetometer is installed on the ship's mast to take advantage of the ship's attitude sensor. Such a sensor is required for other instruments such as multibeam echosounders (e.g., Isezaki 1986; Seama et al. 1993; Korenaga 1995). The towed vector magnetometer has no specific correction for the vehicle magnetization: the only limitation is the high cost of any accurate attitude sensor, which one may hesitate to install in a towed (and easily lost) fish. Conversely, the STCM is affected by the strong magnetic effect of the ship, which should be adequately modelled and removed. Installing the fluxgate sensors on a mast, the extreme point of the ship, allows the observations to be explained adequately (to first order) by the following model.

Vector magnetic field measurements onboard a ship are the sum of the ambient geomagnetic field at the vessel location and the induced and remanent magnetic fields of the ship expressed as (Isezaki 1986):

$$\mathbf{B}_{\mathbf{obs}} = \underline{R}\,\underline{P}\,\underline{Y}\,\mathbf{B} + \underline{A}\,\underline{R}\,\underline{P}\,\underline{Y}\,\mathbf{B} + \mathbf{B}_{\mathbf{p}} \qquad (4.22)$$

where $\mathbf{B}_{\mathbf{obs}}$ is the observed magnetic field vector, $\underline{R}$, $\underline{P}$ and $\underline{Y}$ are the three matrices of rotation due to the roll, pitch and yaw, respectively (see Fig. 4.3), $\mathbf{B}$ is the ambient magnetic field vector, and $\underline{A}$ is the magnetic susceptibility tensor of the ship for a given location of the sensor. Finally, $\mathbf{B}_{\mathbf{p}}$ corresponds to the remanent magnetic field vector of the ship, and $\underline{A}\mathbf{B}$ is the field vector due to the ship's induced magnetic moment. R, P, and Y are given by the attitude sensor measurements. Once $\underline{A}$ and $\mathbf{B}_{\mathbf{p}}$ are known, it becomes possible to determine the ambient geomagnetic field $\mathbf{B}$ from the measurements $\mathbf{B}_{\mathbf{obs}}$ (Isezaki 1986).

To determine $\underline{A}$ and $\mathbf{B}_{\mathbf{p}}$, the usual technique is to acquire calibration data at a location where the ambient geomagnetic field $\mathbf{B}$ does not vary much and can be approximated by the IGRF field model. Specific navigation maneuvers called 'figures of eight' are carried out: they consist of a two consecutive narrow circles of opposite direction, i.e., a clockwise and a counterclockwise loops. The loops in opposite directions result in opposite ship roll, a way to sample the widest possible range of relative orientations of the ship and the ambient field. A large range of relative orientations insures a better constrained determination of $\underline{A}$ and $\mathbf{B}_{\mathbf{p}}$ by least-squares inversion of the calibration loop data (Isezaki 1986; Seama et al. 1993; Korenaga 1995). A faster alternative to figures of eight for ships equipped with bow thrusters is to undertake 360° rotations. Whereas Isezaki (1986) used the IGRF models to assign a value to $\mathbf{B}$ (so with uncertainty), Lesur et al. (2004) performed their own absolute measurements during the rotation, directly estimating the field strength assuming that the bulk susceptibility of the vessel is isotropic. The latter is true for a fibreglass boat, but not for steel research vessels. The accuracy of such an approach reaches 0.2° in declination, 0.05° in inclination and 10 nT in total intensity values. The accuracy of STCM measurements is not better than several tens of nT and can be improved by filters applied to improve the signal to noise ratio (Korenaga 1995).

STCM has been widely used by Japanese research vessels for the last 20 years (Isezaki 1986; Seama et al. 1993; Korenaga 1995) and are getting more popular in Korea (Lee and Kim 2004), France and Germany (König 2006). A difficulty is that, how ever carefully the calibration loops and the reduction of the data are performed, noise still affects the data, because the model used to estimate the ship's magnetization is so simplistic. For instance, the viscous remanent magnetization acquired by a ship sailing on the same heading for a long time will result in a slow and systematic variation of the anomalies, easily removed with a linear regression. Other more complex effects involve Foucauld currents in the ship, a conductive body moving in the Earth's magnetic field. For these reasons, the STCM measurements are only relative estimates of the geomagnetic vector useful for crustal anomaly studies, whereas the proton precession and Overhauser magnetometers provide absolute values of the field amplitude suitable for geomagnetic studies. The two types of measurements are complementary.

4.16.2 Deep-Sea Magnetic Observations

Sea-surface magnetic observations, typically acquired more than 2000 m above the magnetized sources, lack sufficient resolution to address some scientific problems. Here 'resolution' does not means the resolution of the instrument but the ability of the recorded signal (the magnetic anomaly) to detect a given variation of the causative physical property (the magnetization of a source body). Sea-surface anomalies barely resolve the longest wavelengths of geomagnetic field intensity as recorded by the oceanic crust (e.g., Canda and Kent 1992a, 1992b; Gee et al. 1996; Bouligand et al. 2006). Simple forward modelling easily demonstrates that the details of these variations or the depiction of ore deposits on the seafloor in association with hydrothermal vents, for instance, are beyond the reach of these data (e.g., Tivey and Dyment 2010).

The magnetic field created by a point source decays as $\frac{1}{r^3}$, where r is the distance to the source body (ln $(\frac{1}{r})$ in the case of a 2D problem, i.e., a line source seen as a point source in cross section). The only way to significantly improve the resolution of the magnetic signal caused by a source bodies is to reduce the distance to these bodies. For marine magnetics, this means evolving from sea-surface to deep-sea measurements.

4.16.2.1 Procedures

There are two ways to get magnetic profiles closer to the seafloor: either towing a magnetometer behind a depressing weight (deep tow magnetometer), or attaching a magnetometer to a deep-sea vessel, either a manned submersible, a Remotely Operated Vehicle (ROV) or an Autonomous Underwater Vehicle (AUV). Deep tow magnetometers are most often operated at between 200 m and 1000 m above the seafloor (depending the depth and roughness of the seafloor, the desired speed of the ship, and the confidence on the navigation), whereas deep-sea vehicles are generally used up to about 50 m above the seafloor. In both cases, the magnetometer must be placed in a pressure case adapted to the operation depth. Both require slow speeds: about 1.5–2 knots for a deep tow instrument, depending on the water depth and the altitude above seafloor of the measurements; 0.5–1 knot for deep-sea vehicles, depending on the type of vehicle and the depth of the dive, compared with the usual 10–12 knots of most oceanographic vessels. For this reason, and because of the higher level of technology required for such experiments, deep-sea magnetic measurements are expensive and sparse.

Another difficulty common to every deep-sea experiment is accurate positioning of the instrument. Unlike the sea-surface magnetometer, towed 200 to 300 m behind the ship and quite easy to locate with reasonable accuracy, the deep tow magnetometer has a cable several kilometers long. Its positioning requires either a depthmeter – to compute an estimated position assuming that the cable is not bending much and currents are not deviating the instrument laterally from the ship's profile – or a beacon emitting acoustic signals to the ship's Ultra Short BaseLine (USBL) receiver, if such a positioning system is available, or a combination of both for better results. The deep-sea vehicle is usually located by a Long BaseLine (LBL) positioning system – implying the mooring of beacons prior to the experiment – or by a USBL system as well. In both cases, the position of the ship (for USBL) and, to a lesser extent, of the beacons (location of moorings, for LBL) are well known from GPS. Detailed surveys by submersibles or ROVs rely on both acoustic positioning and dead-reckoning navigation; in addition, they also use artificial markers provisionally set up on the seafloor at the beginning of the survey and regularly revisited during the survey to avoid any drift in navigation. The accuracy of such navigation is similar to that of GPS, i.e., a few tenths of meters or better, whereas that of deep-tow magnetometers may be closer to a few hundred of meters - probably better if only relative accuracy along the profile is considered.

Deep tow magnetometers can be either autonomous, i.e., a magnetometer, a pack of batteries, and a recording device is towed at the end of a passive cable, or connected to the ship by a conducting cable which provides power to the instrument and real-time data transfer to the ship. Although the latter is far better for unlimited autonomy and real-time control of the instrument (i.e., to insure that the instrument is properly operating and to get the depth of the instrument for safer monitoring of the cable length and the ship's speed), conducting cables are rather expensive and are not readily available on all research vessels. The major difficulty in operating a deep tow magnetometer is with altitude control, and loss of instruments after collision with the seafloor is not uncommon.

Most deep tow magnetometers are scalar devices–proton precession, Overhauser, or the less accurate but cheaper, easier to operate and often adequate magnetoresistive instruments (e.g., Lenz 1990). Fluxgate magnetometers are sometime used to provide the magnetic field intensity, without any specific attempt to obtain the vector components. A deep tow vector magnetometer, quite similar in principle to the surface towed vector magnetometer described above, has been constructed and sucessfully operated on the East Pacific Rise (EPR, Yamamoto et al. 2004, 2005).

Due to its proximity to the seafloor and for safety reasons, it is impossible to tow a scalar magnetometer behind a deep-sea vessel. The magnetometer has to be attached to the hull of the vessel, at the most extreme position as possible, and should therefore be a vector magnetometer, i.e., three orthogonal fluxgate sensors. The method to correct for the magnetic effect of the vessel is similar to the one described above for STCM. Loops which can be used for calibration are spontaneously performed by submersibles like DSS Nautile of IFREMER (because of its slightly unbalanced weight). ROVs can easily be stopped in the mid-water column, far from both the magnetic sources of the seafloor and the ship, to achieve 360° rotations using their lateral thrusters. AUVs can sail calibration loops, if their magnetic effect is large enough to require a correction.

Topography and altitude variations dominate the magnetic signal recorded by a deep-sea vessel, and have a significant effect on deep tow measurements. Modeling and filtering methods (e.g., Guspi 1987; Hussenoeder et al. 1995; Honsho et al. 2009) help to extract the signal of interest, i.e., seafloor magnetization variations.

4.16.2.2 Some Applications

Despite their cost, a significant number of deep-sea magnetic experiments have been carried out for specific societal or scientific purposes.

One of the first cruises to use a deep-tow magnetometer was undertaken to find the wreck of the sunken submarine Thresher in the Northwest Atlantic (Heirtzler 1964, and references therein). The instrument was towed at a depth of ~3000 meters and an altitude of 20–25 m above the seafloor. Although the accuracy of their proton magnetometer TF measurements was 3 nT, they estimated the true error to be ~10 nT considering the error on sensor position.

Many deep-sea magnetic experiments have taken place at mid-ocean ridges, as part of the effort to explore them. Klitgord et al. (1975) performed several deep tow profiles across the EPR. Macdonald et al. (1983) demonstrated the outward dipping slope of the polarity boundaries, which results from the combination of lava piling and seafloor spreading, by considering measurements at different altitudes above the EPR. This was later confirmed by direct measurements on the Blanco Fracture Zone (Tivey et al. 1998a). Gee et al. (2000) and Pouliquen et al. (2001a, b) have shown from deep tow measurements on the EPR and the Central Indian Ridge, fast and intermediate spreading center respectively, that the oceanic crust is confidently recording not only geomagnetic polarity reversals but also the geomagnetic intensity variations. Honsho et al. (2009) extended this observation to the magmatic areas of slow spreading centers from submersible observations on the Mid-Atlantic Ridge. These observations are allowing high resolution dating of the seafloor using geomagnetic intensity variations in the well-constrained Brunhes and Matuyama sequences. Tivey et al. (1998b) have been able to map the thickness of a recent lava flow from its magnetic signature as recorded by AUV ABE of WHOI on the Juan de Fuca Ridge. Conversely, Shah et al. (2003) have used the same AUV to map the ultrafast EPR at 18°S and found a magnetic low interpreted as depression as the signature of hot dykes, as well as lobes that may mark different lava flows erupted under different geomagnetic paleointensities.

Other important features that exhibit magnetic signature at deep-sea vessel altitudes are active and fossil hydrothermal sites (e.g., Tivey and Dyment 2010). Sites lying on a basaltic basement are associated with a negative magnetic anomaly, i.e., the titanomagnetites are altered to titanomaghemites and non magnetic minerals under the effect of pervasive hydrothermal fluid circulation (Tivey et al. 1993; Tivey and Johnson 2002). Conversely, sites lying on ultramafic rocks such as site Rainbow on the Mid Atlantic Ridge are associated with a strong positive anomaly (Dyment et al. 2005), possibly the result of new magnetic minerals created by serpentinization (magnetite) or by sulfide deposition and accumulation (pyrrhotite). These results suggest deep-sea magnetics is a suitable method to detect and characterize fossil hydrothermal vents and evaluate the mining potential of such ore deposits on the seafloor.

Deep-sea magnetics data have also been collected over passive margins, for instance on the peridotite ridge off Galicia (e.g. Whitmarsh and Miles 1995). Sibuet et al. (2007) used deep-tow magnetic measurements to suggest that serpentinization of outcropping mantle at some oceanic margins could generate parallel magnetic lineations similar to seafloor spreading anomalies. Another successful application is seamount magnetism, where Gee et al. (1988) mapped the non-uniform magnetization of Jasper seamount with sufficient resolution to better constrain paleomagnetic poles than would have been done with surface magnetic measurements. In general, deep-sea magnetic measurements are combined with other geological and structural information to determine an equivalent magnetization distribution (given a magnetized source geometry), to be compared to rock magnetic property measurements (Macdonald et al. 1979; Gee et al. 1988; Ravilly et al. 2001; Honsho et al. 2009).

4.17 Conclusions for Marine Magnetics

Several aspects of marine magnetic observations have been reviewed, trying to emphasize the evolution of errors associated with such data. The scalar measurement is very large dataset and covers all northern parts of oceans very well, but which shows gaps in the southern oceans. Most of these data were acquired between 1960 and 1980. Even though early data are affected by different kind of errors (such as no correction for the external field), one can retrieve the true magnetic anomaly value by applying Comprehensive Models. Also, instrumental errors have been considerably reduced with improvements in scalar magnetometers such as Overhauser sensors. Vector measurements are becoming more common in scientific marine campaigns mainly because specific sailing techniques like 'Figures of Eight' and data processing now allow the initial problems of sensor orientation and ship noise contribution to be overcome. Finally, to map the small-wavelength magnetic anomalies over oceanic areas, deep-sea magnetic measurements have been undertaken for the last 20 years. The results of such observations have considerably improved our vision of the shallow crust's magnetization, and it is now a field of research in its own right.

4.18 General Conclusion

In this manuscript we try to give an overview of the magnetic data acquisition and processing techniques for both airborne and marine surveys. These techniques have constantly evolved since they appeared at the turn of the 20th century. Whereas a century ago researcher were trying to acquire data at "higher and higher" altitudes for the sake of complete coverage of large areas, nowadays "lower and lower" altitudes are aimed for the sake of higher resolution. Major technological improvements in instrumentation for both the acquisition of magnetic measurements and navigational data, favored higher accuracy and resolution field mapping. For aeromagnetics, this implies that the survey have to be flown at very low altitudes, whereas for marine magnetic data acquisition, measurements have to be closer to the ocean floor. The Unmanned Aerial Vehicle (UAV) and both the Autonomous Underwater Vehicle (AUV) and Remotely Operated Vehicle (ROV) are seen as solutions for acquiring safely and efficiently such data. The developments of these techniques will continue in the future as: (1) Large areas are still to be surveyed, particularly over the Oceans; (2) Significant efforts are still required to patch together the existing surveys; (3) intermediate wavelength magnetic anomalies ($\sim$500 km) are not yet properly mapped.

Acknowledgements The authors collectively would like to thank the reviewer, Kathryn Whaler, and acknowledge her work that greatly improved the quality of manuscript. MH would like to warmly acknowledge the contributions of K. Allek, N. Bournas, J. Luis in compiling and processing aeromagnetic data and P. Mouge for his help. T. Ishihara kindly accepted to review a preliminary version of the part devoted to marine magnetics. YQ also acknowledges him and M. Catalãn for precious references that helped to build and write the marine magnetics sections. Novatem, and Geometrics companies kindly provided illustrations used in this article.

References

Achache J, Cohen Y, Unal G (1991) The french program of circumterrestrial magnetic surveys using startospheric balloons. EOS Trans Am Geophys Union 72:97–101

Acuña MH, Scearce CS, Seek JB, Scheifele J (1978) The magsat vector magnetometer-a precision fluxgate magnetometer for the measurement of the geomagnetic field, Technical report, NASA/GSFC TM 79656

Acuña MH, Connerney JEP, Ness NF, Lin RP, Mitchell D, Carlson CW, McFadden J, Anderson KA, Reme H, Mazelle C, Vignes D, Wasilewski P, Cloutier P (1999) Global distribution of crustal magnetization discovered by the mars global surveyor MAG/ER experiment. Science 284(5415):790–793

Airo ML (2002) Aeromagnetic and aeroradiometric response to hydrothermal alteration. Surv Geophys 23:273–302

Allan T (1969) A review of marine geomagnetism. Earth Sci Rev 5:217–254

Allek K (2005) Traitement et interpretation des donnees aeromagnetiques acquises au-dessus des regions de tindouf et de l'eglab (SO de l'Algerie): impact sur l'exploration du diamant, Master's thesis, Universite des Sciences et de la Technologie Houari Boumediene, USTHB

Allek K, Hamoudi M (2008) Regional-scale aeromagnetic survey of the south-west of algeria: a tool for area selection for diamond exploration. J Afr Earth Sci 50:67–78

Anderson B, Longacre M, Quist P (1999) Comparison of a new marine magnetometer system to high-resolution aeromagnetic data a case study from offshore Oman. In: Proceedings of the SEG International Exposition and 70th Annual Meeting, Houston

Anderson DE, Pita AC (2005) Geophysical surveying with georanger uav. Am Inst Aeronaut Astronaut 50:67–78

Balsley JRJ (1946) The airborne magnetometer, Preliminary report 3, 8p., U.S. Geological Survey, Washington, DC

Barton CE (1997) International geomagnetic reference field: The seventh generation. J Geomagnetic Geoelectric 49: 123–148

Bell WE, Bloom AL (1957) Optical detection of magnetic resonance in alkali metal vapor. Phys Rev 107(6): 1559–1565

Bhattacharyya BK (1969) Bicubic spline interpolation as a method for treatment of potential field data. Geophysics 34:402–423

Bhattacharyya BK (1971) An automatic method of compilation and mapping of high-resolution aeromagnetic data. Geophysics 36(4):695–716

Binder AB (1998) Lunar prospector: overview. Science 281(5382):1475–1476

Blakely RJ (1988) Curie temperature isothermal analysis and tectonic implications of aeromagnetic data from Nevada. J Geophys Res 93(B10):11817–11832

Blakely R, Cox A, Iufer E (1973) Vector magnetic data for detecting short polarity intervals in marine magnetic profiles. J Geophys Res 78:6977–6983

Bouligand C, Dyment J, Gallet Y, Hulot G (2006) Geomagnetic field variations between chrons 33r and 19r (83–41 ma) from sea-surface magnetic anomaly profiles. Earth Planet Sci Lett 250(3–4):541–560. doi:10.1016/j.epsl.2006.06.051

Bournas N (2001) Interpretation des donnees aerogeophysiques acquises au-dessus du hoggar oriental (algerie), PhD thesis, Universite des Sciences et de la Technologie Houari Boumediene, USTHB

Bournas N, Galdeano A, Hamoudi M, Baker H (2003) Interpretation of the aeromagnetic map of eastern hoggar (algeria) using euler deconvolution, analytic signal and local wavenumber methods. J Afr Earth Sci 37:191–205

Boyce J, Reinhardt E, Raban A, Pozza M (2004), Marine Magnetic Survey of a Submerged Roman Harbour, Caesarea Maritima, Israel. Int J Naut Arch 33(1):122–136. doi:10.1111/j.1095-9270.2004.010.x

Bozzo E, Colla A, Caneva G, Meloni A, Caramelli A, Romeo G, Damaske D, Moeller D (1994) Technical procedures for aeromagnetic surveys in antarctica during the italian expeditions (1988–1992). Ann Geophys XXXVII(5):1283–1294

Braginski SI, Roberts PH (1995) Equations governing convection in earth's core and the geodynamo. Geophys Astrophys Fluid Dyn 79(1):1–97

Briggs IC (1974) Machine contouring using minimum curvature. Geophysics 39:39–48

Bullard E, Mason R (1961) The magnetic field astern of a ship. Deep Sea Res 8:20–27

Cady JW (1990) Alaska as a frontier for aeromagnetic interpretation. In: Geologic applications of modern aeromagnetic surveys. U.S. Geological Survey Bulletin 1924:75–84

Cande S, Kent D (1992a) A new geomagnetic polarity time scale for the late Cretaceous and Cenozoic. J Geophys Res 97:13917–13951

Cande S, Kent D (1992b) Ultrahigh resolution marine magnetic anomaly profiles: a record of continuous paleointensity variations. J Geophys Res 97:15075–15083

Chapman S, Bartels J (1940) Geomagnetism, vol. II. Clarendon Press, Oxford, p 633

Chiappini M, Meloni A, Boschi E, Faggioni O, Beverini N, Carmiciani C, Marson I (2000) Shaded relief magnetic anomaly map of italy and surrounding marine areas. Ann Geophys 43(5):983–989

Cohen TJ, Lintz PR (1974) Long term periodicities in the sunspot cycle. Nature 250:398–399

Cohen Y, Menvielle M, Le Mouël J (1986) Magnetic measurements aboard a stratospheric balloon. Phys Earth Planetary Inter 44:348–357

Cordell LE, Hildenbrand TG, Kleinkopf MD (1990) Notes of discussion group on the midcontinent. In: Geologic applications of modern aeromagnetic surveys. U.S. Geological Survey Bulletin 1924:90–91

Courtillot V, Le Mouël J, Mayaud P (1977) Maximum entropy spectral analysis of the geomagnetic activity index aa over 107-year interval. J Geophys Res 82(19):2641–2649

Dehmelt HG (1957) Slow spin relaxation of optically polarized sodium atoms. Phys Rev 105(5):1487–1489

Dobrin MD, Savit CH (1988) Introduction to geophysical prospecting. Mc Graw Hill, New York, NY, p 867

Doll WE, Gamey TJ, Beard LP, Bell DT (2006) Airborne vertical magnetic gradient for near-surface applications. The Leading Edges 25(1):50–53

Doyle HA (1987) Geophysics in Australia. Earth Sci Hist 6(2):178–204

Dyment J, Tamaki K, Horen H, Fouquet Y, Nakase K, Yamamoto M, Ravilly M, Kitazawa M (2005) A positive magnetic anomaly at Rainbow hydrothermal site in ultramafic environment. Eos Trans AGU 86(52, Fall Meet Suppl): 21–08

Engels M, Barckhausen U, Gee J (2008) A new towed marine vector magnetometer: methods and results from a Central Pacific cruise. Geophys J Int 172:115–129. doi:10.1111/j.1365-246X.2007.03601.x

Eve AS (1932) A magnetic method for estimating the height of some buried magnetic bodies. Trans Geophys Prospect Am Inst Mining Metallurgical Eng 101:200–215

Eve AS, Keys DA (1933) Applied geophysics in the search for minerals. Cambridge University Press, Cambridge

Ferris JK, Vaughan APM, King EC (2002) A window on west Antarctic crustal boundaries: the junction between the antarctic peninsula, the filchner block and weddell sea oceanic lithosphere. Tectonophysics 347:13–23

Finlay C, Maus S, Beggan C, Hamoudi M, Lowes FJ, Olsen N, Thèbault E (2010) Evaluation of candidate geomagnetic field models for igrf-11. Earth Planets Space submitted 1–18

Foster MR, Jines WR, van der Weg K (1970) Statistical estimation of syetematic errors at intersections of lines of aeromagnetic survey data. J Geophys Res 75:1507–1511

Frost BR, Shive PN (1986) Magnetic mineralogy of the lower continental crust. J Geophys Res 91(B6):6513–6521

Gee J, Schneider D, Kent D (1996) Marine magnetic anomalies as recorders of geomagnetic intensity variations. Earth Planet Sci Lett 144:327–335

Gee J, Tauxe L, Hildebrand J, Staudigel H, Lonsdale P (1988) Nonuniform Magnetization of Jasper Seamount. J Geophys Res 93(B10):12159–12175

Gee J, Cande S (2002) A surface-towed vector magnetometer. Geophys Res Lett. doi:10.1029/2002GL015245

Gee J, Cande S, Hildebrand J, Donnelly K, Parker R (2000) Geomagnetic intensity variations over the past 780 kyr obtained from near-seaoor magnetic anomalies. Nature 408:827–832

Germain-Jones D (1957) Post-war developments in geophysical instrumentation for oil prospecting. J Sci Inst 34:1–8

Gopal BJ, Sarma VN, Rambabu HV (2008) Real time compensation for aircraft induced noise during high resolution airborne magnetic surveys. J Ind Geophys Union 8(3):185–189

Grant F (1970) Statistical models for interpreting aeromagnetic data. Geoexploration 35(2):293–302

Grant F (1972) Review of data processing and interpretation methods in gravity and magnetics, 1964–71. Geophysics 37(2):647–661

Grant F (1985a) Aeromagnetics, geology, and ore environments; i, magnetite in igneous, sedimentary and metamorphic rocks—an overview. Geoexploration 23:303–333

Grant F (1985b) Aeromagnetics, geology and ore environments; ii, magnetite and ore environments. Geoexploration 23: 335–362

Grauch VJS, Millegan P (1998) Mapping intrabasinal faults from high-resolution aeromagnetic data. The Leading Edges 17(1):53–56

Guspi F (1987) Frequency-domain reduction of potential field measurements to a horizontal plane. Geoexploration 24: 87–98

Hamoudi M, Cohen Y, Achache J (1998) Can the thermal thickness of the continental lithosphere be estimated from magsat data. Tectonophysics 284:19–29

Hamoudi M, Thèbault E, Lesur V, Mandea M (2007) Geoforschungszentrum anomaly magnetic map (gamma): a candidate model for the world digital magnetic anomaly map. Geochem Geophys Geosyst 8(6):1–13

Hanna WF (1990) Some historical notes on early magnetic surveying in the U.S. geological survey. In: Geologic applications of modern aeromagnetic surveys. U.S. Geological Survey Bulletin 1924:63–73

Hansen RO (1993) Interpretive gridding by anistropic kriging. Geophysics 58(10):1491–1497

Hardwick CD (1984a) Important design considerations for inboard airborne magnetic gradiometers. Geophysics 49(11):2004–2018

Hardwick CD (1984b) Non-oriented cesium sensors for airborne magnetometry and gradiometry. Geophysics 49(11): 2024–2031

Hawkins WB (1955) Orientation and alignment of sodium atoms by means of polarized resonance radiation. Phys Rev 98(2):478–486

Heezen B, Ewing M, Miller E (1953) Trans-Atlantic profile of total magnetic intensity and topography, Dakar to Barbados. Deep Sea Res 1:25–33

Heiland CA (1929) Geophysical methods of prospecting: Principles and recent successes. Q Colo School Mines XXIV(1):47–77

Heiland CA (1935) Geophysical mapping from the air: its possibilities and advantages. Eng Min J 136: 609–610

Heiland CA (1940) Geophysical exploration. Prentice-Hall, New York, NY

Heiland CA (1953) Method of and apparatus for aeromagnetic prospecting U.S. Patent 2659859

Heirtzler J (1964) Magnetic measurements near the deep ocean floor. Deep Sea Res 11:891–898

Hemant K, Thèbault E, Mandea M, Ravat D, Maus S (2007) Magnetic anomaly map of the world: merging satellite, airborne, marine and ground-based magnetic data sets. Earth Planetary Sci Lett 260(1–2):56–71

Hildenbrand TG, Raines GL (1990) Need for aeromagnetic data and a national airborne geophysics program. In: Geologic applications of modern aeromagnetic surveys. U.S. Geological Survey Bulletin 1924:1–5

Hill M (1959) A ship-borne nuclear-spin magnetometer. Deep Sea Res 5:309–311

Hofman-Wellenhof B, Legat K, Wiener M (2003) Navigation: principles of positioning and guidance. Springer, New York, NY

Honsho C, Dyment J, Tamaki K, Ravilly M, Horen H, Gente P (2009) Magnetic structure of a slow spreading ridge segment: insights from near-bottom magnetic measurements on board a submersible. J Geophys Res. 114 B05101:1–25 doi:10.1029/2008JB005915

Hood L, Zakharian A, Halekas J, Mitchell D, Lin R, na MA, Binder A (2001) Initial mapping and interpretation of lunar crustal magnetic anomalies using lunar prospector magnetometer data. J Geophys Res 106(E11):27825–27839

Hood P (1990) Aeromagnetic survey program of Canada, mineral applications, and vertical gradiometry, in Geologic Applications of Modern Aeromagnetic Surveys

Hood P (2007) History of aeromagnetic surveying in Canada. The Leading Edges 26(11):1384–1392

Hood P, Sawatzky JP, Kornik LJ, McGrath PH (1976) Aeromagnetic gradiometer survey, white lake, Open File 339, Geological Survey of Canada

Horsfall KR (1997) Airborne magnetic and gamma-ray acquisition. J Aust Geol Geophys 17(2):23–30

Hrvoic D (2007) SeaSPY technical application guide, marine magnetics corporation education

Hussenoeder SA, Tivey MA, Schouten H (1995) Direct inversion of potential fields from an uneven track with application to the Mid-Atlantic Ridge. Geophys Res Lett 22:3131–3134. doi:10.1029/95GL03326

Isezaki N (1986) A new shipboard three-component magnetometer. Geophysics 51(10):1992–1998

Jensen H (1965) Instrument details and applications of a new airborne magnetometer. Geophysics XXX(5): 875–882

Jones E (1999) The earth's magnetic field at sea, in marine geophysics. Wiley, Chichester, pp 162–197

Kastler A (1954) Optical methods of atomic orientation and of magnetic resonance. J Opt Soc Am 47(6):460–465

Keating P (1993) The fractal dimension of gravity data sets and its implication for gridding. Geophys Prospect 41: 983–994

Keating P (1995) A simple technique to identify magnetic anomalies due to kimberlite pipes. Explor Mining Geol 4:121–125

Kittel C (2005) Introduction to the solid state physics. Wiley, San Francisco, CA

Klitgord K, Mudie J, Huestis S, Parker R (1975) An analysis of near-bottom magnetic anomalies: sea-floor spreading and the magnetized layer. Geophys. J R Astron Soc 43: 387–424

König M (2006) Processing of shipborne magnetometer data and revision of the timing and geometry of the Mesozoic break-up of Gondwana. Rep Polar Res 525:137 pp.

Korenaga J (1995) Comprehensive analysis of marine magnetic vector anomalies. J Geophys Res 100(B1):365–378

Korhonen JV (2005) Airborne magnetic method: Special features and review on applications, In: Airo ML (ed) Aerogeophysics in Finland 19722004: methods, system characteristics and applications, vol 39. Geological Survey of Finland, special paper Espoo, Finland edn. pp 77–102

Korhonen JV, Koistinen T, Elo S, Saavuori H, Kaariainen J, Nevanlinna H, Aaro S, Haller LA, Skilbrei JR, Solheim D, Chepik A, Kulinich A, Zhdanova L, Vaher R, All T, Sildvee H (1999) Preliminary magnetic and gravity anomaly maps of the fennoscandian shield 1:10, 000, 000. Geological Survey of Finland 27 Special paper, pp 173–179

Korhonen J, Fairhead J, Hamoudi M, Hemant K, Lesur V, Mandea M, Maus S, Purucker M, Ravat D, Sazonova T, Thèbault E (2007) Magnetic anomaly map of the World, Scale 1:50,000,000, 1st edn. Commission for the Geological Map of the World, UNESCO edn.

Langel R (1982) The magnetic earth as seen from magsat, initial results. Geophys Res Lett 9(4):239–242

Langel R, Ousley G, Berbert J, Murphy J, Settle M (1982) The magsat mission. Geophys Res Lett 9(4):243–245

Langel R, Hinze W (1998) The magnetic field of the Earth's lithosphere. Cambridge University Press, Cambridge

Laughton A, Hill M, Allan T (1960) Geophysical investigations of a Seamount 150 miles North of Madeira. Deep Sea Res 7:117–141

Le Mouël JL (1969) Les elements du champ magnetique terrestre, PhD thesis, Faculte des Sciences de l'Universite de Paris

Lee S, Kim S (2004) Vector magnetic analysis within the southern Ayu Trough, equatorial western Pacific. Geophys J Int 156:213–221

Leger JM, Bertrand F, Jager T, Prado ML, Fratter I, Lalaurie JC (2009) Swarm absolute scalar and vector magnetometer based on helium 4 optical pumping. Procedia Chem 1: 634–637

Leliak P (1961) Identification and evaluation of magnetic field sources of magnetic airborne detector equiped aircraft. Ins Radio Eng Trans Aerospace Navigational Electron 8: 95–105

Lenz J (1990) A review of magnetic sensors. Proc IEEE 78(6):973–989

Lesur V, Clark T, Turbitt C, Flower S (2004) A technique for estimating the absolute vector geomagnetic field from a marine vessel. J Geophys Eng 1:109–115

Lesur V, Wardinski I, Rother M, Mandea M (2008) GRIMM: the GFZ reference internal magnetic model based on vector satellite and observatory data. Geophys J Int 173(2): 382–394. doi:10.1111/j.1365-246X.2008.03724.x

Lesur V, Wardinski I, Asari S, Minchev B, Mandea M (2009) Modelling the Earth's core magnetic field under flow constraints. Earth Planet Space 62(6):503–516

Logachev AA (1947) The development and applications of airborne magnetometers in the u.s.s.r. Geophysics (trans: Russian Hawkes HE) 11:135–147

Luis JF (1996) Le leve aeromagnetic des acores, PhD thesis, Institut de Physique du Globe de Paris, IPGP

Luis JF, Miranda JM (2008) Reevaluation of magnetic chrons in the north atlantic between 35n and 47n: Implications for the formation of the azores triple junction and associated plateau. J Geophys Res 113(B10105):1–12

Luis JF, Miranda JM, Galdeano A, Patriat P, Rossignol JC, Mendes-Victor LA (1994) The acores triple junction evolution since 10 ma from aeromagnetic survey of the mid-atlantic ridge. Earth Planetary Sci Lett 125: 439–459

Lum CW (2009) Coordinated searching and target identification using teams of autonomous agents, PhD thesis, University of Washington

Lum CW, Rysdyk RT, Pongwunwattana A (2005) Autonomous airborne geomagnetic surveying and target identification, in AIAA Infotech@Aerospace Conference

Macdonald K, Kastens K, Spiess F, Miller S (1979) Deep tow studies of the Tamayo Transform Fault. Marine Geophys Res 4:37–70

Macdonald K, Miller S, Luyendyk B, Atwater T, Shure L (1983) Investigation of a Vine-Matthews magnetic lineation from a submersible: the source and character of marine magnetic anomalies. J Geophys Res 88:3403–3418

Macmillan S, Maus S (2005) International Geomagnetic Reference Field-the tenth generation. Earth Planets Space 57:1135–1140

Macmillan S, Maus S, Bondar T, Chambodut A, Golovkov V, Holme R, Langlais B, Lesur V, Lowes F, Luhr H, Mai W, Mandea M, Olsen N, Rother M, Sabaka TJ, Thomson A, Wardinski I (2003) The ninth-generation international geomagnetic reference field. Phys Earth Planetary Inter 140:253–254

Malahoff A, Feden R, Fleming H (1982) Magnetic Anomalies and Tectonic Fabric of Marginal Basins North of New Zealand. J Geophys Res 87(B5):4109–4125

Mason R (1958) A magnetic survey off the west coast of the United States between latitudes 32° and 36°N, longitudes 121° and 128°W. Geophys J 1:320–329

Mason R, Raff A (1961) A magnetic survey off the west coast of North America, 32°N to 42°N. Geol Soc Am Bull 72: 1259–1265

Maus S, McLean S, Dater D, Luhr H, Rother M, Mai W, Choi S (2005) Ngdc/gfz candidate models for the 10th generation internationale geomagnetic reference field. Earth Planets Space 57:1151–1156

Maus S, Sazonova T, Hemant K, Fairhead JD, Ravat D (2007) National geophysical data center candidate for the world digital magnetic anomaly map. Geochem Geophys Geosyst 8(6):10

Maus S, Macmillan S, McLean S, Hamilton B, Thomson A, Nair M (2009) The us/uk world magnetic model for 2010–2015, Technical Report NES-DIS/NGDC, NOAA

McCafferty AE, Van Gosen BS (2009) Airborne gamma-ray and magnetic anomaly signatures of serpentinite in relation to soil geochemistry. Appl Geochem 24: 1524–1537

Merkouriev S, DeMets C (2006) Constraints on Indian plate motion since 20 ma from dense Russian magnetic data: Implications for Indian plate dynamics. Geochem Geophys Geosyst Q02002. doi:10.1029/2005GC001079

Merkouriev S, DeMets C (2008) A high-resolution model for EurasiaNorth America plate kinematics since 20 ma. Geophys J Int. doi:10.1111/j.1365-246X.2008.03761.x

Merrill R, McElhinny M (1983) The earth's magnetic field: its history, origin, and planetary perspective. Academic, London

Miles PJ, Partner RT, Keeler KR, McConnel TJ (2008) Unmanned airborne vehicle geophysical surveying US 2008/0125920 A1

Millegan P (2005) Broader spectrum, fewew folks-gavity and magnetics. The Leading Edges 24(S1):36–41

Milligan PR, Franklin R (2004) Magnetic anomaly map of Australia, 4th edn. Geoscience Australia, Canberra, Scale 1:5,000,000

Milligan PR, White A, Heinson G, Brodie R (1993) Micropulsation and induction array study near ballarat victoria. Explor Geophys 24(2):117–122

Minty BRS (1991) Simple micro-levelling for aeromagnetic data. Explor Geophys 22:591–592

Mittal PK (1984) Algorithm for error adjustment of potential field data along a survey network. Geophysics 49(4):467–469

Muffly G (1946) The airborne magnetometer. Geophysics 11:321–334

Nabighian M, Grauch V, Hansen R, LaFehr T, Li Y, Peirce J, Phillips J, Ruder M (2005) The historical development of the magnetic method in exploration. Geophysics 70(6):33–61. doi:10.1190/1.2133784

Nazarova K, Tsetkov YA, Heirtzler J, Sabaka TJ (2005) Balloon geomagnetic survey at stratospheric altitudes, In: Reigber C, Luhr H, Schwintzer P, Wickert J (eds) Earth magnetic field. Springer, New York, NY, pp 273–278

Ness NF (1971) Interaction of the solar wind with the moon. Phys Earth Planetary Inter 4:197–198

Nielsen OV, Petersen JR, Primdahl F, Brauer P, Hernando B, Fernandez A, Merayo JMG, Ripka P (1995) Development, construction and analysis of the 'oersted' fluxgate magnetometer. Meas Sci Technol 6:1099–1115

Nielsen OV, Brauer P, Primdahl F, Risbo T, Jorgensen JL, Boe C, Deyerler M, Bauereisen S (1997) A high-precision triaxial fluxgate sensor for spece applications: layout and choice of materials. Sens Actuators A59:168–176

O'Connell MD, Smith RS, Vallee MA (2005) Gridding aeromagnetic data using longitudinal and transverse gradients with the minimum curvature operator. The Leading Edges 24:142–145

Olsen N (2002) A model of the geomagnetic field and its secular variation for epoch 2000 estimated from Oersted data. Geophys J Int 149:454–462

Olsen N, Lühr H, Sabaka TJ, Mandea M, Rother M, Toffner-Calusen L, Choi S (2006) CHAOS—a model of the Earth's magnetic field derived from CHAMP, Oersted, and SAC-C magnetic satellite data. Geophys J Int 166:67–75

Olsen N, Mandea M, Sabaka TJ, Toffner-Clausen L (2009) Chaos-2 —a geomagnetic field model derived from one decade of continuous satellite data. Geophys J Int 179: 1477–1487

Overhauser AW (1953) Polarization of nuclei in metals. Phys Rev 92:411–415

Packard M, Varian R (1954) Proton gyromagnetic ratio. Phys Rev 93:941–947

Pang X, Lintz C (2009) Study on aircraft magnetic compensation based on fir model, In: International Symposium on Intelligent Information Systems and Applications IISA'09 Quindao, China

Pariso J, Rommevaux C, Sempere JC (1996) Three-Dimensional Inversion of Marine Magnetic Anomalies: Implications for Crustal Accretion along the Mid-Atlantic Ridge (28°–31°30′N). Mar Geophys Res 18:85–101

Parker RL, Huestis S (1974) The Inversion of Magnetic Anomalies in the Presence of Topography. J Geophys Res 79(11):1587–1593

Parkinson BW, Enge PE (1996) Differential gps, In: Zarchan P, Parkinson B, Spilker J Jr, Axelrad P, Enge P (eds) Chapter 1, American Institute of Aeronautics and Astronautics., pp 3–49.

Parsons LW, Wiatr ZM (1962) Rubidium vapour magnetometer. J Sci Instrum 39:292–300

Paterson NR, Reeves CV (1985) Applications of gravity and magnetic surveys: The state-of-the-art in 1985. Geophysics 50(12):2558–2594

Pouliquen G, Gallet Y, Dyment J, Patriat P, Tamura C (2001a) A geomagnetic record over the last 3.5 million years from deep-tow magnetic anomaly profiles across the Central Indian Ridge. J Geophys Res 106:10941–10960

Pouliquen G, Gallet Y, Dyment J, Patriat P, Tamura C (2001b) Correction to A geomagnetic record over the last 3.5 million years from deep-tow magnetic anomaly profiles across the Central Indian Ridge. J Geophys Res 106:30549

Pulz E, Jackel KH, Linthe HJ (1999) A new optically pumped tandem magnetometer: principles and experiences. Meas Sci Technol 10:1025–1031

Purucker M, Whaler K (2007) Crustal magnetism. In: Schubert G (ed) Treatise of geophysics, vol 5. Chapter 6, Elsevier, Amsterdam:195–237

Quesnel Y, Catalán M, Ishihara T (2009) A new global marine magnetic anomaly data set. J Geophys Res B04106. doi:10.1029/2008JB006144

Raff A, Mason R (1961) A magnetic survey off the west coast of North America, 40°N to 52.5°N. Geol Soc Am Bull 72:1259–1265

Ravat D, Hildebrand T, Roest W (2003) New way of processing near-surface magnetic data: the utility of the comprehensive model of the magnetic field. Leading Edge 22: 784–785

Ravilly M, Horen H, Perrin M, Dyment J, Gente P, Guillou H (2001) NRM intensity of altered oceanic basalts: a record of geomagnetic paleointensity variations? Geophys J Int 145:401–422

Reeves CV (1993) Limitations imposed by geomagnetic variations on high quality aeromagnetic surveys. Explor Geophys 24:115–116

Reeves CV (2005) Aeromagnetic surveys: principles, practice and interpretation. e-book published by Geosoft

Reford MS (1980) Magnetic method. Geophysics 45(11): 1640–1658

Reford MS (2006) Gradient enhancement of the total magnetic field. Leading Edge 25(1):59–66

Reford MS, Sumner JS (1964) Aeromagnetics. Geophysics XXIX (4):482–516

Reid AB (1980) Aeromagnetic survey design. Geophysics 45(5):973–976

Reigber C, Luhr H, Schwintzer P (2002) Champ mission status. Adv Space Res 30(2):129–134

Reigber C, Schwintzer P, Luhr H (1999) The champ geopotential mission. Bolletino di Geofisica Terica e Applicata 40: 285–289

Reynolds RJ, Schlinger CM (1990) Notes of discussion Group on Rock Magnetics. In: Geologic applications of modern aeromagnetic surveys. U.S. Geological Survey Bulletin 1924:99–101

Reynolds RL, Rosenbaum JG, Hudson MR, Fishman NS (1990) Rock magnetism, the distribution of magnetic minerals in the earth's crust , and aeromagnetic anomalies. In: Geologic applications of modern aeromagnetic surveys

Rice JAJ (1993) Automatic compensator for an airborne magnetic anomaly detector U.S. Patent 5182514

Ridsdill-Smith TA, Dentith MC (1999) The wavelet transform in aeromagnetic processing. Geophysics 64:1003–1013

Rigoti A, Padilha AL, Chamalaun FH, Trived NB (2000) Effects of the equatorial electrojet on aeromagnetic data acquisition. Geophysics 65(2):553–558

Ripka P (1996) Noise and stability of magnetic sensors. J Magnet Magnetic Mater 157–158:424–427

Sabaka TJ, Olsen N, Langel R (2002) A comprehensive model of the quiet-time, near-earth magnetic field: Phase 3. Geophys J Int 151(1):32–68

Sabaka TJ, Olsen N, Purucker ME (2004) Extending comprehensive models of the Earth's magnetic field with Oersted and CHAMP data. Geophys J Int 159:521–547

Sapunov V, Denisov A, Denisova O, Saveliev D (2001) Proton and Overhauser magnetometers metrology. Control Geophys Geodesy 31(1):119

Schouten H, Denham C (1979) Modelling the oceanic magnetic source layer, In: Talwani M, Harrison C, Hayes D (eds) Deep drilling results in the Atlantic Ocean: ocean crust, American Geophysics Union, series 2 Washington, DC:151–159

Seama N, Nogi Y, Isezaki N (1993) A new method for precise determination of the position and strike of magnetic boundaries using vector data of the geomagnetic anomaly field. Geophys J Int 113:155–164

Sexton JL, Hinze WJ, R.B. von Frese, Braile LW (1982) Long-wavelength aeromagnetic anomaly map of the conterminous united states. Geology 10:364–369

Shah A, Cormier M, Ryan W, Jin W, Sinton J, Bergmanis E, Carlut J, Bradley A, Yoerger D (2003) Episodic dike swarms inferred from near-bottom magnetic anomaly maps at the southern East Pacific Rise. J Geophys Res 2097. doi:10.1029/2001JB000564

Sibuet J, Srivastava S, Manatschal G (2007) Exhumed mantle-forming transitional crust in the Newfoundland-Iberia rift and associated magnetic anomalies. J Geophys Res. 112 B06105 doi:10.1029/2005JB003856

Sichler B, Hékinian R (2002) Three-dimensional inversion of marine magnetic anomalies on the equatorial Atlantic Ridge (St. Paul Fracture Zone): Delayed magnetization in a magmatically starved spreading center? J Geophys Res. 107(B12):2347 doi:10.1029/2001JB000401

Slack H, Lynch V, Langan L (1967) The geomagnetic gradiometer. Geophysics 32:877–892

Smith K (1997) Cesium optically pumped magnetometers : Basic theory of operation, Technical Report M-TR91, GEOMETRICS

Stolz R, Zasarenko V, Schulz M, Chwalla A, Fritzsch L, Meyer HG, Kostlin EO (2006) Magnetic full-tensor squid gradiometer system for geophysical applications. The Leading Edges 25(2):178–180

Sundberg K, Lundberg H (1932) Magnetism. Trans Am Inst Mining Metallurgical Eng

Supper R, De Ritis R, Motschka K, Chiappini M (2004) Aeromagnetic anomaly images of volcano and southern lipari islands (aeolian archipelago, italy). Ann Geophys 47(6):1803–1810

Talwani M, Dorman J, Worzel J, Bryan G (1966) Navigation at sea by satellite. J Geophys Res 71:5891–5902

Tarlowski C, Simonis F, Whitaker A, Milligan R (1992) The magnetic anomaly map of australia. Explor Geophys 23(2):339–342

Tarlowski C, Mc Ewin AJ, Reeves CV, Barton CE (1996) Dewarping the composite aeromagnetic anomaly map of australia using control traverses and base stations. Geophysics 61(3):696–705

Telford WM, Geldart LP, Sheriff RE (1990) Applied geophysics, 2nd edn. Cambridge University Press, Cambridge

Tivey M, Dyment J (2010) The magnetic signature of hydrothermal systems in slow spreading environments, In: Rona P, Devey C, Dyment J, Murton B (eds) Diversity of hydrothermal systems on slow spreading ocean ridges, Am Geophys Union Monogr, Series 188 Washington, ISBN 978-0-87390-470-8

Tivey M, Johnson H (2002) Crustal magnetization reveals subsurface structure of Juan de Fuca Ridge hydrothermal vent fields. Geology 30:979–982

Tivey M, Rona P, Schouten H (1993) Reduced crustal magnetization beneath the active sulfide mound, TAG Hydrothermal Field, Mid-Atlantic Ridge 26°N. Earth Planet Sci Lett 115:101–116

Tivey M, Johnson H, Fleutelot C, Hussenoeder S, Lawrence R, Waters C, Wooding B (1998a) Direct measurement of magnetic reversal polarity boundaries in a cross-section of oceanic crust. Geophys Res Lett 25:3631–3634 (1998a)

Tivey M, Johnson H, Bradley A, Yoerger D (1998b) Thickness measurements of submarine lava ows determined from near-bottom magnetic field mapping by autonomous underwater vehicle. Geophys Res Lett 25:805–808

Tohyama F, Nishio Y, Yamagishi H, Yamagami T (2007) Geomagnetic field observation using fluxgate magnetometer system onboard balloons in antarctica. Proc Schl Eng Tokai Univ., Ser. E 32:19–25

Tsetkov YA, Belkin VA, Kanonidi KD, Kharitonov AL (1995) Physico-geological interpretation of the anomalous geomagnetic field measured in the stratosphere. Phys Solid Earth (English Translation) 31(4):329–332

Urquhart W (2003) Airborne magnetic surveys: Past, present and future, Canadian Exploration Geophysics Society (KEGS).

Vacquier V, Raff A, Warren R (1961) Horizontal displacements in the oor of the northeastern Pacific Ocean. Geol Soc Am Bull 72:1251–1258

Vacquier VV (1946) Apparatus for responding to magnetic field U.S. Patent 2406870

Van Den Bossche P, Coles S, Murrell D, Madotyeni Z (2004) Maritime wreck surveys: Search for the wreck of the Dutch East India Company slave ship, MEERMIN, wrecked in 1766 in Struisbaai, South Africa, Technical Report 0182, Council for Geoscience and Iziko Museums, Project 0463

Verhoef J, Roest W, Macnab R, Arkani-Hamed J (1996) Members of the Project Team, Magnetic anomalies of the Arctic and North Atlantic Oceans and adjacent land areas, Technical Report GSC Open File 3125, parts a and b (CD-ROM and project Report), Geological Survey of Canada, Dartmouth, Nova Scotia, 225pp

Vine F, Matthews D (1963) Magnetic anomalies over oceanic ridges. Nature 4897:947–949

Waters G, Phillips G (1956) A new method of measuring the Earth's magnetic field. Geophys Prosp 4:1–9

Whitham K, Loomer EI (1957) Irregular magnetic activity in northern Canada with special reference to aeromagnetic survey problems. Geophysics 22:646–659

Whitham K, Niblett ER (1961) The diurnal problem in aeromagnetic surveying in Canada. Geophysics XXVI(2): 211–228

Whitmarsh R, Miles P (1995) Models of the development of the West Iberia rifted continental margin at 40°30′N deduced from surface and deep-tow magnetic anomalies. J Geophys Res 100(B3):3789–3806

Williams PM (1993) Aeromagnetic compensation using neural networks. Neural Comput Appl 1:207–214

Wyckoff RD (1948) The gulf airborne magnetometer. Geophysics 13:182–208

Yamamoto M, Seama N, Isezaki N (2004) Genetic Algorithm inversion of geomagnetic vector data using a 2.5-dimensional magnetic structure model. Earth Planets Space 56:217–227

Yamamoto M, Seama N, Isezaki N (2005) Geomagnetic paleointensity over 1.2 Ma from deep-tow vector magnetic data across the East Pacific Rise. Earth Planets Space 57: 465–470

Zhana YX (1994) Aeromagnetic anomalies and perspective oil traps in china. Geophysics 59(10):1492–1499

Zimmerman JE, Campbell WH (1975) Tests of cryogenic squid for geomagnetic field measurements. Geophysics 40(2): 269–264

Zonenshain LP, Verhoef J, Macnab R, Meyers H (1991) Magnetic imprints of continental accretion in the ussr and adjacent areas. EOS Trans Am Geophys Union 72(29): 305–310

Chapter 5

Instruments and Methodologies for Measurement of the Earth's Magnetic Field

Ivan Hrvoic and Lawrence R. Newitt

Abstract In modern magnetic observatories the most widely used instrument for recording magnetic field variations is the triaxial fluxgate magnetometer. For absolute observations, the declination-inclination magnetometer, in conjunction with a proton precession or an Overhauser magnetometer, is the norm. To meet the needs of users, a triaxial fluxgate must have a resolution of 0.01 nT. It must also have good temperature and long-term stability. Several sources of error can lead to degradation of the data, temperature variations and tilting of the sensors being among the most important. The declination-inclination magnetometer consists of a single-axis fluxgate sensor mounted on a nonmagnetic theodolite. With care, most sources of error can be eliminated, and an absolute accuracy of better than 0.1 arcmin is achievable. Proton precession and Overhauser magnetometers make use of the quantum-mechanical properties of protons and electrons to determine the strength of the magnetic field. The Overhauser magnetometer is rapidly supplanting the proton magnetometer (0.1 nT once per second sensitivity) because it can sample the field much more rapidly and precisely (0.01 nT once per second). Potassium magnetometers, which belong to the family of optically pumped magnetometers, are an attractive alternative to Overhauser magnetometers, especially when used in a dIdD instrument.

I. Hrvoic (✉)
GEM Systems Inc., 135 Spy Court, Markham, ON, Canada L3R 5H6
e-mail: Info@gemsys.ca

Abbreviations

AMOS	Automatic Magnetic Observatory System (Canada)
ASMO	Automatic Magnetic Observatory System (USA)
AUTODIF	Automated DIM
BMZ	Balance magnetometrique zero
CARISMA	Canadian Array for Realtime Investigations of Magnetic Activity
CCD	charge coupled device
dIdD	(delta Inclination/delta Declination)
DIM	Declination-inclination fluxgate magnetometer
DMI	Danish Meteorological Institute
EDA	Electronic Design Automation
EPR	Electron Paramagnetic Resonance
GAUSS	Geomagnetic Automated System
GPS	Global Positioning System
IAGA	International Association of Geomagnetism and Aeronomy
IGRF	International Geomagnetic Reference Field
KASMMER	Kakioka Automatic Standard Magnetometer
LEMI	The Laboratory of Electromagnetic Innovations
MACCS	Magnetometer Array for Cusp and Cleft Studies
MRI	Magnetic resonance imaging
NIM	The National Institute of Metrology (CHINA)
NIST	The National Institute of Standards and Technology (USA)
NMR	Nuclear magnetic resonance
NPL	National Physical Laboratory (U.K.)
PCs	Personal computers

M. Mandea, M. Korte (eds.), *Geomagnetic Observations and Models*, IAGA Special Sopron Book Series 5,
DOI 10.1007/978-90-481-9858-0_5, © Springer Science+Business Media B.V. 2011

PDAs	Personal Digital Assistants
ppm	proton precession magnetometer
ppm	parts per million
QHM	Quartz horizontal magnetometer
THEMIS	Time History of Events and Macroscale Interactions during Substorms
TCXO	Temperature Compensated Crystal Oscillator
TPM	Torsion photoelectric magnetometer
TMS	tetra methyl silane
UCLA	University of California at Los Angeles' fluxgate magnetometer
VNIIM	D.I. Mendeleyev Institute for Metrology (RUSSIA)

5.1 Introduction

Instruments to measure the Earth's magnetic field at magnetic observatories fall into two categories: those that measure the temporal changes in the field on a continual basis without regard to the absolute accuracy of the observation, and those that measure the absolute value of the magnetic field at an instant in time. For almost a century and a half, the photographic variometer, or magnetograph, was the primary instrument for recording temporal fluctuations in the magnetic field. Today, the triaxial fluxgate is the instrument most widely used for this task, although some observatories use a suspended magnet system that produces an electrical output. A wide variety of instruments have been used to measure the absolute value of the magnetic field: the induction magnetometer, the QHM (quartz horizontal magnetometer), the BMZ (balance magnétométrique zéro), the declinometer, the declination-inclination magnetometer (DIM), and scalar magnetometers such as the proton precession magnetometer and the Overhauser magnetometer. Although the QHM and declinometer are still in use, they have been replaced by the DIM and ppm or Overhauser magnetometer at most observatories.

Magnetometers, in particular the fluxgate and scalar, have a wide variety of uses outside the observatory environment. The ppm/Overhauser is used to calibrate other magnetometers. It is an essential tool for mineral and oil exploration, and has scientific applications in the fields of volcanology and archeology. Numerous arrays and chains of triaxial fluxgates have been deployed for studying the rapid variation magnetic field and solar terrestrial interactions. Arrays of fluxgates, often in conjunction with telluric sensors for measuring the electric field, have been used for studies of crustal conductivity. Fluxgates and scalar magnetometers have been installed aboard ships, aircraft, and satellites for mapping the magnetic field near and above the Earth's surface. Fluxgate and scalar magnetometers also have a wide range of non-scientific applications (Gordon and Brown, 1972). These include: submarine detection, weapons and vehicle detection, navigation, non-destructive testing of materials and many more.

In this chapter we will concentrate primarily, but not exclusively, on those instruments that are used in a modern observatory setting: the triaxial fluxgate, the DIM and various forms of scalar magnetometers. Fluxgate magnetometers suitable for observatory use are also suitable for magnetometer arrays. We will describe the basic theory behind each instrument, its mode of operation, and the development of ancillary equipment for storage and telemetry, its relative and/or absolute accuracy and the related sources of error. Short descriptions of other magnetometers, that were once popular or that may be popular in the future, are also given.

5.2 Fluxgate Magnetometer

Since the invention of the fluxgate magnetometer in the 1930s, more than 100 different variations of the instrument have been designed, using different core configurations and different core materials (Jankowski and Sucksdorff, 1996). In part, this is a reflection of the myriad of commercial and scientific uses to which fluxgates have been put, as detailed above. However, for observatory use, instruments are needed that have both high sensitivity and good long and short term stability, as discussed in the next section. Only two designs are currently capable of meeting these requirements: those using ring core sensors and those using double core sensors. (A third design, the so-called race-track sensor, is intermediate between these two). The fluxgate mechanism is discussed in Section 5.2.2.

5.2.1 Instrument Standards and Sources of Error

In 1986, the first IAGA Workshop on Magnetic Observatory Instruments was held in Ottawa. Of the 27 instruments tested or exhibited at the workshop, seven were triaxial fluxgates (Coles, 1988). Reading the results of the comparison between instruments (Coles and Trigg, 1988), leads us to make the following observations:

1. Do not always believe the manufacturer's specifications, especially the temperature coefficient.
2. The sensitivity and noise of most instruments were approximately, 5–10 μV/nT and 0.1 nT respectively. These were almost identical to values given by Stuart (1972) in his review of magnetometry circa 1970. The one exception was the Narod ring core fluxgate whose low noise performance enabled a resolution of 0.01 nT (Narod, 1988).
3. Thermal and mechanical stability were major problems (Coles and Trigg, 1988). Again, there had been little apparent progress since Stuart (1972).

One result of the Workshop was the development of specifications for an ideal observatory variometer (Trigg, 1988). The consensus reached by those at the workshop is given in Table 5.1 Also shown in the table are current INTERMAGNET standards, denoted by table footnote a (St-Louis, 2008).

Note that no mention is made of the absolute accuracy of the system. The magnetometer is considered a variometer. INTERMAGNET requires an absolute accuracy of 5 nT in definitive data, that is, in data that are corrected by adding baseline values obtained from absolute observations. Stability is an important factor for both relative accuracy and absolute accuracy. (See Appendix 1 for a discussion of baselines and absolute and relative accuracy.)

The geomagnetic time spectrum spans over twenty orders of magnitude, from millions of years to fractions of a second (Constable 2007). Traditionally, magnetic observatories have been concerned with the part of the spectrum that includes secular, solar cycle, diurnal and magnetic storm variations. These cover eight orders of magnitude, from centuries to minutes. The magnetic variations of the spectrum in this range are relatively large, from about 1 nT at 1-min to about 1000 nT at about 1000 years. Relatively noisy magnetometers with low resolution (~0.1 nT) are adequate for recording this part of the spectrum.

There is currently a great deal of interest in the space science community in studying fluctuations in the one-second to 1-min range. Most of the major magnetometer chains and arrays (THEMIS, MACCS, CARISMA to name only a few) now record data at one-second intervals. However, the amplitude of the geomagnetic spectrum decreases by two orders of magnitude in the band between 1-min and 1 s, which means that magnetometers are required whose

Table 5.1 Specifications of an ideal magnetometer (after Trigg, 1988)

Rugged	Mean time before failure 24 months	Passband	DC to 1 Hz DC to 0.1 Hz[a]
Reliable	mean time to repair 1 day	Noise	0.03 nT in passband
Protected against	lightning, humidity, RF interference	Linearity	0.1% at full scale
Power	<100 W, uninterruptible	Timebase	1 s month^{-1} 5 s month^{-1} [a]
Resolution	0.1 nT 0.1 nT [a]	Sampling rate	10 Hz
Dynamic range	>±3000 nT (8000 high latitude 6000 elsewhere) [a]	Measurement interval	5 s 1.0 s [a]
Stability	0.25 nT per month 5 nT per year [a]	Temperature coefficient	Sensor <0.1 nT/°C Console <0.1 nT/°C 0.25nT/°C [a]
3 component sensor construction	orthogonal within ±30′ Stable to 0.3″ /month 0.3″/°C	Tilt sensors	Resolve 1″ (every 10 min) Stability 1″/month

[a] denotes INTERMAGNET standards

noise characteristics and resolution (sensitivity) exceed the current standards. In 2005, INTERMAGNET conducted a survey of users of magnetic observatory data to ascertain the required resolution and timing accuracy of one-second data (Chulliat et al. 2009). Although the response to the survey was not large, there appears to be a consensus that a resolution of 0.01 nT and a timing accuracy of 10 ms meet the current needs of the scientific community. These requirements were adopted by INTERMAGNET with one important revision. The resolution requirement was revised to 1 pT. (Chulliat et al. 2009). INTERMAGNET is currently working towards developing standards, based on these requirements, for the recording of one-second data at its observatories.

5.2.2 Fluxgate Mechanism

A fluxgate magnetometer is a device for measuring magnetic field by utilizing the non-linear characteristics of ferromagnetic materials in the sensing elements (Aschenbrenner and Goubau, 1936). All fluxgate sensors use cores with high magnetic permeability that serve to concentrate the magnetic field to be measured (Evans, 2006). We will describe the operation of a sensor with a linear twin core (Fig. 5.1). A winding through which the excitation current is applied, is placed around each core. In a twin core sensor, the cores are wound so that they are excited in opposite directions. The excitation current must be large enough to drive the cores into saturation; typically, currents an order of magnitude larger than theoretically necessary are used. The output signal is obtained from a second winding, that encircles both cores.

When the core is not saturated (the excitation current, I, is zero), the core's relative permeability, μ_r, is maximum; this concentrates the ambient field within the core, producing a magnetic flux, Φ, that is μ_r times larger than the field in a vacuum. When a current I, is fed into the winding it creates a magnetic field, H_s, that is strong enough to saturate the core. The permeability becomes close to that of a vacuum and the flux collapses. It recovers during the next half cycle of the excitation signal, only to collapse again when the core saturates. The sense, or pick-up coil, detects these flux changes, which occur at twice the frequency of the excitation signal since there are two flux collapses during each cycle. In the absence of an external field the saturations are symmetrical and the sensor coil will pick up only odd harmonics. The presence of an external magnetic field (to be measured) disturbs this symmetry creating even harmonics, the second harmonic being dominant. Even harmonics are a measure of the applied magnetic field. In general, the sense coil will pick up all harmonics. This can be problematic since the odd harmonics (generated by the excitation current) are much larger than the even ones Using a two core sensor (Fig. 5.1), in which the excitation phase is oppositely directed in each core, solves this problem since the induced voltage produced by the excitation winding is cancelled by the phase reversal. This also holds true for ring core and racetrack sensors.

The signal from the sense coil is fed to a phase sensitive detector referenced to the second harmonic.

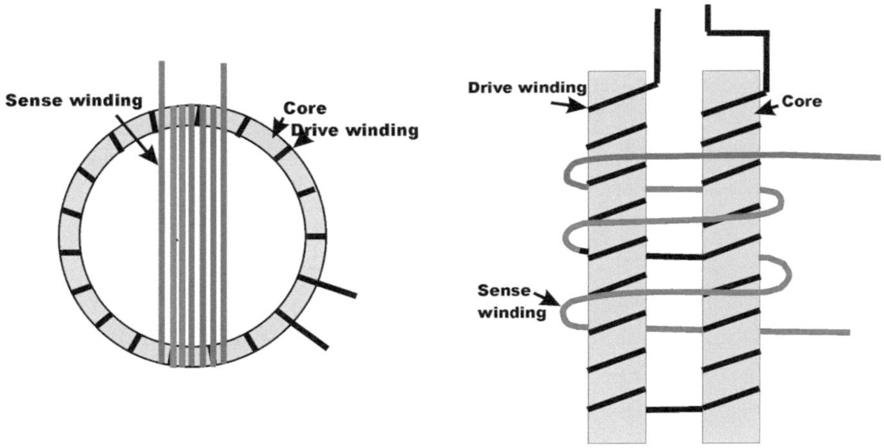

Fig. 5.1 Ring core sensor on the *left*, twin core sensor on the *right*

Because fluxgate sensors work best in a low field environment (Stuart, 1972), most fluxgate magnetometers use negative feedback so that the sensor essentially operates as a null detector. To raise the precision of the measurements, the sensor may be placed inside bias coils that cancel most of the Earth's magnetic field (Fig. 5.2).

Primdahl (1979) derived an equation for the voltage output of the sense coil in terms of changes in the permeability of the core μ_r. The Earth's magnetic induction component, B, induced along the core axis, produces a magnetic flux, $\Phi = BA$, in a core of cross-sectional area A. As described above, when the permeability, μ_r, changes, the flux changes, inducing a voltage in the sense coil.

$$V_s = nA\frac{dB}{dt} \tag{5.1}$$

where n is the number of turns in the sense coil. Inside the core, the field is given by

$$B = \frac{\mu_r B_e}{[1 + D(\mu_r - 1)]} \tag{5.2}$$

Fig. 5.2 Triaxial fluxgate sensor with bias coils which cancel out the Earth's magnetic field so that the sensors can operate in a low field environment

where B_e is the external magnetic induction. From these expressions, Primdahl (1979) derived the basic fluxgate equation

$$V_s = \frac{nAB_e(1 - D)\left(\frac{d\mu_r}{dt}\right)}{[1 + D(\mu_r - 1)]^2} \tag{5.3}$$

D is called the demagnetizing factor. It can be seen from this equation that the output voltage is produced by the change in permeability and that the demagnetizing factor plays an important role in determining the signal size. The demagnetizing factor is highly dependent on the shape and size of the core.

Stuart (1972), in his exhaustive review of the state of magnetometry circa 1970, pointed out several engineering problems that had to be overcome to achieve a fluxgate magnetometer with the precision, accuracy, and stability necessary for observatory deployment. These include the need to eliminate other harmonics without causing phase distortion of the second harmonic, and the need to ensure that the fundamental excitation voltage should not contain a second harmonic component. Gordon and Brown (1972) state bluntly: "Properly designed optimized electronics are axiomatic for best low-level fluxgate response." Other difficulties that must be addressed by the manufacturer of an observatory-quality fluxgate include the following:

1. The requirements for a very low noise sensor.
2. The presence of zero offset, which means that the sensor does not give a zero output in a zero field.
3. The variability in the output due to changes in temperature. Temperature affects the instrument in several ways. The coil characteristics may be dependent on temperature; temperature differences may cause strain on the mechanical system; temperature may also affect electronic components.

Thermal stability is of major concern to those who use fluxgate magnetometers in an observatory setting. To achieve an absolute accuracy of better than 5 nT for each datum requires both frequent calibrations as discussed in Part 5.3, and a magnetometer that has good long-term stability. We should add that in addition to thermal stability, mechanical stability is also required—the sensor must not move (tilt). The manufacturers of magnetometers attempt to achieve thermal stability in a variety of ways. In the Narod magnetometer sensor, for example, all sensor materials are chosen to have similar thermal expansion coefficients; the sensor also has a temperature feedback loop (Narod and Bennest, 1990). Similarly, sensors produced by the LEMI company are fabricated from ceramic glass which has a near zero thermal expansion factor (Korepanov, 2006). In the sensor of the widely used UCLA magnetometer, the ring core is placed in a hermetically sealed container filled with paraffin oil (Russell et al. 2008), and on deployment in the field the sensor is buried. According to the authors, this reduces the effect of temperature to 2 nT seasonally and 0.1 nT diurnally. Narod and Bennest (1990) claim a thermal stability of 0.1 nT per degree for the sensor and 0.2 nT per degree for the electronics. The LEMI sensors have a thermal stability of less than 0.2 nT per degree. The temperature dependency of the LEMI sensors is also linear, so that corrections for temperature may be possible, since the instruments have thermal sensors imbedded in the sensor and the electronics. The problem of temperature can be circumvented by keeping both sensor and electronics in a thermostatically controlled, constant temperature environment. The effect of temperature is reduced to an amount that is less than the error in the absolute observations. It has been found, however, that the on/off cycling of some temperature controllers produces noise in the data. Therefore, any thermostatically controlled system should be thoroughly tested before it is installed in an observatory.

Tilt is another problem that can seriously affect the output of a fluxgate magnetometer. If the sensor is mounted on a pillar that moves for some reason—the freeze-thaw cycle, the wet-dry cycle, changes in temperature—then the orientation of the fluxgate sensor assembly will change, and the sensors will no longer measure the three magnetic field components that they are supposed to measure. This is not a serious problem if the tilting progresses slowly and remains small. It is then manifested as a slow drift in one or more of the magnetic field components for which corrections can be applied by adding baseline values derived from absolute observations. Absolute observations are made once per week at most observatories. If tilting progresses rapidly and non-linearly, so that significant changes occur on a time scale shorter than 1 week, aliasing can occur, which means it is impossible to obtain the true value of the magnetic field components on a minute-to-minute basis. One way to eliminate the problem of tilt is to place the sensor assembly in a tilt-compensating suspension. Trigg and Olsen (1990) describe the suspension developed for use with the Narod magnetometer sensor (Fig. 5.3). Rasmussen and Kring Lauridsen (1990) describe the suspension developed for the larger DMI (Danish Meteorological Institute) magnetometer sensor. Note that tilt-compensating suspensions cannot compensate for a rotation of the sensor due to twisting of the pillar.

5.2.3 Data Collection and Telemetry

Early fluxgate magnetometers existed in an analogue world. They provided a voltage that produced an analog output on a chart recorder, so the processing and use of the information were virtually identical

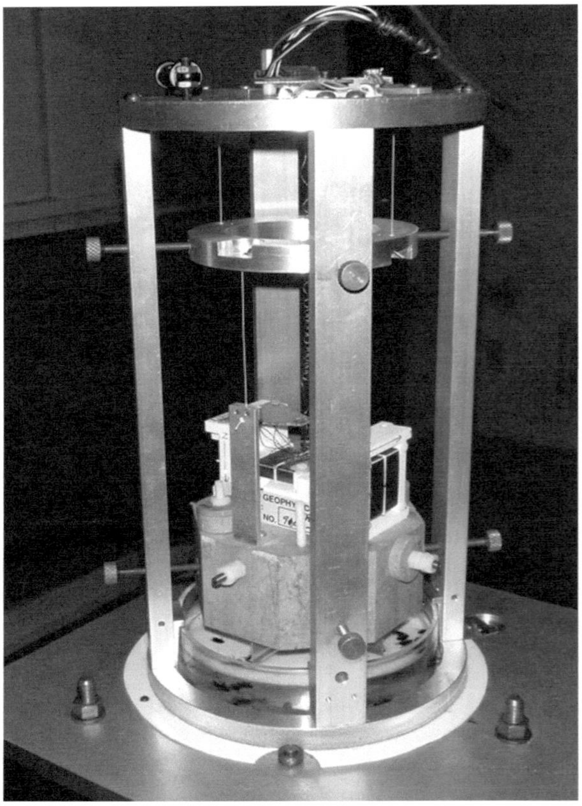

Fig. 5.3 Narod ring core sensors mounted in a tilt-reducing suspension

to those of a photographic magnetogram. When the computer age arrived, a few forward-thinkers saw the potential in a magnetometer-computer partnership, which meant recording data digitally, or alternatively digitizing analog records afterwards. Alldredge and Saldukas (1964) described one of the first magnetometer systems designed with a digital output that recorded on magnetic tape. The system, called ASMO, was also capable of transmitting data over a phone line to a remote receiving centre. In 1969, the first AMOS (Automatic Magnetic Observatory System) was deployed in Canada (Fig. 5.4). It was designed to record digitally on 200 bpi tape. The system also featured an innovative telephone verification system that enabled an operator to communicate with the AMOS and to diagnose system operating problems remotely (Delaurier et al. 1974).

Digital recording was not for the faint of heart. Delaurier et al. (1974) wrote: "Such problems as power failures, electronic device breakdowns, mechanical troubles and other unpredictable difficulties can cause data gaps, bad coding, parity errors and irregular physical record length." Thus, it became necessary to develop a suite of editing programs to deal with a class of errors that had never existed in the analogue era. But there was no going back since scientists had already discovered that having digital data made it possible to use a wide range of analytical tools that enabled them to extract much more information than they were able to obtain from photographic records or hourly mean tables.

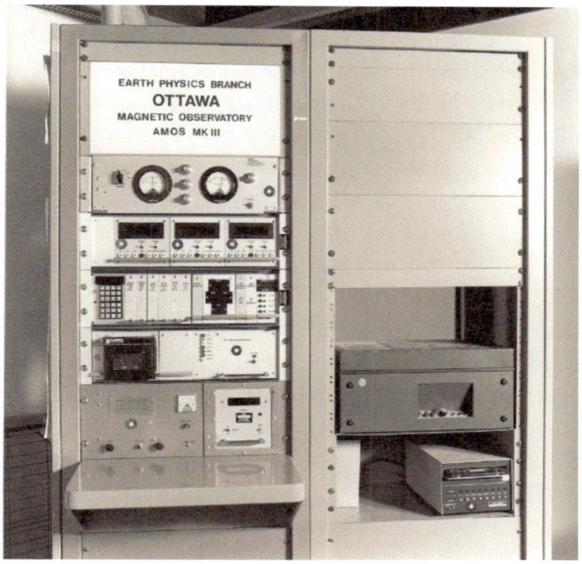

Fig. 5.4 The AMOS Mk 3 is a second-generation automated observatory system deployed at Canadian observatories during the 1980s

It soon became apparent that magnetic tape was not a suitable medium for recording geomagnetic data, especially at remote observatories where there was little control over the cleanliness of the environment in which the tape drive was located. Thus, observatory operators were quick to embrace alternative storage devices. Many different types of data collection platforms have been developed or tried: personal computers (PCs), personal digital assistants, or PDAs (Merenyi and Hegymegi, 2005), WORM (write once read many) drives, Zip drives, and many others. Important criteria for any data acquisition system are robustness and stability, low power consumption, and a user friendly interface.

At many observatories, the PC has become the centre of the magnetometer system. All magnetometers feed data into the PC, which controls the operation of each of them. Data can be stored on the PC's hard disk as well as on peripheral storage devices. The PC also controls the telemetry of data via satellite or the internet (see Fig. 5.5). To achieve the timing accuracy that users require (10 ms), most observatories use a timing system based on the GPS (global positioning system).

5.3 Declination-Inclination Fluxgate Magnetometer

The declination-inclination fluxgate magnetometer (commonly called the DI-flux or DIM) is the instrument of choice for doing absolute observations of the magnetic field. Although INTERMAGNET does not forbid the use of other instruments, the technical manual does state that a DIM and a proton precession/Overhauser magnetometer are an increasingly popular combination (St-Louis, 2008). In fact, all INTERMAGNET observatories use the DIM/Overhauser combination as their primary absolute instruments. Although Jankowski and Sucksdorff (1996) do describe other instruments in the *Guide for Magnetic Measurements and Observatory Practice*, they state that the DIM combined with a ppm "is the recommended pair of absolute instruments". At

Fig. 5.5 Instruments typical of a modern magnetic observatory. Clockwise from *lower right*: computer for data storage and system control. Fluxgate sensor in a tilt reducing suspension; Overhauser magnetometer and sensor; fluxgate magnetometer electronics; satellite transmitter

the XII IAGA Workshop on Geomagnetic Observatory Instruments, Data Acquisition and Processing held at Belsk in 2006, all 29 instruments that took part in the instrument comparison session were DIMs (Reda and Neska, 2007).

However, some observatories still use classical declinometers and QHMs, (quartz horizontal magnetometers) so we describe them briefly in a later section. Those who want more details can find them in Jankowski and Sucksdorff (1996) or Wienert (1970).

Although the use of the DIM may be almost universal today, its acceptance by the magnetic observatory community was slow in coming. An early version of the instrument was developed by Paul Serson who used it in 1947 and 1948 in a field survey to determine the position of the North Magnetic Pole (Serson and Hannaford, 1956). Its first use in a Canadian observatory dates from 1948, and by 1970 it was in use at all Canadian observatories. However, the instrument is not even mentioned in Wienert's (1970) *Notes on Geomagnetic Observatory and Survey Practice*. In 1978, the Institut de Physique du Globe in France developed a version of the instrument for use at its high southern latitude observatories (Bitterly et al. 1984). The motivation for this development was the same as that of Serson. The use of classical instruments, such as QHMs, becomes extremely difficult at high latitudes because of the weakness of the horizontal component of the magnetic field. After they had made 127 comparisons to the standard instruments at Chambon-la-Forêt Observatory, Bitterly et al. (1984) concluded that the instrument was stable, with no apparent long-term drift and that its accuracy was better than 5″ of arc for both D and I.

5.3.1 Observing Procedure

A DIM consists of a fluxgate sensor mounted on the telescope of a magnetically clean theodolite and the associated electronics (Fig. 5.6). The fluxgate sensor is mounted with its magnetic axis parallel to the axis of the theodolite's telescope. In practice, there will always be a misalignment which results in a collimation error. Fortunately, this, as well as most other errors, can be eliminated by proper observational procedure.

There are two methods of observation possible: the null method and the residual method. Both methods require that the sensor be placed in four positions that cancel the collimation and offset errors. Both methods also require that total intensity be recorded simultaneously with the observations of D and I. We shall describe the null method first. The telescope is set in the horizontal plane and the alidade is rotated until the output of the magnetometer is zero. This indicates that the sensor is aligned perpendicular to the horizontal component of the magnetic field. The angle at which this occurs is noted. The alidade is then rotated roughly 180° and finely adjusted until zero output is achieved

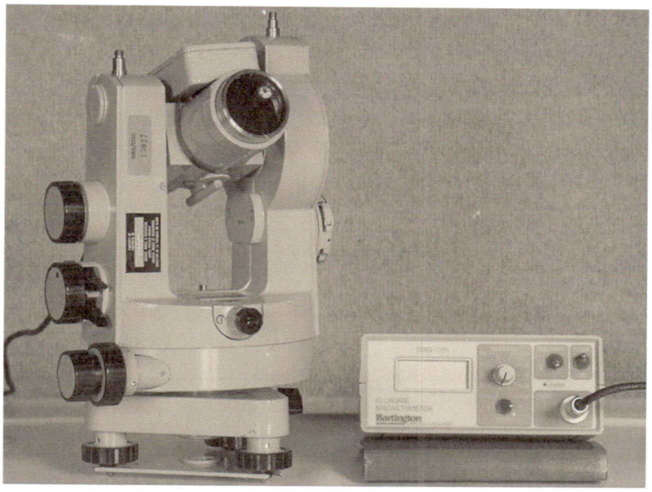

Fig. 5.6 A declination inclination magnetometer. The instrument shown here consists of a Zeiss-Jena 010 theodolite and a Bartingrton 01H single axis fluxgate

again. Next, the telescope is inverted and two more positions at which the output is zero are found. The average of these four values (call it *A*) gives the direction of the horizontal magnetic field in some arbitrary frame of reference. To get the declination (D) we must compare *A* to the direction of true north. At an observatory this is done by sighting a reference mark (*B*) whose true bearing is known (*Az*). Then

$$D = A - (B - Az) \tag{5.4}$$

To measure inclination, the telescope is first aligned in the magnetic meridian, the direction of which is usually obtained from the previous declination observation. (The inclination is actually quite insensitive to misalignment in the meridian, so an approximate value is often sufficient, as discussed in Section 5.3.2) Then, two positions at which the output is zero are found, one with the sensor above the telescope, one with the sensor below. The alidade is rotated exactly 180°, and the positions of two more nulls are recorded. The inclination is derived from these four values.

The residual method follows the same basic procedure except that the position of the telescope is not adjusted to give a zero output. Instead, it is set to some convenient value near the null position and the value of the magnetic field component is read off the magnetometer's meter.

Since the magnetic field will vary over the length of time required to observe in all four positions, it is important to null the meter or read the residual in sync with the observatory's triaxial fluxgate. This will allow changes in the field to be taken into account during post-observation processing. To compute baselines for components of the magnetic field other than *D* and *I*, values of total intensity are obtained from the observatory's ppm or Overhauser magnetometer.

Detailed instructions for observing with a DIM in an observatory setting are given by Jankowski and Sucksdorff (1996). Newitt et al. (1996) give instructions for the use of the instrument in a field setting. Both of these Guides may be obtained by contacting the Secretary-General of IAGA[1]. The theory behind the operation of the instrument is given by Kring Lauridsen (1985) and Kerridge (1988). The former deals with the residual method; the latter deals with the null method.

5.3.2 Instrumental Accuracy and Sources of Error

One important function of the first International Workshop on Magnetic Observatory Instruments was a comparison of absolute instruments. Six of the nine instruments so compared were DIMs (Newitt et al. 1988). The most obvious, and the best, way to compare

[1] Secretary General of IAGA's email is iaga_sg@gfz-potsdam.de

two instruments is to make simultaneous measurements on two pillars. Since an observation contains an error component that is dependent on the observer, two sets of observations should be made with the observers exchanging places after the first set. The differences in *D* and in *I* between the two pillars must be known precisely. If they are not known, then the instruments must be interchanged, and another two sets of observations carried out.

This method is obviously not practical when the number of instruments is large. As an alternative, each set of observations from each instrument can be used to calculate spot baseline values, (as discussed in Appendix 1) for the observatory's triaxial fluxgate magnetometer. The baseline values are then compared to determine differences in the DIMs. This method of comparing instruments is dependent on one major assumption: the observatory fluxgate magnetometer must be stable or at worst must vary only very slowly with time. At the first IAGA Workshop, this assumption was found to be invalid. Newitt et al. (1988) wrote: "it is obvious that the Ottawa AMOS does not have sufficient temperature stability to allow comparisons with an accuracy of a fraction of a nanotesla." Nevertheless, "under adverse conditions baselines can be determined with an accuracy of 1 to 2 nT." The authors also felt that an accuracy of better than 1 nT would be achievable under more favourable observing conditions.

Comparisons of absolute instruments have been made at all subsequent Observatory Workshops. Since the fifth workshop, in 1992, the instruments have been exclusively DIMs. The results of these comparisons have been published in the proceedings of each workshop, but it is difficult to compare the results from one workshop with those of another since the statistics were seldom computed in the same manner. However, the published results of the workshops indicate that a skilled observer using a magnetically clean instrument can obtain an absolute accuracy of a few arc-seconds, or roughly 1 nT.

Several factors can contribute to the error in an observation made with a DIM. Most of these can be reduced to zero by proper procedure and care. Potentially serious sources of error are discussed below:

1. Magnetization in the theodolite can lead to large errors, but it is well-known in the observatory community that theodolites said to be non-magnetic must nevertheless be checked. This is a source of error that can and should be totally eliminated.
2. Movement of the sensor with respect to the telescope will result in an error. This is a problem that can be detected by routinely calculating the collimation angles from the four D readings or the four I readings (see, for example, Jankowski and Sucksdorff, 1996), so the problem can be detected easily and fixed.
3. Large vertical magnetic gradients are a source of error, especially if the gradients are non-linear. The theories developed by Kring Lauridsen (1985) and Kerridge (1988) assume that the magnetic field is the same regardless of the position of the sensor. If there is a vertical gradient, the field will be different in the sensor up and down positions. Experiment has shown that if the gradient in the vertical field is linear, the observational procedure will eliminate its effect. However, this will not be the case for non-linear gradients. It is normal practice to choose a site for a magnetic observatory with low magnetic gradients; Jankowski and Sucksdorff (1996) recommend gradients be less than 1 nT/m, both horizontally and vertically. For such observatories, gradient errors are a non-issue. However, perfect sites cannot always be found, and observatories have and must be built in areas where the gradient is higher than desirable. Observations made with a DIM at such observatories may contain an error of unknown size due to gradients. Some field or repeat station observations made with a DIM are also likely to contain gradient errors.
4. The measurement of declination with a DIM requires referencing the observed value to a known azimuth. An error in the azimuth will result in a systematic error in the declination. In the field, azimuth has traditionally been determined by sun observations, with an accuracy of roughly 1 arcmin (Newitt et al. 1996). North-seeking gyros have also been used (see, for example, Kerridge, 1984), and the use of GPS is now becoming quite common. At a magnetic observatory, a professional surveyor can be brought in to determine the azimuth of the reference mark to a very high degree of accuracy, so this should not be a source of error.
5. There is a very real potential for error when sighting the reference mark. A large temperature contrast between the observatory building and the outside

will cause an apparent erratic motion of the reference mark when viewed through an open window. Viewing through glass can lead to a systematic error due to the index of refraction of the glass, unless the sight line is at right angles to the window. The human factor also comes into play here. In a test carried out at Ottawa Magnetic Observatory, three observers made a series of sightings on the azimuth and noted the angle. Differences in the angles recorded by the observers were as large as 12 arcsec.

6. Failure to set the telescope in the magnetic meridian will introduce an error in inclination. However, both Kerridge (1988) and Kring Lauridsen (1985) state that the positioning is not very critical. Coles (1985a) worked out an analytical expression for computing the true inclination when the theodolite is not aligned with the magnetic meridian:

$$\cos^2 I = \frac{\cos^2 I'}{\cos^2(D - D') + \cos^2 I' \times \sin^2(D - D')} \tag{5.5}$$

where I is the true inclination I' is the observed inclination $D - D'$ is the angle between the true and assumed magnetic meridian.

Table 5.2 gives errors for a few values of I' and $D - D'$. For values of inclination typical of Europe, (55° to 70°) aligning the telescope to within 5′ of the true magnetic meridian will lead to errors in inclination of about 1 or 2 arcsec.

7. Improper leveling of the theodolite will lead to errors in declination. This is another error that is completely preventable if a theodolite with a gravity-oriented vertical scale is used. Even if the base of the theodolite is slightly off-level (by less than 4 arcmin) the vertical scale will indicate the true angle of the telescope relative to the horizontal. The telescope can then be placed in the horizontal by setting the vertical scale to exactly 90° or 270° before each reading. If a theodolite without this feature is used, this source of error becomes much more important. Coles (1985b) has developed analytical expressions for the leveling error. These lead to the following rule of thumb: The error in declination is approximately equal to four times the leveling error.

8. Magnetic disturbances are another potential source of error. In theory, both the null and the residual methods should be immune to the effects of disturbances because readings are synchronized with the sampling cadence of the variometer. In practice, the ability to null the instrument or to read the display with a timing error of less than 1 s depends on the frequency content of the magnetic field variations and the skill of the observer. It is always a good idea to avoid observing during disturbed periods. However, at high latitudes this is almost impossible. In such situations, the accuracy of the baselines determined from observations can be improved by taking several observations.

Table 5.2 Error in inclination when telescope is not set in the magnetic meridian

Inclination (degrees)	Azimuth error (degrees)	Inclination error (min)
85	1	0.05
85	10	4.60
70	1	0.17
70	5	4.20
40	1	0.26
40	2	1.03
40	5	6.45
10	1	0.09
10	2	0.36
10	5	2.24

5.4 Scalar (Quantum) Magnetometers

Scalar magnetometry is an offspring of nuclear magnetic resonance (NMR) and electron paramagnetic resonance. Development in this field goes back to the early twentieth century. Studies of the then newly discovered spin of electrons and some nuclei led to NMR spectroscopy, which allowed scientists to decipher structural formulae of complex chemicals, follow chemical/physical processes etc. NMR experiments are done in artificial, strong, and well known magnetic fields. Powerful, medical tools, based on NMR, such as magnetic resonance imaging (MRI) enable us to see the inner details of the human body. MRI has many applications a cancer detection is one of them.

Scalar magnetometers reverse the above experiments. Using a chemical or an elemental vapour of known composition for the sensor enables measurement of the applied magnetic field. Along with nuclear magnetic resonance, scalar magnetometers have been a phenomenal success. First of all, they allow measurements to be made while in motion since the measurements are only very weakly dependent on the sensor orientation. Their unsurpassed absolute accuracy is in the parts per million range. Sensitivities have reached the fT (femtotesla, 10^{-15} T) range with some claims that are orders or magnitude better.

5.4.1 Background Physics

Quantum magnetometry is based on the spin of subatomic particles: nuclei, usually protons, and unpaired valence electrons (Abragam 1961, Slichter 1963, Kudryavtsev and Linert 1996).

Magnetic dipoles are produced by the spin of charged particles precessing around the magnetic field direction. The precession has an angular frequency (Larmor frequency) ω_0

$$\omega_0 = \gamma_n \mathbf{B} \tag{5.6}$$

γ_n is a gyromagnetic constant (not always a constant) and **B** is a magnetic induction or flux density.

Scalar magnetometers measure magnetic induction **B** and not magnetic field **H**. Units of measurement (nanotesla) are units of **B** and not **H**. However **B** and **H** are in vacuum or air tied by a constant μ_o:

$$\mathbf{B} = \mu_o \mathbf{H} \; \mu_o = 4\pi \; 10^{-7} \frac{\text{Vs}}{\text{Am}} \tag{5.7}$$

The gyromagnetic constant, γ_n, is well known only for protons in water (IAGA recommendation: http://www.iugg.org/IAGA/iaga_pages/pubs_prods/value.htm).

$$\gamma_p = 0.2675153362 \; \gamma_p/2\pi = 0.0425763881 \tag{5.8}$$

This precision makes it possible to measure **B** with high sensitivity and accuracy depending on the spectral line width (or length of decay time T_2, see below), the value of γ_n and the signal/noise ratio of the precession frequency signal.

Spinning dipoles orient themselves in the applied magnetic field creating a weak nuclear or electron paramagnetism.

The macroscopic magnetization due to the polarized particles is (Abragam 1961):

$$M = \frac{N\gamma_n^2 h^2/4\pi^2}{4kT\mu_o} B \tag{5.9}$$

N is the number of particles; μ_0 is the magnetic permeability of vacuum; T is the absolute temperature, h and k are constants. Magnetization is collinear with the magnetic field direction and proportional to the number of particles in the sensor, the square of the gyromagnetic constant, and the applied magnetic induction, and is inversely proportional to the absolute temperature. The dynamics of magnetization is described by two time constants: T_1 and T_2. Placed in a magnetic induction, **B**, the magnetization will reach equilibrium exponentially with time constant T_1; turned 90° away from the direction of **B**, the magnetization **M** will precess around it, its amplitude decaying with the time constant T_2:

T_2 may be shortened by inhomogeneity in the magnetic field.

By irradiating the assembly of spins with a magnetic field of Larmor frequency, absorption of energy by particles at the lower energy level can equalize the two populations. This is called saturation. Saturation will obviously eliminate the magnetization M.

5.4.1.1 Polarization

Magnetization M due to the spin of protons/electrons is just too small to produce detectable signals. To improve this, one needs to increase M by polarizing the particles. From Eq. (5.9) one can do that four different ways:

a) by reducing the absolute temperature to a few degrees Kelvin
b) by placing the sensor in a strong "polarization" field for a time interval comparable with T_1
c) by increasing the sensor volume so that there are more particles (N)
d) selecting particles with higher gyromagnetic constant γ_n directly or indirectly

There is one additional way:

e) by the optical pumping of valence electrons so as to manipulate the misbalance of populations of the two energy levels (Alexandrov, Bonch-Bruevich 1992, Happer 1972)

Of these five methods, (a) is impractical and (c) has an easily realized practical limit. (b) An auxiliary DC polarization magnetic field is used in proton precession magnetometers. Fields of a few hundred Gauss are created by sending a polarizing current of some fraction of an Ampere through the sensor coil. After polarization, the same coil serves as the pick-up coil for the precession signal. Polarization must be carried out at approximately right angles to the ambient field. Once the polarization field is removed, the newly formed magnetization will find itself in the plane of precession. It will precess and decay to thermal equilibrium (i.e., in noise) with the time constant T_2. T_2 determines the time interval available for measuring the precession frequency. In liquids and gases T_2 may reach several seconds, while in solids it is milliseconds. This is the reason nuclear (proton and Overhauser) magnetometers use liquid sensors while optically pumped use vapours. (d) Overhauser effect magnetometers deal with a mixture of protons and unpaired electrons in the so called free radicals (Kurreck, et al. 1988).

Polarization of unpaired electrons in thermal equilibrium is about 660 times greater than that of protons. By placing electrons in local fields of Nitrogen nuclei in the molecules of nitroxide free radicals, this ratio is increased to over 30,000. Although only part of this polarization can be transferred to protons, gains in polarization of the protons in thermal equilibrium of over 1000 times are possible. Transfer of electron thermal equilibrium polarization to protons happens when we saturate the electron spectral line by irradiating the sample by an appropriate RF magnetic field of Larmor frequency of free electrons. Transfer dynamics is again determined by the time constant, T_1. The transferred polarization is colinear with the magnetic field direction, and is static. To create a precession signal, one applies a strong, short magnetic pulse (90° or $\pi/2$ pulse) to rotate the magnetization into the plane of precession. A steady state precession signal can also be achieved by applying a weak rotating magnetic field of the proton precession frequency in the plane of precession. (e) Optical pumping deals with vapours of elements in the first column of the periodic table of chemical elements (alkali metals) namely Potassium, Rubidium and Cesium (Lithium and Sodium are chemically too active). These elements have one, unpaired electron in the valence shell and in a vapour form they are ready for electron paramagnetic resonance (EPR). Helium 4, which is in the second column of the periodic table, has two electrons in the valence shell. It can be "prepared" for EPR by applying a weak discharge that lifts one of the two electrons into a metastable state, but only for a very short time, (few microseconds). The return of the electron from the metastable state eliminates the atom from the process. As a result of this depolarization, the spectral line of Helium 4 is wide, some 70 nT.

All alkali metals need to be heated in a vacuum to some 45–55°C to achieve the proper density of vapour.

5.4.2 Proton Precession Magnetometer

The proton magnetometer was the first of the scalar magnetometers. Packard and Varian (1954) patented the method and in the 1960s newly formed Geometrics brought out an instrument that read the Earth's magnetic field to about 1 nT sensitivity. Barringer and Scintrex followed, all with hard wired electronics. EDA ventured into geophysics and brought out the first computerized proton magnetometer. Geometrics, Scintrex and GEM Systems followed suit, each with some improvements. Sensitivities of 1 nT or 0.5 nT were standard. Proton magnetometers were first used in magnetic observatories in the late 1960s.

The proton precession magnetometer was the standard scalar magnetometer up to about the mid 1980s (pre-Overhauser times). In slow mode, with readings in three seconds or so, a sensitivity of 0.1 nT can be achieved. With faster readings (one second is now becoming a standard at INTERMAGNET observatories) perhaps 0.25 nT is achievable with about 0.5 s polarization and 0.5 s reading time.

For the highest absolute accuracy and long term stability, the frequency reference of the Larmor frequency counter must be of adequate stability. A calibrated Temperature Compensated Crystal Oscillator (TCXO), or GPS timing is needed.

A relatively large polarization current (approximately 0.5 A) will polarize protons but will also generate large stray magnetic fields that may interfere with a nearby fluxgate or similar vector magnetometers.

Standard sensor liquid (kerosene) is a mixture of chemicals and its chemical shift[2] is not well known. This will degrade the achievable absolute accuracy. Alternatively, a liquid with known chemical shift can be used, such as methanol with a chemical shift of 3.6 parts per million, benzene 7.4 ppm, acetone 2.2 ppm (Pouchert 1983). Water has a chemical shift of about 5.6 ppm in relation to a reference tetra methyl silane (TMS).

To maximize signal strength the proton magnetometer sensor coils are usually immersed in the liquid. Some use toroidal sensors that are omni-directional and contain the polarizing magnetic field completely, i.e., they will not interfere with the other nearby magnetic sensors (fluxgates or similar). Toroidal sensors are not as efficient as directional sensors with two immersed coils wound in opposition to eliminate far away sources of interference. When installing that kind of sensor one needs to make sure the polarizing magnetic field is at right angles to the magnetic field of Earth. It is best to point the coil axis East-West. The angles are not critical though. It is of utmost importance to install the sensor in a field that is as homogeneous as possible for two reasons:

(a) Any movement of the sensor will change the readings due to local gradients and add to the noise and/or long term drift.
(b) Inhomogeneity, if excessive (over few hundred nT/m), may shorten exponential decay of the precession signal and reduce the time of measurement and, as a consequence the sensitivity.

Proton magnetometers as well as pulsed Overhausers, do not have measurable $1/f$ or low frequency noise, an excellent feature for long term monitoring.

At present, with the requirement for one-second measurements with a sensitivity of 0.1 nT or better, proton magnetometers are becoming marginal for observatory measurements. However they are still used extensively in mineral exploration and elsewhere.

[2] Chemical shifts are due to configuration of the sensor liquid molecules, their nuclear properties, orbital influences of electrons and their span is about 10 parts per million or about 0.5 nT in a field of 50,000 nT.

5.4.3 Overhauser Magnetometers

The Overhauser method has become the standard for magnetic observatories around the world. In essence, an Overhauser magnetometer is a proton magnetometer with all its valuable features plus numerous extras: better signal strength with better sensitivity, less power consumption, no DC polarization and its stray fields, and no significant interruption in measurement. Low power RF polarization allows for concurrent measurement so the measurement of the magnetic field is near-continuous (about 30 ms gap every second).

The use of a simple, chemically pure sensor liquid (methanol or similar) allows for fine tuning of the gyromagnetic constant by taking its chemical shift into account. This gives high absolute accuracy and long term stability.

The determination of the strength of the magnetic field (magnetic induction) using a proton or Overhauser magnetometer is carried out as follows: The precession frequency signal is sufficiently amplified and all zero-crossing times are measured precisely. From a set of zero-crossing times one determines the average period of the precession frequency. The reciprocal of the average period is the precession frequency which is then divided by the gyromagnetic constant for protons.

The major advantage of this method is the ease with which one can obtain readings of a desired resolution, and the ability to choose a sampling rate.

Sensitivities of commercially available Overhauser magnetometers are in the 10–20 pT range for a one second reading interval. The maximum practical rate of readings is 5 s^{-1}.

For high absolute accuracy, the Overhauser magnetometer, like the proton magnetometer, needs a higher stability TCXO adjusted to proper nominal frequency. GPS time, accurate to 1 μs, can be used for calibration. Omnidirectional sensors are available. Directional sensors must point in a direction that is at right angle to the magnetic field direction.

5.4.4 Time of Reading

In scalar magnetometers, determination of the magnetic field is achieved by timing the zero-crossings of the precession signal over a period of time. For a reading rate of once per second the period of integration is about one second. The time of measurement could be shorter than one second if the sensor experiences an excessively large magnetic gradient that shortens the decay of the signal, reduces the sensitivity, and changes the timing of the reading

To determine the true time of reading, i.e., average time t_0 of the signal zero-crossings, one needs to know the times of the first and last zero-crossings.

The precision with which the first and last zero-crossing can be determined depends on the precession frequency or magnetic field strength. At 50,000 nT the precession frequency is about 2,019 Hz

The average period is then 1/2019 s or 0.469 msec and this is the precision of the determination of time of any zero-crossing. If zero-crossings can be taken every half a period, the uncertainty will be 0.2345 msec instead. With the uncertainty of the last zero-crossing added linearly, the overall worst-case uncertainty becomes 0.469 msec.

For low magnetic fields, say 25,000 nT, this will double to about 0.938 msec, while for strong fields it will be reduced. The delay between triggering and the start of taking zero-crossings is up to 30 msec in Overhauser magnetometers. If rounded time of reading t_0 (full second) is required, then the triggering should be at 15 msec before the 0.5 s mark.

Optically pumped magnetometers (potassium) do not have this uncertainty in measuring as their precession frequency exceeds that of proton magnetometers by about 160 times.

5.4.5 *Optically Pumped Magnetometers*

Optically pumped magnetometers are presently quite rare in magnetic observatories. Cesium, Helium 4 and Potassium are available for airborne mineral and oil exploration surveys. Portable models of Cesium and Potassium magnetometers are available for ground mineral and diamond exploration. However, Potassium magnetometers offer good improvements in speed and sensitivity for observatory measurements (Alexandrov, Bonch-Bruevich, 1992). Potassium is the only Alkali metal magnetometer that operates on a single narrow EPR spectral line. This not only maximizes its sensitivity but it ensures a minimum heading error and very high absolute accuracy comparable with the absolute accuracy of Overhauser or ppm. Sub-pT sensitivities for once per second readings are routinely achievable.

5.5 Use of Scalar Magnetometers for Component Determination

The dIdD method of measuring the components of the magnetic field was first proposed by Alldredge (1960). The dIdD system consists of a proton or Overhauser or Potassium magnetometer centered within two orthogonal coil systems that are aligned to be perpendicular to the ambient magnetic field direction in horizontal and vertical planes. High degree of orthogonality of the two bias coils can easily be achieved. Positive and negative bias currents are applied to each coil system in turn and biased total fields are measured; the ambient unbiased field is also measured. From these five readings one can calculate the total intensity and the angles between the magnetic field vector and the axis in which the system is aligned: dD and dI. If the orientations of the coils are known, D and I can be computed.

$$dD = \sin^{-1}\frac{D_p^2 - D_m^2}{4F\cos I\sqrt{\frac{D_p^2+D_m^2}{2} - F^2}} \tag{5.10}$$

$$dI = \sin^{-1}\frac{I_p^2 - I_m^2}{4F\sqrt{\frac{I_p^2+I_m^2}{2} - F^2}} \tag{5.11}$$

D_p and D_m , I_p and I_m are biased magnetic fields while F is the unbiased field. With known I, D, and F, all components can be computed.

Theoretically, a dIdD system is an absolute instrument; it is very weakly affected by temperature, has no zero offset and is linear over a complete range of measurement. However, it is subject to changes in orientation, which means that in practice it is at best a quasi-absolute instrument. Nevertheless, if installed on a good solid pillar, a dIdD can be used to improve the determination of baselines for an observatory's triaxial fluxgate under the assumption that its drift will be slower and more linear than the drift of the fluxgate. Thus baseline values are determined for the dIdD on a periodic basis (usually once per week) using a DIM. The corrected dIdD values are then used to compute baselines for the fluxgate magnetometer on a minute by minute basis.

A traditional problem with the dIdD has been the length of time required to take an observation—up to 25 s when using a Proton magnetometer. This is too long since an active magnetic field can change substantially in 25 s. The problem of excessive time of measurement can be overcome by replacing the proton magnetometer with an Overhauser or Potassium magnetometer. A dIdD equipped with an Overhauser magnetometer can complete a sequence of measurements in one second (0.2 s each segment); a dIdD equipped with a Potassium magnetometer can measure five times per second (0.04 s per segment). A Potassium dIdD can achieve a sensitivity of one arc second measuring once per second. Although switching from one bias to another requires a delay for transients to die out, the time required is so short that either instrument can be considered virtually continuous.

A variation of dIdD proposed by Alpár Körmendi (2008) of Hungary has four sensors installed in toroidal bias coils and working under constant bias (I_p, I_m, D_p, and D_m) and supplemented by unbiased measurements. All components of the measurement are now collected concurrently and the main weakness of the standard dIdD—sequential measurements—is eliminated. However, the proposed system has its own weaknesses. It is expensive since it requires five sensors instead of one. The sensors are not in exactly the same magnetic field. The bias fields are not exactly equal, and it is difficult to make them orthogonal.

5.6 Automated Absolute Observations

Newitt (2007) lists six elements of observatory operations that must be fully or partially automated before an observatory can truly be called automated: data collection, data telemetry, data processing, data dissemination, error detection and absolute observations. For institutes that run remote magnetic observatories and for those who desire to put observatories (as opposed to variometer installations) in remote locations, the automation of absolute observations is of particular importance, since it would remove the necessity of having a trained observer on site. At present, there have been only three serious attempts to automate absolute observations.

An automated vector ppm is described by Auster et al. (2007, 2009). The instrument is equipped with a telephoto lens and a CCD camera, which, along with the accompanying imaging software, are used to determine the measurement direction with respect to a known azimuth. To determine H, Z, and D, the vector ppm is rotated around its vertical axis. The misalignment between the rotation axis and the true vertical axis is measured with tilt sensors. The rotation angle is monitored by a rotary encoder system. Measurements are made every 30 degrees. When the final measurement is made, the software automatically calculates the component values. Tests performed since 2006 indicate that it is possible to keep error to about 2 nT.

GAUSS (Geomagnetic Automated System) is an instrument in which a three component fluxgate sensor is rotated about a very stable, very well defined axis. All three components are recorded in three different positions, from which the magnetic field along the axis of rotation can be calculated. (Hemshorn et al. 2009, Auster et al. 2007). The instrument consists of a turntable on which are mounted a pair of support prisms for the three component fluxgate. All movements are carried out by piezoelectric motors. Angles are measured by encoders that have an error of 1 arcsec. A telescope focuses a laser beam which points along the measurement direction. The fluxgate must be linear over the entire range of the geomagnetic field (0 to ±64000 nT), so an instrument originally designed for space applications is being used. The instrument determines the field intensity in two horizontal directions. A ppm supplies the additional information required to determine the full vector field. The instrument, in its present form, was installed in the Niemegk magnetic observatory in April, 2008 for long term testing.

AUTODIF is an automated DIM that has been in development since the late 1990s. It was demonstrated at the Belsk Workshop in 2006 (Van Loo and Rasson, 2007) and rigorously tested at the Boulder/Golden Workshop in 2008 (Rasson, von Loo and Berrami, 2009). It is designed to reproduce the measurement sequence of a manually operated DIM. The telescope of the theodolite is replaced by a laser and split photo cells which are used to align the device in a known meridian by reflecting the laser beam off a corner cube reflector back onto the photo cell. Non-magnetic piezoelectric motors are used to move the sensor about the horizontal and vertical axes. The angles are measured by custom electronic optical encoders. An electronic bubble level mounted on the alidade provides reference to the horizontal. A lap top and a microcontroller control the instrument. In-house testing has shown that the instrument can achieve an angular accuracy of $0.1'$, which is comparable to that which can be obtained by a skilled observer. When tested at the Boulder Workshop the results were not as good; errors were about 0.2′. However, the system was being tested under environmentally challenging conditions (strong winds) which may have accounted for a large part of the difference. Further testing showed that the weak point of the system was the ultrasonic motor which was unreliable and had a short lifetime. A new motor has been found, but its mode of operation had led to a complete redesign of the system (Rasson et al. 2010). The MKII version was shown at the Changchun Magnetic Observatory Workshop in September 2010.

Although all three of these instruments have given results that agree closely with those obtained by manual observations, long-term reliability under adverse conditions must yet be demonstrated. We may be in the enviable position of having a choice of auto-absolute instruments that work on three different principles.

5.7 Other Magnetometers

In the following sections we describe briefly other magnetometers still in use at some magnetic observatories.

5.7.1 Declinometer

The classical declinometer employs a magnet suspended from a long, torsionless fibre so that it is free to align itself in the direction of the horizontal magnetic field. A mirror is affixed to the magnet perpendicular to the magnetic axis of the magnet. This assemblage is mounted on a non-magnetic theodolite in such a way that the mirror can be sighted through the telescope of the theodolite. The observational procedure is in theory quite simple. The theodolite is turned until the telescope is aligned perpendicular to the mirror. The direction of the magnetic meridian (A) is then read from the theodolite's base. Next, the reference mark is sighted and the angle (B) is read from the theodolite's base. Knowing the true bearing of the reference mark (Az), one can calculate the declination using Eq. (5.4).

In practice, this seemingly simple procedure becomes much more complicated for two reasons. First, it is impossible to attach a mirror exactly 90 degrees to the magnetic axis of the magnet. Second, a truly torsionless fibre does not exist. Thus, a real observation with the declinometer involves using two magnets with different magnetic moments, and taking observations with each magnet in upward and downward positions. Declination is now calculated using Eq. (5.12):

$$D = A1 + c(A1 - A2) - (B - Az) \tag{5.12}$$

where A1 is the average of the magnetic meridian values obtained using the first magnet in the up and down positions; $A2$ is the average value using the second magnet, and c is a coefficient related to the torsion in the fibre that must be determined experimentally (see Jankowski and Sucksdorff, 1996).

Declinometers can also be used to measure horizontal intensity using the classical method of oscillations and deflections developed by Gauss (Wienert, 1970).

5.7.2 Quartz Horizontal Magnetometer

The quartz horizontal magnetometer (QHM) is a simple instrument for measuring the horizontal intensity of the magnetic field. It consists of a tube from which a magnetic-mirror assembly is suspended by a fibre, and a telescope that fits onto an opening in the tube (Fig. 5.7). The instrument may be mounted on a

Fig. 5.7 Quartz horizontal magnetometer

specially designed base or on some other theodolite base using an appropriate adapter. Measuring horizontal intensity with the QHM is straightforward. The theodolite is turned a complete number of half-turns, such that the magnet is moved from the meridian position (A_o) by an angle of at least 45° (A_+). The angle is recorded and then the theodolite is rotated in the opposite direction (A_-). H can then be calculated from the following formula (Jankowski and Sucksdorff 1996):

$$H = C/(1 - k_1 t)(1 - k_2 H \cos \Phi) \sin \Phi \qquad (5.13)$$

$\Phi = (A_+ - A_-)/2$ and $C = 2\pi\,\tau/M$, where τ is the torsion constant of the fibre and M is the magnetic moment of the magnet. C must be determined experimentally by comparison observations. The temperature dependence of C is given by the first term in the denominator; k_1 is the temperature coefficient and t is the temperature. The second term gives the effect of induction on the magnet. Here H refers to an approximate value of the horizontal magnetic field component such as one would obtain from the IGRF. Great accuracy is not necessary since k_2 is typically in the range from 0.0002 to 0.0008. Both k_1 and k_2 are determined experimentally. Because the three constants do change with time, the QHM cannot be considered a true absolute instrument. However, the constants are extremely stable; k_1 and k_2 are considered constant for about 10 years; C must be redetermined after about 2 years (Wienert, 1970).

Observational errors should not exceed a couple of nanoteslas between periodic calibrations.

5.7.3 Torsion Photoelectric Magnetometer

The torsion photoelectric magnetometer (TPM) is an example of turning a classical instrument into one that can satisfy the requirements of modern science. The TPM consists of a suspended magnet and mirror system enhanced to give voltage as an output. This is accomplished by reflecting the light beam onto a pair of phototransformers which transform the angle of deviation into a voltage. The output is amplified and fed to a negative feedback winding which acts to keep the mirror stationary. Thus, the current in the negative feedback circuit is a measure of the strength of the magnetic field component.

Although the TPM uses a classical suspended magnet system as its field detector, the use of negative feedback enables the system to record more rapid variations than a photographic variometer using the same suspended magnet. The sensitivity is also improved. The TPM has good long-term stability (a few nT per year) and a resolution of about 0.01 nT (Jankowski and Sucksdorff 1996).

5.7.4 Kakioka KASMMER System

Kakioka Observatory in Japan has an interesting setup (KASMMER) for the measurement of components (Tsunomura et al. 1994). Besides standard three component fluxgate magnetometers, three biased scalar magnetometers measure three components of magnetic field concurrently.

Fansleau-Braunbek bias coils are positioned so as to cancel two components of the magnetic field that are not measured, leaving the third one for scalar measurement. Compensation of unwanted components is not overly critical as they are at right angles to the measured component and residual addition to the measured component is suppressed by vectorial addition of one large and two small residual fields.

The latest KASMMER uses custom designed continuous Overhauser magnetometers with one-second recordings.

Available space within bias coils with homogeneous magnetic field makes this set-up somewhat marginal for sensitivity. Proton and Overhauser magnetometers have reduced sensitivity in low magnetic fields. Proper replacement of continuous Overhauser magnetometers with Potassium magnetometers would more than eliminate this weakness.

5.8 Looking Forward

Magnetometry is a mature discipline, so it is unlikely that anything equivalent to the invention of the fluxgate, proton precession, Overhauser and Potassium magnetometers will take place anytime soon. It is more likely that small incremental advances will be made towards the goal of a truly stable vector magnetometer.

Advances will be made in power reduction, data storage and telemetry with the aims of increased automation and cutting costs.

The coming of age of the automated absolute instrument means that it is now possible to consider once again the possibility of a true underwater observatory. All that is needed is a means of determining the direction of true north, which can be accomplished using a north seeking gyroscope (Rasson et al. 2007).

The requirement for 1 pT resolution and sensitivity means constructing a sensor with noise less than or equal to 1 pT/sqrt(Hz)@1 Hz. Sensor noise is dependent on the quality of the material from which the core is made. The supply of material from which many of the low noise fluxgate sensors currently in use were made is almost exhausted, and the probability of obtaining more is small. The challenge, therefore, is to come up with new materials for making low noise sensors. Some manufacturers have made considerable progress in developing new materials and now claim that they can achieve a noise level of 5 to 7 pT at 1 Hz. Theoretical studies show that it should be possible to reduce noise by at least another order of magnitude, to under 100 fT (Koch et al. 1999). In addition, new methods of processing the output signal (e.g., time domain signal extraction) are also under development. In fact, fluxgate magnetometers with noise levels below 1 pT have already been built (Vetoshko et al. 2003), but their cost and complexity make the use impractical for geomagnetic purposes.

DIdDs equipped with Potassium magnetometers may prove to be an attractive alternative to fluxgates. With the recent developments in global positioning (GPS) there are now opportunities to precisely orient bias coils of d*I*d*D* in vertical (determination of *I*) and horizontal East-West direction (determination of *D*). With this orientation d*I* and d*D* become *I* and *D*.

Inexpensive GPS boards and antennas are now available to differentially determine position within 0.5 cm. With two antennas separated by a distance of approximately 20 m, the uncertainty in alignment will be 5×10^{-4} radians or $1.72'$. For observatory use this set-up still needs initial calibration. However for some field use especially in directional drilling for oil and minerals this sensitivity and accuracy is more than adequate. Having both a fluxgate and a dIdD operating at the same time with the same sampling interval and hopefully the same filtering would provide an excellent check on data quality.

Experimental supersensitive Potassium gradiometer installations are now in operation at few selected observatories.

Appendix 1: Accuracy and Baselines

The different applications to which magnetometers may be put require either absolute accuracy or relative accuracy. Absolute accuracy is required for a wide variety of scientific investigations: studies of secular variation, main field morphology, fluid flow in the core; long-term external field variations, from Sq to solar cycle, to name a few. Only relative accuracy is required for most other types of investigations: studies of magnetic storms, sub-storms, pulsations. Spatial surveys over a small area carried out for mineral exploration may often need only relative accuracy.

To define absolute and relative accuracy let us consider an observation of the magnetic field (or one of its components), $F(t)$, where t refers to the time of observation. This observation will not normally equal the true value of the magnetic field which we will call $F_T(t)$. The difference between the true and the observed values, δ, is composed of a systematic error, ξ, and a random error, ε

$$\delta(t) = F_T(t) - F(t) = \xi(t) \pm \varepsilon. \qquad (5.14)$$

Note that the systematic error, ξ, and as a consequence, δ, is a function of time. When absolute accuracy is required, that is, when there is a requirement for δ to be close to zero, periodic calibration observations are made to determine the value of $\xi(t)$. These calibration observations (normally called "absolute" observations) are carried out at least once per week. It is assumed that the change in $\xi(t)$ between calibration observations is small and linear, so that values of $\xi(t)$, referred to as baseline values, can be interpolated for all observations of F that fall between the times of the absolute observations. If this is not the case, then aliasing will occur, leading to spurious information for time scales shorter than 2 weeks. Under these assumptions, the absolute value at time t is

$$F_a(t) = F(t) + \xi_a(t) \qquad (5.15)$$

where $F_a(t)$ is the absolute value of the magnetic field and our best estimate of the true value. The absolute error, then, is

$$\delta_a(t) = F_T(t) - F_a(t) = \xi(t) - \xi_a(t) \pm \varepsilon \qquad (5.16)$$

The most recent guide for magnetic observatories (Jankowski and Sucksdorff, 1996) claim that at the best observatories absolute accuracy of better than 1 nT can be achieved. The INTERMAGNET Technical Manual (St-Louis, 2008) gives a more realistic figure of 5 nT.

We now consider the case where observations are made at the same position at two different times t_1 and t_2. The observed difference, $F(t_1) - F(t_2)$ will differ from the true difference $F_T(t_1) - F_T(t_2)$ by the amount $\delta_r = \xi(t_1) - \xi(t_2) \pm \sqrt{2}\varepsilon$. This is the relative error. The actual size of the systematic error ξ is unimportant. What is important is that it be constant over the time interval t_1 to t_2.

Appendix 2: Absolute Accuracy of Scalar Magnetometers

Most manufacturers of scalar magnetometers have ignored the question of their long term stability since it is of little importance for most usages. Temperature compensated crystal oscillators of stabilities of ± 1 ppm over −40 to +40°C temperature range and over 1 year aging are commercially available and recommended for observatory work. For long term stability and accuracy one needs to take into account several additional factors (Hrvoic 1996):

1. Gyromagnetic constant
2. Frequency reference
3. Details of taking zero crossings of the precession frequency
4. Phase stability of the precession frequency (proton and Overhauser magnetome ters)
5. Chemical shift of the Overhauser/ppm sensor liquid

While points 2–4 are engineering problems that can be resolved by proper design of the electronics, the gyromagnetic constant is "given" to us by National Standards Associations like NIST(USA), NPL (U.K.), VNIIM (Russia), NIM (China).

The value of the gyromagnetic constant and the tolerances are updated periodically. In the past we even had "western" and "eastern" γ_p determined by NIST, NPL and VNIIM and strong and weak fields γ_p. γ_p is traditionally computed for water as a solvent. For different solvents, the value must be corrected due to "chemical shifts" caused by molecules of the solvent.

Water has chemical shift relative to TMS (tetramethylsilane) of 5.6 parts per million i.e., under the same conditions the precession frequency of protons in water will be 5.6 ppm lower that that of TMS; Methanol has a shift of about 3.6 ppm i.e., in the same magnetic field it will give 2 ppm higher frequency. Since $B = \frac{\omega_o}{\gamma_p}$, we need to increase γ_p for methanol by 2 ppm. The uncorrected error would be 0.1 nT at 50,000 nT magnetic field.

Any magnetic inclusions in the sensor will change the local field and influence the measurement. Paramagnetic and diamagnetic materials like housing, copper wire etc may also influence the measurement and reduce absolute accuracy. And so can the gradient over sensor volume.

Presently one can attain an absolute accuracy of a fraction of a part per million with properly designed scalar magnetometers (Overhauser or Potassium). This, at a minimum, requires a thermostated crystal oscillator or Rubidium/Cesium frequency standards and very careful design of sensors and electronics.

Acknowledgments We wish to thank Jean Rasson and Barry Narod and the anonymous reviewer for their valuable input.

References

Abragam A (1961) The principles of nuclear magnetism. Oxford at the Clarendon Press, Ely House, London

Alexandrov E, Bonch-Bruevich V (1992) Optically pumped atomic magnetometers after three decades. Opt Eng 31(4):711

Alldredge LR (1960) A proposed automatic standard magnetic observatory. J Geophys Res 65:3777–3786

Alldredge LR, Saldukas I (1964) Automated standard magnetic observatory. JGR 69:1963–1970

Aschenbrenner H, Goubau G (1936) Ein Anordnung zur Registreirung rascher magnetischer Stoerungen. Hochfrequenztechnik Elektroakustik 47:177–181

Auster HU, Mandea M, Hemborn A et al. (2007) GAUSS: A geomagnetic automated system. In: Reda J (ed) XII IAGA Workshop on Geomagnetic Observatory Instruments, Data Acquisition and Processing. Publs Inst Geophys Pol Acad Sc c-99:398

Auster V, Hillenmaier O, Kroth R et al. (2007) Advanced proton magnetometer design and its application for absolute measurements. In: Reda J (ed) XII IAGA Workshop on Geomagnetic Observatory Instruments, Data Acquisition and Processing. Publs Inst Geophys Pol Acad Sc c-99:398

Auster V, Kroth R, Hillenmaier O et al. (2009) Automated absolute measurement based on rotation of a proton vector magnetometer In: Love JJ (ed) Proceedings of the XIIIth IAGA Workshop on geomagnetic observatory instruments, data acquisition, and processing: U.S. Geological Survey Open-File Report 2009–1226

Bitterly J, Cantin M, Schlich R et al. (1984) Portable magnetometer theodolite with fluxgate sensor for earth's magnetic field component measurements. Geophys Surv 6: 233–240

Chulliat A, Savary J, Telali K and Lalanne X (2009). Acquistion of 1-second data in IPGP magnetic observatories. In: Love JJ (ed) Proceedings of the XIIIth IAGA Workshop on geomagnetic observatory instruments, data acquisition, and processing: U.S. Geological Survey Open-File Report 2009–1226

Coles RL (1985a) To determine the expression for correcting I readings when the theodolite was not set in the magnetic meridian. Unpublished report, Geological Survey of Canada

Coles RL (1985b) Analytical expressions for the errors introduced by an off-level adjustment of a theodolite used in the determination of magnetic declination and inclination. Unpublished report, Geological Survey of Canada

Coles RL (ed) (1988) Proceedings of the international workshop on magnetic observatory instruments. Geological Survey of Canada Paper 88–17

Coles RL, Trigg DF (1988) Comparison among digital variometer systems. In: Coles RL (ed) Proceedings of the international workshop on magnetic observatory instruments. Geological Survey of Canada Paper 88–17

Constable C (2007) Geomagnetic spectrum, temporal. In: Gubbins D, Jerrero-Bervera E (ed) Encyclopedia of Geomagnetism and Paleomagnetism, Springer, Dordrecht, The Netherlands

Delaurier JM, Loomer EI, Jansen van Beek G et al. (1974) Editing and evaluating recorded geomagnetic field components at Canadian observatories. Publications of the Earth Physics Branch, 44(9)

Evans K (2006) Fluxgate magnetometer explained http://www.invasens.co.uk/FluxgateExplained.PDF Accessed 10 Nov 2009

Gordon DL, Brown RE (1972) Recent advances in fluxgate magnetometry. IEEE Trans Magnetics MAG-8:76–83

Happer W (1972) Optical pumping. Rev Mod Phys 44

Hemshorn A, Pulz E and Mandea M (2009) GAUSS: Improvements to the geomagnetic automated system. In: Love JJ (ed) Proceedings of the XIIIth IAGA Workshop on geomagnetic observatory instruments, data acquisition, and processing: U.S. Geological Survey Open-File Report 2009, 1226

Hrvoic I (1996) Requirements for obtaining high accuracy with proton magnetometers. In: Rasson JL (ed) Proceedings of the VIth workshop on geomagnetic observatory instruments data acquisition and processing. pp 70–72 Institut Meteologique du Belgique, Brussels. Also available at www.gemsys.ca

Jankowski J and Sucksdorff C (1996) Guide for magnetic measurements and observatory practice. IAGA

Kerridge D (1984) Determination of true north using a Wild GAK-1 gyro attachment. British Geological Survey Geomagnetic Research Group Report No. 84–16

Kerridge DJ (1988) The theory of the fluxgate theodolite. British Geological Survey Geomagnetic Research Group Report No. 88/14

Koch RH, Deak, JG and Grinstein (1999) Fundamental limits to magnetic field sensitivity of flux-gate (sic) magnetic field vectors. Appl Phys Lett 75:3862–3864

Korepanov V (2006) Geomagnetic instruments for repeat station surveys. In: Rasson J, Delipetrov (ed) Geomagnetism for Aeronautical Safety. pp 145–166, Springer, Dordrecht, The Netherlands

Körmendi A (2008) Private communication alpar@freemail.hu

Kring Lauridsen E (1985) Experiences with the DI-fluxgate magnetometer inclusive theory of the instrument and comparison with other methods. Danish Meteorological Institute Geophysical Papers R-71

Kudryavtsev AB, Linert W (1996) Physico-Chemical Applications of NMR, World Scientific, River Edge, New Jersey

Kurreck H, Kirste B, Lubitz W (1988) Electron Nuclear Double Resonance Spectroscopy of Radicals in Solution, VCH Publishers, New York, New York

Merenyi L, Hegymegi L (2005) Flexible data acquisition software for PDA computers. In: XI IAGA Workshop on Geomagnetic Observatory Instruments, Data Acquisition and Processing, Kakioka Magnetic Observatory, Kakioka, Japan

Narod B (1988) Narod Geophysics Ltd. ringcore magnetometer S-100 version. In: Coles RL (ed) Proceedings of the international workshop on magnetic observatory instruments. Geological Survey of Canada Paper 88–17

Narod B, Bennest JR (1990) Ring-core fluxgate magnetometers for use as observatory variometers. Phys Earth Planet Int 59:23–28

Newitt LR (2007) Observatory automation. In: Gubbins D, Jerrero-Bervera E (ed) Encyclopedia of Geomagnetism and Paleomagnetism, Springer, Dordrecht, The Netherlands

Newitt LR, Barton CE, Bitterly J (1996) Guide for magnetic repeat station surveys. IAGA

Newitt LR, Gilbert D, Kring Lauridsen E et al. (1988) A comparison of absolute instruments and observations carried out during the IAGA workshop. In: Coles RL (ed) Proceedings of the international workshop on magnetic observatory instruments. Geological Survey of Canada Paper 88–17

Packard ME, Varian R, (1954) Phys Rev 93:941

Palangio P (1998) A broadband two axis flux-gate magnetometer. Ann Geofis 41:499–509

Pouchert C (1983) Aldrich Library of NMR Spectra, Aldrich Chemical Co. P.O. Box 355, Milwaukee WI53201 U.S.A

Primdahl F (1979) The fluxgate magnetometer. J Phys E Sci Instrum 12:241–253

Rasson J, van Loo S, Berrani N (2009) Automatic diflux measurements with AUTODIF. In: Love JJ (ed) Proceedings of the XIIIth IAGA Workshop on geomagnetic observatory instruments, data acquisition, and processing: U.S. Geological Survey Open-File Report 2009–1226

Reda J, Neska M (2007) Measurement session during the XII IAGA workshop at Belsk. In: Reda J (ed) XII IAGA

Workshop on Geomagnetic Observatory Instruments, Data Acquistion and Processing. Publs Inst Geophys Pol Acad Sc c-99(398):7–19

Rasmussen O, Kring Lauridsen E (1990) Improving baseline drift in fluxgate magnetometers caused by foundation movement, using a band suspended fluxgate sensor. Phys Earth Planet Int 59:78–81

Russell CT, Chi PJ, Dearborn DJ et al. (2008) THEMIS ground-based magnetometers. Space Sci Rev.doi:10.1007/s11214-008-9337-0

St-Louis B (ed) (2008) INTERMAGNET technical reference manual, version 4.4. http://intermagnet.org/publications/im_manual.pdf. Accessed 21 Nov 2009

Serson PH and Hannaford WLW (1956) A portable electrical magnetometer. Can J Technol 34:232–243

Slichter CP (1963) Principles of Magnetic Resonance, Harper & Row, New York, New York

Stuart WF (1972) Earth's field magnetometry. Rep Prog Phys 35:803–881

Trigg DF (1988) Specifications of an ideal variometer for magnetic observatory applications. In: Coles RL (ed) Proceedings of the international workshop on magnetic observatory instruments. Geological Survey of Canada Paper 88–17

Trigg DF, Olson DG (1990) Pendulously suspended magnetometer sensors. Rev Sci Instrum 61:2632–2636

Tsunomura S, Yamataki A, Tokumoto T, Yamada Y (1994) The New System of Kakicka Automatic Standard Magnetometer. Mem Kakioka Magn Observ 25(1,2)

Van Loo S, Rasson J-L (2007) Presentation of the prototype of an automated DIFlux. In: Reda J (ed) XII IAGA Workshop on Geomagnetic Observatory Instruments, Data Acquistion and Processing. Publs Inst Geophys Pol Acad Sc c-99(398):7–19

Vetoshko PM, Valeiko VV, Nikitin PI (2003) Epitaxial yttrium iron garnet film as an active medium of even-harmonic magnetic field transducer. Sensor Actuator A 106:270–273

Wienert KA (1970) Notes on geomagnetic observatory and survey practice. UNESCO, Brussels

Chapter 6

Improvements in Geomagnetic Observatory Data Quality

Jan Reda, Danielle Fouassier, Anca Isac, Hans-Joachim Linthe, Jürgen Matzka, and Christopher William Turbitt

Abstract Geomagnetic observatory practice and instrumentation has evolved significantly over the past 150 years. Evolution continues to be driven by advances in technology and by the need of the data user community for higher-resolution, lower noise data in near-real time. Additionally, collaboration between observatories and the establishment of observatory networks has harmonized standards and practices across the world; improving the quality of the data product available to the user. Nonetheless, operating a high-quality geomagnetic observatory is non-trivial. This article gives a record of the current state of observatory instrumentation and methods, citing some of the general problems in the complex operation of geomagnetic observatories. It further gives an overview of recent improvements of observatory data quality based on presentation during 11th IAGA Assembly at Sopron and INTERMAGNET issues.

6.1 Introduction

The network of surface geomagnetic field observatories was begun to be formed in the first half of the nineteenth century, its initiators being Carl Friedrich Gauss and Alexander von Humboldt. Since that time, the methodology of geomagnetic observations has changed a lot, both the magnetic field recording and the absolute measurements. From simple photographic recording and tedious absolute measurements we came to digital recordings with suspended variometers and absolute measurements by proton magnetometers and non-magnetic theodolites with fluxgate sensors. However since Gauss's time, geomagnetic observatories are monitoring variations in the Earth's magnetic field following the same principle. All are performing separate absolute measurements and variometer recordings.

The systematic development of the methodology of magnetic observations resulted in a huge improvement of their quality. This concerns the strictly measurable parameters, such as noise of magnetic sensors, resolution of the recordings, accuracy of absolute measurements, accuracy of time setting, as well as less measurable ones, such as the ways the data are published and made available to the users. In spite of this tremendous progress, a magnetometer able to make automatic absolute measurement of magnetic field variations is still not commercially produced. We are still using two types of instruments: one for variation recordings, and the other for linking the recorded variations with the absolute values of the field components.

The data provided by geomagnetic observatories should satisfy two criteria: the observational series should be continuous, and the data should be accurate and recorded in a magnetic clean environment. A statistically effective way to avoid gaps is to make recordings by two independent instrument sets. The matter of accuracy is more complicated. The accuracy that is necessary to measure pulsations of a few nT in amplitude should differ from that related to magnetic storms, and differ again if we are monitoring the secular variations. The accuracy for pulsation recording has typically an order of magnitude of a few pT, it is limited mainly by the noise of the magnetic sensors and by the noise of quantization. Magnetic storms have

J. Reda (✉)
Institute of Geophysics, Polish Academy of Sciences, Warsaw, Poland
e-mail: jreda@igf.edu.pl

M. Mandea, M. Korte (eds.), *Geomagnetic Observations and Models*, IAGA Special Sopron Book Series 5,
DOI 10.1007/978-90-481-9858-0_6,

amplitudes of hundreds of nT, they are usually measured with an accuracy of a few nT. The main factors influencing this range are the precision of the magnetic sensor orientation as well as the precision of scale value determination. For observation of secular variations, an accuracy of 0.5–2 nT is achieved generally, which is mainly determined by the quality of absolute measurements and by base line stability.

The improvement of accuracy may be due to various factors. Some of these are a consequence of a general technological progress. An example is the manner how the magnetic field has been recorded, which evolved from photographic recording to digital recording with high resolution. The improvement of accuracy may be also forced by the needs of users. For instance, the magnetic field declination has been (and still is) widely used in navigation, as well as for military purposes Hence, there was a need for maps containing information on declination, and consequently a need for increasing accuracy in geomagnetic field measurements. At present, new practical applications of the magnetic field data are evolving. An example is the directional geological drilling related to oil and gas mining (Reay et al. 2005).

The quality of geomagnetic observations is strongly affected by anthropogenic factors. These manifest themselves as artificial perturbations in the natural magnetic field variations, sometimes in the form of jumps of the measured magnetic field elements, undesirable with a view to the study of secular variations. The sources of perturbations can be nearby, as for instance maintenance work in the observatory or constructions in the vicinity. Still worse are the perturbations due to currents flowing in the ground from, e.g., electric railroads or power lines, notably DC ones. Unfortunately, such perturbations may be effective over tens of kilometers.

And most importantly, because of the broad band of signals of interest it has to be a reasonable guarantee that a suitable magnetic environment would persist for at least 50 years. This means that managing institutes, local and international scientific community might look to how to keep each existent observatory running and how to adopt the changes of the technology which are certain to be required: high resolution low drift magnetometers and compact, simple data logging and processing systems (Isac et al. 2009).

To be successful, the geomagnetic investigations should be accompanied by information describing these data, called metadata. These should include information facilitating the assessment of accuracy of the data and other necessary or useful details, such as instrumentation, data formats, and the like.

6.2 Quality of Recording of Geomagnetic Variations

The variations of the Earth's magnetic field are monitored by means of various types of variometers. There is no current instrument with the stability, the resolution and the dynamic range to continuously measure the magnetic field across all bands of interest, so at present, all existing variometers are relative instruments. They are not able to generate absolute values—their readings refer to a base line value, which has to be determined by means of absolute measurements.

As the Earth's magnetic field is a vector, it is common that magnetometers include multiple single sensor elements to describe the variations of the vector. A variety of variometers based on differing physical principles have led to single sensor instruments, with or without auxiliary equipment to resolve the complete vector information of the Earth's magnetic field variations.

The quality of the data of an observatory depends significantly on the suitable combination of manual absolute measurements and continuous variometer recordings. For absolute measurement matters refer to Section 6.3 "Quality of observations of secular variations". The variometer recordings are periodically calibrated to absolute level by reference to manual, absolute measurements and the derivation of the variometer's baseline values. Since baseline values are usually daily spot values, the variation of a variometer's baseline between absolute measurements is modelled by interpolation (typically polynomials or splines). Base line stability is an important parameter for the evaluation of an observatory, although further quality parameters can be derived, depending on the variometer type.

6.2.1 Physical Principles of Variometers

6.2.1.1 Fluxgate Magnetometers

The fluxgate magnetometer uses the principle of the transformer working in saturation. An alternating current of a frequency f flows through the primary (excitation) coil. The core consists of a material of high permeability. If an external field exists, the signal in the secondary (pick-up) coil will consist of the frequency f and higher harmonics as well. The second harmonic is related to the intensity of the magnetic field component aligned to the core direction. The electronic unit of the magnetometer separates the second harmonic and generates a voltage output signal. Commonly, a feedback coil is used to maintain the core in zero-field. The feedback current is converted to voltage using a stable resistor and the voltage digitised by means of an analogue-to-digital converter (ADC).

Two different designs of fluxgate sensors exist:

- Bar core
- Ring core

Bar core sensors are "direction–true". They record a more accurate real measurement value of the component which it is aligned to. Readings of ring core sensors also contain information of the different components. However, the noise level of ring core magnetometers is smaller than that of bar core ones.

The fluxgate magnetometer is useful to gain vector information of the Earth's magnetic field strength. Usually three sensors are aligned orthogonally. Two of the sensors are oriented in the horizontal plane, while the 3rd one is perpendicular to the horizontal plane. One of the two horizontal sensors may be oriented to the true North or to magnetic North. Both constellations allow the determination of unique component information of the Earth magnetic field. Some observatories use a design of only two fluxgate sensors oriented to monitor the variations of declination (D) and inclination (I). Additionally a scalar magnetometer (proton or optically pumped) is operated to obtain the complete vector. Fluxgate magnetometers are relative instruments. Their base line values have to be calibrated by absolute measurements.

Although fluxgate magnetometers are routinely sampled at higher data rates, the commonly available data sets are one-minute. There is presently a demand from the user community for higher time resolution data and observatories are being encouraged to disseminate one-second data. One-second data require a special design of fluxgate magnetometers (Korepanov et al. 2009), or more conventional fluxgate sensors with modern data loggers (Chulliat et al. 2009; Shanahan and Turbit 2009). A low noise level and specially designed filters for the wanted signal are necessary.

6.2.1.2 Photoelectric Feed-Back Magnetometers

Torsion variometers were used in observatories for more than 100 years. Brunelli et al. (1960) described a magnetometer based on torsion variometers with photoelectric converters. The use of photoelectric conversion causes that the magnet's deflections in response to the magnetic field changes are transformed into the electric current changes. The widely known example of such construction is torsion photoelectric magnetometer PSM developed and produced at the Institute of Geophysics, Polish Academy of Sciences, Poland (Jankowski et al. 1984). The PSM magnetometer is based on quartz variometers designed by Bobrov (1962), in which quartz fibres are attached to both sides of the magnet, so that the magnet is suspended in the fibre.

Three Bobrov variometers combined with electronic feedback can easily be mounted orthogonally in a relatively small space. So a component monitoring of the Earth's magnetic field is possible. PSM magnetometers are successfully operated at all of the Polish magnetic observatories and at some other international ones.

Torsion photoelectric magnetometers are also relative instruments. Their base line values have to be calibrated by absolute measurements. Due to their impulse response they have a limited sampling rate. Nevertheless they can be used to record 1-second samples.

6.2.1.3 Vector Variometers Based on Scalar Magnetometers

Scalar magnetometers such as proton magnetometers and optically pumped magnetometers can be adapted by means of special constructed coils to monitor components of the Earth's magnetic field vector. Proton magnetometers make use of proton precession a nuclear physical effect. The principle originally does not generate a continuous measurement signal. The instruments can generate their readings only intermittently—by means of a change of polarization and frequency measurement. Overhauser proton magnetometers make use of spin coupling between the protons and electrons to significantly improve the signal-to-noise ratio. Some versions are able to generate a continuous output signal, but their absolute accuracy is reduced. Vector proton magnetometers were constructed from the 1980s onward.

Optically pumped magnetometers replaced the proton magnetometers in vector magnetometers from about 1998 onward. Exploiting the relationship between magnetic field strength and electron energy states, the advantages of optically pumped magnetometers are their continuous output signal, higher resolution and a much better signal-to-noise ratio.

The output signal of proton and optically pumped magnetometers is a frequency related to the scalar value (total intensity) by the so called gyromagnetic ratio, which depends only on natural constants. The Consultative Committee for Electricity and Magnetism (CCEM) within the Bureau International des Poids et Mesures (BIPM) provides the values of the gyromagnetic ratios for proton and optically pumped magnetometers. The last proton gyromagnetic ratio γ_p changes were adopted in 1991 (Rasmussen 1991) and 2009 (Mohr et al. 2008).

To monitor components of the Earth's magnetic field vector specially constructed coils are needed to compensate one or more components. The first vector proton magnetometers were designed to monitor the horizontal or the vertical intensity (*H* or *Z*) and the magnetic East component. These instruments were almost exclusively used for absolute measurements. Later on (around 1995) constructions monitoring the inclination (*I*) and declination (*D*) emerged, called d*I*d*D* (deltaI–deltaD). Finally the proton sensors were replaced by optically pumped ones (Hegymegi et al. 2004).

d*I*d*D* variometers can take readings of the 3 field components *F* (total intensity), *D* (declination) and *I* (inclination) although not concurrently. The currents in the two coils have to be individually switched on and also reversed. So the instrument is limited in its sampling rate. Due to the prolonged sample time, monitoring of 1 s samples by means of a d*I*d*D* magnetometer is at present not possible.

Most significantly, proton magnetometers, Overhauser magnetometers and optically-pumped magnetometers are capable of measuring the total intensity absolutely. Theoretically also the declination and inclination can be obtained absolutely but this requires a very exact orientation of both compensation coil axes with respect to the geographical coordinate system. This is impossible in practice. Furthermore the alignments of the coils with respect to each other and the geographic orientation may change with the time. So the instrument is at most quasi-absolute. Its base lines have to be adjusted by means of absolute measurements as well.

6.2.2 Practical Aspects of Variometer Operation

Fluxgate and photoelectric compensated torsion variometers are temperature dependent. Proton and optically pumped magnetometers are temperature independent. So vector magnetometers based on the latter two instruments should be temperature independent as well, but this is not the case due to the used coils, even if a thermally very stable material is used for the construction of the coil system. So all vector variometers are subject to temperature dependence. Ideally, vector magnetometers are maintained at a stable temperature, but where a magnetometer is operated under varying temperature conditions, temperature dependency may be compensated by using temperature coefficients.

Temperature coefficients have to be determined by means of special thermal test equipment (Csontos et al. 2007). Fluxgate sensors and electronic units have to be tested separately, because both have different temperature coefficients. This is also the case for torsion photoelectric magnetometers. For the thermal correction of the recordings of both these variometers the temperature has to be recorded separately for the sensor and the electronic unit, if both are placed in different rooms. The electronic units of proton and optically pumped magnetometers are temperature independent, because their time base for the frequency measurement is usually temperature compensated. So only the thermal coefficients of the coil systems of scalar magnetometer based vector variometers have to be known and considered.

The correction for the thermal behaviour of variometers is not ideal. It is much better to run the variometers under stable temperature conditions. In case of fluxgate and photoelectrically compensated variometers both the sensor and the electronic unit should be operated in the same temperature stabilized room, if it offers enough space to achieve a suitable distance between them to avoid interference.

All types of variometers require stable pillars for the placement of their sensors. The sensors have to be correctly adjusted to monitor the components of the field vector. Any movement of the sensor influences the long-term quality of the observatory data. Several manufacturers produce fluxgate sensors with suspension which compensates pillar tilts. Such a suspension system was also fitted to a d*I*d*D* variometer (Hegymegi et al. 2004).

Observatory data need to be exactly referred to the time of their monitoring. So a suitable time base is necessary. Radio clock or satellite based time synchronization of data loggers is widely used (Linthe 2004). The readings taken during the absolute measurements have to be exactly referred to the time as well. So to accurately reference the absolute measurements to the variometer recordings, adjusted precise reference clock is of critical importance.

Associated with the call of data users for higher time resolution observatory data, improvements in timing accuracy have also been specified. To generate 1 s observatory data accurate, timely sampling is of high importance. The evaluation of timing accuracy is non-trivial, however Rasson el al. (2009) describe a testing method and apparatus.

6.2.3 Quality Detection of Variometers

6.2.3.1 Base Line Behaviour

The quality of observatory data depends to a large extend on the quality of the variometer used. The base line of the variometer is adjusted by means of absolute measurements, which are carried out periodically. The quality of a variometer can be evaluated by means of its base line plot for all its recorded components. Figure 6.1 shows the base line plot of the main variometer of Niemegk observatory. The variometer is a suspended triaxial fluxgate magnetometer FGE, produced by the Danish Meteorological Institute Copenhagen (now produced at DTU Space, Technical University of Denmark, Copenhagen). The absolute measurements are performed by means of a DI-flux Zeiss theodolite THEO 010B equipped with a Bartington fluxgate magnetometer MAG01H. Absolute total intensity measurements are carried by means of the Overhauser proton magnetometer GSM19.

The black dots depict the results of the absolute measurements; the purple lines are the adopted base

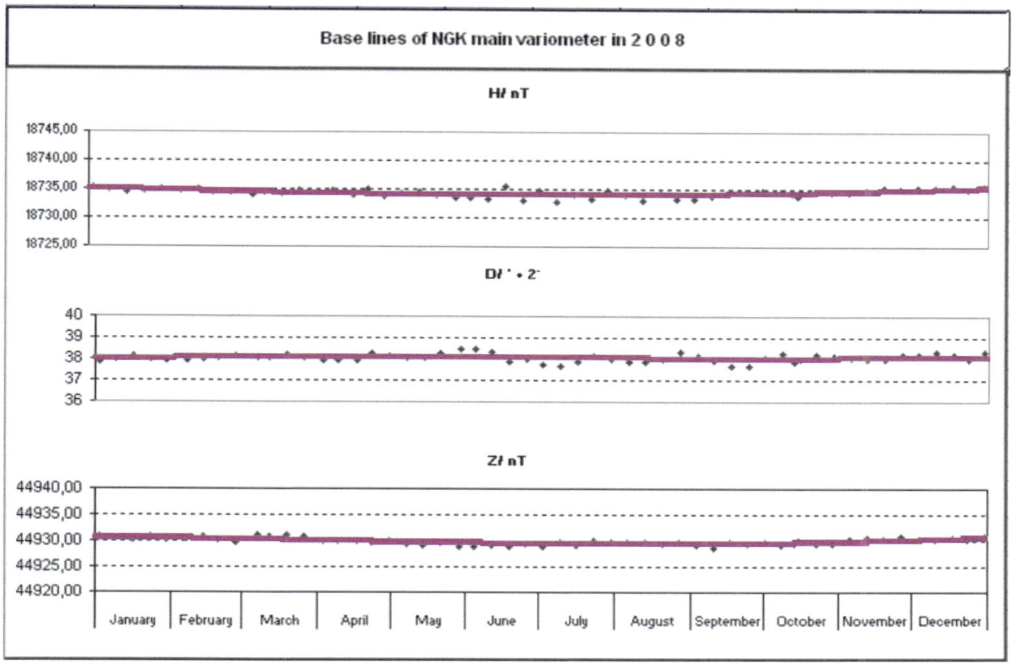

Fig. 6.1 Base lines of the Niemegk observatory main variometer of 2008

lines of the three components H, D and Z. In this example, a 3rd order polynomial is used for determining the base line. The vertical scale of 20 nT is often used by observatories, it is adequate for the most baselines. The base line plot includes two types of information; the base line variation and the uncertainty of the absolute measurements.

6.2.3.2 Delta-F Check

Further quality control is possible on the basis of certain special conditions existing at an observatory. The use of modern data processing hardware and software enables an immediate quality check of the observations. Suitable time intervals of recordings (for example the last 24 h) can be plotted at the computer screen. Malfunctions and perturbations on the recordings can be conveniently recognised. If the observatory operates a scalar magnetometer independently of the vector magnetometer, then ΔF (the difference between the total field calculated from the vector recordings and the recorded total intensity), can be displayed as well. A straight horizontal ΔF line plot depicting a constant value shows a good variometer behaviour. The constant value represents the offset between the F level at the position of the scalar magnetometer sensor and the position, where the absolute measurements are carried out (absolute measurement pillar, which is the place to which all the observatory data refer). Figure 6.2 shows an example of Surlari (SUA) observatory.

The ΔF plot is much more sensitive than plots of single components for observatory problems. Jumps, spikes or drifts indicate problems of the base line values, scale values and internal or external magnetic perturbations. Spikes are caused in most cases by external or internal perturbations. The length of the spike corresponds to the time interval when any ferromagnetic material, for example a vehicle, was placed too close to the variometer or when DC current has caused interference on the instrument. Jumps or long-term drifts in ΔF can indicate a base line problem of the variometer. But for the indication of long-term drifts a longer time interval of the plot is necessary. If ΔF represents a similar behaviour as the diurnal variation of one or more of the field components, surely a problem of the scale values exists.

A plot of ΔF can be used to identify problems in a vector variometer, but since ΔF is calculated as the square-root of the sum of the squares of the components, the sensitivity of a ΔF plot is dependent on the magnitude of the components aligned with the variometer sensors. For example, for an HDZ-oriented

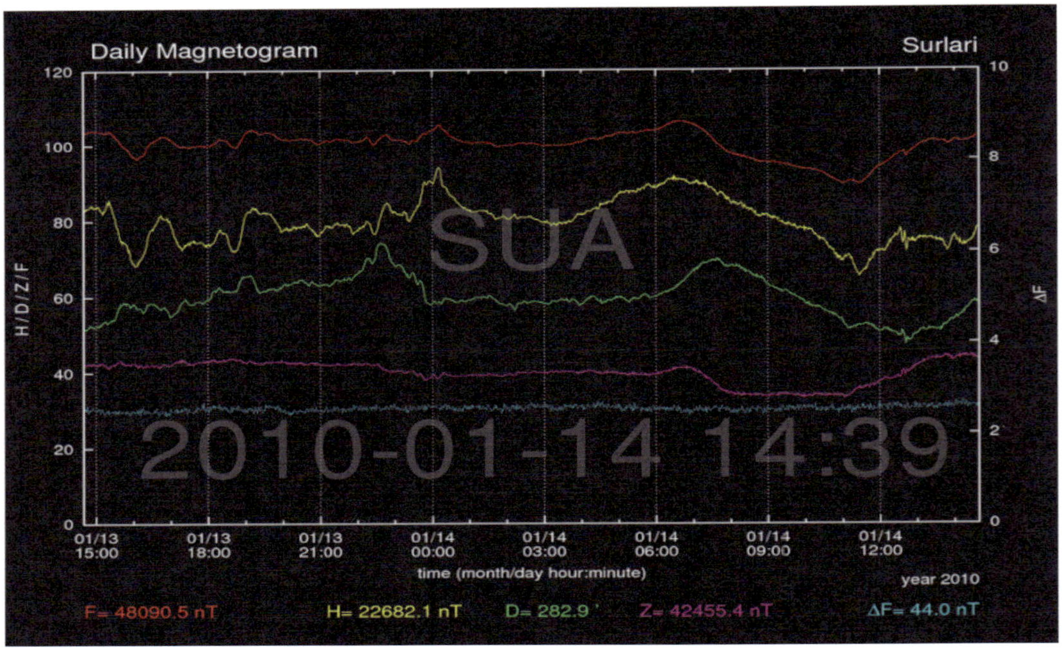

Fig. 6.2 Plot of the most recent 24 h of Surlari observatory displaying the vector variometer components H, D and Z, the independently recorded total intensity F and ΔF calculated as $\Delta F = \text{sqrt}\,(H^2 + Z^2) - F$

variometer in mid-latitude, ΔF is most sensitive to problem with the Z-sensor, less so for H and errors on the D-sensor are generally undetectable. Even if the variometer is oriented geographically (one of the horizontal sensors is oriented to the true North), the geographic East component sensor (Y) contributes due to its small value in comparison to both the other components very little to ΔF at most of the locations of magnetic observatories (southern Africa is an exception where declination varies between $-8°$ and $25°$). Only a rotation of the sensors in the horizontal plane by $45°$ against the geographic orientation may balance the contribution of the horizontal components within ΔF. The disadvantage of this orientation is a non-trivial alignment of the sensor and a complicated coordinate transformation in all the observatory data processing.

The ΔF-error-detection method is only applicable if a triaxial magnetometer is used and if an independent scalar F recording is operated. So only in case of triaxial magnetometers without any application of a scalar magnetometer this method can be used. In case of observatories using DIF orientation of their instruments (vector variometers based on scalar magnetometers and further designs deriving components from a scalar F recording) this method is impossible to be used because the scalar F recording is not independent.

6.2.3.3 Inter-Comparison with Other Magnetometers

If an observatory operates more than one variometer, an inter-comparison can be carried out. It is possible to detect problems of one of the variometers as base line jumps or drifts, scale value errors or internal or external perturbations. Especially useful is the existence of variometers based on different measurement principles or produced by different manufacturers. A very fortunate situation exists, if the observatory operates three vector variometers and three scalar magnetometers continuously. In that case the instrument, which causes the problem, can be uniquely identified. Figures 6.3 and 6.4 are examples of daily

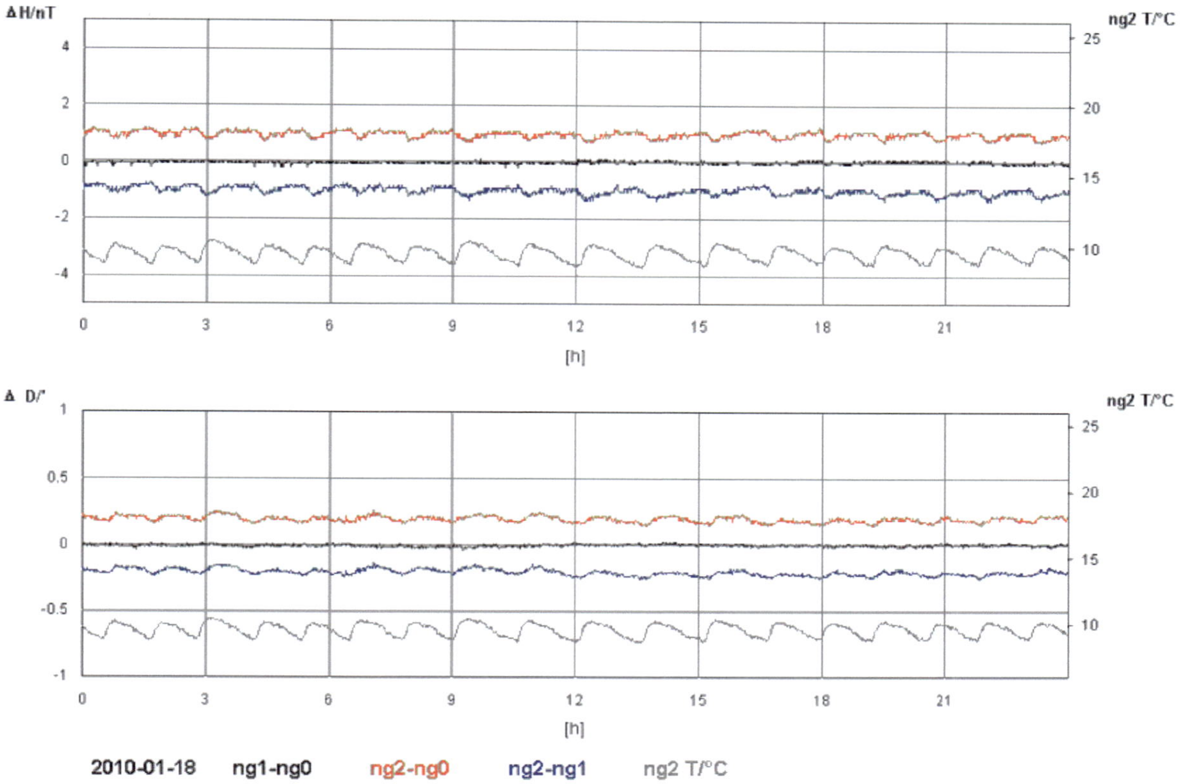

Fig. 6.3 Difference plots of the three Niemegk observatory systems ng0, ng1 and ng2: H (*upper panel*) and D (*lower panel*). The ng2 temperature is plotted at both panels in grey

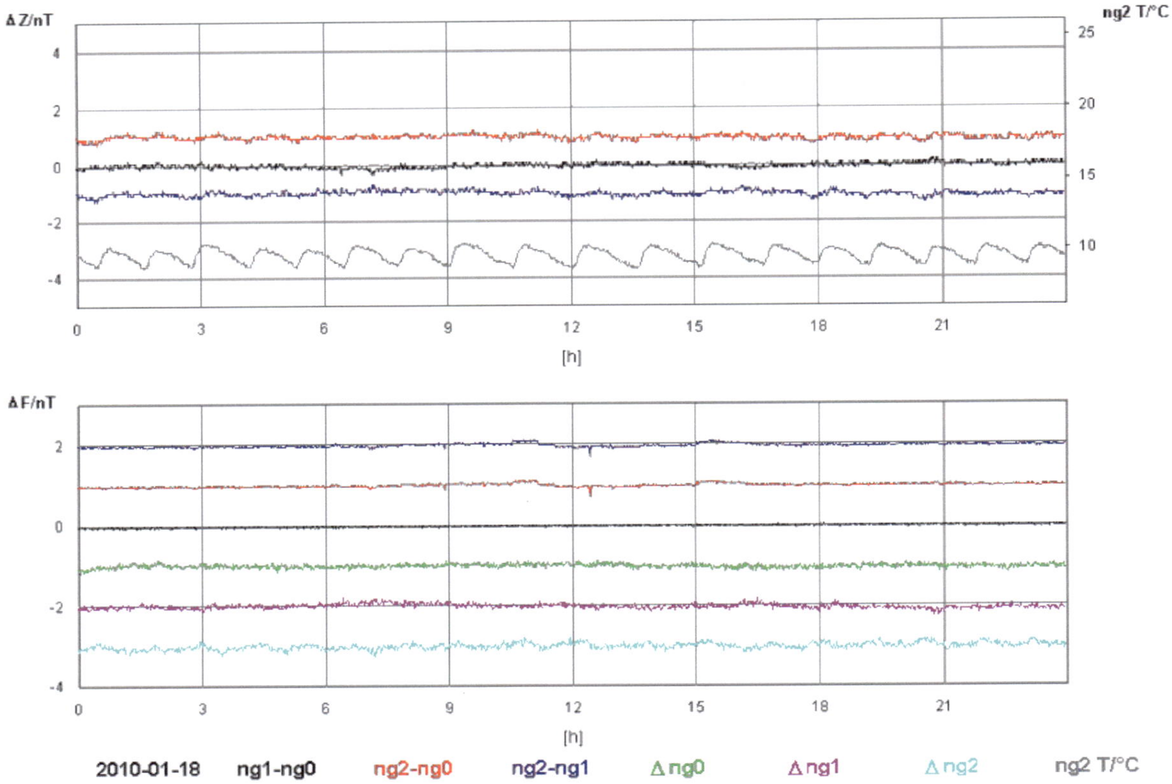

Fig. 6.4 Difference plots of the three Niemegk observatory systems ng0, ng1 and ng2: *Z* and ng2 temperature plot (*upper panel*) and *F* recordings and ΔF plot (*lower panel*)

difference plots of all three variometer systems at Adolf Schmidt Niemegk observatory called ng0, ng1, ng2.

Each system consists of a triaxial fluxgate magnetometer FGE and an Over-hauser proton magnetometer GSM90-F1 produced by GEM Systems, Markham, Canada. Figure 6.3 shows the differences of the horizontal intensities (*H*) in the upper panel and declination (*D*) differences in the lower panel. At Fig. 6.4 in the upper panel the vertical intensity (*Z*) differences are depicted, while in the lower panel the total intensity (*F*) differences are plotted. The colours of the plots at Figs. 6.3 and 6.4 refer to the systems that are compared, and are identical for all components (*H*, *D*, *Z* or *F*). Black refers to ng1–ng0, red to ng2–ng0 and dark blue to ng2–ng1. The additional colours used in the lower panel are used for the $\Delta F = \text{sqrt}\,(H^2 + Z^2)$-$F$ plots of any system: green for ng0, purple for ng1 and light blue for ng2. The following conditions exist for the systems:

– The vector and scalar variometers of ng0 and ng1 are placed in the variometer house under very stable thermal conditions (less than 0.1°C temperature variation per day).
– The vector and scalar variometers of ng2 are located in a small non-magnetic hut thermo-electrically heated, but having a temperature hysteresis of about 1.5°C. The hut is not very well situated with respect to magnetic perturbations in comparison to the variometer house.

The following conclusions can be drawn from the plots in the figures:

– The single vector components *H*, *D* and *Z* of ng0 and ng1 behave completely identical (black lines in Fig. 6.3 and upper panel of Fig. 6.4). The differences are straight lines of very low noise.
– The same can be said about for the total intensity *F* recordings (black line in the lower panel of Fig. 6.4).

- The green, purple and light blue lines show in general a good quality of the base lines and the scale values of all the systems as displayed by stable ΔF plots. Both ng0 and ng1 show very low noise levels.
- The red and dark blue lines in Fig. 6.3 and upper panel of Fig. 6.4 show a sinusoidal behaviour coinciding with the grey lines, which depict the temperature plot of the hut, in which ng2 is located. So the temperature variation can be clearly identified as the reason for the not satisfying difference plot including the ng2 vector components. The most significant thermal influence is on both the horizontal components; its influence on the vertical component is hardly recognised.
- The light blue line at the lower panel of Fig. 6.4 also shows the sinusoidal behaviour, which is caused by temperature variations in the hut where ng2 is located.
- The red and the dark blue line in the lower panel of Fig. 6.4 show spikes and slight jumps. They are caused by the movement of ferromagnetic materials in the vicinity of the F sensor of this system, which is placed outside of the ng2 hut to avoid interferences.

These plots offer a very useful diagnostic procedure of the quality of the observatory data.

A further example of the use of variometer intercomparisons for observatory quality matters is the correction of erroneous scale values. Surlari observatory (SUA) operates three triaxial vector variometers: The fluxgate FGE, a further fluxgate produced by Bartington, England and a torsion photoelectric magnetometer PSM. Only the FGE (HDZ orientation) and Bartington (X, Y, Z orientation) systems were compared. The horizontal components of the Bartington were subsequently converted into H and D. The components H and Z did not show any significant differences, but D clearly showed the diurnal variation of the geomagnetic field. The upper panel of Fig. 6.5 shows the D recording of FGE in black and the Bartington

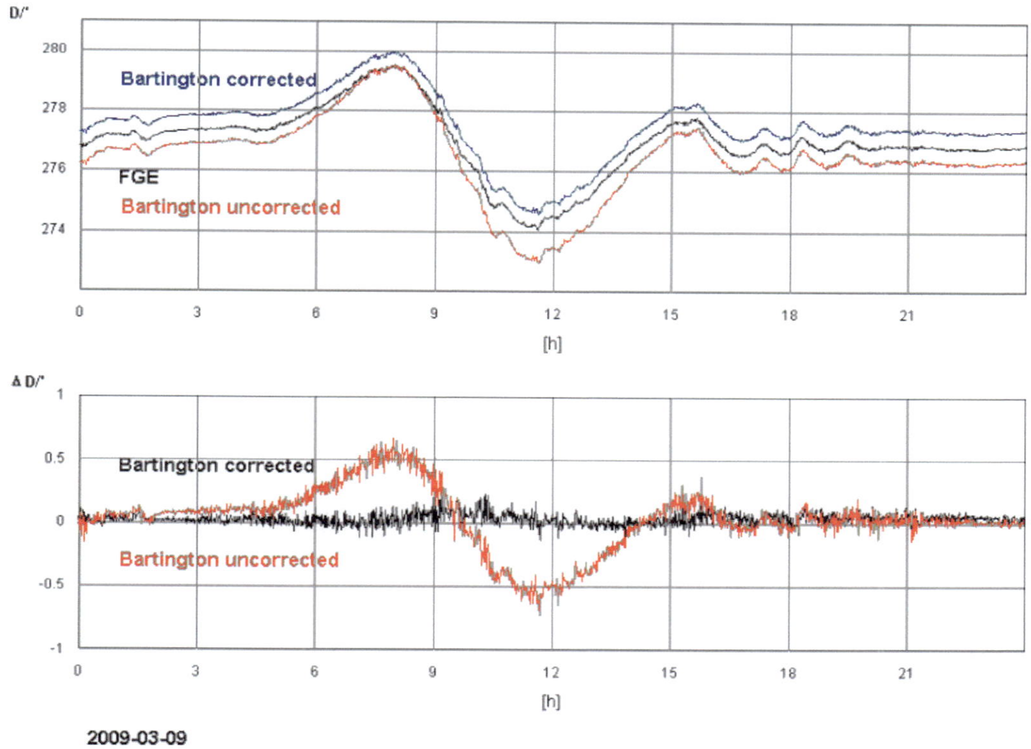

Fig. 6.5 *Upper panel*: D recording of FGE (*black*), Bartington uncorrected D recording (*red*) and Bartington scale value corrected D recording (*blue*). Lower panel: Difference of FGE and scale value corrected Bartington D recordings (*black*) and difference of FGE and uncorrected Bartington D recordings (*red*)

D recording in red. The lower panel of Fig. 6.5 shows the differences of both the *D* recordings in red. The difference coincides clearly with the diurnal variation.

A scale value correction was determined from the differences of the *D* variations and applied to the Bartington *D* recordings. The result is shown at Fig. 6.5. The black plot line depicts the FGE *D* recording, the red one shows the uncorrected Bartington *D* recording while the blue plot represents the scale value corrected Bartington *D* recording in the upper panel. The lower panel compares the difference between the FGE *D* recording and the uncorrected Bartington *D* recording (red plot line) with the FGE *D* recording and the scale value corrected Bartington *D* recording (black plot line). The improvement is clearly to be seen. The noise in the difference plots is caused by the unsynchronized time base of the Bartington data logger. The FGE logger is GPS synchronized, while the Bartington system uses the PC clock, which is manually synchronized from time to time.

A further possibility of variometer inter-comparison is the check of the data of one observatory by means of the data of neighbouring observatories. This procedure is practiced by INTERMAGNET to check the definitive data (see Section 6.5). Similar methods were applied in the check of historical hourly mean values (Korte et al. 2007).

6.3 Quality of Secular Variation Observations

One of the aims of geomagnetic observatories is the monitoring of secular variations. The observatories are also often a standard for regional (national) repeat stations, in which the observation of secular variations is the only objective. The time scale of these observations is tens to hundreds of years. The longer the time series and the greater the absolute accuracy of such observations, the more valuable they are.

The improvement in the quality of data from the whole network of observatories and repeat stations over the world leads to a greater accuracy of global geomagnetic field models, such as IGRF, or magnetic maps, e.g., WMM. In 2001, for instance, IAGA made a decision to enlarge the number of spherical harmonic co-efficients of the IGRF model from 10 to 13 (Maus et al. 2005). Detailed monitoring of secular variations brought about, among other things, a discovery of jerk-type secular variations. Further studies of the jerk-type events make it necessary to investigate secular variations in greater detail (Chambodut and Mandea 2005), thus enhancing the accuracy of geomagnetic observations.

In order to monitor secular variations at observatories, it is necessary to rely on the results of absolute measurements. The accuracy of the data produced in an observatory is based on the accuracy of the absolute measurements (Mandea 2009). The baseline values are calculated by subtracting the result provided by the variometer from the absolute measurement result. Since the zero-value (offset) of the variometer is adopted in an arbitrary manner, the baseline value itself has no physical meaning. However, the scatter of baseline values and the baseline stability provide important information on the correctness of observations. The scatter of the measurement should be close to the absolute measurement accuracy. If this is the case, and the frequency of absolute observations is high enough to adequately sample any variation in the baselines, then we can state with high probability that the recording of variations is reliable, and the scale values, thermal coefficients and the orientation of sensors are determined correctly. The long-term behaviour of baseline values is evidence of the variometer's operation stability, and stable baselines make the monitoring of secular variations much easier. The baseline plot, with frequent absolute measurements, small scatter, and small drift are an indication of the good quality of data (McLean et al. 2004), and good performance of the geomagnetic observatory. In the INTERMAGNET network, the presently required accuracy is ± 5 nT (St-Louis 2008), but many observatories supply data that are even of higher standards and accuracy. The recording of geomagnetic field variations should be stable enough so that the baseline values between the consecutive absolute measurements are nearly the same. Typically, the absolute measurements are made once or twice a week. The frequency of these measurements depend on the variometer characteristics, the stability of the piers and installation and logistical considerations (St-Louis 2008).

Absolute measurements are usually done by means of a DI-fluxgate magnetometer (the fluxgate sensor is fixed to the telescope of a non-magnetic theodolite) and a proton magnetometer. The DI-fluxgate magnetometer measures the magnetic field direction relative to the horizontal plane (inclination) and the angle between the local magnetic north and the geographical north (declination). The measurement requires manual operation and takes about 20 min. per measurement. The observation procedure eliminates unknown parameters such as sensor offset, collimation angles and theodolite errors and relies heavily on operator skill and conscientiousness. The total geomagnetic field value is measured by a proton magnetometer, these measurements are far simpler. The use of optically pumped magnetometers is also considerable, but indeed they are used very rarely for absolute total intensity measurements in magnetic observatories.

Here it is worth mentioning the idea of automatically performing absolute measurements. This idea was born a long time ago, necessitated by the operation of seafloor observatories and unmanned observatories at remote places. Two teams are developing instruments based on different technical ideas. The Geomagnetic AUtomated SyStem—GAUSS (Auster et al. 2007) was derived from the calibration principle of spaceborn fluxgate magnetometers. The other instrument AUTODIF is based on the AUTOmation of the DI-Flux theodolite (Rasson et al. 2009b). Further attempts have been made to operate instruments for absolute vector field measurements. Examples of them are: deltaD-deltaI dIdD quasi absolute variometers (Csontos et al. 2007, Chambodut and Schott 2009) and the construction of a quasi-absolute optically pumped vector magnetometer (Pulz et al. 2009). These new systems aim to be reliably working automated absolute geomagnetic instruments for worldwide use. However it is worth noticing that such instruments are not widely used in geomagnetic observatories, where still two types of instruments are used: one for variation recordings, and the other for linking the re-corded variations with the absolute values of the field components.

The responsibility for the quality of geomagnetic data lies with the personnel of an observatory and its host organisation (McLean et al. 2004). Experience shows, that a very efficient way to improve the quality of observations is to facilitate exchange between the personnel of different organisations. This concerns the comparisons of instruments for magnetic measurements as well as the comparisons of data. A very good example is the international comparisons of absolute instruments and the technical and scientific presentations and discussions during the IAGA Workshops for Geomagnetic Observatories. Another example from the pre-internet era is the comparison of 02 h and 11 h UT momentary values practiced for some decades (since 1955) by many European observatories. The main objective of these comparisons was to detect fluctuations and unnatural jumps of absolute levels of the observatory standard, related to absolute measurements. This concerned such problems as defective absolute instruments, incorrect adjustments, magnetic impurities in the immediate vicinity of the pier, electromagnetic interferences (Schulz and Gentz 1998).

Nowadays, the possibility of comparing data between observatories has grown considerably. The observatories belonging to the worldwide INTERMAGNET network are sending, in the near-real time (within 72 h from recording), their data denoted as reported, to the GIN (Geomagnetic Information Node) centres. These data are immediately available for users. The final data, having a definitive status (DD—Definitive Data) are placed, once the year ends, on the INTERMAGNET web server (http://www.intermagnet.org). Moreover, the INTERMAGNET issues a DVD (formerly a CD-ROM) containing the definitive data from all the observatories belonging to this network.

Comparisons of instruments and data, and first of all the implementation of proton and DI-flux magnetometers, has motivated most of the observatories to discontinue the monitoring of secular variations according to their own absolute levels of the observatory, and replace them by the observations of absolute levels of IAGA standard. Nowadays researchers in the field of geomagnetism know the difference between these two notions only from the literature, e.g., from book by Wienert (1970).

A very important factor contributing to the improvement of the quality of geomagnetic observations is the demand of the users; it is particularly so when the data are within the sphere of interest of large research projects such as, for instance, the Swarm satellite mission, which needs accurate data with a relatively short delay. A similar expectation concerning data is related to the calculation of Dst and AE indices (Baillie et al. 2009). It is due to the users' pressure that the preparations are under way to define a type of data that would be close in accuracy to the definitive data, yet available much sooner. This new type of data, named quasi-definitive (QD), was proposed by Chulliat et al. (2009) during the 11th IAGA Assembly. According to the presented postulates, the QD data are to be published within 30 days after recording, their accuracy being very similar to that of the definitive data. According to Chulliat et al. (2009), the data should be subject to the following verifications:

- visual inspection of the quasi-definitive baseline,
- check of the continuity between the current quasi-definitive and last years definitive data,
- check of scalar residuals ΔF (see above),
- visual inspection of all components at different time scales.

Geomagnetic observatory data from the Southern Hemisphere and the oceans are particularly very valuable for modelling of the secular variation because of the sparse network in these regions. The important task to support data quality from these regions is addressed by many institutes that operate geomagnetic observatories and by the project INDIGO, whose main aim is to establish new observatories or to improve existing ones, enabling them to fulfill INTERMAGNET requirements (Rasson et al. 2009a).

6.4 External Factors Disturbing Observations of the Geomagnetic Field

Geomagnetic observatories are supposed to provide a geomagnetic field record, from which the sources of the geomagnetic field can be studied. These sources are electric currents in the Earth's core (source of the main field), in the ionosphere and magnetosphere, as well as electric currents induced within the Earth from the aforementioned time varying magnetic fields. As disturbing signals we consider all contributions that mask the natural signals. All natural and disturbing signals have a certain time scale and spatial scale that might help identify them. For the study of a specific natural signal in geomagnetic observatory records, the most problematic effects would come from a disturbing signal with a similar time scale/frequency content.

An additional magnetic field is the crustal field or observatory bias, arising from the crustal magnetization. Jankowski and Sucksdorff (1996) recommend to choose observatory locations such that the observatory data is representative for a large area. and that areas with gradients in the crustal field or a laterally inhomogeneous electric conductivity are to be avoided. An observatory bias can hardly be avoided. Any long term changes in the observatory bias are difficult to separate from secular variation (true main field changes) by observatory data alone, but independent satellite-derived main field models allow to determine the observatory bias and potential changes with time (Mandea and Langlais 2002; Macmillan and Thomson 2003). A special problem arises, when the source rock to an observatory bias is exposed to daily or seasonal temperature changes, which then could change the rock magnetization and the resulting magnetic field. Such effects are suspected to influence some observatories. For instance, these effects have been reported for observatories located on volcanic islands like Martin de Vivies-Amsterdam Island (IAGA code AMS) or Port Alfred (IAGA code CZT) where the base lines display an annual variation clearly linked to the annual variation of ambient temperature (pers. comm. Jean-Jacques Schott). Pillar differences (e.g., from local gradients in the magnetic field due to strongly magnetized volcanic rocks), when not properly taken into account, can lead to problems when scalar magnetic field measurements are involved, like for the ΔF-error-detection method or the calculation of baselines. Jankowski and Sucksdorf (1996, on page 129) demonstrate this with an (arbitrary) example, where, in a field strength of roughly 50.000 nT, two pillars have a pillar difference of $\Delta X = 100$ nT. These pillars are subject to a variation of 1000 nT in X, leading to a deviation of 2 nT in the pillar difference ΔF between the two pillars (but ΔX, ΔY and ΔZ stay constant, irrespective of the field variation).

The aforementioned laterally inhomogeneous electric conductivity, e.g., due to inhomogeneity in the earth's crust or a nearby ocean, will also influence geomagnetic field recordings. Some scientific studies might want to investigate these effects, others will be hampered. For a more detailed discussion see chapter 9 in Jankowski and Sucksdorff (1996).

Before a geomagnetic observatory is established, it should also be checked, if the location is affected by disturbing anthropogenic magnetic fields. These could come from close-by magnetic objects or, more likely, from electric currents. Often, geomagnetic observatories were established in a suitable environment, but the increasing urbanization and industrialization of an area could cause disturbing signals at a later time. Therefore, the man-made noise at a geomagnetic observatory has to be checked regularly. A good possibility to do so is the comparison with neighbouring observatories (e.g., Korte et al. 2007), and the occasional monitoring of high frequency magnetic field changes.

Electric power with 50 or 60 Hz usually does not pose a problem since it can be efficiently low-pass filtered for geomagnetic observatory data, where the band-width of interest is below 1 Hz (however, an AC disturbing field could cause problems in extreme cases, e.g., when strong enough to saturate the fluxgate sensor). Additionally, in many technical installations, the phase and the neutral wire run in the same cable, such that a large part of their magnetic field cancel each other. Railways and some power lines conduct the return current through the ground. These currents can spread out significantly and then become problematic to geomagnetic observations in a greater region, especially when they are DC-currents. There are cases documented where the effect is observed in 3 km (Curto et al. 2009) and in 5 km distance, theoretical predictions recommend a minimum distance of 30 km between DC-railways and geomagnetic observatories, depending on geological and meteorological parameters influencing the conductivity (Pirjola et al. 2007). In some countries, railways operate at 16 2/3 Hz and because of return currents spreading far out from the rails and the frequency being closer to 1 Hz than the 50 or 60 Hz discussed above, they could be more relevant disturbing signals. Magnetic fields with 16 2/3 Hz have been observed at Fürstenfeldbruck observatory (IAGA code FUR), and an additional analog low-pass filter is used to suppress them (pers. comm. Martin Beblo and Martin Feller). The railway lines in the area around Niemegk observatory (IAGA code NGK) are more distant to NGK (10 to 20 km) than in the example from FUR. Still, a significant 16 2/3 Hz signal can be observed by induction coils at NGK. However, this signal is not degrading the geomagnetic recordings, which sufficiently behave like a low-pass filter (pers. comm. Hans-Joachim Linthe).

The influence of DC-railways on geomagnetic observatory data has been discussed recently for the Spanish Ebro observatory (IAGA code EBR) by Curto et al. (2009). This previously magnetically undisturbed observatory was affected by railway electrification in 1973, a situation that improved after changes to the railway system in 1997. Although the railway electrification being DC, the main effect on the magnetograms is in the form of spikes. These spikes correspond to arrivals or departures at the train station closest to EBR. An algorithm to remove these spikes from 1-min digital data was developed (Curto et al. 2009), however the authors advocate the relocation of the geomagnetic recordings to a place further away from the railway. Pirjola et al. (2007) and Lowes (2009) have described the problems expected from DC railways to magnetic measurements and treated the problem in a more general and mathematical way, giving insight into the technical specifications of railway electrification as well as formulas to predict their magnetic disturbing fields. A method called "remote reference", which was developed for magnetotelluric measurements, is currently evaluated to study the influence of DC railways on geomagnetic observatories (pers. comm. Anne Neska).

Two single wire, DC power lines were recently found to cause problems at the Danish geomagnetic observatory Brorfelde (IAGA code BFE) and will be discussed in the following. A more detailed account of this investigation can be found elsewhere (Matzka et al. 2009; Fox Maule et al. 2009a, 2009b). During the magnetically very quiet recent years, suspicious disturbances with DC character (time scales >1 h) became obvious in the differences in the geomagnetic recordings between BFE and a neighbouring Danish variometer station or the next geomagnetic observatory (Wingst, IAGA code WNG, Germany). These disturbances were simultaneous in the *H* and *Z* component, and have a very characteristic timing (starting and stopping at the full hour) and must be attributed to disturbances in the recordings at BFE (Fig. 6.6).

The disturbance in H was expected to originate from an underground East-West current sheet running below BFE. This telluric current could be confirmed by magnetotelluric measurements of the electric field in the ground, which had the same temporal pattern as the disturbances in the magnetic record. The disturbance in the *Z* component of BFE was thought to come from a current in a distant, long, horizontal cable. This cable was thought to be a power line that is feeding the current sheet below BFE with its return current. Taking this very simple model of a reasonable current system and taking into account the observed geometry of the disturbance, it was clear that the sought after power line must lie to the South of BFE. To identify

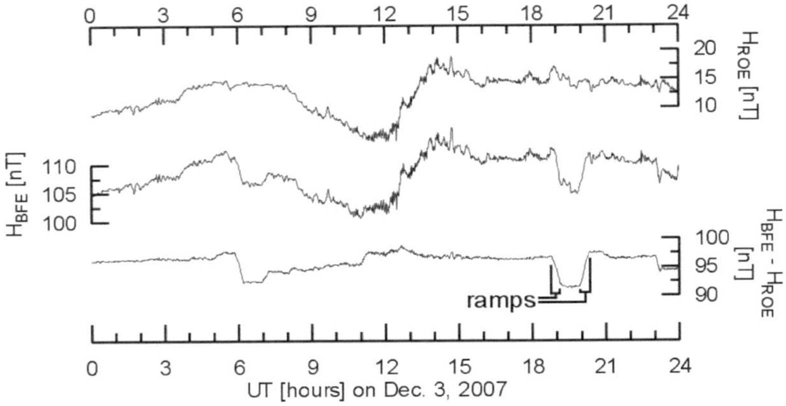

Fig. 6.6 24 h recording of the geomagnetic *H* component in BFE and at the Danish variometer station ROE, as well as the difference between the two recordings BFE minus ROE (*H* is given with an arbitrary offset for BFE and ROE). The natural signals at both stations are similar. A disturbing signal is superimposed on BFE recordings, that is characterized by ramps that lead to a change in the level of the BFE recordings

this power line, a variometer was installed at various locations to the South of BFE, recording for a few days at each location. In these recordings, the amplitude of the disturbance in the *Z* component increased with increasing distance from BFE, indicating that the variometer locations were getting closer and closer to the source. A change in sign of the disturbing signal in the *Z* component between two variometer station locations would be proof that the power line is running between those two locations. In fact, the variometer was set up at several locations, ranging from hundreds of meters to tens of kilometers distance to BFE without crossing the power line. The power line in question was finally identified by matching the time series of the disturbance measures at BFE with the time series of power transmission in the Kontek DC power line (1500 A, 400 kV, connecting Denmark and Germany since 1995), which causes up to about 4 nT in *H* and 2 nT in *Z*. (The geometry of the Kontek power line turned out to be more complex then the simple straight cable assumed in the initial model and lies to the South-East of BFE and is mostly North-South running.) Employing the same method, another DC power line (the Baltic Cable, 1335 A, 450 kV, connecting Sweden and Germany since 1994) was identified to also influence the BFE recordings, however to a lesser extend. The recordings at BFE are almost all the time affected by the power cables. In fact, the only time in Fig 6.6 when both power lines are turned off is the short period between 19 and 20 UT that is surrounded by "ramps", leading to a significantly lower value for the H component.

Both power lines are single wire and send the electric current through an underwater cable in one direction only. They utilize the Baltic Sea to conduct the return current. The temporal pattern of power transmission for the two power lines affects minute means, hourly daily and monthly means (e.g., in periods of several months where the power lines are shut off for maintenance). Its effect on the K index calculated at BFE was confirmed to be small, but non-negligible. Similar effects on the daily variation or secular variation are obviously present, but have not yet been studied in great detail.

In summary, it can be expected that DC railways are currently the most likely source of disturbance for geomagnetic observatories. DC power lines are used to transmit power through the sea, to connect otherwise separated power grids, or for very long distance transmission. The geomagnetic observatory community should keep an eye on how the technology and infrastructure for electric power distribution develops in the future.

6.5 Inspection of Reported and Final Geomagnetic Data, Aims of Verification of Data

Since geomagnetic observatories changed over to digital recording, data processing plays an important rule in the observatory practice. Besides the pure measurement instruments computers became necessary components of the observatory equipment. IAGA considered this fact for example in their observatory workshops, held successfully every 2 years by including the issue of data processing. The introduction of data processing into the operation of magnetic observatories was on the one hand necessary to the ongoing reduction of observatory personnel. On the other hand it opened the possibility of more economic operation of observatories and more effective quality check of measurements, recordings and data.

In the era of the Internet, the preliminary time series (INTERMAGNET's reported data) acquired in geomagnetic observatories are often available in near-real time, while the final absolute time series (definitive data) are disseminated with many months delay, being subject to many checks.

The reported data are usually used in applications where the reliable representation of higher-frequency magnetic field variations is more important rather than absolute levels or secular variation. This concerns, e.g., the forecasts of magnetic activity, radiowave propagation, or space weather. In the case of reported data, it is not possible to verify them prior to dissemination. The present systems of data recording and transmitting, with appropriate Internet application, make it possible to follow the data in close-to-real time. Careful monitoring of the previously sent data and implementation of appropriate correcting measures enable us to improve the quality of data and reduce the number of gaps in the records. Such a procedure is of particular importance in the case of unmanned observatories where the planned visits of the observers are rare. An example of internet application for current monitoring of data from a network of USGS (U.S. Geological Survey) observatories is shown in Fig. 6.7 (Finn and Berarducci 2009).

Another example is the Internet application available at the address http://rtbel.igf.edu.pl which makes it possible to graphically compare the curves from two observatories, as well as to monitor the ΔF difference (Nowozynski and Reda 2007). This application can be used for data in the INTERMAGNET format, both IMFV1.22 and IAF.

In any case, owing to the present-day computer technology, the current data can be verified in a continuous manner. When the observatory has several re-cording sets, any suspicious differences between the

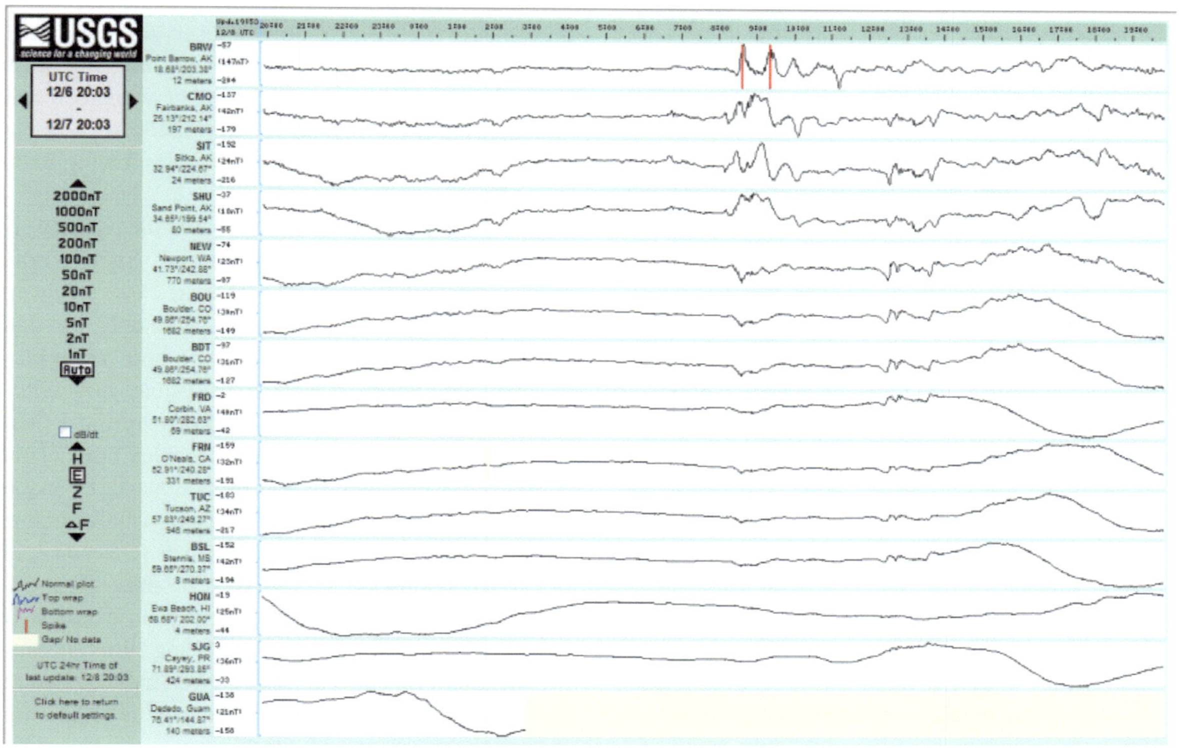

Fig. 6.7 An internet application for on-line monitoring of data from observatories belonging to the USGS (http://geomag.usgs.gov/realtime/)

recording sets should give an automatic alert. The verification of data quality may be faster and more reliable when the observatory records the total intensity F_s by means of a proton magnetometer or optically pumped magnetometer. A comparison of F_s with the total field F_V calculated from the recorded X, Y, and Z (or H, D, and Z) is a very good indication of the data quality. The observation of the ΔF difference ($\Delta F = F_V - F_S$) helps to identify problems such as:

- perturbation by magnetic objects placed too close to the sensors, in particular those changing their location,
- improperly determined scale values,
- effect of temperature changes on the variometer sensors,
- non-orthogonality of variometer sensors.

The methods of checking the recording sets have been described in detail in the book IAGA guide for magnetic measurements and observatory practice" (Jankowski and Sucksdorff 1996). The book contains, among other things, practical advice for daily, weekly or monthly observational routines.

The final, definitive data, prepared in the form of absolute time series, are a flag-ship product of each geomagnetic observatory, so the utmost accuracy of absolute levels is demanded. In INTERMAGNET observatories, the final data are accompanied by files containing information on the instruments, personnel, baseline values, annual means from the current and past years, etc. Prior to publication, all these data should go through several checks, such as:

- Visual inspection of daily, monthly and yearly variations, to detect, first of all, spikes, jumps, or major artificial perturbations. Special care should be given to the curves at the boundaries between days, months and years.
- Verification of baseline values.
- Comparison of the basic recording set with a spare one if exists.
- Verification of the ΔF difference.
- Verification of internal consistency of data, e.g., the consistency of annual means with the time series.
- In case of any doubts, a comparison with data from adjacent observatories.
- Verification of data formats.

Of particular importance is the baseline verification. The inspection should be mainly related to:

- RMS of the residuals between the adopted baseline and the observed baseline values and possible incorrectness in the adopted baseline determinations,
- baseline instabilities and jumps,
- the number of absolute measurements and their even distribution throughout the year.

An example of a baseline plot (made with the IMCDVIEW program discussed later) on the basis of files in the IBFV1.11 format is shown in Fig. 6.8.

It is also important to make a periodic survey of the artificial noise level in geomagnetic observatories. One of the methods of assessing the artificial noise levels is to analyze the rate of changes of the field, dB/dt, for periods of very small geomagnetic activity (Love 2006). The daily dB/dt diagrams for definitive one-minute data of X, Y, and Z can be constructed using the INTERMAGNET page http://www.intermagnet.org/apps/dataplot_e.php?plot_type=db_plot or the IMCDVIEW application. In Fig. 6.9 we present daily dB/dt diagrams of X, Y, and Z of two observatories of similar geomagnetic latitudes but strongly differing in the level of artificial noise in the natural geomagnetic field variations. Such a method of evaluation of artificial noise becomes more and more popular, owing to its simplicity.

To make a fast verification of the quality of data, appropriate software is needed. Such a software is available from the DVD issued every year by INTERMAGNET, which contains not only the data but also the software for inspecting and preliminarily analyzing them. It is a multisystem Java application named IMCDVIEW developed by the British Geological Survey. The software enables most of the above-mentioned quality control actions to be made. It has many options, e.g., it enables a visualization of the one-minute, hourly and daily values (Dawson et al. 2009). This software was used to produce Figs. 6.8, 6.9 and 6.10. In Fig. 6.10 we present an example of the use of this software for visualizing differences in two neighbouring observatories over half a year. In this very case the comparison helped to notice and remove the problem with the Y baseline in one of the observatories.

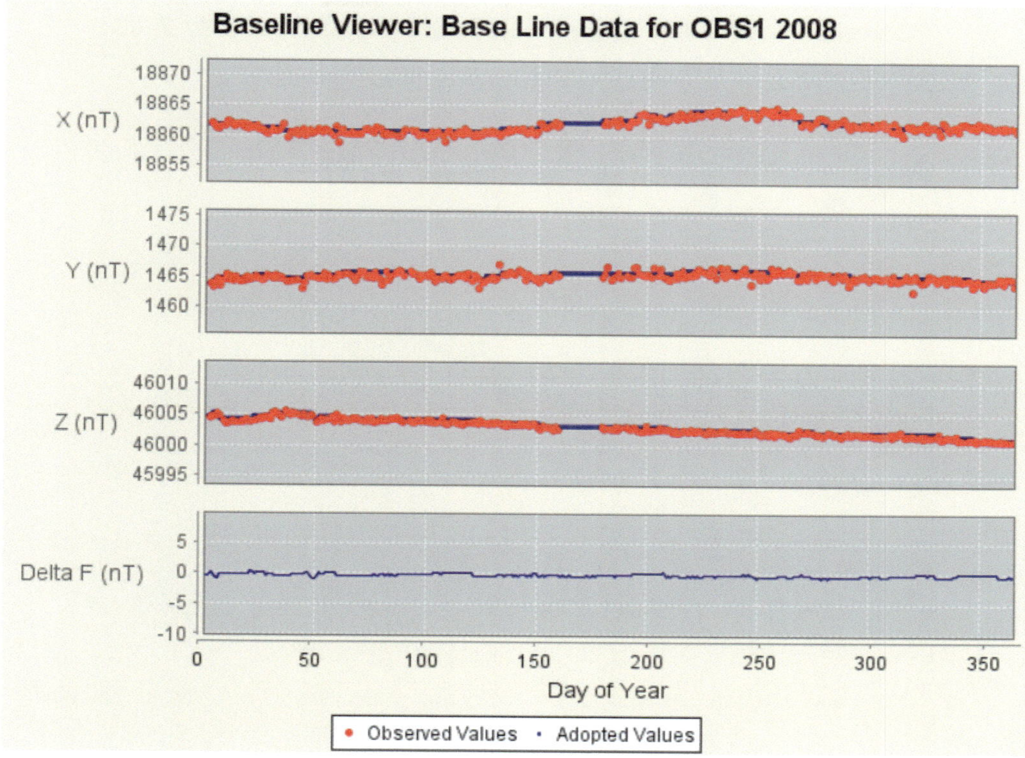

Fig. 6.8 An example of a baseline plot from an observatory with a good baseline. This type of plot can be used for baseline verification

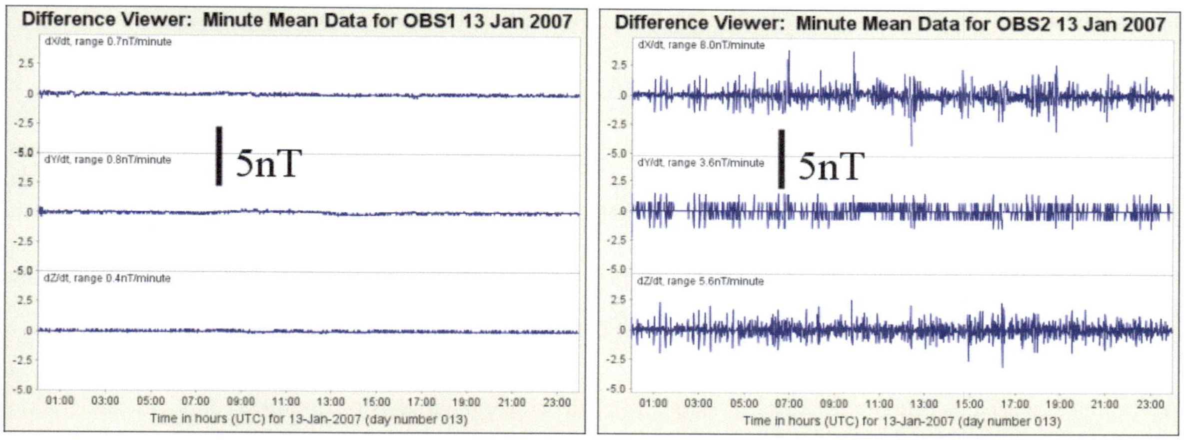

Fig. 6.9 Daily diagrams of dB/dt made with the use of IMCDVIEW software (for two European observatories)

The IMCDVIEW software makes it also possible to inspect the metadata contained on the DVD, as well as to convert the data formats.

It is worth noting that the details of the magnetic observation procedures at each observatory are slightly different. The specific features are related to the equipment or software used, the qualification and experience of the personnel, etc. The observatories determine their own methods of data collecting and verification. Still, the exchange of experience and

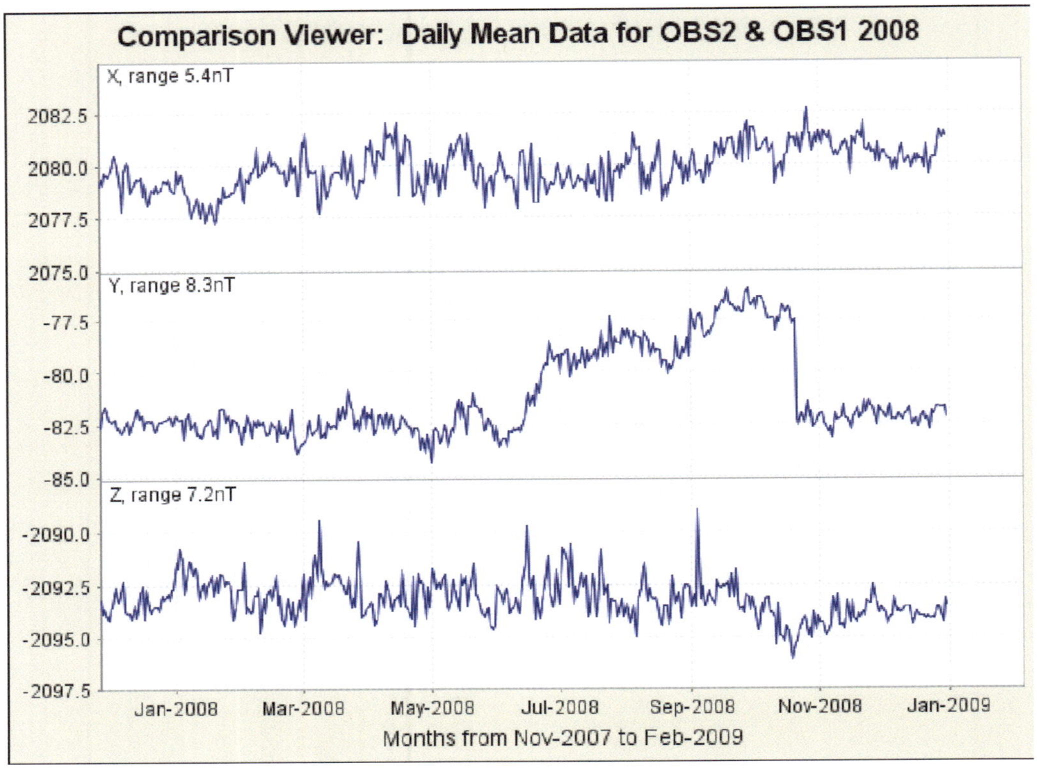

Fig. 6.10 An example for the use of the IMCDVIEW software for making a comparison of two observatories

comparisons of all kinds lead to the eventual refinement of the quality of geomagnetic observations.

The improvement of data quality is augmented by the participation of observatories in the international association IAGA, especially in the world digital magnetic observatories network INTERMAGNET. The observatories gathered in IAGA supply their observational data to the World Data Centres (WDC). The WDCs make various efforts to assure the appropriate data quality norms. It is to be noted, though, that in accordance with the principles described in the "Guide to the WDC System", the final responsibility for the data relies on the data contributors (Rishbeth 1996). For instance, the basic rule at the WDC in Edinburgh is that no corrections to the supplied data be introduced, except correcting the formatting errors and typos. Among other things, this means that no thorough spike-type error verification is made and such errors are not removed. This concerns also discontinuities in the baselines. If the limited verification detects some errors, the WDC in Edinburgh offers help to improve the data quality. However, the subsequent publication of corrected data is in the responsibility of the data contributor (Dawson et al. 2009).

INTERMAGNET's policy on data quality is to a large extent similar to that of WDCs. However, in the case of INTERMAGNET, the minimum quality requirements have been specified, e.g., the minimum accuracy is ±5 nT for definitive data. The observatories trying to be included to this network have to inform what type of instruments they use for magnetic field observations and provide a sample of baseline values over at least a year (St-Louis 2008). The observatories which do not satisfy the requirements are not accepted to the network. INTERMAGNET offers help in preparation of definitive data for publication. This concerns, first of all, the verification of data; the observatories are informed of the faults, in case some are noticed. The verification is mainly related to the data formats, but also, in a limited scope, to the baseline values and perturbations. If the definitive data are not accepted, they are not allowed to be published on the INTERMAGNET internet webpage and the DVD. In extreme cases, when the quality of data is unacceptable and this situation lasts over consecutive years, the observatory may be deprived of the status of INTERMAGNET network member.

6.6 Metadata and Data Quality

6.6.1 What is Metadata?

Any form of data is of little practical application without some knowledge of the source and characteristics of the data, whether that metadata is explicit or implicit. The metadata associated with a set of data may be rudimentary—such as the type of data, the time and location of the recording—or may be more specific—sensor type, processing methods, etc.—but any use of a data set will draw on the associated metadata, hence it is vital to the data user that data and their associated metadata are coherent, accurate and accessible.

The need to record metadata along with geomagnetic data has always been recognised, however the metadata fields and the manner in which the metadata have been published have varied with time and with the institute publishing the data. With the demise of the published observatory yearbook and the move to centralised electronic data sources, the problem of data and metadata becoming disassociated and inaccessible is now beginning to be addressed. Reay et al. (2010) describe how metadata have been published historically and how there is currently a concerted international effort to set metadata standards aimed at giving consistency to the fields of metadata recorded and also to the form in which metadata is published.

6.6.2 Metadata in Geomagnetism

One of the basic purposes of metadata is to provide a data user with elementary information for a particular data set. For geomagnetic observatory data, the metadata is likely to contain time and location of observation (latitude, longitude, elevation, observatory name and IAGA code), orientation of components and data units.

However, a digital sample of the magnetic field at a particular point and time will have been through a series of processes that manipulate the sample, including instrument response, sampling technique, scaling, baseline fitting, post processing, re-orientation, reformatting and distribution. Where any of these processes has an effect on the value of the sample, there is also an effect on the quality with which that data point represents the original magnetic field. A complete metadata record would detail or provide reference to all processes impacting data quality, from the instrument transfer function to the algorithms used to derive the published data products and how these change with time. Ideally, metadata provides traceability for a data sample to the original data and should be applied to all data products, including geomagnetic indices. As an example of the relevance of metadata to the quality of a data set, Martini and Mursula (2006) describe the difficulty in identifying the source of an inconsistency in the Inter-Hour Variability index of Eskdalemuir Observatory data in the absence of a documented method of calculating the hourly values.

In many applications of observatory data, such as down-hole navigation by the oil and gas industry, data error bounds are regarded as significant as the data itself. Metadata for a data sample could usefully include comprehensive quality indices such as absolute error bounds and estimates of noise levels, but an advantage of digitised metadata over observatory yearbooks is that quality factors such as absolute observations, baseline residuals and inter-comparisons between instruments can more readily be published and manipulated, allowing data quality indices, such as error bounds, to be derived.

If digitised metadata are to mirror the information historically contained in yearbooks, then these records may also include reference to operational practice at observatories such as observatory publications, staff lists, non-geomagnetic activities, etc. Of more direct relevance to data quality, metadata may record man-made or natural signal contained in the data where these have been identified by the data producer. It may be the case that the originating field contains signals (such as those caused by ground-motion from earthquakes) that are either of interest to or regarded as unwanted noise by a data user. Both Kakioka Observatory (Tokumoto and Tsunomura, 1984) and Brorfelde Observatory (Matzka et al. 2009, Fox Maule et al. 2009a, 2009b) have documented the effect of earth-leakage from local or regional high-current DC electricity supplies on magnetic recordings.

A complete metadata record would document such events and provide detail of any post-processing technique applied to the data designed to minimise these artificial signals, thus providing a resource for future reference.

Generally, it should be the purpose of metadata to record sufficient information on factors that have had an influence on the quality of a data sample such that a prospective data user will be able to assess the quality of the sample and its suitability for their application. Hence, data quality and metadata are fundamentally linked.

However, it is not currently common practice to explicitly quantify all factors influencing data quality in metadata and it is most generally the case that data quality is implied by the standards to which a data set conforms. Even so, data standards may not be explicitly recorded in the metadata, but are often implied in the data; as in the case when data are sourced from an organisation such as INTERMAGNET. As data are easily redistributed to WDCs and other data sources (e.g., SuperMAG) the link between the data and the quality metadata is frequently broken along with the user's reference to the data standards. Indeed, data quality may be adversely affected by the redistribution if data are transformed to an alternate coordinate system or data values are truncated to conform to a spe-cific data format.

It is worth noting that data standards are frequently loosely defined (such as those for INTERMAGNET one-minute data) and do not comprehensively describe all of the processes and parameters affecting the quality of that data. This is in part deliberate, to satisfy the requirement for a standard while encompassing the existing broad range of observatory designs, instrument types and procedures. However, this results in an incomplete record of the metadata. It is also commonplace for standards to change with time as new instrumentation and data processing practices are accepted by the observatory community, so existing documented standards cannot be expected to be applicable to historic or future data. This underlines the necessity to publish data with a complete metadata record and to maintain the metadata with the data.

In addition, the quality standards of a particular piece of data are not static. With modern communication technologies, it is straightforward to revise and redistribute data, creating multiple versions of a data set, each version with a different set of quality attributes. As an example, INTERMAGNET observatories will transmit preliminary data to meet the needs of users in near real-time, but will reassess instrument baselines and publish definitive data after the calendar year end. With the upcoming *Swarm* satellite mission, there is a new initiative encouraging observatories to publish data within weeks of recording to a level of data quality close to definitive. For quasi-definitive observatory data to be a practical resource to the global geomagnetic field modelling community, observatories will need to be able to estimate and publish the level of data quality (likely to be measured as a statistical distribution of the errors against definitive quality data) along with the data. It is foreseeable that in the near future, observatories will routinely publish multiple versions of data, each with a revision of the baselines and an associated improvement in data quality with time. It will therefore become increasingly important to observatories, data managers and data users that data are published with a full metadata record describing version numbers, publication date and data quality indicators.

6.6.3 Geomagnetic Metadata Standards

Presently, no digital metadata standard exists for geomagnetic observatory data so it is imperative that a common standard is established to meet the needs of the user community and to preserve the information formerly published in observatory yearbooks. The metadata working group being led by the National Geophysical Data Center Boulder and the other World Data Centres for Geomagnetism aims to establish such a standard but, as described in (Reay et al. 2010), is faced with a number of difficulties. Although metadata standards have been well defined for spatial data, none of these can be readily applied to a time-series such as observatory data where all of the metadata fields are also time-variable.

The working group is likely to set a basic standard with potential for future expansion, giving the possibility to add data quality indicators as required. Consideration will be given to the tools allowing data-producing institutes to maintain their own metadata records and also to the mechanisms with which the link between the data and the metadata are preserved through distribution. As data can be manipulated by data centres, so the metadata must also be dynamic and data centres must also be encouraged to maintain the metadata as they do the data, documenting such modifications to the data as reorientation, file format changes and versioning control.

A further consequence of the move away from observatory yearbooks to digital data distribution is the potential loss of a publication date for definitive data. Currently, WDCs issue an annual call for data and INTERMAGNET sets final submission date for yearly data to encourage data producers to publish a definitive revision of the data with finalised baselines and metadata. Although definitive data are occasionally corrected, a definitive data publication date benefits the data producers by imposing a regular epoch when data and metadata are set and published. It also benefits the user community by providing a unique data set that can be applied and referenced to. There are risks that, as observatories routinely revise and republish data to meet the demands for faster distribution, definitive data are no longer published. The observatory community and data centres should continue to encourage the publication of definitive data, along with the associated metadata, and these data should continue to be regarded as those with the highest quality.

References

Auster HU, Mandea M, Hemshorn A, Pulz E, Korte M (2007) Automation of absolute measurement of the geomagnetic field. Earth Planets Space 59(9):1007–1014

Baillie, O Clarke E, Flower S, Reay S, Turbitt, C (2009). Reporting quasi-definitive observatory data in real near time. Presentation during 11th IAGA Assembly, Sopron, Hungary, 502-MON-O0930-1130

Bobrov, YN (1962) Series of quartz magnetic variometers (in Russian). Geomagn Aeronom 2(2):348–356

Brunelli E, Raspopov, OM, Yanovskiy BM (1960) Vysokochuvstvitelnaya varyatsionnaya stantsiya dlya registratsii korotkoperiodnykh kolebaniy magnitnogo polya Zemli. Sb po geofiz priborstr 5

Chambodut A, Mandea M (2005) Evidence for geomagnetic jerks in comprehensive models. Earth Planets Space 57(2):139–149

Chambodut A, Schott, J-J (2009) Geomagnetic Observatory Practice, Instrumentation and Network. Presentation during 11th IAGA Assembly, Sopron, Hungary, 508-FRI-O1330-0632

Chulliat A, Peltier A, Truong F, Fouassier, D (2009) Proposal for a new observatory data product Quasi-definitive data. Presentation during 11th IAGA Assembly, Sopron, Hungary, 502-MON-O0945-1255

Chulliat A, Savary J, Telali K, Lalanne, X (2009) Acquisition of 1-second data in IPGP magnetic observatories. In: Love JJ (ed) Proceedings of XIII IAGA Workshop on Geomagnetic Observatory Instruments, Data Acquisition and Processing. U.S. Geological Survey Open File Report 2009-1226, 54–59

Csontos A, Hegymegi L, Heilig B (2007) Temperature tests on modern magnetometers. In: Reda J (ed) Proceedings of the XIIth IAGA Workshop on Geomagnetic Observatory Instruments, Data Acquisition and Processing, Belsk. Publs Inst Geophys Pol Acad Sc C-99(398):171–177

Curto JJ, Marsal S, Torta JM, Sanclement E (2009) Removing spikes from magnetic disturbances caused by trains at Ebro Observatory. In: Love JJ (ed) Proceedings of XIII IAGA Workshop on Geomagnetic Observatory Instruments, Data Acquisition and Processing. U.S. Geological Survey Open File Report 2009-1226, 60–66

Dawson E, Reay S, Macmillan S, Flower S, Shanahan, T (2009) Quality control procedures at the World Data Centre for Geomagnetism (Edinburgh). Presentation during 11th IAGA Assembly, Sopron, Hungary, 502-MON-P1700-1167

Finn CA, Berarducci AM (2009) USGS magnetic observatory operations: status and planned improvements. Presentation during 11th IAGA Assembly, Sopron, Hungary, 502-MON-O0900-0653

Fox Maule C, Thejll P, Neska A, Matzka J, Pedersen LW, Nilsson A (2009a) Analyzing and correcting for contaminating magnetic fields at the Brorfelde geomagnetic observatory due to high voltage DC power lines. Earth Planet Space 61(11):1233–1241

Fox Maule C, Matzka J, Pedersen LW (2009b) Correcting geomagnetic observations from Brorfelde, Denmark for disturbances caused by nearby power lines. Presentation during 11th IAGA Assembly, Sopron, Hungary, 502-MON-O1330-0108

Hegymegi L, Heilig B, Csontos A (2004) New suspended dIdD magnetometer for observa-tory (and field?) use. Proceedings of the XIth IAGA Workshop on Geomagnetic Observatory Instruments, Data Acquisition and Processing, Kakioka and Tsukuba, Japan, pp 28–33

Isac A, Mandea M, Linthe H-J (2009) Surlari Observatory: where we have been and where we are. Presentation during 11th IAGA Assembly, Sopron, Hungary, 502-MON-O0915-0906

Jankowski J, Marianiuk J, Ruta A, Sucksdorff C, Kivinen M (1984) Long-term stability of a torque balance variometer with photoelectric converters in observatory practice. Geophys Surv 6:367–380

Jankowski J, Sucksdorff C (1996) Guide for Magnetic Measurements and Observatory Practice, IAGA, Warsaw, 235 pp

Korepanov V, Marusenkov A, Rasson J (2009). 1-second INTERMAGNET standard magnetometer—1-year operation. Presentation during 11th IAGA Assembly, Sopron, Hungary, 502-MON-O1145-0155

Korte M, Mandea M, Olsen N (2007) Worldwide observatory hourly means from 1995 to 2003: Investigation of their quality. In: Reda J (ed) XII IAGA workshop on geomagnetic observatory instruments, data acquisition and processing. Publs Inst Geophys Pol Acad Sc C-99(398):275–283

Linthe H-J (2004) MAGDALOG-Data Logger for Magnetic Observatories. Proceedings of the XIth IAGA Workshop on Geomagnetic Observatory Instruments, Data Acquisition and Processing, Kakioka and Tsukuba, Japan, pp 90–94

Love JJ (2006) Examination of INTERMAGNET Observatory Noise, Minutes of the OPSCOM/EXCOM INTERMAGNET MEETING, Warsaw 2006, unpublished.

Lowes FJ (2009) DC railways and the magnetic fields they produce—the geomagnetic context. Earth Planet Space 61(8):i–xv

Macmillan S, Thomson A (2003) An examination of observatory biases during Magsat and Orsted missions. Phys Earth Planet Int 135:97–105

Mandea M, Langlais B (2002) Observatory crustal magnetic biases during Magsat and Orsted satellite missions. Geophys Res Lett 29. doi:10.1029/2001GL013693

Mandea M, Isac A (2011) Geomagnetic Fields, Measurement Techniques. In: Harsh K. Gupta (ed) Encyclopedia of Solid Earth Geophysics, Springer. doi 10.1007/978-90-481-8702-7

Martini D, Mursula K (2006) Correcting the geomagnetic IHV index of the Eskdalemuir Observatory. Ann Geophys 24:3411–3419

Matzka J, Pedersen LW, Fox Maule C, Neska A, Reda J, Nilsson A, Linthe H-J (2009) The effect of high-voltage DC power lines on the geomagnetic measurements at BFE. In: Love JJ (ed) Proceedings of XIII IAGA Workshop on Geomagnetic Observatory Instruments, Data Acquisition and Processing. U.S. Geological Survey Open File Report 2009-1226, 162–170

Maus S, Macmillan S, Chernova T, Choi S, Dater D, Golovkov V, Lesur V, Lowes F, Luhr H, Mai W, McLean S, Olsen N, Rother M, Sabaka T, Thomson A, Zvereva T (2005) The 10th generation international geomagnetic reference field. Phys Earth Planetary Inter 151:320–322

McLean S, Macmillan S, Maus S, Lesur V, Thomson A, Dater D (2004) The US/UK World Magnetic Model for 2005–2010, NOAA Technical Report NESDIS/NGDC-1

Mohr PJ, Taylor BN, Newell DB (2008) CODATA recommended values of the fundamental physical constants: 2006. Rev Mod Phys 80(2):633–730

Nowozynski K, Reda J (2007) Comparison of observatory data in quasi-real time. In: Reda J (ed) XII IAGA Workshop on Geomagnetic Observatory Instruments, Data Acquisition and Processing, Belsk, Poland. Publs Inst Geophys Pol Acad Sc C-99(398):123–127

Pirjola R, Newitt L, Boteler D, Trichtchenko L, Fernberg P, McKee L, Danskin D, Jansen van Beek G (2007) Modelling the disturbance caused by a dc-electrified railway to geomagnetic measurements. Earth Planets Space 59(8):943–949

Pulz E, Jäckel K-H, Bronkalla O (2009) A quasi absolute optically pumped magnetometer for the permanent recording of the Earth's magnetic field vector. In: Love JJ (ed) Proceedings of the XIIIth IAGA Workshop on Geomagnetic Observatory Instruments, Data Acquisition and Processing, Boulder and Golden, Colorado, USA. U.S. Geological Survey Open File Report 2009-1226, 216–219

Rasmussen O (1991). The proton gyromagnetic ratio. IAGA News, No 30, p 78

Rasson J (2009) Testing the time-stamp accuracy of a digital variometer and its data logger. In: Love JJ (ed) Proceedings of the XIIIth IAGA Workshop on Geomagnetic Observatory Instruments, Data Acquisition and Processing, Boulder and Golden, CO. U.S. Geological Survey Open File Report 2009-1226, 225–231

Rasson J, Ameen A, Ashfaque M, Borodin P, Brenes JL, Daudi E, Efendi N, Flower S, Hidayat M, Husni M, Kampine M, Kusonski O, Langa A, Monge I, Mucussete A, Ghulam, MG, Nhatsave A, Santika IKO, Riddick J, Suharyadi D, Turbitt C, Yusuf M (2009a) INDIGO: Better Geomagnetic Observatories where we need them. Presentation during 11th IAGA Assembly, Sopron, Hungary, 502-MON-P1700-0162

Rasson J, van Loo S, Berrami N (2009b) Automatic DIflux measurements with AUTODIF. In: Love JJ (ed) Proceedings of the XIIIth IAGA Workshop on Geomagnetic Observatory Instruments, Data Acquisition and Processing, Boulder and Golden, Colorado, USA. U.S. Geological Survey Open File Report 2009–1226, 220–224

Reay SJ, Allen W, Baillie O, Bowe J, Clarke E, Lesur V, Macmillan S (2005) Space weather effects on drilling accuracy in the North Sea. Ann Geophys 23:3081–3088

Reay S, Nose M, Sobhana A, Sergeyeva N, Kharin E, McLean S, Herzog DC (2010) Data and Metadata: Types and Availability. IAGA Special Sopron Book Series—Geomagnetic Observations and Models (Chapter 7)

Rishbeth H (ed) (1996) Guide to the World Data Center System. General Principles World Data Centers Data Services. Issued by the Secretariat of the ICSU Panel on World Data Centres

Schulz G, Gentz I (1998) Results of the momentary value comparison between European observatories—A summary of the last two decades. VIIth IAGA Workshop on Geomagnetic Observatory Instruments, Data Acquisition and Processing, Niemegk. Scientific Technical Report STR98/21. In: Best A, Linthe H-J (ed), pp 366–374

Shanahan TJG, Turbitt CW (2009) Evaluating the noise for a commonly used fluxgate magnetometer—for 1-second data. In: Love JJ (ed) Proceedings of XIII IAGA Workshop on Geomagnetic Observatory Instruments, Data Acquisition and Processing. U.S. Geological Survey Open File Report 2009-1226, 239–245

St-Louis BJ (ed) (2008) INTERMAGNET Technical Reference Manual, Version 4.4, Available at www.intermagnet.org

Tokumoto T, Tsunomura S (1984) Calculation of Magnetic Field Disturbance Produced by Electric Railway. Mem Kakioka Magnetic Observ 20(2):33–44

Wienert KA (1970) Notes on geomagnetic observatory and survey practice. Published in 1970 by UNESCO, Paris, 217 pp, Series: Earth sciences

Chapter 7

Magnetic Observatory Data and Metadata: Types and Availability

Sarah J. Reay, Donald C. Herzog, Sobhana Alex, Evgeny P. Kharin, Susan McLean, Masahito Nosé, and Natalia A. Sergeyeva

Abstract The availability of magnetic observatory data has evolved rapidly with the transition of observatories from analogue photographic magnetograms to digital electronic recordings, and the advent of the internet for instant global access to information of every sort. Metadata (information about the data) is undergoing its own transformation in order to accompany the rapid and extensive dissemination of these data. This chapter describes the types of data historically and currently produced by geomagnetic observatories and introduces new data types such as one-second and quasi-absolute data recently discussed at the 11th IAGA Scientific Assembly in Sopron, Hungary. We review the availability of these data types from the World Data Centres, INTERMAGNET and other sources. Finally, we discuss developments in metadata describing the current efforts in the geomagnetism community to gather, store and distribute this information about the data to better assist scientific discovery.

7.1 Introduction

Magnetic observatories continuously measure the strength and direction of the Earth's magnetic field (Macmillan, 2007). The availability of data from these observatories has evolved greatly with the transition from the production of analogue photographic magnetograms to digital electronic recordings, and the development of technological capabilities such as the Personal Computer (PC); CD- and DVD-ROMs for the inexpensive archiving of large quantities of data; and the internet for instant global access to information of every sort. Metadata (information about the data) are undergoing their own transformation in order to accompany the rapid and extensive dissemination of these various data sets. Metadata describes the content, quality, originator, and other characteristics of a data set, and supports the proper interpretation and use of the data.

For more than a century, magnetic observatory data were only available on printed materials as analogue traces on photographic paper known as magnetograms, products derived from those traces, such as temporal averages (hourly, daily, etc.) and magnetic activity indices. Metadata consisted of yearbooks in which information about the observatories was provided, and copies of these various data products were reproduced in these books. The launching of the International Geophysical Year (IGY) in 1957 brought with it the beginning of the World Data Centres (WDC) established as central repositories where scientists could go to obtain data from a number of locations worldwide.

In the late 1960s, magnetic observatories began converting to digital electronic equipment and a true revolution in data types and availability was underway. The individual traces could now be kept separate (traces often crossed each other and became confused on magnetograms) and the standard observatory data product changed from hourly means to minute means. Most significantly, these data could now be collected, manipulated, archived and disseminated using computers.

S.J. Reay (✉)
British Geological Survey, Murchison House, West Mains Road, Edinburgh EH9 3LA, UK
e-mail: sjr@bgs.ac.uk

M. Mandea, M. Korte (eds.), *Geomagnetic Observations and Models*, IAGA Special Sopron Book Series 5,
DOI 10.1007/978-90-481-9858-0_7,

The emergence of the PC in the 1980s, and the development of CD-ROMs for data storage saw the beginning of "small science" where individual scientists were able to have large quantities of data available to study for minimal cost. In 1990, for example, the U.S. Geological Survey (USGS) produced the first CD-ROM of one-minute magnetic observatory data for a 5-year period (1985–1989) from the USGS network of (then) 13 stations (Herzog and Lupica, 1992), and served as the model for the production of a series of CD-ROMs by INTERMAGNET containing a global set of observatory data.

During the 1980s and 1990s, significant advances were also made in the timeliness of data delivery through the use of satellites. Data Collection Platforms (DCPs), equipped with satellite transmission systems, began delivering data from observatories to data processing centres, with delays of the order of tens of minutes. An organisation called INTERMAGNET was formed which began transferring data from observatories worldwide using satellites, and which developed Geomagnetic Information Nodes (GINs) where users could obtain preliminary data from the participating stations quickly by means of email requests.

But without question the greatest advancement in data availability has come about from the impact of the World Wide Web (WWW) as a data discovery and distribution tool for the internet. Beginning in 1992, when WWW access became more broadly available to the general public, the growth in use of the WWW for access to data has been astonishing. Today virtually every institution, public and private, has a website with the capability of providing access to their information and data. As magnetic observatories have progressed in their abilities to collect and process data faster and with higher cadences (one-second data, for example), these data have also become ever more readily available. This change has implications for traditional WDC roles of data discovery and delivery, but also for long-term data archive.

In this chapter we will look at some of these developments and advances in data and metadata types and how they have been made available. Using an historical perspective, and with the help of results presented at the 11th IAGA Scientific Assembly held in Sopron, Hungary in 2009, we hope to provide an account of where we are today regarding these topics.

7.2 Data Types

7.2.1 *Printed Media*

From the earliest days of magnetic observatory operations, data were collected on paper records of one sort or another. Variations in the magnetic field were recorded as continuous traces on photographic paper. Observations of the absolute magnitudes of the field components were recorded and provided calibration reference data with which to convert the analogue trace amplitudes into magnetic field values. These observations could then be reduced to produce tables of results at various time-scales: hourly, daily, monthly, annual. In time, range data started to be presented and then activity measures, such as the *K*-index (1939) were developed. Magnetic records also included absolute measurements, information on the baseline used and indications of the most quiet and active days (from 1911).

These records were presented in the form of yearbooks. These yearbooks and photographic analogue traces are archived and are available from different sources and in a variety of formats. Many magnetogram traces are stored on microfilm and microfiche. In this digital age we are keen to preserve these, often, fragile documents and allow greater access for analysis of these historical records. Some have been scanned as images and some have been entered directly to form digitised electronic files.

7.2.1.1 Eye-Observations

The earliest records from magnetic observatories were from eye-observations made manually by observers at set times during each day. For example at Greenwich observatory observations were made at 2-hourly intervals except on "term-days" when measurements were every 5 min. At Colaba observatory observations were made hourly except during disturbed times when they were every 15 min or every 5 min during severely disturbed conditions. Figure 7.1 shows the 5-min eye-observations for Colaba during the September 1859 "Carrington Storm". The laborious nature of these manual observations encouraged the development of automatic recording devices. In 1847 Brooke designed the automatic photographic magnetograph that was to

BOMBAY MAGNETICAL OBSERVATIONS. 169

DISTURBANCE OBSERVATIONS, 1859.

Date and Göttingen Mean Time.	Declination. Large. At Full Time.	Declination. Small. 5 min. after Full Time.	Horizontal Force Magnetometers. Large. 2 min. after Full Time.	Horizontal Force Magnetometers. Small. 4 min. after Full Time.	Thermometers. Large. H. F. M.	Thermometers. Small. S. H. F.	Vertical Force Magnetometer. 2 min. before Full Time.	Vertical Force Magnetometer. Thermometer.
H. M.			Sc. Read. Uncorrected.	Sc. Read. Uncorrected.			Sec. Read. Uncorrected.	
Sept. 1st. 18.15	28.070	29.22	15.27	28.22	83.6	81.0	59.00	82.9
18.30	56.131	56.42	11.52	21.20	84.0	81.2	58.35	83.3
18.45	49.202	49.00	—	—	84.0	81.4	59.45	83.3
19.00	49.202	48.95	3.20	10.22	84.1	81.6	58.75	83.3
19.05	56.063	55.00	2.47	16.80	84.1	81.8	57.55	83.7
19.10	50.368	49.22	3.88	17.20	84.1	81.8	56.40	83.6
19.15	48.859	46.00	7.78	21.10	84.4	81.9	55.60	83.6
19.20	36.548	35.00	13.00	26.40	84.5	81.9	54.86	83.9
19.25	37.940	37.44	13.55	26.50	84.5	81.9	56.20	83.9
19.30	31.707	31.24	13.45	26.20	84.6	82.0	55.75	83.9
19.35	28.687	31.57	12.57	27.50	84.7	82.3	55.25	84.0
19.40	27.727	28.70	12.10	22.75	84.7	82.5	57.62	84.0
19.45	29.991	30.79	13.65	24.40	84.7	82.6	58.22	84.0
19.50	32.667	32.62	14.45	26.20	84.8	82.7	58.02	84.1
19.55	24.846	28.70	14.46	27.65	84.8	82.7	57.48	84.1
20.00	30.196	23.99	16.54	28.80	84.9	83.0	58.55	84.1
20.05	22.168	24.25	15.17	29.35	84.9	83.0	58.25	84.2
20.10	24.914	23.73	16.28	28.55	84.9	83.0	59.28	84.2
20.15	31.088	25.56	15.60	29.07	84.9	83.1	58.64	84.3
20.20	22.101	22.63	14.45	30.13	84.9	83.2	58.15	84.1
20.25	17.298	18.76	17.35	29.25	84.9	83.2	59.35	84.1
20.30	18.379	13.01	17.50	30.90	85.0	83.5	58.00	84.1
20.35	23.130	20.33	13.35	23.30	85.1	83.5	59.45	84.2
20.40	19.767	19.02	13.45	25.20	85.1	83.6	59.55	84.3
20.45	18.670	17.90	14.40	26.15	85.1	83.6	59.00	84.3
20.50	21.071	19.81	15.45	27.88	85.0	83.6	59.60	84.4
20.55	18.327	19.02	16.85	29.65	85.0	83.6	57.50	84.4
21.00	22.787	17.72	13.40	26.25	85.2	83.7	59.85	84.5
21.05	19.356	20.33	13.70	25.50	85.2	83.8	60.62	84.7
21.10	20.522	20.85	12.75	23.60	85.3	83.8	60.40	84.8
21.20	19.356	20.33	12.80	25.50	85.3	83.9	59.85	85.0
21.30	18.944	19.81	14.05	25.40	85.3	84.0	59.60	85.0
21.40	17.778	18.76	13.85	25.70	85.3	84.0	59.50	85.0
21.50	14.554	18.50	14.10	26.00	85.3	84.0	59.40	85.0
22.00	15.445	17.45	13.65	25.70	85.3	84.0	59.88	85.0
22.10	16.955	17.72	13.59	25.45	85.3	84.0	60.15	85.0
22.20	17.641	18.65	13.45	24.90	85.3	84.0	59.85	85.0
22.30	17.572	18.76	13.48	25.15	85.4	84.0	59.95	85.0
22.45	17.641	18.76	14.35	26.07	85.5	84.0	60.00	85.2
23.00	17.778	19.02	13.93	25.50	85.4	83.9	60.25	85.3
23.15	19.013	20.07	14.40	25.75	85.3	83.9	60.62	85.3
23.30	21.758	22.68	14.53	26.85	85.2	83.8	60.65	85.2
23.45	23.198	24.40	14.70	26.65	85.1	83.6	61.00	85.1
Sept. 2nd. 0.00	24.639	25.04	14.64	26.68	85.0	83.5	61.00	85.1
0.15	23.130	25.04	15.40	26.65	85.0	83.5	60.60	85.0
0.30	23.473	26.08	14.92	26.50	84.9	83.4	60.30	85.0
0.45	23.130	24.51	17.23	28.05	84.8	83.3	59.85	85.0
1.00	19.836	21.90	17.43	28.60	84.6	83.2	60.20	85.0
1.15	20.590	20.07	16.55	29.55	84.3	82.8	59.45	85.0
1.30	17.298	20.91	14.97	28.00	84.1	82.8	59.75	85.0
1.45	20.865	22.16	13.10	24.70	83.9	82.8	[illegible]	84.9
2.00	19.836	20.07	12.57	25.00	83.6	82.4	60.50	84.7
0.15	31.707	32.50	12.45	24.60	83.5	82.4	60.60	84.5
0.30	24.363	21.38	12.10	23.45	83.3	82.1	60.60	84.3
0.45	20.453	19.28	10.97	23.50	83.2	82.0	60.60	84.1
3.00	19.767	21.64	10.50	22.00	83.0	82.0	60.20	84.0
3.15	22.305	23.31	10.20	21.20	83.0	82.0	60.75	84.0
3.30	24.159	25.82	9.87	21.30	83.0	82.0	60.65	83.9

43—1859.

Fig 7.1 Eye-observations made with Grubb's magnetometer at Colaba Observatory for 1–2 September 1859. The observation frequency increases from 15 to 5-min observations as the severity of the storm increases (see Tsurutani et al. 2003). (Credit: Indian Institute of Geomagnetism) form the standard technique for observatory operations for over a century (Brooke, 1847).

7.2.1.2 Magnetograms

For more than a century, the changes in the Earth's magnetic field have been recorded at observatories as analogue traces on sheets of photographic paper. Under darkroom conditions, mirrors attached to magnets suspended from quartz filament fibres, known as *variometers*, reflect light from a source through a horizontal lens and onto a sheet of photographic paper generally mounted on a drum that completes a rotation once a day. Time marks on the hour, generated by a light flash, provide a temporal reference. These paper records are known as *normal magnetograms*. There are also two auxiliary types of magnetograms produced at some stations: *storm magnetograms*, with reduced sensitivity, provide better amplitude resolution during periods of enhanced magnetic activity; and *rapid-run magnetograms* are recorded on drums that rotate more than once a day as the drum moved laterally, providing greater time resolution. In addition to the mirrors mounted on the magnets in the variometer, stationary mirrors are also used to reflect light onto the recording drum and produce a fixed (non-varying) trace known as a *baseline*. A schematic illustration of this arrangement can be seen in Fig. 7.2a. An example of a *normal* magnetogram, from Kakioka, Japan can be seen in Fig. 7.2b. Although still in use today at some observatories, these systems have become less common as the magnetic field is more often recorded using digital systems as described in Section 7.2.2.

7.2.1.3 Calibration Data

In addition to the magnetogram traces continuously recording the variation in the magnetic field periodic measurements (generally weekly) of the absolute values of the magnetic field vector components were made using a variety of special instruments. Furthermore, special deflection magnets of known magnetic moment were used to obtain conversion factors, known as *scale values*, to convert the traces from units of distance to units of magnetic field strength. Knowing the absolute values of the field components for a given time interval, one could then combine these with the scale values to derive an absolute value for the baseline. With this, a calibrated value of the magnetic field at any point along the magnetogram trace could then be determined to convert the analogue traces into definitive magnetic values.

7.2.1.4 Hourly Values

After the magnetograms were developed, an observer would use a scaling glass to estimate the amplitude of each trace for each hour from its respective baseline. Using the calibration data, these could then be converted to hourly values of the magnetic field. In the earliest days of observatory operations these hourly values were often spot observations taken every 60 min on the hour. Later, just after the turn of the 20th century, observatories began to estimate the average amplitudes over the whole hour to provide hourly mean values (HMVs). The first HMVs were calculated by Schmidt (1905) for Potsdam observatory with many others following suit shortly afterwards. These HMVs referred to the period of 60 min from the start of the hour, resulting in a mean centred on the following half-hour (i.e., a mean labelled as hour 10 is centred at 10 h 30 min). Convenient for hand-scaling between the hour marks on a magnetogram, this method continued into the digital era, thus ensuring homogeneity throughout the HMV data sets.

Until the advent of digital recording systems HMVs were the primary data product from magnetic observatories. Both the spot hourly values and the HMVs were usually compiled into monthly tables as shown in Fig. 7.3. These tables were published in observatory yearbooks along with the information about the values that would enable distinction between spot and mean values. Unfortunately there were cases of insufficient or incorrect information being recorded, causing the potential for incorrect conclusions to be reached by researchers using these data sets and underlining the importance of clear and accurate metadata (see Section 7.4).

7.2.1.5 Magnetic Activity Indices

In addition to the values of the magnetic field components themselves, efforts were made to develop indicators of the various levels of geomagnetic activity.

(a)

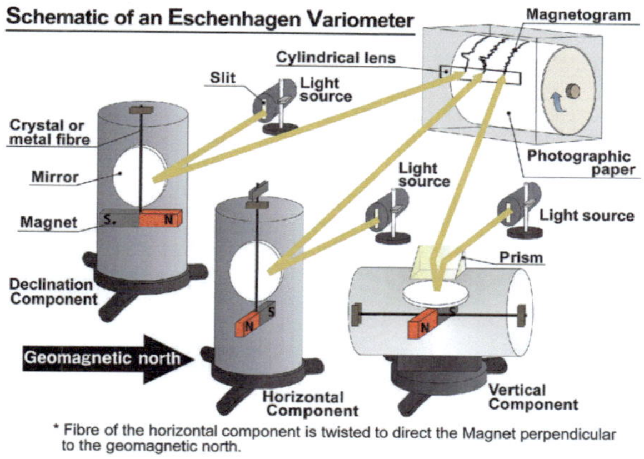

(b)

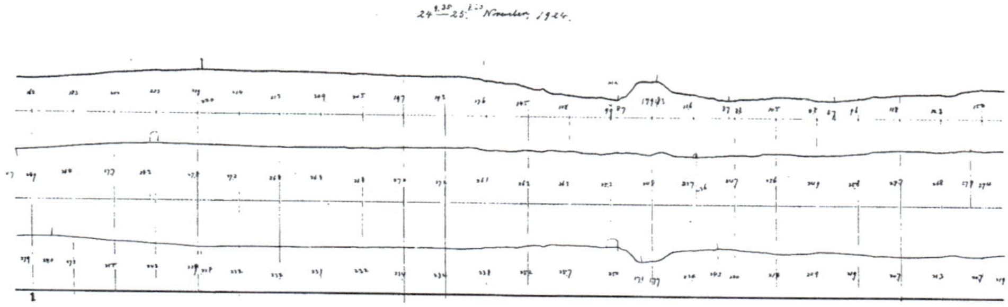

Fig. 7.2 (**a**) Schematic of an Eschenhagen type magnetic variometer (for details see Wienert, 1970) (**b**) Normal-run magnetogram recorded at Kakioka observatory on November 24, 1924. (Credit: WDC for Geomagnetism, Kyoto)

A comprehensive guide to this subject is given in the chapter "Geomagnetic Indices" by Menvielle et al. (2010) within this book and in the classical text by Mayaud (1980). Here, we mention only a few of the most commonly used indices that were historically derived from and originally stored as analogue records.

The 3 h *K*-index is a quantitative measure of local magnetic activity based upon the range of fluctuations in the observatory traces over 3 h. It uses a semi-logarithmic scale from 0 to 9, with 0 indicating completely quiet conditions and 9, highly disturbed conditions. It is intended to measure geomagnetic disturbances outside the normal diurnal quiet time variations and so these had to be accounted for by the experienced hand-scaler. The largest range of either the horizontal component (*H*) or declination (*D*) trace (originally the vertical component (*Z*) was also used) is selected. *K*-indices can also be converted into eight 3 h equivalent linear amplitudes, *ak*, and summed, these provide a local daily amplitude index, *Ak*.

The scaling of *K*-indices at a wide network of observatories enabled the derivation of various *K* based planetary indices adopted by IAGA: *Kp* from which *ap* is derived; *am*; and *aa*. The planetary activity index, *Kp*, is derived from the standardised *K*-indices of 13, mostly mid-latitude, observatories (Bartels et al. 1939). The name *Kp* comes from the German phrase "*p*lanetarische *K*ennziffer", meaning planetary index. It is designed to monitor the influence of the auroral electrojet current system, and the magnetospheric ring current and field-aligned currents. As with *Ak*, *ap* is the 3-hourly equivalent planetary amplitude derived from *Kp* and *Ap* is the daily average of the eight values. *Kp*, *ap* and *Ap* extend back to 1932. Although much used by the external magnetic field communities, these indices are generally considered to have two main weaknesses: one is the poor global distribution of the observatory network used; and the second is that the length of the time series is insufficient for century-long studies. The first was solved by the design of the *am*-index (Mayaud, 1967), which uses a more extensive

FORM C&GS-686e (1-64)

MAGNETOGRAM HOURLY SCALINGS (UNIVERSAL TIME)

U.S. DEPARTMENT OF COMMERCE COAST AND GEODETIC SURVEY GEOMAGNETISM DIVISION

Values are in tenths of mm. and are averages for successive periods of one hour beginning at midnight. Hour 01 of local day is hour 12 of the same universal day. Shrinkage corrections have been applied. Negative values are in red, with minus signs shown.

OBSY.	YEAR	MONTH	ELEMENT
HON	1964	JUNE	H

C	Q or S	Loc. Q	Da.	01	02	03	04	05	06	07	08	09	10	11	12	Da.	13	14	15	16	17	18	19	20	21	22	23	24	SUM
0		T	01	280	258	251	249	240	238	240	235	239	249	259	260	01	276	287	288	289	300	316	320	307	275	249	229	229	6363
0		T	02	227	227	219	210	218	220	221	230	239	240	247	250	02	261	251	257	261	277	289	291	292	298	287	269	255	6036
0	Q	T	03	242	247	239	231	231	231	230	229	231	234	231	228	03	229	229	234	239	252	260	280	290	302	298	288	270	5975
0		T	04	243	232	239	235	228	222	213	203	218	210	218	219	04	214	219	221	226	249	261	285	292	291	283	272	274	5767
0	Q	T	05	269	257	239	239	230	230	230	230	232	232	234	237	05	239	238	232	230	240	262	280	280	271	252	242	258	5683
0	Q	T	06	266	266	255	241	245	253	252	251	240	243	229	219	06	220	240	238	235	261	281	280	270	282	280	262	230	6044
0			07	209	205	209	206	218	226	229	230	232	238	240	242	07	244	243	242	242	250	268	287	339	343	318	281	310	6061
0			08	297	250	245	229	238	241	230	210	157	132	162	190	08	198	202	217	220	250	269	266	252	250	262	252	255	5474
0			09	250	239	231	212	221	200	183	189	188	200	201	208	09	209	213	215	234	280	289	299	326	340	353	380	382	6042
1	S		10	392	362	305	203	028	-210	-258	-138	-0-5	-068	032	049	10	058	058	047	024	050	079	107	111	130	110	071	089	1536
1	S		11	078	073	-005	010	070	071	066	054	070	080	099	110	11	111	128	143	135	152	169	199	221	220	164	150	147	2709
0	S		12	132	125	111	092	093	092	107	140	150	154	150	142	12	173	183	172	177	173	210	200	193	178	169	123	123	3584
0			13	138	152	157	140	129	126	154	148	149	149	189	189	13	162	170	171	173	181	196	211	220	210	179	181	208	4082
0			14	189	180	179	163	148	149	139	156	170	160	160	162	14	160	179	179	181	204	223	220	217	248	264	260	240	4530
0			15	210	195	189	171	175	183	179	177	179	178	180	186	15	200	199	200	210	221	232	250	255	257	224	231	235	4916
0	Q	T	16	220	207	190	185	189	188	182	182	182	191	191	190	16	191	195	192	192	211	228	222	220	230	249	261	249	4937
0		T	17	225	201	188	187	182	182	181	182	190	188	190	190	17	193	196	210	221	240	250	242	242	250	245	240	239	5054
0			18	231	219	211	205	208	211	212	221	229	233	239	221	18	200	210	210	212	240	262	272	250	230	198	176	191	5291
0		T	19	202	208	209	200	198	199	202	208	211	219	213	211	19	211	209	210	210	218	239	259	260	249	221	202	201	5169
0	S		20	205	175	129	101	071	111	151	169	186	199	198	201	20	192	179	159	169	212	227	227	218	219	220	229	239	4386
0	S		21	222	121	140	165	170	170	173	165	180	188	180	183	21	185	197	181	182	192	219	218	189	200	222	220	213	4465
0			22	228	219	200	191	193	205	208	212	210	215	219	221	22	229	229	231	241	230	236	244	239	228	210	192	198	5234
0			23	205	207	191	178	179	163	153	168	171	184	209	203	23	216	206	203	197	209	228	239	239	221	210	200	212	4796
0			24	199	189	192	187	187	193	198	200	208	210	211	210	24	218	222	225	230	240	258	262	267	259	251	252	249	5317
0			25	239	229	230	190	126	141	117	085	138	120	159	158	25	162	168	171	180	193	215	220	214	222	229	247	252	4405
0			26	241	219	188	171	170	190	192	190	202	190	179	199	26	190	190	194	189	202	212	228	221	221	212	206	198	4794
0			27	210	214	210	199	191	192	190	190	192	203	208	220	27	255	257	279	270	292	293	272	249	229	212	210	229	5466
0			28	239	240	232	228	227	220	205	206	210	219	239	268	28	250	228	222	238	229	255	240	239	229	203	193	182	5441
0			29	178	170	178	169	180	193	192	190	183	230	221	200	29	194	192	191	208	221	250	259	256	250	233	219	209	4966
0	Q	T	30	207	208	220	221	211	210	209	210	212	210	212	220	30	221	220	222	228	239	248	250	252	262	258	258	262	5470
			31													31													

SCALED BY	HO PERSONNEL
CHECKED BY	HO PERSONNEL
SIGNS REVERSED BY	HO PERSONNEL
PUNCHED BY	

Preliminary base-line and scale values:

Interval Beginning	Base-line Value	Scale Value

( ) Interpolated

[ ] Significant portion of hour interpolated.

☐ No record; or no values available because of faulty record.

[ ] Scaling uncertain because of magnetic storm.

<> Record off sheet for part or all of hour; if value is given, curve was estimated for missing part.

* Derived from ________ Mgph., converted to Normal Mgph.

MONTHLY SUM (720)	149993
MONTHLY MEAN	208
DAYS WITH GAPS:	

USCOMM-DC 6469-P63

Fig. 7.3 A table of hourly mean values from Honolulu observatory for the *H* component in June 1964. Each *column* shows the hour in UT and each *row* indicates the day of the month. These values are summed at the end of each row and a monthly sum and mean are produced. The table also indicates the five internationally quiet (*Q*) and disturbed (*S*) days

network of observatories and is available from 1959. The latter was solved by the design of a much simpler *K* based planetary index, the *aa*-index (Mayaud, 1972). Using only two near-antipodal observatories it could be extended back further in time to 1868.

Another important measure of magnetic activity is the *D*isturbed *s*torm *t*ime, *Dst*-index (Sugiura, 1964). This hourly index is based on the average value of *H* measured hourly, as described in Section 7.2.1.4, at four near-equatorial geomagnetic observatories. It was originally introduced to measure the magnitude of the current which produces the symmetric disturbance field, but is often used to estimate the strength of magnetic storms caused largely by the magnetospheric ring current. This (mainly) westward flowing ring current causes *H* to be depressed from its normal level. *Dst* can be used as a measure of magnetic storm intensity because the strength of the surface magnetic field at low latitudes is inversely proportional to the energy content of the ring current.

7.2.1.6 Yearbooks

Every year the results from a magnetic observatory would be collated and published in the form of a yearbook. Along with photographic magnetometer traces yearbooks are one of the primary sources of printed geomagnetic observatory data. The type of data published in yearbooks has changed over the years as instrumentation has developed and methods were standardised, but all share commonality. Generally, yearbooks contain descriptive text describing the observatory; this may detail its location, housing, staff, instrumentation, and data processing methods. It also contains the final definitive measurements and results from the observatory. As an example, yearbooks from Greenwich observatory originally contained results called 'indicators of the magnetometers'; firstly from eye-observations and then taken directly from the photographic magnetometer records. These were published alongside records of magnetic dip (these days normally referred to as inclination, *I*) and absolute measures of *H*. In later years hourly, daily, monthly and annual means were presented alongside *K*-indices, diurnal ranges, absolute measurements, baseline information and, sometimes, selective magnetograms and descriptive notes of magnetic disturbances. The results for many observatories were often published alongside meteorological observations.

It is worth noting when examining data from the earliest yearbooks that care must be taken regarding time recordings. Prior to the adoption of standard time-zones and Universal Time in 1884, time conventions at each observatory could vary considerably. Some used a local civil time where a day started at midnight whereas others used an astronomical time where a day started at noon or some other local convention (Boteler, 2006).

Presently, the tradition of publishing yearbooks is still practiced by many observatories worldwide. However since advent of digital distribution of results this practice has been discontinued at some observatories. Yearbooks continue to be the best source of valuable metadata on observatory practice, instrumentation, data processing and quality control and the publication of these are encouraged by IAGA (Jankowski and Sucksdorff, 1996).

7.2.1.7 Conservation and Conversion of Printed Media

During classical, analogue operations magnetograms, HMV tables, *K*-index tables, and calibration data made up the core of data types produced at magnetic observatories, although other derived products such as sudden storm commencements (*ssc*) and magnetic pulsations (Matsushita and Campbell, 1967) were also generated at some stations. At some WDCs the various paper products were archived onto microfilm rolls and microfiche cards. WDCs periodically published individual and joint catalogues identifying available data at each of the geomagnetic data centres (e.g., Abston et al. 1985). Figure 7.4 shows the holdings of analogue and digital data at WDC-Kyoto and illustrates that the digital-age of magnetic observatories only began in the mid-1980s. There are vast amounts of data 'locked away' in analogue form that is difficult to access and analyse in this digital era. Special procedures are required to preserve these data and allow electronic access.

Much effort has been made over the years to digitise magnetogram traces (e.g., Curto et al. 1996), but this has proved to be a challenging task. During magnetically active times, generally of most interest to researchers, the traces for the different components often cross each other making it difficult to identify which is which. Nonetheless, at the 11th IAGA Scientific Assembly, the British Geological Survey (BGS) reported on a programme to digitise magnetograms from the UK observatories dating back to 1848 (Clarke et al. 2009). The primary goal of the project is one of conservation, with the capture of the

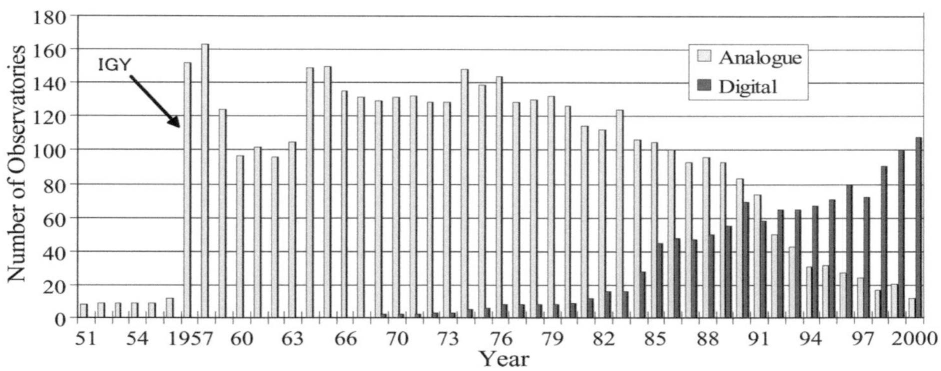

Fig. 7.4 The number of analogue normal-run magnetograms and digital one-minute data held at the WDC-Kyoto. (Credit: WDC for Geomagnetism, Kyoto)

magnetograms as images (Fig. 7.5), which will also enable increased access to the scientific community. The second aim is to develop a semi-automated procedure to convert the images into digital values with a greater time resolution than has previously been reported.

In addition, many of the WDCs discussed in Section 7.3 of this chapter have been involved in a special project funded by an International Council for Science (ICSU) grant. This "Magnetogram Rescue Project" involved WDCs Kyoto, Moscow, and Mumbai and aimed to locate old magnetograms and convert them into digital images (Iyemori, 2005). The results of this project are available from these WDCs.

The observatory community has invested most of its efforts in this area to convert the tables of data in yearbooks and other bulletins (for example, HMVs, *K*-indices) into digital electronic values. Various attempts have been made at this, including the scanning techniques of Optical Character Recognition

Fig. 7.5 Digital image capture of historical magnetograms. (Copyright: BGS/NERC)

(OCR). The unreliability of this procedure, especially for HMV tables (generally due to the high variability of different typefaces and hand-written entries in the tables), caused these efforts to be directed more toward manual entry and checking (see for example, Nagarajan, 2008; Fouassier and Chulliat, 2009; Nevanlinna and Häkkinen, 2010). While the reliability of this manual procedure is very good, it is also labour intensive and can therefore be expensive. As a compromise, and in order to make the data accessible electronically, many data sets have been, and are being, scanned into images that will at least allow for computer manipulation by researchers.

7.2.2 Electronic Media

The earliest digital electronic data from magnetic observatories were recorded in 1969 from two French stations, Dumont d'Urville (DRV) in Antarctica, and Port-aux-Français (PAF) on the Kerguelen Islands in the Indian Ocean. These stations used photoelectric feedback and Cesium vapour magnetometers to sample the northerly intensity (X), easterly intensity (Y) components, and total intensity (F) of the magnetic field at one-minute intervals. The introduction of the triaxial fluxgate magnetometer (see for example, Trigg et al. 1971), and use of analogue-to-digital converters in data loggers by the early 1970s, ushered in the true digital electronic age in magnetic observatory operations. These devices sampled the field multiple times per minute which were then filtered to produce one-minute values. This also was the beginning of one-minute values becoming the standard published magnetic observatory results. Many institutes added proton precession scalar magnetometers at their stations which provided separate measurements of the absolute F, and also served as a reference to compare with the fluxgate vector data. Observatory data quality is discussed in the chapter "Improvements in geomagnetic observatory data quality" by Reda et al. (2010) within this book, where a full description of modern magnetic observatory instrumentation, operations and data processing is provided.

7.2.2.1 Minute Means

Various brands of linear core, three-component fluxgate magnetometers are installed at observatories throughout the world. Originally these instruments recorded an output voltage proportional to the strength of the ambient magnetic field. Using the calibration data, the voltages were converted to magnetic field values. The output from a fluxgate magnetometer can be sampled at various frequencies; initially systems were set to sample every 10 s but 1 s sampling is now more common. Regular manual absolute observations are still required to derive baseline values for the fluxgate data. Combined, these provide definitive data (see for example, Turner et al. 2007). Early on, data were stored on magnetic tape and processed to derive 2.5 min means. Over time these faded from use and one-minute means became the standard. These definitive digital one-minute data form the basis for easy computation of various time averages, including hourly, daily, monthly, and annual means.

As with the analogue records, the delay in the availability of early digital data could be considerable. This was due to several factors including the fact that recording tapes at many stations were left for weeks before they were retrieved. Even today, it requires several months of weekly absolute observations, to enable the production of reliable final baselines, before the final definitive data can be delivered to the WDCs.

With the many advantages that digital data provided, and the proliferation of low-cost computational capabilities, the demand for these one-minute data grew. New DCPs began allowing access remotely through modems, and requests for the data rapidly increased. Many researchers, for example in the space physics community, did not require definitive data as they are only interested in the variations in the magnetic field. The internet, with its email and FTP capabilities, allowed institutions to send preliminary data to users more quickly and efficiently. By 1991, INTERMAGNET had established GINs where a global collection of observatory one-minute data could be retrieved. Demand from industry for observatory data also grew. One example is its use, in near real-time, to improve the accuracy of bore-hole surveys in offshore directional drilling operations carried out by oil and gas companies (Reay et al. 2005). Today, one-minute data are available from a multitude of locations through websites within minutes of the measurements being made.

However, three important points concerning the one-minute data should be emphasised. First, in the early 1990s magnetometers and data collection systems were developed that produced an output in magnetic units rather than voltage units (Narod, 2009). The field measured in this way may appear to be the correct value that one might expect for a given location, but it is not. There are correcting factors, such as zero-level offsets and orthogonality issues that must be corrected for with the use of baselines in the calibration data. With the widespread proliferation of preliminary one-minute data, one could potentially mistake the preliminary data for the definitive data. The preliminary values can be as little as a few to as many as hundreds of nT apart from the definitive values. Nowadays INTERMAGNET clearly separates each type of data on their website and the IAGA-2002 data format explicitly states the type of data both in the filename and the headers. This gives another good example of the power of metadata.

Secondly, early on, the one-minute means were often computed using simple arithmetic mean algorithms, sometimes referred to as a "box-car" mean. However, when converting from analogue to digital, in order to preserve the full information in the analogue signal, it is necessary to sample at least twice the maximum frequency of the analogue signal. Sampling at lower rates introduces erroneous components due to *aliasing*. A numerical filter can minimise the effects of aliasing in the construction of one-minute values from one-second samples. A Gaussian filter is often used for this purpose, although other types meeting the same requirements are also used. INTERMAGNET now requires numerical filtering of one-minute data as a criterion for attaining INTERMAGNET Magnetic Observatory (IMO) certification and encourages the use of a Gaussian filter using 90 one-second samples. The USGS, for example, began applying a Gaussian filter to derive its one-minute data from 1995 and the BGS use a 61-point cosine filter, having adopted this method in 1997 when instruments capable of one-second sampling were first installed in the UK.

Finally, we should mention that minute means have not always been centred on the minute, but rather were often computed from the 60 sec of values prior to the minute. It wasn't until the IAGA meeting in Canberra, Australia in 1979 that the decision was taken to encourage observatories to centre the minute values exactly on the minute (Resolution 12).

7.2.2.2 Hourly, Daily, Monthly, Annual Means

An on-going debate within the geomagnetic observatory community is the question of how HMVs should be computed when data are missing within the hour. The question centres on the accuracy of HMVs when the one-minute data for the hour are not complete. Over recent years a number of researchers have considered what may affect the accuracy of HMVs considering various factors including how many data points are missing, the distribution of these data gaps (random, continuous blocks), the level of magnetic activity and, the magnetic latitude of the observatory. Both Mandea (2002) and Schott and Linthe (2007) concluded that a reliable HMV could be computed if less than 10% of the data were missing. However Herzog (2009) found that during magnetically quiet times, including at a high-latitude station, up to two-thirds of the data could be missing without significant loss of accuracy. Newitt (2009) also found the 10% rule be an over-simplification leading both to the unnecessary rejection and unacceptable inclusion of certain data. He suggested a more statistical approach based on a set maximum permissible error and this was further considered in Marsal and Curto (2009) who studied the effect of a pre-established relative accuracy on HMVs. Love (2009) stated that the level of accuracy set for HMVs could not be better than the <5 nT level set for minute data by INTERMAGNET. In his study, he found that HMVs could satisfy this 5 nT level of accuracy, on average, about 90% of the time if the "10% rule" were applied. Whilst no common consensus yet exists on this issue, an IAGA task force is currently working towards determining a resolution (Hejda et al. 2009).

Questions have also arisen regarding how the daily means should be computed. Differences can arise depending upon whether one computes a daily mean from the average of the 1440 min of the day or from the average of the 24 h of the day. Missing data during the day can give rise to these differences. INTERMAGNET guidance states that a daily mean is calculated from 24 h values and an hourly value is calculated from 60 one-minute values. Monthly and annual means are less sensitive to these differences, although this depends on the amount of missing data and the time distribution of the missing data. These have historically been computed both for all days and for magnetically quiet days using the daily mean values. At the WDC Edinburgh, a note is associated with any annual mean value derived from an incomplete data set. This allows these data to be down-weighted when used in global magnetic field modelling.

7.2.2.3 Digital Magnetic Activity Indices

Although production of digital one-minute data became the standard for observatory operations, the legacy of products produced from the analogue magnetograms, including magnetic indices, continued. The long time series of these products were important to researchers studying the magnetic field and development of "magnetic climatologies" requiring many decades of data.

The derivation of *K*-indices from digital data presented its own unique difficulties. At first the data were plotted and printed on paper thus enabling the index to be hand-scaled in the same way as traditionally done from photographic magnetograms. There was, however, a strong incentive to obtain *K*-indices automatically using computer algorithms. A comparison between the various proposed algorithms was organised by an IAGA Working Group on geomagnetic indices, and four computer-derived methods were found to provide acceptable results when compared to the hand-scaled method. The four methods, discussed in Menvielle et al. (1995), were approved at the IUGG General Assembly in Vienna in 1991. Although many institutes gradually switched to digital derivation, some have continued to use the traditional hand-scaling method.

The convenience and flexibility of digital electronic media provided the opportunity to not only derive existing indices digitally, and in a more timely manner, but also to devise new magnetic activity indices, such as the *AE*-index and *PC*-index.

The *AE*-index is the *a*uroral *e*lectrojet index that is derived from a set of 12 observatories located in the auroral zone latitudes of the northern hemisphere (Davis and Sugiura, 1966). Plots of the *H*-component for each station are superimposed together to form a collection of traces. This collection produces upper and lower bounds of the range of magnetic activity for the group. The maximum positive deviation of the *H*-traces produces an upper bound (*AU*), and the maximum negative deviation produces a lower bound (*AL*). *AE* is the difference between the two and provides a global measure of the magnetic activity caused by ionospheric currents within the auroral oval. It correlates well with magnetospheric sub-storms that can last for several hours. This index is attractive because it is easy to compute digitally, and can be used to provide a measure of magnetic activity over any time scale desired. It also has known limitations; for example during very large storms the auroral electrojet will move equatorward of AE stations and its strength will no longer be measured (e.g., Akasofu, 1981; Feldstein et al. 1997). This index was endorsed by IAGA in 1969 (Resolution 2).

The *PC*-index is the *p*olar *c*ap (dimensionless) index designed to measure geomagnetic activity over the polar cap regions using only a single station located near the geomagnetic North and South poles (Troshichev and Andrezen, 1985). These polar stations are Qaanaaq (previously Thule) in Greenland for the northern hemisphere and Vostok in Antarctica for the southern hemisphere. This index is used to describe the principal features of the solar wind and the Interplanetary Magnetic Field (IMF), and indicates the total energy input into the magnetosphere. This was recommended to IAGA in Birmingham (1999) and endorsed in Hanoi (2001). The endorsement was later suspended due to different procedures being used in the derivation of the index in each hemisphere (*PCN* and *PCS*). At Sopron, Stauning et al. (2009) outlined a unified procedure now in use for both indices.

7.2.2.4 One-Second Data

With advances in magnetometers and other hardware systems and circuitry, it was inevitable that observatories would begin collecting and storing one-second data. Currently data reported from observatories are most commonly one-second values recorded from a magnetometer sampling at 1 Hz. Unlike one-minute and hourly mean data these are spot values and the accuracy of the timing (to the millisecond) is not well controlled. Providing true one-second mean data (obtained by filtering higher frequency data) is a challenging endeavour requiring sophisticated magnetometer specifications, timing accuracy, sampling restrictions, physical and numerical filtering processes, data processing and storage techniques. The greatest obstacle to overcome is instrumental noise which at 1 Hz is typically 10 pT $(\sqrt{\text{Hz}})^{-1}$, greater than geomagnetic signal at this frequency.

At the IAGA conference in Sopron, several authors reported on issues regarding developments regarding one-second data. Korepanov et al. (2009) presented the results of tests with a new magnetometer, developed at the Lviv Center of Institute for Space Research in the Ukraine, which meets the standards for one-second data set by INTERMAGNET. Their magnetometer accomplished a 1 pT resolution matching the standard proposed by INTERMAGNET. The sensitivity threshold is close to 1 pT, and the damping of higher frequency ambient noise, especially that of power line harmonics has been achieved. Dourbes and Conrad observatories are currently using this particular magnetometer. Worthington et al. (2009a) reported on an effort to estimate the noise levels and timing accuracy of the fluxgate magnetometers and A/D converters for the one-second data being collected at the USGS, compared to the standards proposed by INTERMAGNET. The results showed that the USGS observatories studied have noise levels of 0.01–0.02 nT and the timing is close to the 10 ms accuracy proposed by INTERMAGNET in 2008.

Others observatories around the world are investigating this issue and developing new techniques and instrumentation to provide one-second data (for example Chulliat et al. 2009a; Shanahan and Turbitt, 2009; Worthington et al. 2009b). Whilst this is a new development for most, the Japan Meteorological Agency has been recording one-second data at Kakioka observatory for about 25 years and at Memambetsu and Kanoya observatories for over 10 years (Minamoto, 2009).

A difficulty with one-second data is the question of what format to use for distribution. The IAGA-2002 format (see Appendix 1) allows for the possibility of one-second data, but only provides for a resolution of 0.01 nT or 10 pT, and this is insufficient for the INTERMAGNET proposal to record one-second data to 1 pT resolution. Nonetheless, it was decided at the INTERMAGNET meeting in Sopron (2009), that one-second preliminary data would be distributed from the website. IMOs will be requested to submit all one-second data that they hold. Initially the IAGA-2002 data format is to be used for these data, despite the resolution problem. Several possibilities for a more satisfactory format were discussed at the Sopron meeting, including: the development of a new format using XML; the development of a new format using the CDF (Common Data Format) [1] currently used in space physics and magnetic satellites communities; or modifying the current IAGA-2002 format.

7.2.2.5 Quasi-Definitive Data

The significant time delay between the availability of preliminary data and definitive data from observatories prompted the proposal of a new data type at the 11th IAGA Scientific Assembly in Sopron. Preliminary data are accessible almost immediately in many cases, but the definitive data can take a year or more to become available. Field modellers have a need for what has been termed "quasi-definitive data," defined as "data corrected using temporary baselines shortly after their acquisition and very near to being the final data of the observatory" (Chulliat et al. 2009b). Such baseline corrected observatory data are useful to modellers to quickly detect geomagnetic jerks (Chulliat et al. 2010), to test IGRF-11 candidate models (Chulliat, 2009c; Finlay et al. 2010) and for the validation of level-2 products within the context of the upcoming Swarm mission (Friis-Christensen et al. 2006), expected to be launched in 2012. The appeal of this data to modellers is apparent, but one concern with this proposal is the notion of the values being "very near" to that of the final definitive data. One standard proposed at the INTERMAGNET meeting following IAGA was that the data should be within 5 nT of the final definitive values, although verification will only be possible following production of the definitive data. Recent studies suggest observatories should be able to meet this requirement, but confirmation of the accuracy achieved will always be after the fact.

One study presented in Sopron reported on work carried out at the BGS and demonstrated that quasi-definitive data can be produced in near real-time (Baillie et al. 2009). BGS use piecewise polynomials to estimate baselines, and these are extrapolated to provide the daily baseline values used to construct quasi-definitive data in near real-time. For the five BGS INTERMAGNET observatories, comparisons were made between the *X*, *Y* and vertical

(Z) component quasi-definitive hourly mean values, available on a next day basis, and the final definitive hourly mean values from 2000 to 2007. The distribution of differences showed that for all observatories the quasi-definitive data were within 5 nT of the final data 94–99% of the time. Peltier and Chulliat (2010) also recently proposed a method for producing quasi-definitive data every month and demonstrated that the difference between quasi-definitive and definitive data was less than 1 nT, well within INTERMAGNET standard, for nine observatories having very different baseline characteristics. At Sopron, IAGA encouraged magnetic observatories to produce baseline-corrected quasi-definitive data shortly after their acquisition (Resolution 5) (IAGA, 2009).

7.3 Data Availability

7.3.1 World Data Centres for Geomagnetism

During the IGY, the ICSU created a number of WDCs that were designed to collect, catalogue, archive and distribute geophysical and solar data sets from a few centralised locations (ICSU Panel on WDCs, 1996). There was, and continues to be, a clear need for a few central repositories where scientists can go to obtain data in a consistent format and avoid catastrophic and irreversible losses of collected data (Mandea and Papitashvili, 2009). Originally, WDCs were set up in the United States (WDC-A), Russia (WDC-B), Europe (WDC-C1) and Asia-Oceania (WDC-C2). This has expanded to other regions, notably China (WDC-D) in 1988. WDCs were originally established to manage and preserve data from the physical sciences, such as geomagnetism, oceanography, and meteorology. Since these beginnings WDCs have expanded greatly in both the disciplines they support and the ways in which they meet the scientific community's requirements. Today, WDCs are more broadly environmental and include such disciplines as biodiversity and ecology, soils, and land processes. There are now more than 50 WDCs located in 12 different countries, as shown in Fig. 7.6. Geomagnetic observatory data, geomagnetic indices, survey data, models and other sources are maintained at several of these WDCs.

WDCs are funded and maintained by their host countries on behalf of the international scientific community. The WDCs operate under the guidelines of an ICSU panel, and from the outset have championed full and open access with data made available free of change, or for the nominal costs of reproduction. WDCs are encouraged to exchange data with each other and there are informal arrangements between data centres holding geomagnetic data to exchange new data sets received. WDCs are also expected to assure a reasonable standard of data quality and documentation. The ultimate responsibility however for data quality lies with the data contributor and not the WDC (ICSU Panel on WDCs, 1996). Recently Korte et al. (2007) identified a number of inconstancies in the hourly mean data sets held at the (then) WDC for Geomagnetism in Copenhagen. It is preferable that data producers re-examine their data sets and update the WDC database accordingly. However in some cases WDCs have, and do, make changes to the data held within. For example, Dawson et al. (2009) reported in Sopron on WDC for Geomagnetism, Edinburgh's efforts to correct simple typographical and formatting errors to the data sets. In this case it is not necessary to return the data to the individual observatory to correct. Improving the metadata associated with data sets, noting any errors, and if changes are made, is perhaps the first step to improving the quality of the data held.

7.3.1.1 World Data Centre for Geomagnetism, Edinburgh

The WDC for Geomagnetism, Edinburgh [2] was established in 1966 at the Institute of Geological Sciences in Sussex, which later became the BGS, and moved to its current location in Edinburgh in 1977. BGS, which is part of the Natural Environment Research Council (NERC), a publicly-funded agency, had concentrated on gathering data primarily for use in global magnetic field modelling—mainly annual mean data from the worldwide observatory network. Whereas, the WDC in Copenhagen, which was hosted by the Danish Meteorological Institute (DMI), gathered one-minute and hourly observatory data. WDC-Copenhagen provided access to these data sets online via a 'Data Catalogue' website. In 2007, BGS agreed

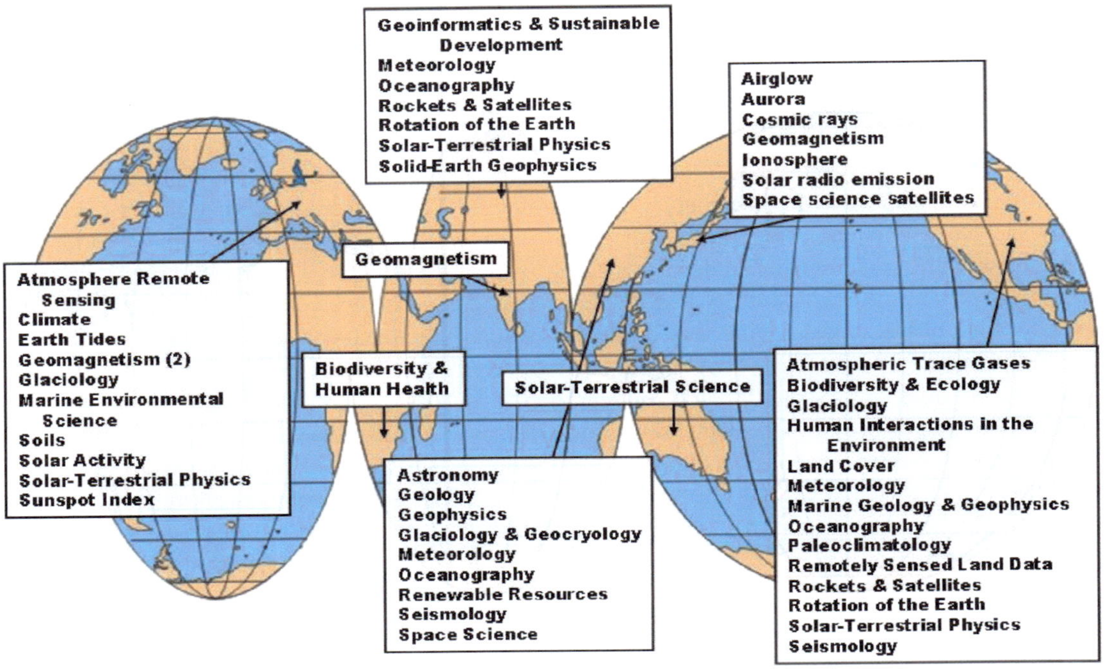

Fig. 7.6 Locations and scope of the various World Data Centres worldwide. (Redrawn from original courtesy of Prof B. Minster, past Chair of the ICSU Panel on WDC)

to take over responsibility for these data sets and the operation of this 'Data Catalogue' website.

Data now held at the WDC-Edinburgh include digital data from geomagnetic observatories worldwide (Fig. 7.7) including one-minute mean values from 1969; hourly values from 1883; and annual means from 1813. Also available are uncorrected one-second data from 2000 onwards from UK observatories and digital versions of some yearbooks. Data from land, marine and aeromagnetic surveys and repeat stations worldwide from 1900 onwards are available, as well as charts and computations of main field models, including the World Magnetic Model (WMM) and International Geomagnetic Reference Field (IGRF) model. Further digital data include definitive magnetic activity indices (*K, Kp, ap, Ap, aa, Aa, Cp and C9*); estimated real-time planetary indices (ap_{est}, Ap_{est}, aa_{est} and Aa_{est}) and solar activity indices (*SSN* and *F10.7*) [3]. In addition its analogue data holdings include shipborne declination data from 1590 onwards; archived magnetograms for several UK stations from 1850; and a library containing yearbooks, expedition memoirs, original survey observations and similar items. Data are available online, by anonymous FTP or by request.

7.3.1.2 World Data Center for Geomagnetism, Kyoto

In 1957, the Faculty of Science, Kyoto University was assigned to establish the WDC-C2 for Geomagnetism [4] since it was well-known for its achievements in geomagnetic research. This led to the foundation of the World Data Archive for Geomagnetism in the Kyoto University Library in December of the same year. In 1976, the Solar Terrestrial Physics Subcommittee of Science Council of Japan adopted the resolution that the existing data archive should be managed by a new research institution. As a result, on April 18, 1977, the Data Analysis Center for Geomagnetism and Space Magnetism (DACGSM) was established as a new institution in the Faculty of Science, Kyoto University and an associate professor was assigned to it. For over 30 years, DACGSM has been operating the WDC for Geomagnetism, Kyoto and providing a leading data service.

WDC Kyoto holds geomagnetic field data in form of normal-run magnetograms, rapid-run magnetograms, hourly digital values, one-minute digital values, and one-second digital values. Kyoto also derives

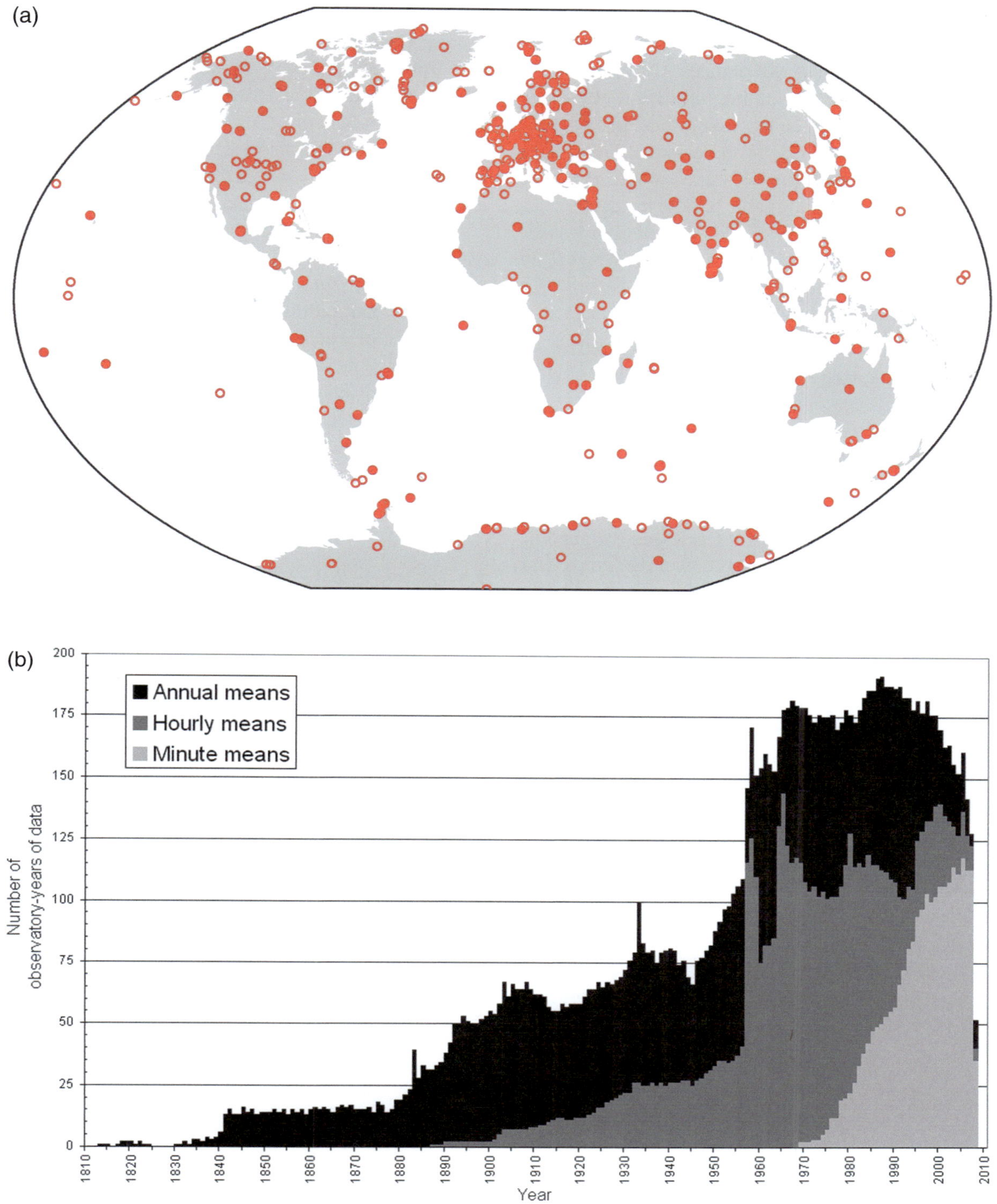

Fig. 7.7 (**a**) Locations of geomagnetic observatories worldwide with data held in the Edinburgh WDC. *Solid dots* indicate operational observatories, *open dots* are sites that are currently closed. (**b**) The number of data holdings at the World Data Centre (Edinburgh) for various time resolutions. (Copyright BGS/NERC)

geomagnetic indices, including the *AE*-index and the *Dst*-index, both of which are IAGA sanctioned indices, as well as the *SYM*/*ASY* indices (Iyemori, 1990). Recent advances in computing have made it possible for this data centre to collect and display geomagnetic field data in real-time from more than 30 observatories. Magnetograms have been copied onto microfilm or microfiche and are available back to the early 1900s. Converted digital images of old magnetograms resulting from the ICSU "Magnetogram Rescue Project" are also available from the Kyoto WDC [5].

7.3.1.3 World Data Center for Geomagnetism, Copenhagen

The WDC for Geomagnetism, Copenhagen is maintained by the National Space Institute at the Technical University of Denmark (DTU Space) [6] since 2010. It was established at the DMI for the IGY in 1957. Its Geomagnetic Data Master Catalogue for geomagnetic observatory hourly means and one-minute means moved to WDC Edinburgh in 2007.

WDC-Copenhagen currently holds digital data for the *PCN*-index derived from Qaanaaq geomagnetic observatory, and a mirror of the GFZ Potsdam ftp-server, home of the *Kp*-index. It has substantial analogue holdings and catalogues of these holdings, in the form of searchable PDF files, are available online [7]. This includes approximately 4300 years of magnetograms from about 280 geomagnetic observatories preserved on microfilm or microfiche; about 3200 years of tabulated hourly means from about 230 observatories; nearly 790 years of quick-run or pulsation magnetograms from about 90 observatories; approximately 1400 records of special events from about 90 different observatories; nearly 3300 years of *K*-indices from about 180 observatories; and on the order of 100 years of *Q*-index for about 22 observatories. The WDC-Copenhagen holdings also include data from 406 observatories mainly from the IGY onwards, also data from the Second Polar Year, 1932–1933. Most of the data are stored on microfilm, microfiche, and in printed publications.

7.3.1.4 World Data Centre for Geomagnetism, Mumbai

The Indian Institute of Geomagnetism (IIG), Navi Mumbai houses the WDC for Geomagnetism, Mumbai [8]. The centre is situated on the campus of old Colaba Magnetic Observatory, Bombay, which operated from 1841 to 1905. India's participation in the study of the earth's magnetism dates back to the period 1834–1841, when the country joined the Gottingen Magnetic Union. Full-fledged functioning of the WDC for Geomagnetism started at the IIG in 1991 under the recommendation of ICSU and the WDC Panel.

The centre holdings include magnetograms, geomagnetic hourly values, yearbooks of various observatories, digital data on magnetic tape, diskettes and CD-ROMs, microfilms and microfiche of the magnetograms, solar geophysical data bulletins containing solar rotation numbers; *Kp*, *Ap*, *Cp* and Zürich sunspot numbers, and indices such as *Dst*, *AE*, *AU*, *AL*, and *aa*. The centre is equipped with ample facilities for data retrieval, and these data are available in machine readable form upon written request or personal visit to the WDC.

7.3.1.5 World Data Center for Solar-Terrestrial Physics, Moscow

The WDC system in Russia consists of two facilities. The WDC-B was one of the original data centres established in the USSR by the Academy of Sciences of the USSR in 1957 to support the IGY. In 1971, the WDC for Solid Earth Physics (WDC for SEP) in Moscow was established as a part of WDC-B, and is maintained by the Geophysical Center of the Russian Academy of Sciences (RAS). WDC for SEP has concentrated on gathering data about the main magnetic field and its secular variations. A considerable part of these data are stored as publications on paper and microfilm, but some are also available in digital electronic form from their website [9]. Data stored at the WDC for SEP, Moscow include digital annual means values from 1813 onwards; publications of original land, marine and aeromagnetic survey observations; catalogues of measured values of the geomagnetic field elements, secular variation data, and charts of the geomagnetic field elements and its anomalies.

The other facility is the WDC for Solar-Terrestrial Physics (STP), also in Moscow [10]. The WDC for STP holds analogue data (mainly on microfilm and microfiche) from 1957 for the worldwide network of 264 magnetic stations as normal-run magnetograms, rapid-run magnetograms, hourly-, minute- and daily-mean values, as well as global magnetic activity indices and various publications. All standard geomagnetic data stored in the WDC for STP, Moscow, are accessible on the Moscow SPIDR (Space Physics Interactive Data Resource) website and mirrored worldwide (see Section 7.3.5.1).

Additional data in non-standard formats are available on the website of the WDC for STP. Digital electronic data from the WDC for STP website include data from magnetic observatories in Russia and the Former Soviet Union; hourly-mean values for 38 observatories mainly from the IGY onwards; one-minute values from 41 observatories mainly from 1983 onwards; global magnetic activity indices (*aa*, *Kp*, *Ap*, *AE*, *Dst*, *Pc*); digital images of magnetograms beginning from 1957; *ssc* from 1868 onwards; and a catalogue of geomagnetic Pc1 pulsations at the Borok and Mirny observatories from 1957–1992. Participation in the ICSU "Rescue of Magnetograms Project" resulted in digital images of magnetograms from nine observatories covering over 100 observatory-years worth of data.

7.3.1.6 World Data Center for Solar-Terrestrial Physics, Boulder

WDC-A was one of the original data centres established in the United States by the Coast & Geodetic Survey (C&GS) in 1957 to support management of the full suite of IGY data. The National Geophysical Data Center (NGDC) was created in 1965 to assume responsibility for long-term management of the geophysical data, including the WDCs for Geomagnetism, Gravity, and Seismology. NGDC moved to Boulder, CO from Washington, DC in 1972, assuming responsibility for the WDC-A for Solar Activity hosted by the University of Colorado. Currently, NGDC operates two WDCs; the WDC for Solar-Terrestrial Physics (WDC-STP) [11] and the WDC for Geophysics and Marine Geology (WDC-GMG) [12].

The WDC-STP holds data sets relating to solar activity, the space environment, ionosphere, thermosphere, geomagnetism, and cosmic rays. Geomagnetic observations at magnetic observatories are maintained as one-minute, hourly, daily, monthly, and annual mean values. In addition to the observatory data, the WDC-STP also manages magnetic variation, repeat observation, and land survey data, satellite data and global indices of magnetic activity. The WDC-GMG manages magnetic data collected at sea, aeromagnetic data, and develops and distributes magnetic field models, including the joint US/UK WMM and IAGA IGRF. The marine and airborne magnetic data are available through the GEODAS (Geophysical Data System) online system at [13]. Digital data are available online through their websites or by FTP. The WDC-STP also supports the SPIDR data portal that allows internet users to access, browse, display, and analyse STP data.

7.3.1.7 World Data Centre for Solar-Terrestrial Science, Sydney

The WDC for Solar-Terrestrial Science [14] is operated by IPS (Ionospheric Prediction Service) Radio and Space Services, a program within the Bureau of Meteorology of the Australian Government. The centre was established in 2000 and operates in Sydney, Australia. It contains ionospheric, magnetometer, spectrograph, cosmic ray data and solar images that are available for download via FTP. Currently WDC-Sydney holds magnetometer data for ten locations in Australia and two in Pakistan.

7.3.1.8 World Data Center for Geophysics, Beijing

The WDC for Geophysics [15] is operated by the Institute of Geology and Geophysics, Chinese Academy of Sciences. It was established in 1988 and is based in Beijing, China. With the support of the institute, Chinese Academy of Sciences and Ministry of Science and Technology, they have established online access to scientific data, such as geomagnetic data, gravity data, geoelectric field data and seismic wave data. This WDC holds one-minute and *K*-index data for Beijing Ming Tombs (BMT) observatory. Much effort has also been made to rescue historical magnetograms from 1877–1962 from Sheshan magnetic observatory. This collection has now been microfilmed and some magnetograms have been scanned and made available from their website (Peng, 2007).

7.3.2 INTERMAGNET

INTERMAGNET stands for *Inte*rnational *R*eal-time *Mag*netic Observatory *Net*work. Its objective, to establish a "global network of cooperating digital magnetic observatories, adopting modern standard specifications for measuring and recording equipment, in order to facilitate data exchange and the production of geomagnetic data products in close to real-time" (St-Louis et al. 2008) was achieved and further improvements continue to be the goal. The origin of the organisation and current operation are described in detail by Kerridge (2001) and Rasson (2007). Since its inception in 1986 INTERMAGNET has become the *de facto* organisation for setting standards for the highest quality observatory operations. Data types are defined as *R*eported (the raw data without any corrections), *A*djusted (data that have some modifications made to it), and *D*efinitive (the final published data), and these codes are used in the headers of the adopted data exchange formats. Instrument standards, such as accuracy and resolution were also established. Observatories that meet these standards and requirements are known as INTERMAGNET Magnetic Observatories, or IMOs.

In 1991, the USGS proposed that INTERMAGNET produce a CD-ROM of definitive one-minute data from IMOs, based upon a model the USGS had published (Herzog and Lupica 1992). Since then, INTERMAGNET has produced CD-ROMs, and now DVD-ROMs, annually. For 2007, the DVD-ROM contains data for 104 observatories from 41 countries. In 1997 a website was created [16] and this is now an important portal for the availability of both definitive and preliminary worldwide observatory data.

Following the 11th IAGA Scientific Assembly in Sopron, INTERMAGNET management resolved to make both one-second and quasi-definitive data available from all IMOs that are able to provide it (S Flower, personal communication, 2009). It was decided to restore all archived preliminary data to the website. The ability to provide commercial protection of data by time-limited distribution from the website will be introduced. The 'gold standard' for near real-time data delivery was agreed to be at 2–3 min delay and a discussion session was held to try to identify techniques that might be used to implement this standard.

7.3.3 World Data System

In 2006 ISCU established an ad-hoc Strategic Committee on Information and Data (SCID) to advise ISCU on the future organisation and direction of its activities in relation to scientific data and information. The SCID considered input from three ICSU Interdisciplinary Bodies: the WDC, the Federation of Astronomical and Geophysical Data analysis Services (FAGS) and the Committee on Data for Science and Technology (CODATA).

Despite the success of the WDCs, the review highlighted some basic problems that needed to be addressed. There was no real "system" in the sense of a managed or coordinated effort between the WDCs. While the ICSU WDC panel provided guidance, the WDCs have, for the most part, operated as individual institutions rather than a single entity, and the concept of data interoperability was only minimally implemented. There are also large disparities between WDCs in the developed countries and those in developing countries, no central comprehensive directory or catalogue of data available at the WDCs and an increasingly urgent problem with old data holdings being at risk of decay, lost, or discarded (ICSU 2008).

In its review the SCID also considered the operation of the FAGS and CODATA. The FAGS [17] was formed by ISCU in 1956 as part of IGY. Their principal purpose has been to encourage the analysis of long-term data sets and produce data products for the scientific community. The review by SCID identified similar issues to the WDCs with the services being very much a product of history rather than by design. CODATA were established by ICSU in 1966. It provides a cross-disciplinary focus on scientific data and aims to improve the quality and accessibility of scientific data and the methods by which those data are acquired, managed, and analysed.

Following the review by the SCID, at the 2008 ICSU General Assembly, it was agreed to abolish the ICSU Panel on WDCs and create a World Data System (WDS) incorporating former WDCs with the FAGS, and other centres and services [18]. Those WDCs and FAGS currently in existence must now apply to ICSU to re-establish membership as a part of the WDS. The WDS will provide a coordinated and strategic response to the data needs of the global scientific community with international initiatives such as the International Polar Year (IPY).

The WDS will transition from a set of standalone WDCs, and individual Services, to a common globally interoperable distributed data system, while incorporating new scientific data activities. The new system will have advanced interconnections between data management components for disciplinary and multidisciplinary scientific data applications. WDS will have a broader disciplinary and geographic base than the previous ICSU bodies and will strive to become a worldwide 'community of excellence' for scientific data. Eventually the WDS plan is to implement a Global Data System of Systems (GDSS) that should be interoperable with current and emerging data systems such as Global Earth Observations System of Systems (GEOSS) [19] (Minster et al. 2009).

7.3.4 The International Service of Geomagnetic Indices

The International Service of Geomagnetic Indices (ISGI) is a collaborative international service established by IAGA and falling within the FAGS. ISGI collects, derives, validates, maintains and distributes time-series of geomagnetic indices and other geomagnetic data products. The service is also responsible for the publication and distribution of any IAGA Bulletins containing information on geomagnetic indices.

The main ISGI operations are carried out at the Laboratoire Atmosphères, Milieux, Observations Spatiales (LATMOS), Paris (official derivation of *aa*, *am*, *Km*). The other ISGI collaborating institutes are: GeoForschungsZentrum (GFZ), Potsdam (official derivation of *Kp*, *Ap* and international quiet and disturbed days); Observatorio del Ebro, Roquetes (official derivation of *ssc* and solar flare effects (*sfe*)); and WDC-C2 for Geomagnetism, Kyoto (official derivation of *Dst* and *AE*). Data availability for the latter is described in more detail in Section 7.3.1.2 and further information on the other institutes follows.

7.3.4.1 LATMOS, France

Established in 2009 following a reorganisation of scientific organisations in France, LATMOS has now replaced Centre d'étude des Environnements Terrestre et Planétaires (CETP) as the main ISGI host. As well as having the primary responsibility for the derivation of the *aa*, *am* and *Km* indices, LATMOS also maintains the ISGI database of the IAGA endorsed indices and operates the ISGI publication office and the main ISGI website [20]. As well as providing online access to the ISGI data, the website contains valuable information (metadata) on the derivation and meaning of the indices. The definitive indices available online are: *aa*, *Kpa* (from 1868); *am*, *Kpm* (from 1959); *Dst* (from 1957); AU, AE and AL (from 1957); *ssc* (from 1868); international quietest days (from 1932); and *Kp* (from 1932). Monthly bulletins are also produced, which contain various provisional indices and include graphs of musical diagrams of *Km* and *aa*. These are circulated by post as well as being made available on the website.

7.3.4.2 GeoForschungsZentrum, Germany

Helmholtz-Center Potsdam, the German Research Center for Geosciences or GFZ is the national German research centre for the Earth sciences. The Adolf-Schmidt-Observatory Niemegk, belonging to GFZ, derives and maintains the *Kp, ap* and *Ap* indices as well as the lesser known planetary indices, *Cp* and *C9*. Definitive indices are produced twice a month and quick look versions are provided in near-real time. As well as the planetary indices, GFZ's ISGI responsibilities include the selection of the international quietest days (*Q-days*) and most disturbed days (*D-days*) of each month, which are classified using the *Kp* indices. These, the aforementioned planetary indices and *ssc* data provided by Observatorio del Ebro are all available on the GFZ website [21] in various formats, including ASCII tables, histograms and Bartels music diagrams. They are also available to download directly from the GFZ anonymous FTP site.

7.3.4.3 Observatorio del Ebro, Spain

The Ebro Observatory is a research institute founded by the Society of Jesus in 1904 to study the Sun-Earth relationships. It is a non-profit organisation associated with the Spanish Research Council (CSIC). Since 1975, Ebro Observatory was entrusted by IAGA to host the International Service on Rapid Magnetic Variations. It has responsibility for collecting, creating and publishing the official IAGA lists of rapid variations, which consist of *ssc* and *sfe* records. Annual lists of *ssc* data are available online from 1868 and *sfe* data from 1995 [22]. Monthly updates are provided to LATMOS and GFZ for further distribution.

7.3.5 Other Data Resources

7.3.5.1 Space Physics Interactive Data Resource

The SPIDR [23] is a data portal designed to allow a solar terrestrial physics customer to intelligently access and manage historical space physics data for integration with environment models and space weather forecasts. SPIDR is a distributed network of synchronous databases, web-portals and web-services. There are SPIDR sites in Boulder, Paris, Nagoya, Sydney, Beijing, Kiev, and Cape Town. SPIDR databases include geomagnetic one-minute, hourly means and annual means, as well as global geomagnetic and solar indices.

7.3.5.2 Observatory Operator's Websites

Many institutions operating magnetic observatories in the modern era have their own websites. This is an additional resource providing access to geomagnetic observatory data and data products to the user community. Since the advent of digital data distribution in the 1990s, observatories might not necessarily submit their data to a WDC as recommended by IAGA. It is recognised (Mandea and Papitashvili, 2009) that the traditional "push data" approach (to a WDC for example) is gradually being replaced by a "pull data" approach via data mining techniques such as virtual observatories (VO).

7.3.5.3 Variometer Networks

In addition to the data resources mentioned thus far, where full-field definitive observatory data and activity indices can be obtained, there are also a large number of variometer stations located around the globe providing additional data resources. Unlike magnetic observatories, variometer stations record only the short-term variations in the magnetic field and not the absolute level. These instruments complement the network of magnetic observatories and these measurements are used especially in the study of solar-terrestrial science.

There is a large amount of data (past and present) from many different magnetometer chains or networks around the world. A worldwide collaborative effort to pull these together was initiated and a single source of ground based magnetometer data is now available at SuperMAG [24]. This data portal website provides access to variometer data from more than 200 stations worldwide. Data are available from 1997 and can be viewed and downloaded online as linear plots and maps.

SuperMAG relies on acquiring data from a variety of sources including magnetometer networks such as IMAGE [25] (International Monitor for Auroral Geomagnetic Effects) and CARISMA [26] (Canadian Array for Realtime Investigations of Magnetic Activity). IMAGE is a network of 31 magnetometer stations operated by 10 institutions from Estonia, Finland, Germany, Norway, Poland, Russia and Sweden. CARISMA operate an array of 25 stations throughout Canada. INTERMAGNET also contributes data to SuperMAG since magnetic observatories can double as variometer stations. There are dozens of similar magnetometer networks located across the globe and are too numerous to mention here. Most of these are collaborative ventures in their own right, relying in turn on the individual institutes operating the stations.

7.4 Metadata and Metadata Standards

Metadata are an increasingly important aspect of geomagnetic data provision, both for the WDCs and for other data providers. Metadata are information about data; they are the "who, what, when, where, why and how" of a data set. They describe the content, quality, originator, and other characteristics of a data set that help users understand the nature of that data set and how to use it. As time passes, and personnel, instrumentation, and data processing procedures change, it is important that current and future generations have sufficient information about the data to enable them to independently understand and use them. The objective is to provide documentation that ensures that the values contained in a data set have the necessary reference information, and that even non-experts in a discipline will be able to use the data properly.

As Buneman (2005) notes, the web has radically changed the way scientific research is carried out, with more rapid and varied access to data, but this has produced new issues in maintaining the scientific record. As data can readily be copied and altered we need some means of verifying authorship and data provenance for researchers. For data producers too the issue of data citation is a growing concern. Wayne (2005a; 2005b) demonstrates the many benefits of metadata but also some of the obstacles to its production. Adhering to a metadata standard can be complex and time-consuming and there may be little immediate, tangible benefit to data producers. However the potential benefit is great.

The subject of metadata was raised at the IAGA conference in Sopron, and at the INTERMAGNET meeting that followed. Reay et al. (2009) highlighted the need for good data provenance to assist scientific examination of geomagnetic data sets. Better metadata would lead to improvements in the curation of data at the WDCs; addressing any inconsistencies seen in data sets and providing a clear 'paper-trail' of any transformations or corrections to data. It would also provide clear quality assurance to researchers, assisting data selection for global field models. In addition; common, well developed metadata improves the "discoverability" of data through the many metadata clearinghouses.

Svalgaard and Cliver (2007) discuss the current issues of data quality within WDCs and note that metadata about, especially, historical data is sorely lacking. They also point out the current difficulties in feeding back any corrections into the WDCs. If comprehensive metadata records were associated with a dataset a researcher could trace all the data processing steps applied to a data set, more easily identifying any inconsistencies. This would also provide a means for archivists to document any subsequent corrections made.

Consider another scenario: say, if you wanted to gather all geomagnetic data recorded by a fluxgate magnetometer, spanning 1990–2000, located within N30–50 latitude range. Currently this would require a significant amount of investigation to determine which observatories this applied to and where to get this data. If this simple metadata was recorded and stored in a central database in a known metadata standard this type of query would take seconds.

At the Sopron assembly, IAGA recognised the importance of metadata preservation in supporting geophysical studies (Resolution 7) and encouraged relevant agencies to support the generation, preservation and dissemination of metadata to ensure the future usability of these data for interdisciplinary studies. A session on metadata has been proposed for the next IUGG General Assembly in 2011.

INTERMAGNET has assigned a representative to work with groups in Japan, the UK, the US, and elsewhere on the issue of station-level metadata requirements and proposals. INTERMAGNET will notify the IMOs that a proposed standard would be forthcoming, and they should prepare to incorporate a new element into the observatory operations. It was also suggested that observatories place more effort on generating yearbooks with the omission of magnetogram plots if that would facilitate their production. The WDCs are also considering this issue and a wiki has been established to further coordinate discussions on this matter.

There are many issues to be considered in the development of a metadata standard for magnetic observatories. In the following sections we discuss some of the types of metadata currently available, or in development, that might provide a solution to capturing this important information resource.

7.4.1 Magnetic Observatory Metadata

7.4.1.1 Yearbooks

Yearbooks provided the earliest form of metadata for geomagnetic observatories, and are still being produced by many institutions today. These contain a permanent record of the status of the observatory for a given year and generally contain information about the station location, instrumentation, contact persons and record significant events or changes to the observatory throughout the year. These are published alongside the actual data products reported from an observatory.

Traditionally these were produced as printed books and these are stored in libraries, institutes and WDCs across the globe. Nowadays yearbooks are often produced in electronic PDF format that can be found from the websites of some institutions, WDCs, and INTERMAGNET [27]. Additionally some efforts are being made to scan historical yearbooks allowing electronic access to these important records.

Whilst the metadata held in yearbooks are often precisely what we wish to capture in a metadata standard, the format that this information is held is not currently standardised and is not suitable for data discovery applications. Principally, observatory operators should be encouraged to record this metadata, in whatever form, and then those within the geomagnetism community engaged in data curation must consider how to transfer this into an established metadata format. Improvements in data mining may be a way forward in this respect.

7.4.1.2 INTERMAGNET

The best source for digital observatory metadata, at least for those observatories within the organisation, is INTERMAGNET. Each year INTERMAGNET publish annual definitive observatory results on a CD/DVD-ROM. Alongside the data a *Readme* file is supplied for each observatory detailing basic information on location, instrumentation, sampling, filtering, contacts etc. This ASCII text file is in a standardised format with mandatory and optional fields. As these metadata records are directly related to a single published dataset they also provide temporal metadata that is challenging to capture otherwise.

Beyond this published metadata there is much implicit metadata associated with classifying an observatory as an IMO. If an observatory meets INTERMAGNET standards you can infer much information about its data processing, quality assurance, data distribution and instrumentation standards. However to a non-expert most of this information is not readily apparent: to access this type of metadata you would need to refer to the Technical Reference manual (St-Louis et al. 2008) held separate to the data.

7.4.1.3 Metadata Standards

Despite the wealth of metadata, in yearbooks and from INTERMAGNET, there is currently no metadata standard that can adequately describe data from magnetic observatories. With metadata, standardisation is important because it provides a common set of formats, terminology and definitions that facilitates the use and exchange of metadata digitally. It provides uniformity and consistency over time, enabling the development of tools and resources for metadata exchange and manipulation. One might say that metadata standards are to metadata what data formats are to data.

Considering what form a metadata standard for magnetic observatory data may take, we can suggest some of the basic information that would be required. These include, but are not limited to:

- *Contact Information*—The name of the responsible institute that produces the data, including addresses and personal contacts.
- *Data Description*—A description of the type of data, including a narrative summary of the nature of the data and its possible applications and a list of entities and attributes relevant to the data.
- *Station Description*—A description of the station in question, including coordinates, elevation, and possibly photographs and maps.
- *Instrumentation*—A description of the types of instruments in use at the observatory, and the nature of the data that are retrieved from them.
- *Data Processing*—A description of the processes and methodology used to process the data from instrument recordings to the final definitive values. This should include a description of data formats and how the numerical values are to be interpreted.
- *Data Quality*—A general assessment of the quality of the data set considering completeness, accuracy, quality control methods and so on.
- *Data Distribution*—A description of how and from where the data may be acquired, particularly through the WDCs.

To help us towards a metadata standard for geomagnetic observatory data we must look toward those standards currently in use for other, similar, datasets. Metadata standards established for geospatial data may act as such a guide. While these standards address many of the basic requirements of our metadata (for data discoverability, distribution, etc.) they lack the scope to deal with parameters that change with time. For an observatory nothing is fixed for all time: instrumentation, processing techniques, responsible institutions, staff contacts, data types, data quality, and even the observatory location can change. How to address this issue is one the greatest challenges facing metadata standardisation.

7.4.2 FGDC Standard

One of the first standards produced for geospatial data was developed by the Federal Geographic Data Committee (FGDC) in the US [28], which was designed primarily as the template for storing and distributing geographic spatial metadata. The FGDC standard consists of seven information sections:

1. *Identification*—Basic information about the data set and originator.
2. *Data Quality*—General assessment of the quality of the data set.
3. *Spatial Data Organisation*—Used to represent data set spatial information.
4. *Spatial Reference*—Reference frame for data set coordinates.
5. *Entity and Attribute*—Details about the information content.
6. *Distribution*—Distributor information and how to obtain the data set.
7. *Metadata Reference*—Metadata responsible parties and current status.

This standard is complex, requiring the data provider to fill in many metadata fields to comply. These are often in a terminology that is unclear to non-experts in the standard. The FGCD standard was developed primarily for data products that have a single publication date, such as maps, and consequently has limitations with regard to parameters that change in time—an important factor for observatory metadata. One would have to write a separate record for each change that occurs.

The NGDC has produced a modified standard for observatory data based on FGDC standard, omitting sections that are not applicable to magnetic observatories (Fischman et al. 2009). Within SPIDR, authorised institute personnel can now enter metadata for their own observatories using this modified FGDC template. SPIDR provides online forms corresponding to the various sections of the modified FGDC standard, from which observatory-level metadata records in the form of XML files are generated. While this goes some way to capture the required metadata it is still overly complex for observatory data providers to complete, does not solve the problem of tracking observatory histories, and has not been well received by the geomagnetic community as yet.

7.4.3 ISO-19115 Standard

In 2003 the International Standards Organisation (ISO) Technical Committee 211 on Geographic Information/Geomatics (ISO/TC 211) released the metadata standard ISO-19115. This combined various elements of national geospatial metadata standard including the FGDC (USA), CEN/TC287 (Netherlands), and ANZLIC (Australia and New Zealand), as well as contributions from other members of the Open Geospatial Consortium [29].

The FGDC mandatory fields and ISO core metadata fields are quite similar, although ISO-19115 offers the possibility of more detail, and includes special coverage of raster and imagery information. While the FGDC standard has seven informational sections, ISO-19115 has 14 comprising more than 400 individual metadata elements. As with the FGDC standard, most of these are optional and only a few of the available elements are likely to be used. While this ISO-19115 standard offers greater complexity and detail, it has the same issues as the FGDC standard with no capacity to track time-changing parameters at observatories.

7.4.4 SPASE Data Model

In 2006 NASA funded five new VOs to cater for different aspects of solar system science. A VO is a complex distributed environment with a goal to provide a single point of discovery of data and related resources (Merka et al. 2008). Groups, including the Space Physics Archive Search and Extract (SPASE) consortium [30], are defining metadata standards to aid in archiving and sharing of information resources from VOs.

The SPASE data model provides enough detail to allow a scientist to understand the content of the solar-terrestrial data products together with essential retrieval and contact information (King et al. 2010). The SPASE metadata schema divides the solar-terrestrial environment into a limited set of 12 resource types. Most commonly used are Display Data, Numerical Data, Granule, Instrument, Observatory and Person. These resources are interconnected to create a network of resources and the relationship between these resources fully describes a dataset. VOs can harvest these resource descriptions and allow users to find data.

As the SPASE data model manages the interrelationship between resources, such as observatory and instrument, it may prove to be better placed to manage the issue of time-varying information in magnetic observatory data.

7.4.5 XML

XML stands for eXtensible Markup Language and was designed to store and transmit structured data in plain text. It allows users to define their own vocabulary using tags to mark-up data in a similar in way to how HTML is used to mark-up web pages. XML has become the de facto standard for the exchange of information, in part, because of this flexibility. A user can develop any set of properties to define the elements of a data set and describe that data set using corresponding tags. This is known as creating an XML *schema*. Data and metadata can be stored in a document conforming to that XML schema, and be exchanged and read by both computers and humans.

For example, consider an XML document used to represent an instance of an instrument type in operation at a magnetic observatory:

```
<instrumentTypeUsageInstance>
      <beginDate>1975-05-09</beginDate>
      <vectorMagnetomterType>EDA Fluxgate magnetometer</vectorMagnetomterType>
      <endDate>1981-12-31</endDate>
</instrumentTypeUsageInstance>
```

The angle brackets surround a tag name, with the backslash used to indicate the end of an XML element. In this example *instrumentTypeUsageInstance* is an element that identifies a period of time when a particular type of instrument was in use. The *vectorMagnetometerType* element describes the type of instrument in use and the time-specific elements *beginDate* and *endDate* define the boundaries of the period when this instrument was in use. Note also that XML elements may be nested to reflect the structure of the data.

By constructing an XML schema that fully encapsulates magnetic observatory metadata we begin to define a type of metadata standard. This standard would be able to describe time-varying aspects of an observatory's operation by defining appropriate elements, which is an advantage over using current established standards. By using XML as a base, we can define the metadata and the method of storage and exchange at the same time.

XML is primarily used to exchange information between computers. XSLT (Extensible Stylesheet Language: Transformations) is a tool that allows XML to be transformed into various formats such as plain text, HTML or into XML of a different schema which may be more appropriate for the end-user [31]. For example the SPIDR system stores metadata in XML files and uses XSLT to transform this information into HTML for display to the user.

7.4.6 Databases

Another possibility being pursued at the NGDC, and reported on in Sopron, is the construction of a database schema to store the relevant observatory parameters with time histories of these parameters as fields in the database tables (Mabie et al. 2009). For example, changes in instrumentation from one time period to another (start time/stop time) can be stored in the database tables, and retrieved when a user requests metadata on that instrumentation within a given time-frame. The advantage of this approach is that web pages can be developed that provide selectable metadata categories (e.g., contact information, changes to instrumentation, etc.) to accompany data sets requested by users. If a user requests data from a certain observatory for a given interval, it can be designed so that the user may also request the metadata on instrumentation, institution address, data quality, or any other metadata category deemed of interest to the user. A menu of choices could be designed so that the user could select the kind of metadata of importance to them. The user would not have to search for the metadata separately, but could download it as an accompanying file with the data itself.

Whilst the XML and database approach may allow for the complete capture of observatory metadata this is still one step away from an established international metadata standard. Translation algorithms will be required to convert metadata held in a local form or within a magnetic observatory database/XML schema into an accepted standard when one is established.

7.5 Metadata Distribution

Once a metadata standard has been established there are many other issues to be addressed. We must consider how this metadata is stored and managed and where it can be accessed from. Should multiple sites distribute the metadata, and how would we handle consistency between sites? How would we collect and ingest metadata records? Although observatory operators will have the required information it may require a third-party to translate this information into the established metadata standard.

Beyond establishing a metadata standard these and many other issues must be addressed in time. For now, we will look at some selected examples of how metadata records for various geophysical data are distributed. All these metadata portals require a manual process to obtain the metadata separate from acquiring the data itself. They all lack the desirable feature of obtaining metadata simultaneously with data.

7.5.1 SPIDR VO (USA)

The SPIDR is a distributed network of synchronised web-accessible databases, developed by the NGDC, which provides users with access to current and historical STP data and metadata. SPIDR allows users to login as a guest or register with a username and password. After selecting the *geomagnetic view* option, the user can then select *view/modify stations metadata*. After selecting an observatory the user is presented with the modified FGDC metadata record for that observatory.

Each institution can assign a person responsible for maintaining and updating the metadata associated with the data from their own observatories. That person registers with SPIDR, requesting permission to access and edit the metadata records for those observatories.

As this system is based on the modified FGDC standard it has the same drawbacks regarding observatory metadata as discussed in Section 7.4.2.

7.5.2 GeoMIND (Europe)

GeoMIND (*Geo*physical *M*ultilingual *I*nternet-*D*riven Information Center) [32] is an internet site dedicated to information about geophysical systems in Europe. It is a consortium of 12 organisations from nine different countries, and provides metadata on geophysical data holdings from these countries in a consistent and seamless way, in whatever language the user selects. The website allows a user to identify a data set and its general content, and then be informed about how to obtain the data itself. This system allows for searching and editing of metadata records, as well as the importing and exporting of records in a variety of formats.

Whilst geomagnetic monitoring is included as a geophysical record within GeoMIND it is not currently well populated with metadata. The metadata are available on the web in the form of XML files conforming to ISO-19115. Only basic contact, citation and data distribution metadata are currently covered.

7.5.3 GEOMET (Australia)

GEOMET is the name of a database developed by Australian Geosciences that holds metadata records in the ANZLIC standard. ANZLIC (Australia New Zealand Land Information Council) is the metadata standard adopted by Australia and New Zealand that provides a schema required for describing geographic information, products, and services. It is similar to the FGDC standard in the parameters it defines, giving information about the identification, geographical boundaries, the quality, the spatial and temporal limits, spatial reference, and distribution of digital geographic data. ANZLIC assigns a unique identifier for each metadata record, which includes the usual metadata parameters such as an abstract describing the data set in general terms; geographic bounding coordinates; data quality; and contact information. ANZLIC has been made compatible with the ISO-19115 standard for geospatial metadata. Again, as this is similar to the FGDC and ISO-19115 standards, it will suffer from the same issues concerning geomagnetic observatory data.

7.5.4 IUGONET (Japan)

IUGONET (Inter-university Upper atmosphere Global Observation NETwork) [33] is a cooperative program in Japan between the National Institute of Polar Research, Tohoku University, Nagoya University, Kyoto University, and Kyushu University to expand the global radar, magnetometer and optical observation equipment and ground-based network in order to enhance each agency's effective use of various observational data sets.

Because of the diversity of the data from upper atmosphere observations, the program is developing a database for the metadata, which will include observation times, location, equipment, data formats, etc. to be made available on the internet. The metadata will make the data understandable to all researchers, and promote interdisciplinary research by a variety of different agencies, including those in fields other than that of upper atmosphere science. IUGONET will handle metadata in XML in the IUGONET format based on SPASE. Figure 7.8 shows an example of such metadata for the HF radar.

7.5.5 GeoNetwork (Open Source)

The GeoNetwork project [34] is an open-source and standards-based application for managing and delivering spatially referenced information and other resources via the internet. The software is intended to provide access to a wide variety of data, metadata,


```
<?xml version="1.0" encoding="UTF-8"?>
<Spase lang="en" xmlns="http://www.spase-group.org/data/schema"
  xmlns:xsi="http://www.w3.org/2001/XMLSchema-instance"
  xsi:schemaLocation="http://www.iugonet.org/data/resouces/iugonet-
1.0.0.xsd">
  <Version>1.0.0</Version>
  <NumericalData>
    <ResourceID>
      spase://IUGONET/NumericalData/SuperDARN/HOK/common_erg_cdf
    </ResourceID>
    <ResourceHeader>
      <ResourceName>
        SuperDARN Hokkaido HF radar, common mode data distributed by
        ERG-SC
      </ResourceName>
      <ReleaseDate>2009-04-01T00:00:00</ReleaseDate>
      <ExpirationDate>2199-12-31T23:59:59</ExpirationDate>
      <Description>
        Common mode data generated by SuperDARN Hokkaido HF radar. Data
        files are distributed in the CDF format through ERG-SC
      </Description>
      <Contact>
        <PersonID>spase://IUGONET/Person/Nozomu.Nishitani</PersonID>
        <Role>PrincipalInvestigator</Role>
      </Contact>
      <Contact>
        <PersonID>spase://IUGONET/Person/Kanako.Seki</PersonID>
        <Role>DataProducer</Role>
      </Contact>
      <Contact>
        <PersonID>spase://IUGONET/Person/Tomoaki.Hori</PersonID>
        <Role>MetadataContact</Role>
      </Contact>
    </ResourceHeader>

    ......
```


Fig. 7.8 An example of metadata archived by the metadata database of IUGONET. All metadata are described as XML in the IUGONET format based on the SPASE model and other kinds of information, obtained from multiple disciplines, and organised and documented in standardised ways. It provides the ability to manage and administer access to geospatial databases containing data and related metadata. It provides support for ISO-19115, FGDC and Dublin Core metadata standards. It also incorporates various services such as access to ESRI-based servers running the ArcIMS protocol. There are many agencies that use the GeoNetwork open source system to manage their metadata, including the Group on Earth Observations [35].

7.6 Conclusion

In this chapter we have reviewed the range of geomagnetic observatory data and indices that are available to the scientific community. The types and availability of this data are summarised in Table 7.1. We have seen that the type of data recorded and supplied by observatory operators is not fixed for all time. Extracting information from historical printed data products are of interest to those examining trends in long time-series data and current efforts worldwide to digitally capture this important information resource should be lauded. Furthermore, exciting developments in the supply of one-second and quasi-absolute data should have a significant impact on the range and type of scientific analysis that can be achieved.

Curation of these important data sets for our and future generations is of utmost importance. This task has been carried out with great care for over 50 years by the WDCs providing long-term data stewardship and free and open access to all. This is an interesting

Table 7.1 A summary of the type of geomagnetic data available and where this data is available from

		Data Sources													
		WDC-Edinburgh	WDC-Kyoto	WDC-Copenhagen	WDC-Mumbai	WDC-STP Moscow	WDC-SEP Moscow	WDC-STP Boulder	WDC-GMG Boulder	WDC-Sydney	WDC-Beijing	INTERMAGNET	ISGI- LATMOS	ISGI-GFZ	ISGI- Ebro
Digital Observatory	One-Second	■	■	-	-	-	-	-	-	■	-	-	-	-	-
	One-Minute	■	■	-	-	■	-	■	-	-	■	■	-	■	■
	Hourly Values	■	■	-	■	■	-	■	-	-	-	■	-	-	■
	Annual Means	■	-	-	-	-	■	■	-	-	-	■	-	-	-
Indices	Planetary (*K-derived*)	■	■	■	■	■	-	■	-	-	-	-	■	■	-
	Storm (*Dst, ASY/SYM*)	-	■	-	■	■	-	■	-	-	-	-	■	-	-
	Auroral (*AE*)	-	■	-	■	■	-	■	-	-	-	-	■	-	-
	Polar (*PC*)	-	-	■	-	■	-	-	-	-	-	-	-	-	-
	SSC/SI	■	-	-	-	■	-	-	-	-	-	-	■	■	■
	SFE	■	-	-	-	-	-	-	-	-	-	-	-	-	■
	Solar	■	-	-	■	-	-	■	-	■	-	-	-	-	-
	Other	■	■	-	-	■	-	■	-	■	-	■	-	-	-
Survey Data	Repeat	■	-	-	-	-	-	-	■	-	-	-	-	■	-
	Land/Air/Marine	■	-	-	-	-	■	-	■	-	-	-	-	-	-
Analogue	Magnetograms	■	■	■	■	■	-	-	-	-	■	-	-	-	-
	Digitised Magnetograms	■	■	-	■	■	-	-	-	-	-	-	-	-	-
	Yearbooks	■	■	-	■	-	-	-	-	-	-	-	-	-	-
	Other	■	-	■	-	-	■	-	-	-	-	-	-	-	-
Global Magnetic Field Models		■	■	-	-	-	■	-	■	-	-	-	-	-	-
Digital Metadata Records		■	■	-	-	-	-	■	-	■	-	■	■	-	-

time for data centres as we move towards a new World Data System to replace WDCs and provide a data service for the 21st century. INTERMAGNET and ISGI also play an important role in data availability, both in encouraging the establishment and improvement of standards at observatories and in the standardisation and distribution of data sets.

As data are now recorded and distributed electronically, important information about these data sets can become disassociated from the actual data. These metadata are vital to gain a full understanding of the data and any treatment applied to it. IAGA have recently recognised the importance of recording metadata and there are numerous efforts within the geomagnetism community to try and establish a way of capturing and distributing this important resource. This is one of the most pressing and challenging tasks in geomagnetic data curation today; but if we can achieve consensus and standardisation the scientific community will reap the benefits.

Acknowledgments The authors wish to thank David M. Clark, Dr. Hans-Joachim Linthe, Dr. Jürgen Matzka and Dr. Susan Macmillan for their valued contributions. We would like to thank Ellen Clarke for her critical review and many helpful additions to the manuscript. We would also like to thank Prof. Mioara Mandea and Dr. Monika Korte for the opportunity to prepare this chapter. This paper is published with the permission of the Director of BGS (NERC).

Glossary

ANZLIC—Australia New Zealand Land Information Council
BGS—British Geological Survey
CARISMA—Canadian Array for Realtime Investigations of Magnetic Activity
CDF—Common Data Format
CETP—Centre d'étude des Environnements Terrestre et Planétaires
CODATA—Committee on Data for Science and Technology
CSIC—Consejo Superior de Investigaciones Científicas (Spanish Research Council)
DACGSM—Data Analysis Center for Geomagnetism and Space Magnetism
DCP—Data Collection Platform
DMI—Danish Meteorological Institute
ERSI—Environmental Systems Research Institute
FAGS—Federation of Astronomical and Geophysical data analysis Service
FGDC—Federal Geographic Data Committee
FTP—File Transfer Protocol
GEODAS—Geophysical Data System
GIN—Geomagnetic Information Nodes
GFZ—GeoForschungsZentrum
HMV—Hourly Mean Value
IAGA—International Association of Geomagnetism and Aeronomy
ICSU—International Council for Science
IGRF—International Geomagnetic Reference Field
IGY—International Geophysical Year
IIG—Indian Institute of Geomagnetism
IPY—International Polar Year
IMAGE—International Monitor for Auroral Geomagnetic Effects
IMO—INTERMAGNET Observatory
INTERMAGNET—International Real-time Magnetic Observatory Network
ISGI—International Service of Geomagnetic Indices
ISO—International Standards Organisation
IUGONET—Inter-university Upper atmosphere Global Observation NETwork
LATMOS—Laboratoire Atmosphères, Milieux, Observations Spatiales
NERC—Natural Environment Research Council
NGDC—National Geophysical Data Center
OCR—Optical Character Recognition
SEP—Solid-Earth Physics
sfe—Solar Flare Effects
SPASE—Space Physics Archive Search and Extract
SPIDR—Space Physics Interactive Data Resource
ssc—Sudden Storm Commencements
STP—Solar-Terrestrial Physics
USGS—United States Geological Survey
VO—Virtual Observatory
WDC—World Data Centre
WDS—World Data System
WMM—World Magnetic Model
WWW—World Wide Web
XML—eXtensible Markup Language
XSLT—eXtensible Stylesheet Language Transformations

Appendix 1: Data File Formats

With the advent of digital data records it was soon recognised that a common data format for storing and presenting geomagnetic observatory was required. Without data standards it would be impossible for users of worldwide observatory data to make use of the vast collection of data available from many different institutions and countries.

Over time different data formats have been established with different motivations behind each approach. The most prevalent standard for one-minute data is currently IAGA-2002 whereas WDC Exchange Format is still often used for hourly mean data.

World Data Centre Exchange Format

In the 1980s a common file format for dissemination of geomagnetic data was developed. This WDC Exchange Format [36] was a continuation from the format for punched card and was designed to make maximum use of the limited RAM and disk capacity of computers at that time. It is not a convenient format for the user and there is no space for any metadata. For hourly means each value is decomposed into a tabular base value (shared by all hourly values for a given day), and a tabular value (Fig. 7.9). For each day the tabular base and the 24 values are presented with a daily mean value at the end of the row. For minute means each value in an hour is displayed along one horizontal row with an hourly mean value given at the end.

INTERMAGNET GIN Dissemination Formats

INTERMAGNET defined a number of format standards for the dissemination of data from their GINs [37]. There are standards for one-minute (IMFV1.22), hourly (IHFV1.01) and daily (IDFV1.01) data. For example, for one-minute means, data are organised on a day-file basis. One file contains 24 one-hour blocks, each containing 60 minutes worth of values. Blocks are padded with 9's if incomplete. Information is encoded in ASCII and there is only very limited metadata included e.g., noting if data are reported, adjusted or definitive.

```
ESK DEC0109 335 00 HDZF R EDI 03473568 000000 RRRRRRRRRRRRRRRR
 174920 -022201  463428 495316   174931 -022205  463427 495320
 174937 -022207  463427 495321   174928 -022201  463427 495318
 174925 -022201  463427 495317   174931 -022209  463427 495320
...
ESK DEC0109 335 01 HDZF R EDI 03473568 000000 RRRRRRRRRRRRRRRR
 174923 -022227  463424 495313   174922 -022228  463424 495314
 174921 -022230  463424 495314   174924 -022232  463424 495313
 174924 -022232  463424 495313   174924 -022232  463423 495314
...
ESK DEC0109 335 02 HDZF R EDI 03473568 000000 RRRRRRRRRRRRRRRR
 174928 -022221  463420 495312   174926 -022223  463419 495312
 174926 -022225  463420 495311   174924 -022223  463420 495310
 174923 -022223  463420 495311   174924 -022221  463421 495311
```

IAGA-2002 Format

The IAGA ASCII Exchange Format [38] was adopted in 2001. It is used as a data exchange format for geomagnetic data (samples and means) from observatories and variometer stations at cadences from milliseconds up to and including monthly means. This ability to accommodate data of different cadence is a major strength of this file format. It is also flexible and can account for files containing different durations of data (e.g. a month file of one-minute data) with its well-defined file naming scheme. The file names themselves provide critical metadata and within each file there are twelve mandatory file header

Tabular bases **D, H & Z mean values for hour 1, day 1, month 1, 2008 are 345 deg 47.0 min, 12575 nT and 50161 nT**

```
LRV0801D01   20 345 470 436 429 418 425 447 459 447 437 434 432 423 413 411 404 412 415 418 423 427 433 440 462 441 432
LRV0801D02   20 345 440 437 436 435 430 435 428 433 436 437 436 432 418 406 403 406 415 412 415 423 429 434 439 443 427
...
LRV0801H01   20 120 575 571 584 582 589 592 595 595 595 593 591 589 589 589 589 589 591 593 592 593 592 591 590 590 589
LRV0801H02   20 120 592 586 585 588 590 592 594 596 595 596 593 590 588 587 588 591 594 597 595 592 594 591 591 588 591
...
LRV0801Z01   20 500 161 144 148 166 179 180 177 176 179 180 181 182 183 183 185 186 186 185 185 184 185 188 189 182 178
LRV0801Z02   20 500 182 181 182 184 183 182 183 182 181 182 182 182 183 183 184 185 185 185 188 191 189 189 186 184 184
```

Fig. 7.9 WDC hourly mean value format, with the tabular base values highlighted records for further metadata. These include the type of data (definitive, provisional, variation), if the data are instantaneous or a mean, and how the values are centred. It also has an optional, variable length comment field for more descriptive metadata associated with the dataset. The data records are presented in a fixed four-column ASCII format which is convenient for the user to read and manipulate.

```
Format                 IAGA-2002                                   |
Source of Data         Danish Meteorological Institute             |
Station Name           Narsarsuaq                                  |
IAGA CODE              NAQ                                         |
Geodetic Latitude      61.160                                      |
Geodetic Longitude     314.560                                     |
Elevation              4                                           |
Reported               XYZF                                        |
Sensor Orientation     DIF                                         |
Digital Sampling       0.01 seconds                                |
Data Interval Type     Filtered 1-minute (00:30 - 01:29)           |
Data Type              Definitive                                  |
# This area is where the data source or distributor can include    |
# any additional information needed for proper use of data.        |
DATE       TIME         DOY     NAQX      NAQY      NAQZ      NAQF |
2001-03-13 00:00:00.000 072   10800.11  -6100.23  53381.51  54801.12
2001-03-13 00:01:00.000 072   10800.31  -6100.20  53381.51  54801.12
2001-03-13 00:02:00.000 072   10801.11  -6101.23  99999.00  54801.12
2001-03-13 00:03:00.000 072   10803.12  -6100.23  99999.00  54801.12
```

Appendix 2: Internet Links

1. http://cdf.gsfc.nasa.gov
2. http://www.wdc.bgs.ac.uk
3. http://www.geomag.bgs.ac.uk/data_service/home.html
4. http://wdc.kugi.kyoto-u.ac.jp
5. http://wdc.kugi.kyoto-u.ac.jp/film
6. http://www.space.dtu.dk/English.aspx
7. http://www.space.dtu.dk/English/Research/Scientific_data_and_models/World_Data_Center_for_Geomagnetism.aspx
8. http://www.wdciig.res.in
9. http://www.wdcb.ru/sep
10. http://www.wdcb.ru/stp/index.en.html
11. http://www.ngdc.noaa.gov/stp/WDC/wdcstp.html
12. http://www.ngdc.noaa.gov/mgg/wdc/wdcgmg.html
13. http://www.ngdc.noaa.gov/mgg/geodas/trackline.html
14. http://www.ips.gov.au/World_Data_Centre
15. http://gp.wdc.cn
16. http://www.intermagnet.org
17. http://www.icsu-fags.org
18. http://wds.geolinks.org
19. http://www.earthobservations.org/documents/geo_brochure.pdf
20. http://isgi.latmos.ipsl.fr
21. http://www-app3.gfz-potsdam.de/kp_index/index.html
22. http://www.obsebre.es/php/geomagnetisme/variaciorap.php
23. http://spidr.ngdc.noaa.gov/spidr
24. http://supermag.jhuapl.edu
25. http://www.geo.fmi.fi/image
26. http://bluebird.phys.ualberta.ca/carisma
27. http://www.intermagnet.org/Yearbooks_e.php
28. http://www.fgdc.gov/metadata
29. http://www.opengeospatial.org
30. http://www.spase-group.org
31. http://www.w3schools.com/xsl/default.asp
32. http://www.geomind.eu/portal/md_search.jsf
33. http://www.iugonet.org/en
34. http://geonetwork-opensource.org
35. http://www.geoportal.org/web/guest/geo_home
36. http://www.wdc.bgs.ac.uk/catalog/format.html
37. http://www.intermagnet.org/FormatData_e.php
38. http://www.ngdc.noaa.gov/IAGA/vdat/iagaformat.html

References

Abston CC, Papitashvili NE, Papitashvili VO (1985) Combined international catalog of geomagnetic data, Report UAG-92, U.S. National Geophysical Data Center

Akasofu S-I (1981) Relationships between the AE and Dst indices during geomagnetic storms. J Geophys Res 86: 4820–4822

Baillie O, Clarke E, Flower S, Reay S, Turbitt C (2009) Reporting quasi-definitive observatory data in near real time. 11th IAGA Scientific Assembly, Abstract Book 94 pp

Bartels J, Heck NH, Johnston HF (1939) The three-hour-range index measuring geomagnetic activity. J Geophys Res 44:411–454

Boteler DH (2006) Comment on time conventions in the recordings of 1859. Adv Space Res 38:301–303

Brooke C (1847) Description of apparatus for automatic registration of magnetometers and other meteorological instruments by photography. Phil Trans R Soc London 137:69–77

Buneman P (2005) What the Web Has Done for Scientific Data, and What it Hasn't. In Advances in Web-Age Information Management: 6th International Conference,

Web Age Information Management 2005, Hangzhou, China, October 2005 Proceedings, pp 1–7

Chulliat A (2009c) Evaluation of IGRF-11 candidate models, secular variation. From IGRF-11: Progress reports, candidate models, evaluations and test models http://www.ngdc.noaa.gov/IAGA/vmod/EVALS/Eval_Chulliat.pdf

Chulliat A, Peltier A, Truong F, Fouassier D (2009b) Proposal for a new observatory data product: quasi-definitive data. 11th IAGA Scientific Assembly, Abstract Book 94 pp

Chulliat A, Savary J, Telali A, Lalanne X (2009a) Acquisition of 1-Second Data in IPGP Magnetic Observatories. Proceedings of the XIIIth IAGA Workshop on Geomagnetic Observatory Instruments, Data Acquisition, and Processing: U.S. Geological Survey Open-File Report 2009–1226, pp 54–59

Chulliat A, Thébault E, Hulot G (2010) Core field acceleration pulse as a common cause of the 2003 and 2007 geomagnetic jerks. Geophys Res Lett 37:L07301. doi:10.1029/2009GL042019

Clarke E, Flower S, Humphries T, McIntosh R, McTaggart F, McIntyre B, Owenson N, Henderson K, Mann E, MacKenzie K, Piper S, Wilson L, Gillanders R (2009) The digitization of observatory magnetograms. 11th IAGA Scientific Assembly, Program Abstract 43 pp

Curto JJ, Sanclement E, Torta JM (1996) Automatic measurement of magnetic records on photographic paper. Comput Geosci 22:359–368

Davis TN, Sugiura M (1966) Auroral electrojet activity index AE and its universal time variations. J Geophys Re 71:785

Dawson E, Reay S, Macmillan S, Flower S, Shanahan T (2009) Quality Control Procedures At The World Data For Geomagnetism (Edinburgh). 11th IAGA Scientific Assembly, Program Abstract 44 pp

Feldstein YI, Grafe A, Gromova LI, Popov VA (1997) Auroral electrojets during geomagnetic storms. J Geophys Res 102:14223–14236. doi:10.1029/97JA00577

Fischman D, Denig WF, Herzog D (2009) A Proposed Metadata Implementation for Magnetic Observatories. Proceedings of the XIIIth IAGA Workshop on Geomagnetic Observatory Instruments, Data Acquisition, and Processing: U.S. Geological Survey Open-File Report 2009–1226, pp 82–85

Finlay CC, Maus S, Beggan C, Hamoudi M, Lesur V, Lowes FJ, Olsen N, Thébault E (2010) Evaluation of candidate geomagnetic field models for IGRF-11. Earth Planets Space (in press)

Fouassier D, Chulliat A (2009) Extending backwards to 1883 the French magnetic hourly data series. Proceedings of the XIIIth IAGA Workshop on Geomagnetic Observatory Instruments, Data Acquisition, and Processing: U.S. Geological Survey Open-File Report 2009-1226, pp 86–94

Friis-Christensen E, Lühr H, Hulot G (2006) Swarm: A constellation to study the Earth's magnetic field. Earth Planets Space 58:351–358

Hejda P, Herzog D, Linthe HJ, Mandea M, Schott J-J, Svalgaard L (2009) On the derivation of hourly mean values with incomplete data. 11th IAGA Scientific Assembly, Abstract Book 94 pp

Herzog DC, Lupica CW (1992) Magnetic observatory data on CD-ROM. Eos Trans Am Geophys Union 73:236

Herzog DC (2009) The effects of missing data on mean hourly values. Proceedings of the XIIIth IAGA Workshop on Geomagnetic Observatory Instruments, Data Acquisition, and Processing: U.S. Geological Survey Open-File Report 2009-1226, pp 116–126

IAGA (2009) IAGA News 46, http://www.iugg.org/IAGA/iaga_pages/pubs_prods/Newsletters/IAGA_News_46.pdf

ICSU Panel on World Data Centers (1996) Guide to the World Data Center System. http://www.ngdc.noaa.gov/wdc/guide/wdcguide.html

ICSU (2008) Ad hoc Strategic Committee on Information and Data. Final Report to the ICSU Committee on Scientific Planning and Review

Iyemori T (1990) Storm-time magnetospheric currents inferred from mid-latitude geomagnetic field variations. J Geomagnetics Geoelectrics 42:1249–1265

Iyemori T, Nose M, McCreadie H, Odagi Y, Takeda M, Kamei T, Yagi M (2005) Digitization of Old Analogue Geomagnetic Data. Proceedings of PV-2005: Ensuring Long-term Preservation and Adding Value to Scientific and Technical data, Royal Society, Edinburgh, UK. http://www.ukoln.ac.uk/events/pv-2005/pv-2005-final-poster-papers/014-poster.doc

Jankowski J, Sucksdorff C (1996) Guide for magnetic measurements and observatory practice. International Association of Geomagnetism and Aeronomy, Warsaw, Poland

Kerridge DJ (2001) INTERMAGNET: Worldwide near-real-time geomagnetic observatory data, Proc. ESA Space Weather Workshop, ESTEC, Noordwijk, The Netherlands, http://www.esa-spaceweather.net/spweather/workshops/SPW_W3/PROCEEDINGS_W3/ESTEC_Intermagnet.pdf

King T, Thieman J, Roberts DA (2010) SPASE 2.0: a standard data model for space physics. Earth Sci Infor. doi:10.1007/s12145-010-0053-4

Korte M, Mandea M, Olsen N (2007) Worldwide Observatory Hourly Means from 1995 to 2003: Investigation of Their Quality. In: Reda J (ed) XII IAGA Workshop on Geomagnetic Observatory Instruments, Data Acquisition and Processing Belsk, 19–24 June 2006, Publications of Institute of Geophysics Polish Academy of Sciences, C-99 (398)

Korepanov V, Marusenkov A, Rasson J (2009) 1-second INTERMAGNET Standard Magnetometer—1-year operation. 11th IAGA Scientific Assembly, Abstract Book 94 pp

Love JJ (2009) Missing data and the accuracy of magnetic-observatory hour means. Ann Geophys 27: 3601–3610

Mabie J, Denig W, McLean S (2009) Geomagnetic Data Management at the NGDC. 11th IAGA Scientific Assembly, Program Abstract 114 pp

Macmillan S (2007) Observatories: an overview. In: Gubbins D, Herrero-Bervera E (eds) Encyclopedia of Geomagnetism and Paleomagnetism. Springer, Netherlands, pp 708–711

Marsal S, Curto JJ (2009) A new approach to the hourly mean computation problem when dealing with missing data. Earth Planet Space 61:945–956

Mandea M (2002) 60, 59, 58, ... How many minutes for a reliable hourly mean? Proceedings of the Xth IAGA Workshop on Geomagnetic Instruments, Data Acquisition, and

Processing, Hermanus Magnetic Observatory, Hermanus, South Africa, pp 112–120

Mandea M, Papitashvili V (2009) Worldwide Geomagnetic Data Collection and Management. Eos Trans Am Geophys Union 90:409–410

Matsushita S, Campbell WH (eds) (1967) Physics of geomagnetic phenomena (2 vols). Academic, New York-London

Mayaud PN (1967) Calcul preliminaire d'indices Km, Kn et Ks ou Am, An, et As, mesures de l'activité magnétique a l'échelle mondiale et dans les hémispheres Nord et Sud. Ann Géophys 23:585–617

Mayaud PN (1972) The aa index: a 100-year series characterizing the geomagnetic activity. J Geophys Res 77: 6870–6874

Mayaud PN (1980) Derivation, Meaning, and Use of Geomagnetic Indices, AGU Geophys Monogr 22

Menvielle M, Iyemori T, Marchaudon A, Nose M (2011) Geomagnetic indices. In: Mandea M, Korte M (eds) Geomagnetic Observations and Models, IAGA Special Sopron Book Series 5, DOI 10.1007/978-90-481-9858-0_7

Menvielle M, Papitashvili N, Hakkinen L, and Sucksdorff C (1995) Computer production of K indices: review and comparison of methods. Geophys J Int 123:866–886

Merka J, Narock TW, Szabo A (2008) Navigating through SPASE to heliospheric and magnetospheric data. Earth Sci Inform 1:35–42

Minamoto Y (2009) Ongoing Geomagnetic field 1-second value measurement by JMA. Proceedings of the XIIIth IAGA Workshop on Geomagnetic Observatory Instruments, Data Acquisition, and Processing: U.S. Geological Survey Open-File Report 2009-1226, pp 190–193

Minster JH, Capitaine N, Clark DM, Mokrane M (2009) Initial Progress in Developing the New ICSU World Data System. American Geophysical Union, Fall Meeting 2009, abstract #IN23A-1059

Narod B (2009) Private communication, Further information can be obtained from Narod Geophysics Ltd., #310, 2475 York Ave., Vancouver, B.C., Canada, V6K-1C9. Tel/fax: 604-732-9083

Nevanlinna H, Häkkinen L (2010) Results of Russian geomagnetic observatories in the 19th century: magnetic activity, 1841–1862. Ann Geophys 28:917–926

Nagarajan N (2008) Retrieval and Distribution of Digital Archives of Photographic Records of Magnetic Variations at Ngri, India. Abstract for XIIIth IAGA Workshop on Geomagnetic Observatory Instruments, Data Acquisition, and Processing http://mines.conference-services.net/resources/328/1322/pdf/IAGA2008_0017.pdf

Newitt LR (2009) The Effects of Missing Data on the Computation of Hourly Mean Values and Ranges. Proceedings of the XIIIth IAGA Workshop on Geomagnetic Observatory Instruments, Data Acquisition, and Processing: U.S. Geological Survey Open-File Report 2009-1226, pp 194–201

Peltier A, Chulliat A (2010) On the feasibility of promptly producing quasi-definitive magnetic observatory data. Earth Planet Space 62(2):e5–e8, doi:10.5047/eps.2010.02.002

Peng F, Shen X, Tang K, Zhang J, Huang Q, Xu Y, Yue B, Yang D (2007) Data-Sharing Work of the World Data Center for Geophysics. Beijing Data Sci J 6: 404–407

Rasson JL (2007) Observatories, INTERMAGNET. In: Gubbins D, Herrero-Bervera E (eds) Encyclopedia of geomagnetism and paleomagnetism. Springer, New York, NY, pp 715–717

Reay SJ, Allen W, Baillie O, Bowe J, Clarke E, Lesur V, Macmillan S (2005) Space weather effects on drilling accuracy in the North Sea. Ann Geophys 23:3081–3088

Reay S, Dawson E, Flower S, Macmillan S, Herzog D (2009) Towards a metadata standard for geomagnetic observatory data. 11th IAGA Scientific Assembly, Abstract Book 94 pp

Reda J, Fouassier D, Isac A, Linthe H-J, Matzka J, Turbitt CW (2011) Improvements in geomagnetic observatory data quality. In: Mandea M, Korte M (eds) Geomagnetic Observations and Models, IAGA Special Sopron Book Series 5, DOI 10.1007/978-90-481-9858-0_7

St-Louis BJ, Sauter EA, Trigg DF, Coles RL (2008) INTERMAGNET Technical Reference Manual, version 4.4 http://www.intermagnet.org/publications/im_manual.pdf

Schmidt, A (1905) Ergebnisse der magnetischen Beobactungen in Potsdam, Veröffentl. des Preuss. Meteol Inst Berlin

Schott J-J, Linthe HJ (2007) The hourly mean computation problem revisited. Proceedings of the XIIth IAGA Workshop on Geomagnetic Instruments, Data Acquisition, and Processing, Institute of Geophysics, Polish Acad Sci Warszawa 135–143

Shanahan TJG, Turbitt CW (2009) Evaluating the Noise for a Commonly Used Fluxgate Magnetometer—for 1-second Data. Proceedings of the XIIIth IAGA Workshop on Geomagnetic Observatory Instruments, Data Acquisition, and Processing: U.S. Geological Survey Open-File Report 2009–1226, pp 239–245

Stauning P, Troshichev O, Janzhura A (2009) The polar cap (PC) index. Present (unified) and past index calculations 11th IAGA Scientific Assembly, Abstract Book 99 pp

Sugiura M (1964) Hourly values of equatorial Dst for IGY. Annals of the International Geophysical Year, vol 35. Pergamon Press, Oxford, pp 945–948

Svalgaard L, Cliver EW (2007) Interhourly variability index of geomagnetic activity and its use in deriving the long-term variation of solar wind speed. J Geophys Res 112:A10111. doi:10.1029/2007JA012437

Trigg DF, Serson PH, Camfield PA (1971) A solid-state electrical recording magnetometer. Department of Energy, Mines, and Resources 41 Ottawa, Canada.

Troshichev OA, Andrezen VG (1985) The relationship between interplanetary quantities and magnetic activity in the southern polar cap. Planet Space Sci 33:415–419

Tsurutani BT, Gonzalez WD, Lakhina GS, Alex S (2003) The extreme magnetic storm of 1–2 September 1859. J Geophys Res 108(A7):1268. doi:10.1029/2002JA009504

Turner GM, Rasson JL, Reeves CV (2007) Observation and measurement techniques. In: Schubert G (ed) Treatise on geophysics, vol 5. Elsevier, Amsterdam, pp 93–146

Wayne L (2005a) Metadata in action. In: Proceedings of the 2005 GISPlanet Conference Proceedings http://www.fgdc.gov/metadata/documents/MetadataInAction.doc (Accessed on 28th July 2009)

Wayne L (2005b) Institutionalize Metadata Before It Institutionalizes You, Federal Geographic Data Committee http://www.fgdc.gov/metadata/documents/InstitutionalizeMeta_Nov2005.doc (Accessed on 28th July 2009)

Wienert KA (1970) Notes of geomagnetic observatory and survey practice. UNESCO Earth Sciences 5

Worthington EW, Sauter EA, Love JJ (2009b) Analysis of USGS One-Second Data. In: Proceedings of the XIIIth IAGA Workshop on Geomagnetic Observatory Instruments, Data Acquisition, and Processing: U.S. Geological Survey Open-File Report 2009–1226, pp 262–266

Worthington EW, White T, Sauter EA, Stewart DC (2009a) Evaluation of 1-Hz data from USGS geomagnetic observatories. 11th IAGA Scientific Assembly, Abstract Book 94 pp

Chapter 8

Geomagnetic Indices

Michel Menvielle, Toshihiko Iyemori, Aurélie Marchaudon, and Masahito Nosé

To the observers

Abstract Geomagnetic indices are a measure of geomagnetic activity, which is a signature of the response of the Earth magnetosphere and ionosphere to solar forcing. They play a significant role in describing the magnetic configuration of the Earth's ionized environment. In the second half of the twentieth century, they have become a key parameter in Solar Terrestrial studies; in the past 15 years, they have become a key parameter in Space Weather, being commonly used to detect and describe Space Weather events. The objective of this chapter is to contribute to a better understanding of the meaning, usefulness, potential and limitations of geomagnetic indices. Standard geomagnetic indices, as well as some newly introduced quantities are considered. We present for each index, or each index family, a short but complete description of the derivation process and a review of the information that the index may provide on the dynamics of, and on the physical processes that take place in the Earth's ionized environment.

Abbreviations

Bx	IMF component along x axis, directed positive towards the Sun
By	IMF component along the y axis, directed positive towards dusk
Bz	IMF component along the z axis, directed positive towards north
CME	Coronal Mass Ejection
D	Declination: angle between the local magnetic field and the geographic north; D is positive when the geomagnetic north is east of geographic north.
DP-1	Disturbance Polar of type 1 current
DP-2	Disturbance Polar of type 2 current
GIC	Ground Induced Currents
GMLAT	*Geo*magnetic *Lat*itude
GSEQ	Geocentric Solar Equatorial coordinate system (x axis is from Earth to Sun, y axis is parallel to solar equatorial plane, z axis is positive northward)
GSM	Geocentric Solar Magnetospheric coordinate system (x axis is from Earth to Sun; z axis is northward in a plane containing the x axis and the geomagnetic dipole axis)
H	geomagnetic field horizontal component along the local geomagnetic north direction, directed positive northward
IGY	International Geophysical Year (July 1, 1957—Dec. 31, 1958)
IMF	Interplanetary Magnetic Field
Intermagnet	*Inter*national *R*eal-time *Mag*netic Observatory *Net*work
LT	Local Time
MLT	Magnetic Local Time
R_E	Earth radius
rms	root mean square
sfe	solar flare effect
Sq	Solar quiet variation
S_R	Solar Regular variation
ssc	storm sudden commencement
ULF	Ultra Low Frequency
UT	Universal Time

M. Menvielle (✉)
LATMOS-IPSL (Laboratoire Atmosphères, Milieux, Observations Spatiales), Université Versailles St-Quentin, CNRS/INSU, Univ. Paris Sud, Boîte 102, 4 place Jussieu, 75252, Paris Cedex 05, France
e-mail: michel.menvielle@latmos.ipsl.fr

M. Mandea, M. Korte (eds.), *Geomagnetic Observations and Models*, IAGA Special Sopron Book Series 5,
DOI 10.1007/978-90-481-9858-0_8,

X	geomagnetic field horizontal component along the geographic north direction, directed positive northward
Y	geomagnetic field horizontal component along the geographic east direction, directed positive eastward
Z	geomagnetic field vertical component, directed positive downward
Institutions	
AARI	Artic and Antarctic Research Institute, St Petersburg, Russia
DMI	Danish Meteorological Institute, Kopenhagen, Denmark
GFZ	GeoForschung Zentrum, Potsdam, Germany
IAGA	International Association of Geomagnetism and Aeronomy
IATME	International Association of Terrestrial Magnetism and Electricity
ISGI	International Service of Geomagnetic Indices
IUGG	International Union of Geophysics and Geodesy
LATMOS	Laboratoire Atmosphères, Milieux, Observations Spatiales, Guyancourt, France
WDC-Kyoto	World Data Center for Geomagnetism, Kyoto, Japan

8.1 Introduction

Indices are widely used in various domains to monitor the evolution of more or less complex phenomena by providing pertinent, reliable, and concentrated information. An index, whether it be geomagnetic or stock market related, is a number that simply represents an event, or a series of events. Each individual value of the index aims at describing the phenomenon under study during a fixed time interval; it is neither a substitute for the original data nor an interpretation of them. The relation between an index value and the original data should be clearly defined, as simple as possible, and the regularity and the homogeneity of the index data series are of great concern.

Geomagnetic indices are a measure of geomagnetic activity, which is a signature of the response of the Earth magnetosphere and ionosphere to solar forcing.

The first attempt to characterise geomagnetic activity was made as early as 1885. It was aimed at estimating geomagnetic disturbances on a daily basis. At that date, the daily range, namely, the daily difference between the highest and the lowest values recorded on one selected geomagnetic component was calculated at the Greenwich observatory using the two horizontal components H (magnetic North) and D (magnetic East).

The so-called "C" character—0, 1, or 2, describing the relative degree of disturbance of the magnetograms—was then introduced, giving rise to the international character C_i that has been calculated for the years 1884–1975 inclusively.[1] However, in spite of the great service provided to the scientific community by the C_i index, its crudeness led to the introduction of new indices that allow an objective monitoring of the irregular variations.

The K index was first introduced by Bartels and co-workers (Bartels et al. 1939—see Section 8.4). K indices from a network of stations were then used to compute "Kp" (Bartels 1949—see Section 8.5.2.1). The Kp index is a "pioneer", which means that it is of crucial importance in the history of geomagnetism, but it remains somewhat imperfect, in particular as a result of the limited number of stations for which data were available at that time, during the cold war and before the International Geophysical Year.

After the International Geophysical Year, it became possible to design new indices enabling a more refined description of the geomagnetic activity, as a result of a better understanding of magnetospheric physics and of improvement in the observatory network: the "Dst" (ring current behaviour; Sugiura 1964—see Section 8.6.1), "AE" (maximum of the auroral electrojet intensity; Davis and Sugiura 1966—see Section 8.3.2), and "am" indices (K-derived planetary activity index; Mayaud 1968—see Section 8.5.2.2) were then proposed. From the K scalings in two almost antipodal stations, Mayaud (1971) also introduced the "aa"

[1] The C figure is the first geomagnetic index defined at an international level: each observer assigned a certain number (0, 1, or 2) to each Greenwich day, by judging the relative degree of disturbance of the magnetogram (resp., quiet, moderately disturbed, or disturbed). The planetary or international magnetic character figure C_i was defined as the mean of the C figures supplied by all the cooperating observatories.

antipodal activity index in order to provide a very long series of geomagnetic activity indices (see Section 8.5.2.4).

Besides, lists of various kinds of magnetic events used to be compiled during the first half of the twentieth century. According to a IAGA resolution passed in 1975, only the lists of "ssc" (storm sudden commencements, Mayaud 1973—see Section 8.6.3) and "sfe" (solar flare effects[2]) are still compiled.

At the current time, these indices are the ones that are acknowledged by IAGA.

More recently, Troshichev et al. (1979; 1988) introduced the "PC" index (polar cap magnetic activity driven by the IMF B_z component—see Section 8.3.1).

All these indices have been designed at a time when observatories were operated using analogue variometers, and their derivation schemes therefore cope with the related requirements. During the last decades of the twentieth century, digital magnetometers were installed in more and more observatories. Methods for computer derivation of K indices were acknowledged by IAGA (Menvielle et al. 1995).

On the other hand, new geomagnetic indices that best utilize the availability of high quality digital data were proposed. They give new insight on geomagnetic activity: "SYM" and "ASY" indices (ring current and field aligned currents; Iyemori 1990—see Section 8.6.2), "IHV" and "IDV" indices (long term variation of geomagnetic activity; Svalgaard et al. 2004; Svalgaard and Cliver 2005—see Section 8.7.1), "αm" and "αa" indices (planetary activity on time intervals shorter than 3 h, Menvielle 2003—see Section 8.5.9), and indices based on geomagnetic pulsations (see Section 8.7.2).

At the turn of the twenty-first century, the Internet revolution opened the way for massive data transfer with short delay. At the same time, indices were recognized as space weather basic data. This resulted in a multiplication of Internet sites where estimated values of IAGA geomagnetic indices were made available within short delays. Since derivation schemes may differ from one site to the other, and from the derivation scheme of the 'official' index, IAGA "*urged the producers of the estimated indices to clearly label them with "est" at the end of each index name to distinguish them from the official IAGA indices*" (Resolution 5, IAGA News 38 1998, p. 42).

This paper presents for each index, or each index family, a short but complete description of the derivation process and a review of the information that the index may provide on the dynamics of, and on the physical processes that take place in the Earth's ionized environment. Our objective is to help the user of geomagnetic indices (not just for statistical studies) not to be in the situation such as that exemplified in the following citation:

> He uses statistics as a drunken man uses a lamp-post: for support rather than illumination. Andrew Lang (1844).

8.2 Physical Background

8.2.1 Basics

The solar wind is a variable supersonic and superalfvenic flow of hot plasma emitted permanently by the Sun and carrying the Sun's magnetic field through the solar system, where it is called the Interplanetary Magnetic Field (IMF). Close to planetary obstacles, the solar wind is slowed down through a bow shock, where it becomes turbulent. In case of magnetized planets such as the Earth, the solar wind stream is deflected around the magnetic field obstacle, compressing it in the dayside, and stretching it into a long tail in the nightside, giving rise to a magnetosphere cavity.

In the case of the Earth's magnetosphere, merging between the IMF and magnetospheric field lines is the main interaction process by which energy, momentum, and plasma are transferred from the solar wind into the Earth's environment. Other mechanisms, such as

[2] A sfe, or magnetic crochet is the sudden perturbation in geomagnetic elements that follows the eruption of a solar flare. sfe events occur when a solar flare points towards the Earth; they are confined mostly to the sunlit hemisphere and are associated with currents that flow primarily in the ionosphere. They are due to the extra ionization produced by X ray and EUV flare radiation. For more details, the reader is referred to, e.g., Curto et al. (1994a, b) and references therein. sfe events are usually noticed on magnetograms at low and mid-latitude stations as an increase in the intensity of the S_R (see Section 8.2.2). Lists of sfe events are compiled by the Service of Rapid Variations on the basis of reports made by observatories from morphological inspection of their magnetograms.

diffusion or viscous interaction, also contribute to this transfer. The magnetopause location of the merging process depends upon the IMF orientation, which is generally expressed in a geocentric reference frame. In the following, we will use the Geocentric Solar Magnetospheric[3] (GSM) coordinate system:

- if the IMF is directed purely southward (IMF-Bz < 0), reconnection occurs with closed magnetospheric field lines on the subsolar dayside magnetopause, forming opened field lines, which are dragged anti-sunward by the magnetic tension at the reconnection site and the solar wind flow;
- if the IMF is directed purely northward (IMF-Bz > 0), reconnection occurs with opened magnetospheric field lines tailward of the cusp, which are first dragged sunward by the magnetic tension, then antisunward by the solar wind flow;
- if the IMF presents also a dawn-dusk component, the reconnection location is still an open question, but it may be shifted away from the subsolar magnetopause. The newly open field lines present a curvature, resulting in an azimuthal component of the convection. Thus, if the IMF is southward and duskward (IMF-By > 0), then the convection shows a poleward component but also a westward component in the northern hemisphere and an eastward component in the southern hemisphere. The convection is reversed if the IMF is southward and dawnward (IMF-By < 0). This effect is known as the Svalgaard-Mansurov effect (Svalgaard 1968, Mansurov 1969).

During the merging process, the solar wind plasma entering the magnetosphere precipitates along magnetic field lines and is simultaneously transported perpendicularly by the convection of magnetic field lines.

These opened field lines pile up in the nightside magnetotail, where magnetic energy is stored. The magnetic configuration of the magnetotail is streched and evolves toward a more and more unstable state. The stored energy in the tail can then be released mainly sporadically during violent episodes called substorms. The stretched opened field lines anti-parallel on both sides of the magnetotail equatorial plane reconnect and plasma present in the tail is released into the nightside ionosphere. The magnetic configuration of the magnetotail returns toward a more stable state, with closed dipolar magnetic field lines dragged sunward by the magnetic tension.

The resulting global magnetospheric convection maps along highly conductive magnetic field lines in the high-latitude ionosphere. The ionospheric convection gives then a condensed view of the general dynamics of the magnetosphere. In case of southward IMF, the ionospheric convection is composed of two cells: with anti-sunward convection at high latitudes, and sunward return convection flow at lower latitudes. However, the dynamics of the Earth's magnetosphere can be very variable and depends upon IMF orientation, solar wind properties on the dayside and substorm activity on the nightside. The resulting ionospheric convection can then be more complex, presenting several cells whose number and shape vary with IMF orientation (see Cowley 1982, for a complete review). At the ionospheric footprints of dayside and nightside reconnected field lines, auroral features are formed, fast ionospheric flows and electric currents are excited and Joule heating is generated.

8.2.2 Electric Currents in the Magnetosphere-Ionosphere System

Several sources of electric currents co-exist in the magnetosphere-ionosphere system. They are directly or indirectly related to the dynamical interaction between the Sun and Earth environments: either via the solar illumination on the terrestrial upper atmosphere or via the solar wind-magnetosphere magnetic interaction.

The first major source of current is the ionosphere dynamo. Solar illumination creates a hot spot in the atmosphere near local noon. It generates a pressure gradient and the atmosphere flows away from the peak pressure. In the conductive regions of the ionosphere, the associated motion of the ionized gas causes charges separation (between electrons and ions) responsible of horizontal currents that mostly flow on the day

[3] In the GSM frame, the x-axis is represented by the Earth-Sun line, directed positive towards the Sun. The y-axis is defined as the cross product of the GSM x-axis and the magnetic dipole axis, directed positive towards dusk. The z-axis is defined as the cross product of the x- and y-axes. The magnetic dipole axis lies within the xz plane.

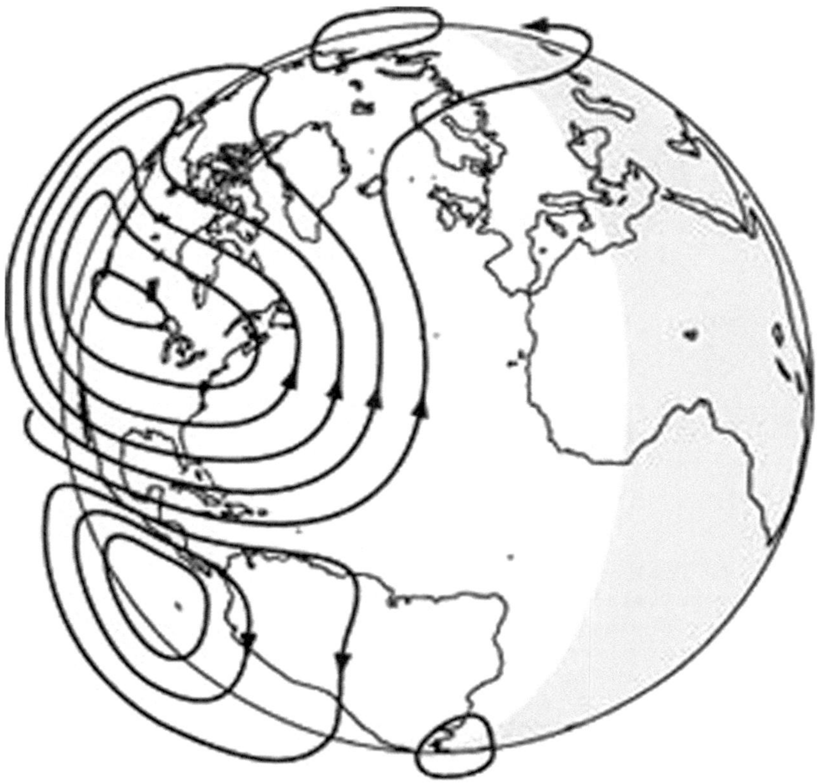

Fig. 8.1 Sketch of the ionosphere dynamo (from http://geomag.usgs.gov/)

side. These currents are sketched on Fig. 8.1; they form two vortices, one in each hemisphere, flowing counter-clockwise in the Northern Hemisphere and clockwise in the Southern Hemisphere. They produce the well-known "solar quiet" S_R[4]ground disturbance.

Due to the anisotropy of the ionosphere, two types of currents co-exist in the 90–120 km height region, the Hall current flowing perpendicular to the electric field and the Pedersen current flowing parallel to the electric field. Close to the dayside magnetic equator, where the Earth magnetic field is almost horizontal, the eastward component of the dynamo electric field drives an eastward Pedersen current and an upward Hall current. As these currents are confined in the 90–120 km region, the upward Hall current is inhibited above and below, and a vertical polarization field has to develop. This polarization field drives a strong eastward Hall current flowing in the 90–120 km altitude region within $\pm 2°$ latitude of the dip equator, known as the equatorial electrojet. The conductivity is thus enhanced in the electrojet region, and it is known as the Cowling[5] conductivity.

The second major source of current is caused by the interaction between the solar wind and Earth planetary magnetic field. As the solar wind particles encounter the Earth's main field, the electrons and ions of the wind are deflected in opposite directions by the Lorentz force. A sheet of electrical current is created, in which the pressure of the solar wind normal to the surface is exactly balanced by the pressure of the Earth's magnetic field just inside the boundary. This current is called the magnetopause current or Chapman-Ferraro current; it flows duskward on the dayside magnetopause, around the two polar cusps

[4] The S_R corresponds to the curve observed during an individual magnetic quiet day; the S_q variation is deduced from S_R curves by averaging them over a given time interval, during which it thus represents the most likely S_R variation.

[5] The Cowling conductivity is then a combination of the Hall and Pedersen conductivities which arise in magnetized collisional plasma, such as the ionosphere.

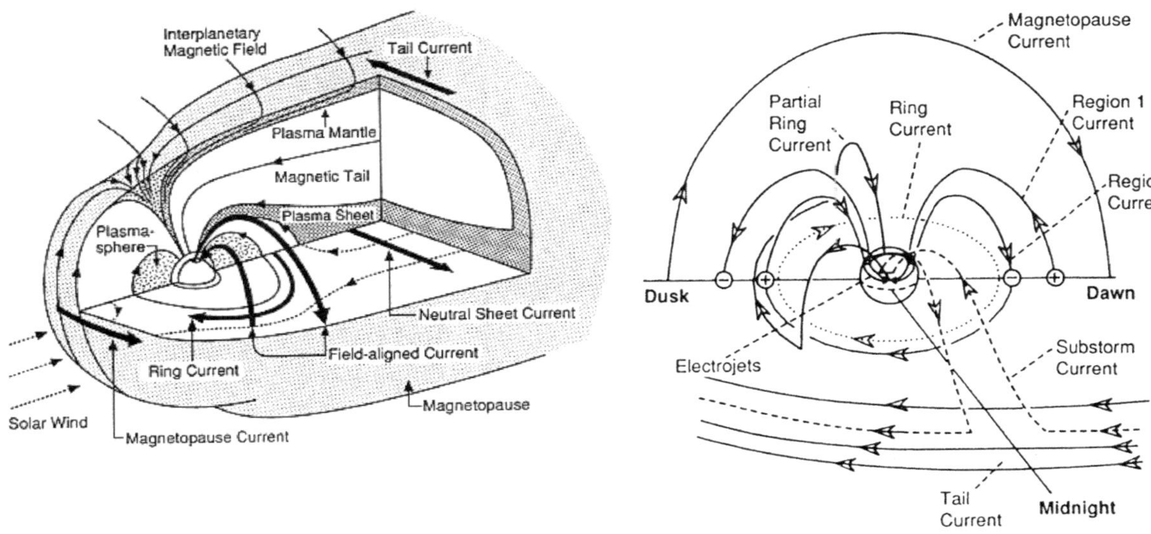

Fig. 8.2 (**a**) Sketch of the magnetosphere. The arrows indicate the current flows associated to the plasma dynamics in the magnetosphere. (**b**) Highly schematic representation of the various current systems linking magnetospheric and ionospheric currents and ultimately responsible for magnetic activity. Not shown are polar-cusp currents, polar-cap closure of the electrojets, and currents associated with IMF-By effects (from McPherron 1995)

(neutral points of the terrestrial magnetic field), and dawnward on the nightside magnetopause where it is called the tail current (see Fig. 8.2). Its magnetic signature at the Earth's surface, or ground effect is a 24 h modulated variation the intensity of which is most important on the dayside. The tail part of the magnetopause current is closed inside the magnetosphere by a current flowing duskward in the magnetic equatorial plane of the magnetotail which separates the Northern and Southern lobes of opposite magnetic field. This current is called the neutral sheet of current. The ground effect of the tail currents is a southward perturbation.

The third major source of current in the magnetosphere is produced by the westward drift of protons and the eastward drift of electrons in the radiation belts region, where the magnetic field is approximately dipolar. This current is known as the ring current; it flows westward around the Earth at a distance between 3 and 6 Earth radius (R_E) and is dominated by the ions, as electrons are rapidly recombined in the atmosphere. Its ground effect is a southward perturbation reducing the strength of the main field, except near the magnetic poles. As the magnitude of the ring current disturbance is proportional to the total energy of drifting particles, this current is strongly enhanced during magnetic storms and substorms when energization of these drifting particles occurs. Moreover, during substorms and the main phase of storms, the drifting ions of the ring current gain energy in the nightside from the cross tail electric field produced by enhanced convection. This causes the ring current to be more intense near dusk, where it is called the partial ring current. It causes an asymmetric pattern of the ground magnetic perturbation, with more southward perturbation at dusk than at dawn.

The origin of the field-aligned currents is mainly explained by magnetospheric convection and is mainly composed of two concentric regions of currents encircling the Earth. They flow from the equatorial magnetosphere to the high-latitude ionosphere along highly conductive field-lines; Fig. 8.2b shows a highly schematic representation of these current systems. The more pole-ward currents are known as Region-1 field-aligned currents and these originate from the anti-sunward convection in the external magnetosphere driven by viscous interaction and reconnection inside the boundary layers between the solar wind and the magnetosphere. The more equator-ward currents are known as Region-2 field-aligned currents and these originate from the divergence of the ring current driven by the azimuthal pressure gradients generated in the magnetospheric ring plasma by the sunward return convection. Currents flow in opposite direction on each side of the noon-midnight plane and also between the two concurrent currents. The effects of the Region-1 and Region-2 currents across the auroral oval cancel each other and cannot be measured on the ground.

The currents flowing in the high-latitude ionosphere are associated with ionospheric convection, which is a direct mapping of the magnetospheric convection:

- the Pedersen currents flow parallel to the convection electric field, i.e., duskward through the polar cap and dawnward on the auroral zones: they close the Region-1 and Region-2 of field-aligned currents. They produce almost no visible ground magnetic perturbation;
- the Hall currents flow perpendicular to the convection electric field, i.e., anti-sunward along the auroral zones and sunward in the polar cap. They cause clear ground perturbations (disturbance polar of type 2 or DP-2 current, see Fig. 8.3). In the auroral oval where the conductivity is high, these currents are confined along very narrow eastward and westward channels called auroral electrojets, and are strongly enhanced during substorms.

Additional magnetosphere-ionosphere current circulations are produced by magnetic reconnection on the dayside and on the nightside. The noon current system is caused by magnetic merging between interplanetary and terrestrial magnetic fields. Its pattern depends upon IMF direction and consequently reconnection geometry and location on the magnetopause. During substorm activity, a midnight current system is also observed. It is called substorm current wedge and it is caused by the divergence of a part of the tail current through the midnight ionosphere, with a downward field-aligned current in the post-midnight sector, a westward enhancement of the Hall current in the midnight sector of the auroral ionosphere (disturbance polar type 1 or DP-1 current) and an upward field-aligned current in the pre-midnight sector (see Fig. 8.3). The DP-1 and DP-2 currents can co-exist during substorms, with DP-1 strongest pre-midnight and DP-2 strongest post-midnight. DP-1 only persists

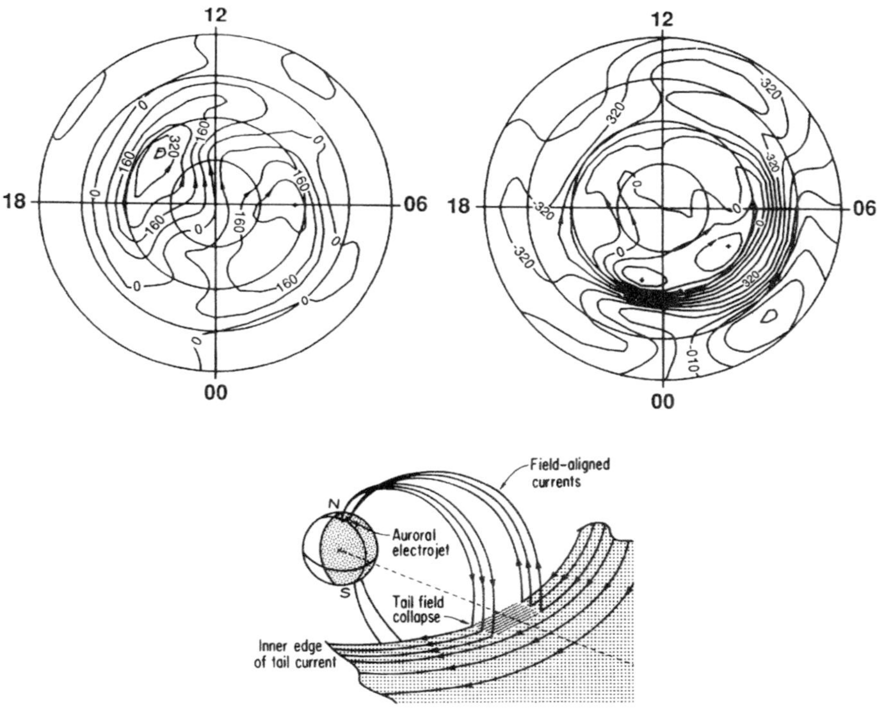

Fig. 8.3 DP-1 (panel a) and DP-2 (panel b) equivalent ionospheric current systems Closed contour lines show the flow lines for an equivalent ionospheric current that produces the observed ground magnetic perturbations. In reality, the currents are three dimensional systems, as illustrated in the case of DP-1 by the perspective view of the substorm current wedge (panel c). (from McPherron 1995)

during the expansion and early recovery phase while DP-2 may be present throughout a substorm. The cause of the substorm current wedge is an important subject of current research.

The transient variations of the geomagnetic field are caused by these electric current systems. They can therefore be considered as the output of a complex highly non-linear magnetosphere-ionosphere filter with the interplanetary conditions at the Earth's location as the input.

One part of the transient variations is due to induced currents in the solid Earth, which depend on the distribution of the conductivity within the Earth. However, it can be shown that the effect of induced currents on the variations of the horizontal geomagnetic field can be described at a first-order approximation by a multiplicative factor. This factor does not differ by more than about 10% from one station to another, except where there are sharp local heterogeneities of the conductivity in the crust and the upper mantle (see, e.g., Menvielle and Berthelier 1991, and references therein).

Magnetic transient variations have a regular component and an irregular one. The words "regular" and "irregular" are taken here in their temporal sense, which means that some variations occur regularly over time while other variations do not. The regular variations are mostly the S_R and S_q variations that result from the ionospheric dynamo. The irregular variations are the magnetic signature of the solar wind forcing: they make up the so-called geomagnetic activity that the geomagnetic indices presented in this chapter aim at describing. Monitoring the magnetic irregular variations at the Earth' surface therefore provides information on the magnetosphere and ionosphere response to its forcing by the solar wind and IMF.

An extensive review of the morphological features of the geomagnetic activity is definitely beyond the scope of this paper. The reader is referred to e.g., Mayaud (1978), Menvielle and Berthelier (1991), and references therein for further details.

Figure 8.4 shows an example of variations of geomagnetic indices during five consecutive days, in August 2003. A storm sudden commencement (ssc, see Section 8.6.3) occurred during the afternoon of August, 17th (UT Day of Year 229). It marks the beginning of a period of intense magnetic activity that lasts about 36 h. The ssc corresponds to a sharp increase of SYM-H values (few tens of nT in few minutes); SYM-H keeps afterward quite high values (about 50 nT) during few hours, then rapidly decreases down to about −130 nT. This decreasing phase is followed by the so-called recovery phase, a few days long period during which SYM-H slowly increases and tends to recover its pre-ssc level. During the decreasing phase and the first part of the recovery phase, geomagnetic activity is intense and all the other indices have high values. Such behaviour of geomagnetic indices is characteristic of geomagnetic storms.

This storm is preceded and followed by periods of magnetic quietness (am values smaller than or equal to 13 nT), or very moderate magnetic activity (am values in the range 13–40 nT). Periods of magnetic quietness are characterized by very low values of all geomagnetic indices.

Geomagnetic activity increases during the second half of August 21st (UT Day of Year 232), and remains intense during the following UT day (am values larger than 60 nT during at least four consecutive 3 h intervals). During this period of intense activity, geomagnetic indices do not behave as they do during the storm, thus indicating that the magnetosphere dynamics is not the same during these two periods of intense magnetic activity. Note that observed negative PC values correspond to periods during which this index has no physical meaning (see Section 8.3.1.2).

8.3 Polar and Auroral Indices

8.3.1 *PC Index*

The polar cap index (PC: PCN: northern; PCS: southern) aims at characterizing the magnetic activity in the polar caps that is driven by the IMF Bz component. Each index (PCN or PCS) basically relies on the use of magnetic variations observed at a single near pole station (PCN: Qaanaaq—formerly known as Thule; PCS: Vostok; see Table 8.1 and Fig. 8.5).

8.3.1.1 History

The idea to define a polar cap index based upon the magnetic activity of the DP-2 current system was first proposed by Troshichev et al. (1979), and the original concept of an index which combines magnetic and

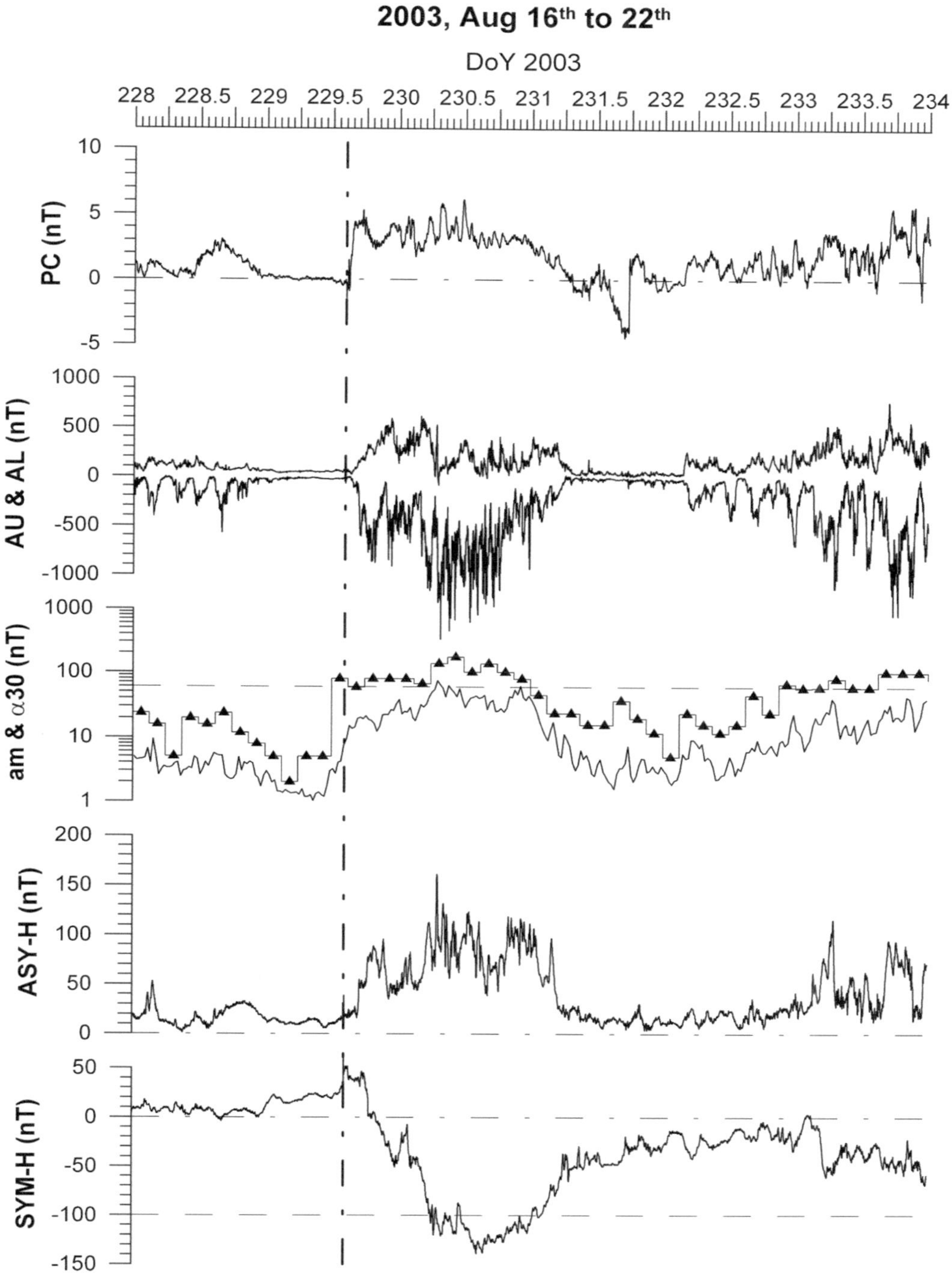

Fig. 8.4 Variation of geomagnetic indices between 2003, August 16th and 22th (UT Day of Year 228 and 234). From *top to bottom*: Polar Cap PCN indices (see Section 8.3.1), Auroral AU and AL indices (see Section 8.3.2), 3 h am (*triangles and step-like curve*) and 30-min αm planetary indices (see Section 8.5), asymmetric (ASY-H) and symmetric (SYM-H) disturbance indices (see Section 8.6.2). The *horizontal dot-dashed lines* (PC, SYM-H and ASY-H indices) correspond to 0 nT; the *horizontal dashed lines* correspond to thresholds used to define intense geomagnetic storms (SYM-H < −100 nT) and intense geomagnetic activity (am > 60 nT during four consecutive 3 h intervals). The vertical axis line corresponds to the storm sudden commencement (*ssc*, see Section 8.6.3). See text for further explanation

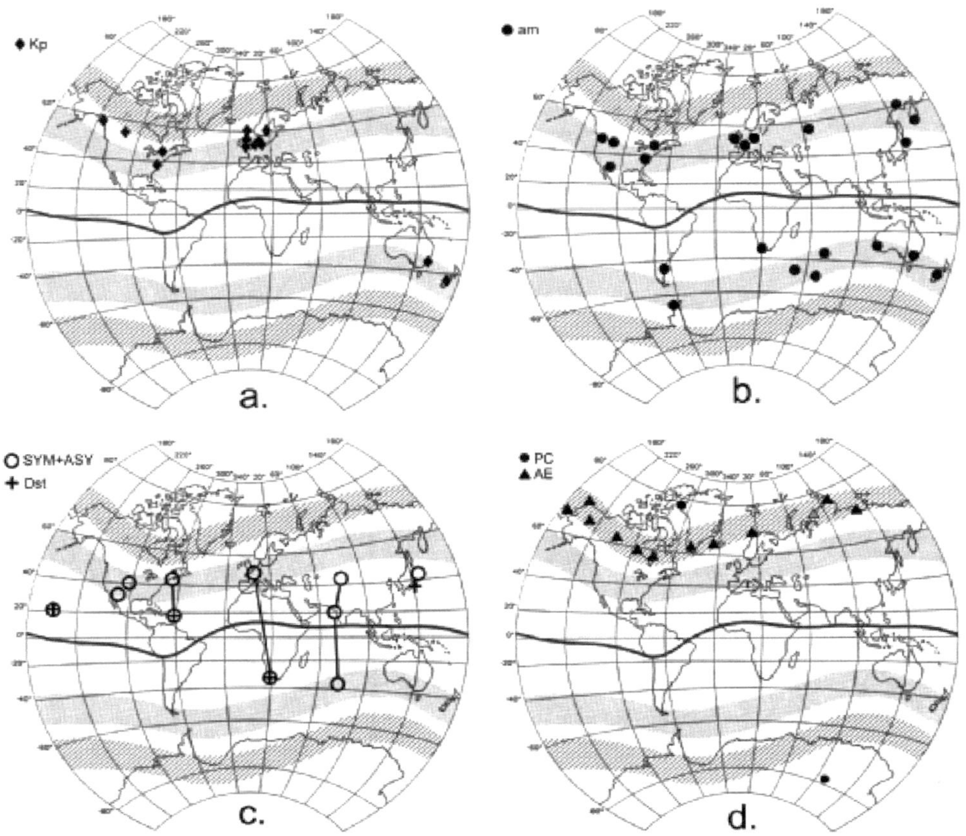

Fig. 8.5 Geographical world maps on which are indicated the positions of stations belonging to the different networks used in deriving geomagnetic indices, at 2005. A solid line indicates the position of the dip equator. The average extension of the auroral zone is sketched by the hatched area, that of the subauroral region by the shaded area (after Berthelier 1993). Panel a: the Kp network. Panel b: the am network. Panel c: the Dst, SYM and ASY networks. The SYM and ASY network stations connected by a solid line are replaced by each other in the index computation, depending on the availability and the condition of the data of the month. Panel d: the AE and PC networks. (After Menvielle and Marchaudon 2007)

Table 8.1 The PC stations

Stations	Corr. Geomag. latitude	Invariant latitude	Magnetic local noon
Qaanaaq	85.4°	86.5°	~14 UT
Vostok	−83.4°	83.3°	~13 UT

solar wind data was first suggested by Troshichev et al. (1988).

Susanne Vennerstrøm was instrumental in the early development of the PC index, including the code, operated by the Danish Meteorological Institute (DMI, Copenhagen), that produces the PCN index (Vennerstrøm et al. 1991; Vennerstrøm et al. 1994); Papitashvili et al. (2001) later fixed a programming error in this code,[6] and also discussed a recognisable daily variation, which is comparable to the seasonal variation, and a solar cycle variation.

The PCS index is computed at the Arctic Antarctic Research Institute (AARI),[7] after its proposal by Troshichev et al. (1988). Although PCN and PCS are

[6] The PCN indices were recalculated once the software bug was corrected. The index is available in both 1 min and 15 min resolution from 1975 until the present, from: http://web.dmi.dk/projects/wdcc1/pcn/pcn.html

[7] The PCS index is available (for registered users) in 15 min resolution from 1978 to Oct. 1992, and in 1 min resolution from Nov. 1992 until the present, from: http://www.aari.nw.ru/clgmi/geophys/pc_req.asp

both defined as described in Section 8.3.1.2, there were however differences between the procedures used for, e.g., secular and daily variations determination, and coefficient computation (see, e.g., McCreadie and Menvielle 2010).

To eliminate any influence of the calculation technique on scientific results a unified method for derivation of the PC index was elaborated at both AARI and DMI (Stauning, Troshichev and Janzhura 2006; Troshichev, Janzhura, and Stauning 2006; 2007a),[8] and new sets of the unified PCN and PCS indices were calculated. The reader is referred to Lukianova (2007) and Troshichev et al. (2007b) for a critical discussion of this method. Stauning (2007) proposed to use the average of the unified PCN and PCS indices as a global polar cap index, the PCC index.

When using the PC index values available on line, it is thus necessary to carefully pay attention to the method used for their derivation, and to refer to the peer reviewed literature for their precise description.

8.3.1.2 Definition of the PC Index

The current PC index is defined as

$$\mathrm{PC} = \frac{\zeta\,(\Delta F_{\mathrm{PC}} - \beta)}{\alpha}. \tag{8.1}$$

where ΔF_{PC} is the magnetic disturbance vector, α and β normalisation coefficients, and ζ a scaling value of 1 m/mV.

ΔF_{PC} is the projection of the actual magnetic disturbance vector (δM, δN) along the direction perpendicular to the DP-2 transpolar current flow:

$$\Delta F_{\mathrm{PC}} = \delta M \sin\gamma \mp \delta N \cos\gamma \tag{8.2}$$

where ΔF_{PC}, δM, and δN are expressed in nT. The disturbance vector is derived from the magnetic measurements made at the station:

$$\begin{aligned} \delta M &= M - M_s - M_d - M_{ss} \\ \delta N &= N - N_s - N_d - N_{ss} \end{aligned} \tag{8.3}$$

where (M, N) denotes magnetic elements pairs, e.g., (H, D): magnetic North and East; (X, Y): geographic North and East. The subscript s denotes secular variation, the subscript d denotes daily regular variations, and the index ss denotes the solar wind sector structure effect (Svalgaard-Mansurov effect, see Section 8.2.1).

The projection angle γ is defined as:

$$\gamma = \lambda + (UT)\,15^\circ \mp \delta_{(H,D)}^{(M,N)} D + \varphi \tag{8.4}$$

with:

$$\delta_{(H,D)}^{(M,N)} \begin{aligned} &= 0 \ \text{ if } (M,N) = (X,Y) \\ &= 1 \ \text{ if } (M,N) = (H,D) \end{aligned}; \tag{8.5}$$

λ is the geographic longitude, D the mean Declination (degrees), UT the universal time, and φ the UT-dependent angle between the direction perpendicular to the DP-2 transpolar current and the noon-midnight local time meridian. When declination is positive (Eastwards), in Eqs. 8.2 and 8.4 we use a "+" for the southern hemisphere and a "−" for the northern hemisphere (Troshichev et al. 1988).

The normalisation coefficients α and β, and the angle φ are obtained through a correlation analysis relating the merging (geoeffective) interplanetary electric field Em and magnetic perturbations projected on various horizontal directions. The direction where correlation is maximal is used for the definition of the index.

Finally, φ is the corresponding angle, and:

$$F_{PC} = \alpha \mathrm{Em} + \beta \tag{8.6}$$

α, β, and φ are defined in a table for each UT hour and calendar month. To obtain the values for times between defined elements a linear variation is assumed.

However, this description is mainly valid during the summer season, when the polar cap is totally illuminated by the Sun, allowing uniform ionospheric conductivities and free circulation of ionospheric currents, and during southward IMF-Bz periods, because in the case of northward IMF (Bz > 0), the Hall currents in the polar cap are reversed and the PC index becomes negative. Then the PC index is not anymore related to the merging electric field that always remains positive.

[8] The full citation is deliberate here so the reader is aware that the papers are by the same authors.

The magnetic contributions to the PC index are more complex during the winter season, when ionospheric conductivities are weak and non-uniform in the dark polar cap. Magnetic perturbations associated with electrojets in the auroral zones and field-aligned currents in the nightside ionosphere, where particles precipitation is present, become important contributions to the PC index. Moreover, substorms periods can lead to an electric field orientation in the near-pole region significantly modified by the deformation of the FACs structure and the appearance of the substorm current wedge (DP-1 current, see Fig. 8.4) in the night-time magnetosphere. This reduces considerably the linear relation between the merging electric field and PC.

As the PC index is calibrated by Em, no variation due to UT or season should appear. This is why PC is parameterized by season (to take into account conductivity differences between illuminated and dark polar caps), by UT (to take into account the rotation with the Earth of the magnetic station) and also by hemisphere (to take into account the different geographic positions of the Qaanaak and Vostok stations). Moreover, solar wind sector structure variations responsible for the Svalgaard-Mansurov effect (see Section 8.2.1) in the magnetosphere are also removed in deriving the PC index. Then, only solar cycle variations will persist: PC is in average higher during solar maximum. Seasonal variations will anyway appear during particular events such as solar wind pressure pulses or magnetic substorms, not directly governed by the merging electric field (see below).

8.3.1.3 Correlation with Interplanetary and Magnetosphere Quantities

As a result of its derivation process, PC is well related to the merging electric field Em, as well as to the polar cap DP-2 ionospheric Hall currents, which are directly controlled by the coupling between the merging electric field and the magnetosphere.

Over 20 years, the PC index has been correlated with various parameters of the solar wind or from the magnetosphere-ionosphere system. The correlations are generally reasonable, being mainly linear or quadratic relationships.

Correlation with Solar Wind, IMF, and Coupling Functions

In their initial studies, Troshichev and Andrezen (1985) and Troshichev et al. (1988) compared the ground magnetic perturbations used to derive the PC index with various solar wind parameters or coupling functions of these parameters (such as IMF-Bz, Vsw.Bz,[9] Em…). All these parameters correlate reasonably well with PC magnetic perturbations, if we allow for a 20 min propagation delay corresponding to the transmission of the solar wind signal between the bow shock and the ionosphere. The maximum correlation is obtained with the merging electric field Em, and this is why PC is calibrated by this coupling function.

In a new method of re-unification of PCN and PCS, Troshichev et al. (2006) showed a linear correlation between Em and PCN-PCS, independent of season and hemisphere. These results allow the validation of the calibration method of the PC indices and the re-unification method between PCN and PCS. The correlation coefficient between PCN-PCS and the merging electric field remains relatively low ($r = 0.6 - 0.65$) but this can be explained by the fact that solar wind parameters are not filtered by the bow shock, the magnetopause and the magnetosphere-ionosphere system and are hardly directly comparable with ionospheric parameters such as PC. More recently, Lyatsky et al. (2007) also made an attempt to correlate the original PCN with slightly more sophisticated solar wind coupling functions than the merging electric field, allowing small improvement in the correlation, especially during the solar minimum period.

Correlation with Solar Wind Pressure

Several studies have attempted to correlate solar wind pressure with the PC index, revealing contradictory results: with significant PC correlation (Lukianova 2003) or weak PC correlation (Huang 2005). The main problem in these studies has been to properly decorrelate the effect of solar wind pressure variations from Em variations. Recent studies obtained with the

[9] Bz is the IMF North-South component, V_{sw} is the solar wind velocity.

re-unified PC technique have confirmed that strong pressure gradients (especially during magnetic storms) have a significant effect on PC and are the second most important factor for PC variation, after the merging electric field Em (Troshichev et al. 2007a).

Contrarily to the merging electric field, the PC index responds only with a few minutes delay to solar wind pressure variations. In case of a solar wind pressure enhancement (decrease), PC displays first a negative (positive) spike of a few minutes followed by a positive (negative) and more progressive enhancement which returns eventually to its basis level even if the solar wind pressure remains high (low). Seasonal effects of the PC response to a solar wind pressure pulse has also been observed by Troshichev et al. (2007), with PC in the summer hemisphere higher than PC in the winter hemisphere, due to higher conductivities in the ionosphere.

These PC signatures have allowed Stauning et al. (2008a) to give an insight into the electrodynamic response of the magnetosphere-ionosphere system to solar wind pressure pulses. In the case of a solar wind pressure enhancement, two small reverse vortices are first generated in the central polar cap, caused by a divergence in field-aligned currents of a magnetopause current excess, caused by the pressure pulse impact, followed by a re-initiation of forward convection with two convection cells flowing antisunward in the polar cap, intensified by the solar wind pressure perturbation circulating around the magnetopause.

Correlation with Cross-Polar Cap Parameters: Electric Field, Potential and Diameter

Several studies have attempted to directly correlate the PC index with other parameters physically describing polar cap properties. The main problem in these studies has been to find reliable data sets to compare PC with, because of the smallness of the database and the diversity of experiments. Thus, electric field experiments and particle detectors onboard the Akebono and DMSP satellites (Troshichev et al. 1996; Troshichev et al. 2000; Lukianova et al. 2002) or more recently, SuperDARN ionospheric HF radars, measuring line-of-sight convection velocity (Fiori et al. 2009), have been used to deduce the cross-polar cap electric field, potential and diameter.

For all these cross-polar cap parameters, quadratic relationships have generally been found with the PC index: generally linear for small PC values with a saturation effect of these parameters appearing at higher PC values. The coefficients of these quadratic relations are very different from one study to the other, at least partly due to the different calibration levels of these experiments. No clear seasonal effect was detected in these studies. Only Lukianova et al. (2002) showed that for disturbed periods (PC above 5), the PC index reaches higher values in the winter polar cap. This effect was explained by Lukianova et al. (2002) by the existence of strong polar cap absorption events caused by solar wind protons bombardment, dramatically increasing conductivities in the dark polar cap.

Correlation with Auroral Parameters: AE, AL, AU Indices, Joule Heating and Auroral Power

Some studies have shown a very strong linear correlation between PC and the AE and AL auroral indices, representing a good characterisation of the auroral electrojets and substorm currents (e.g., Vennerstrøm et al. 1991; Huang 2005; Janzhura et al. 2007; Lyatskaya et al. 2008), but not with the AU index. Strong seasonal effects were observed by Vennerstrøm et al. (1991) with higher correlation between PC and AE-AL during winter and equinox than during summer.

Moreover using re-unified PCN and PCS in both hemispheres, Janzhura et al. (2007) found that isolated magnetic bays and substorms were always preceded by an increase of magnetic activity in the summer polar cap, with the summer PC index running rather independently of the auroral magnetic disturbances, contrary to the behaviour of the AE and AL and winter PC indices which were pretty well correlated. These effects are likely caused by the fact that PC index for the sunlit polar cap (with high ionospheric conductivities) responds mainly to the merging electric field and thus to Hall currents in the polar cap, whereas PC index for the dark winter cap, being limited to by low ionospheric conductivities, responds better to the particle precipitation and field-aligned currents in the auroral zone like the AE and AL indices. Finally, Lyatskaya et al. (2008) found also that the AL index correlated better with the re-unified PC index calculated in the winter hemisphere, but gave another explanation for this effect, by introducing inter-hemispheric field-aligned currents that flow from the summer high-latitude ionosphere and close through the ionosphere in the opposite auroral zone, and thus decrease the field-aligned currents contribution to magnetic disturbances in the summer hemisphere and increase it in the winter hemisphere. Further investigation is probably necessary to conclude on this subject.

Liou et al. (2003) found a correlation between the northern hemispheric auroral power inferred from auroral luminosity acquired from the ultraviolet imager of the Polar satellite and the PCN index, with higher correlation in winter than in summer. They also found that PC correlates with nightside auroral power much better than with dayside auroral power. These seasonal and diurnal effects were again explained in terms of competition between Hall ionospheric convection currents and field-aligned currents contributions caused by variations in ionospheric conductivities.

Finally, Chun et al. (1999) proposed to use PC as a proxy for the hemispheric Joule heat production rate (JH). They found a quadratic relationship between Northern JH estimated from the AMIE technique and PCN and explained it by the fact that PC must be proportional to the polar cap electric field affected itself by a saturation effect. Seasonal differences were also observed and again explained by variations in polar cap conductivity between seasons, but could also be due to a badly constraint estimation of ionospheric conductances in the AMIE technique.

Correlation with the Dst Index

Stauning (2007) found a one-to-one correspondence between enhancements in the re-unified PCC index and the occurrence of global geomagnetic disturbances, such as storms, as determined by the Dst index. As the polar cap index is assumed to provide an indication of the energy input to the magnetosphere while the Dst index is considered to mark the energy stored in the ring current, Stauning (2007) compared the PCC index to the source function Q of the Dst index. This function Q is related to the interplanetary electric field and there is an empirical relation between Q and Dst (Burton et al. 1975). An overall linear correlation was found between Q and PCC, allowing the derivation of an empirical Dst index very close to the experimental one. It is thus possible to get, in near real-time, the Dst index from the PCC index when available.

8.3.1.4 Use and Misuse of PC Index

As the PC index is issued from only one station in each hemisphere, it can be calculated in near real-time. Other important indices, such as auroral indices, the am index, or the Dst index, characterise the energy state of the magnetosphere-ionosphere system; however they are more difficult to derive quickly as they are derived from several magnetic stations. Through the various relationships found between these indices and the PC index, it becomes in theory possible to get access in near real-time to a rough empirical estimation of these other indices. Thus, the PC index is expected to rapidly become important for specification of magnetospheric state and to be useful in scientific and space weather applications.

However, as shown in the previous section, the linear and quadratic relations found between the PC index and the other parameters are often obtained by isolating each source of perturbations in the magnetosphere (isolated substorms, isolated solar wind pressure pulses). In real conditions, the magnetosphere-ionosphere system reacts to a combination of different phenomena (e.g., magnetic storm with strong merging electric field, solar wind pressure pulses, substorms). Then, as stressed by Vennerstrøm et al. (1991), the most serious problem is that several sources can contribute to the PC index and that it is difficult to distinguish between them. An important question can then be raised: are the relations found between PC and other parameters still valid in complex conditions? A recent study by Stauning et al. (2008b) where the PCC/Em ratio was compared with the AL index for magnetically disturbed conditions, showed no PCC/Em variations, which seems to be contradictory with previous studies.

8.3.2 Auroral-Electrojet (AE) Indices

AE indices are acknowledged by IAGA (Resolution 2, IAGA Bulletin 27 1969, p. 123); they are currently routinely produced by WDC for Geomagnetism, Kyoto, Japan, as part of the International Service of Geomagnetic Indices (ISGI); they are made available electronically at the WDC-Kyoto and ISGI Internet sites.[10]

[10] Reference AE values are available on-line at http://wdc.kugi.kyoto-u.ac.jp/ and at http://isgi.latmos.ipsl.fr

8.3.2.1 History

The Auroral-electrojet (AE) index was originally introduced by Davis and Sugiura (1966) as a measure of global electrojet activity in the auroral zone.

After the initial development at the NASA/Goddard Space Flight Center, the calculation of the index was first performed at the Geophysical Institute of the University of Alaska, which published hourly values of the index for the years 1957–1964. The production of 2.5 min values was then made at the Goddard Space Flight Center for the period from September 1964 to June 1968. After these early publications, the index was regularly issued by the World Data Center A for Solar-Terrestrial Physics (WDC-A for STP) in Boulder, Colorado, which published 2.5 min values for the years 1966–1974 and 1.0 min values for 1975 and the first 4 months of 1976.

When it became difficult for the WDC-A for STP to continue producing the AE index, WDC-C2 (operated by the Data Analysis Center for Geomagnetism and Space Magnetism, Faculty of Science, Kyoto University) began to produce the AE index from the International Magnetospheric Study period (1978–1979) onwards. Since then, WDC-C2 for Geomagnetism (renamed WDC for Geomagnetism, Kyoto after 2000) has been publishing 1.0 min values of the AE index.

8.3.2.2 Definition of the AE Indices

The AE index is derived from geomagnetic variations in the horizontal component *H* observed at 12 selected observatories along the auroral zone in the northern hemisphere.

To normalize the data, a base value for each station is first calculated for each month by averaging all the data from the station on the internationally selected five quietest days (Q-days, see Section 8.5.4.1). This base value is subtracted from each value of one-minute data obtained at the station during that month.

Resulting *H* deviations are superimposed, as illustrated by Fig. 8.6. Among the data from all the stations at each given time (UT), the largest and smallest values are then selected. The AU and AL indices are respectively defined by the largest and the smallest values so selected. The symbols, AU and AL, derive from the fact that these values form the upper and lower envelopes of the superposed plots of all the data from these stations as functions of UT. The difference, AU minus AL, defines the AE index, and the mean value of the AU and AL, i.e., (AU + AL) / 2, defines the AO index. The AU and AL indices are intended to express the strongest current intensity of the eastward and westward auroral electrojets, respectively. The AE index represents the overall activity of the electrojets, and the AO index provides a measure of the equivalent zonal current.

The term "AE indices" is usually used to represent these four indices (AU, AL, AE and AO).

A list of the AE stations is compiled in Table 8.2. It should be noted that some of the stations have closed and been replaced by new stations. Cape Wellen was closed in 1996 and was not replaced until the introduction of Pebek in April 2001. Great Whale River observatory was closed in July 1984 and followed by a new station at Poste-de-la-Baleine, in September 1984. Then Poste-de-la-Baleine was replaced by Sanikiluaq in November and December 2007. The locations of the AE stations are shown in Fig. 8.5.

8.3.2.3 Basic Characteristics of AE Indices

It is of interest to examine which station (or which magnetic local time (MLT)) contributes most to the AU and AL indices. Davis and Sugiura (1966) showed that the AU and AL indices during disturbed intervals mostly reflect variations at 1400–2100 MLT and 2300–0500 MLT, respectively. Allen and Kroehl (1975) found that during disturbed intervals, stations located on the nightside make substorm-related AU and AL variations and their contributing peak times are around 1745 MLT for AU and 0315 MLT for AL, which are consistent with the MLT values reported by Davis and Sugiura (1966). During quiet times, stations in the sunlit hemisphere contribute to low-amplitude AU and AL indices and the peak contributions are around 0615 MLT for AU and 1115 MLT for AL.

Diurnal variations (or Universal Time variations) of the AE index have been discussed by many researchers. Allen and Kroehl (1975) showed that both the AU and AL indices during the five internationally selected most disturbed days (*D*-days, see Section 8.5.4.1) in 1970 have larger values around 0900–1800 UT. Basu (1975) reported that the AL index in 1967–1970 is larger on the morning side

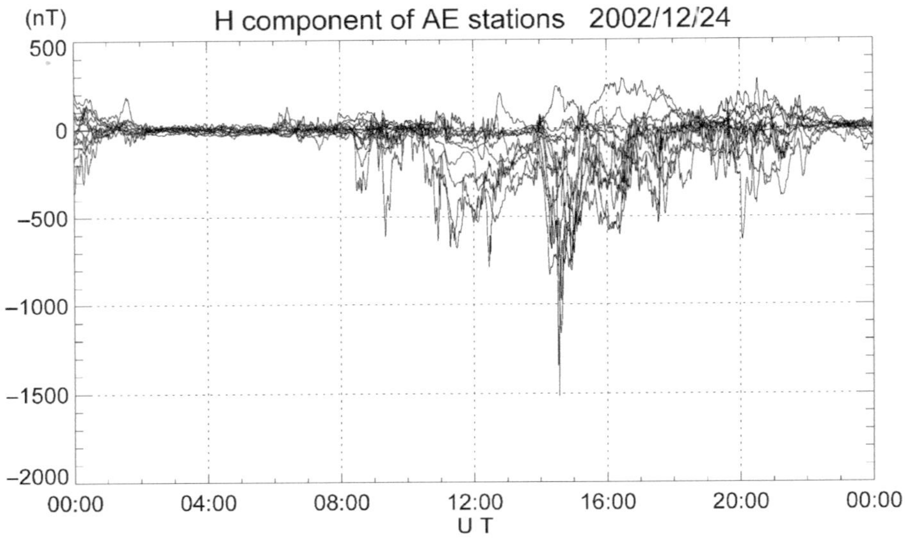

Fig. 8.6 Derivation of the auroral-electrojet indices. *H* deviations from a base value at each AE stations are superimposed: AU is the upper envelope, AL is the lower envelope, AO = (AU + AL)/2, AE = AU − AL

Table 8.2 List of the 12 AE stations. Stations above (below) a double line are currently working stations (old stations). Values of geomagnetic coordinates are given for January 1, 2005 from the IGRF-10 geomagnetic field model

		Geographic		Geomagnetic		
Observatory	Abbrev.	Lat. (°)	Lon. (°)	Lat. (°)	Lon. (°)	Notes
Abisko	ABK	68.36	18.82	66.06	114.66	
Dixon Island	DIK	73.55	80.57	64.04	162.53	
Cape Chelyuskin	CCS	77.72	104.28	67.48	177.82	
Tixie Bay	TIK	71.58	129.00	61.76	193.71	
Pebek	PBK	70.09	170.93	63.82	223.31	Open. in 2001/04
Barrow	BRW	71.30	203.25	69.57	246.18	
College	CMO	64.87	212.17	65.38	261.18	
Yellowknife	YKC	62.40	245.60	68.87	299.53	
Fort Churchill	FCC	58.80	265.90	67.98	328.36	
Sanikiluaq	SNK	56.5	280.8	66.6	349.7	Open. in 2007/12
Narssarssuaq	NAQ	61.20	314.16	69.96	37.95	
Leirvogur	LRV	64.18	338.30	69.32	71.04	
Cape Wellen	CWE	66.17	190.17	62.88	241.36	Clos. in 1996
Great Whale River	GWR	55.27	282.22	65.45	351.77	Clos. in 1984/07
Poste-de-la-Baleine	PBQ	55.27	282.22	65.45	351.77	Open. in 1984/09 Clos. in 2007/11

during summer and on the afternoon side during winter. Mayaud (1980) revealed that both the AU and AL indices for 1968–1974 are enhanced in the afternoon sector during disturbed conditions (AU or |AL| is larger than 50 nT). Ahn et al. (2000a) analyzed the index for 1966–1987, and found that the AL index has peaks at 1300–1800 UT while the AU index becomes larger at 0200–1000 UT and 1400–2000 UT. Cliver et al. (2000) showed that the AE (= AU − AL) index for 1957–1988 has larger values at 0900–1800 UT.

Seasonal/annual variations of the AE index are also considered an important research topic. Allen and Kroehl (1975) found that the AU index is largest during summer and smallest during winter, while the AL index does not show such tendency. Mayaud (1980) and Ahn et al. (2000b) showed that the AU index is largest during summer, and the |AL| index has two peaks around spring and fall. Weigel (2007) also reported similar variations of the AL index. According to a result by Cliver et al. (2000), the AE (= AU − AL) index seems to show the combined effects; that is, a larger AE value appears from March to September.

Solar cycle variations of the AE index are reported by Ahn et al. (2000b), who found that the maximum of AU or AL index does not occur during the year with the maximum sunspot numbers, but during the declining phase of a solar cycle.

8.3.2.4 Relation with Magnetospheric and Ionospheric Physical Quantities

Some studies have related the AE index to the rate of energy dissipation through Joule heating in the ionosphere (Q_J). The first attempt was made by Perreaut and Akasofu (1978), who calculated:

$$Q_J[\mathrm{GW}] = 0.06 \times \mathrm{AE}\ [\mathrm{nT}], \tag{8.7}$$

while Baumjohann and Kamide (1984) derived:

$$Q_J[\mathrm{GW}] = 0.32 \times \mathrm{AE}\ [\mathrm{nT}], \tag{8.8}$$

and found that the correlation coefficient between them was 0.74. A number of similar studies followed and their results are summarized by Østgaard et al. (2002).

The energy which creates disturbances in the auroral region originally comes from the solar wind. Thus, numerous studies have attempted to describe the AE index as a function of the energy input parameter from the solar wind. Iyemori et al. (1979) considered the magnetosphere as a linear system having the solar wind parameter as an input and the AE index as an output. The efficiency of prediction of the AE index by the IMF Bz, or by Vsw·Bs[11] was about 0.6, indicating that a fairly large portion of the AE index can be explained with the linear system. Clauer et al. (1981) also used the linear prediction filtering technique and found that the AL index is well correlated to Vsw·Bs. Shue et al. (2001) examined the effects of the solar wind density on the AU and -AL indices, and found significant correlation between them. A fairly recent study by Li et al. (2007) developed two empirical models to compute the AL index from various solar wind parameters. They provided a good summary of previous works on predicting the AE index. A new approach to predicting the AE index from real-time global MHD simulation has been reported (Kitamura et al. 2008).

[11] Bz is the IMF North-South component, Vsw is the solar wind velocity.

8.3.2.5 Use and Misuse of the AE Indices

The AE index was created such that "The index is a direct measure of the axially nonsymmetric component of Dp (polar disturbance) activity" (Davis and Sugiura 1966, p. 799). Thus, the index is generally considered to be a measure of electrojet activity; the AU and AL indices provide the maximum eastward and westward electrojet currents, respectively. Because of these characteristics, the AE index has been widely used for substorm studies. However, caution should be taken when using the index for the following reasons:

1. The AE stations are distributed in the auroral latitude over an 8° latitude range. Narssarssuaq, Leirvogur, and Barrow are located at 69°–70° geomagnetic latitude (GMLAT), while Tixie Bay is at 61.7° GMLAT and Pebek is at 63.8° GMLAT. This will cause diurnal variations (UT variations) in the values of the index.
2. The AE stations are distributed globally in the longitudinal direction, but the distribution is not uniform. In geomagnetic longitude, the largest separation is 48°, between Abisko and Dixon, and the smallest is 15°, between Barrow and College. This will also cause diurnal variations, similar to (1), as well as underestimation of substorm occurrence and magnitude.
3. The number of AE stations available to report data is sometimes less than 12, because of artificial noise, problems with the magnetometers, or other reasons. In such cases, there is a higher possibility of failure in detecting electrojet enhancement.
4. The index is mainly generated from the ionospheric electrojet current, but magnetospheric currents such as the equatorial ring current also have an effect on the AE index. According to Davis and Sugiura (1966), negative values of the AU index may occur by this effect.

8.3.2.6 What Next for the AE Indices

Recent demands from users include real time or quick derivation of the AE index. In response to these demands, a real time AE index has been available since 1996 at the web-site of the WDC for Geomagnetism, Kyoto.[12] It should be noted that real time values are automatically derived from raw data, resulting in no correction of artificial noise and baseline shifts. Moreover, the number of stations used is usually less than 12, and some auroral electrojet enhancements may not be reflected. Nevertheless, the advantages of the real time AE index far outweigh the disadvantages. The real-time values will thus continue to be provided from the web-site, followed by the provisional AE index with a few month delay, after these are cleaned for scientific use through visual inspection of the professional staff. A more detailed description of the real-time AE index can be found in Takahashi et al. (2004).

An AE index for the southern hemisphere was recently introduced by Weygand and Zesta (2008). The southern AE index is derived from seven stations in the auroral latitude (−60° to −70° GMLAT). The correlation coefficient between the northern (original) and southern AE indices for 7 days in December 2005 was 0.58.

8.4 *K* Index

The *K* index was devised by Bartels et al. (1939) to provide an objective monitoring of the geomagnetic activity, namely the irregular component of the magnetic transient variations that he called "particle variations".

Bartels made in fact a clear distinction between geomagnetic variations arising from solar "wave radiations" and those arising from solar "particle radiations". Because "particle variations" were sometimes evident when there were no visible spots on the Sun, he postulated the existence of so-called "*M* regions" (M for magnetically active), which emitted particle radiation without any visible footprint on the solar surface. It was some decades later that the true nature of *M* regions and their association with coronal holes, and more generally the true nature of the coupling between solar and geomagnetic activity was understood (see Section 8.8).

K index was then routinely used to monitor the magnetic activity at permanent magnetic observatories, as well as at temporary stations. It was extensively analysed and discussed in Mayaud's Atlas of *K* indices (Mayaud 1967) and by Mayaud (1980). A short review of its basic characteristics is given in Menvielle and Berthelier (1991). It is endorsed by IAGA (Resolution 2, IATME Bulletin 11 1940, p. 550).

8.4.1 History

When Bartels and his co-authors devised the *K*-index, computers were rudimentary and digital magnetometers did not exist. The original definition of *K*-indices therefore requires hand scaling on analogue magnetograms.

The question of the derivation of geomagnetic indices from digital data arose with the appearance of digital magnetometers at the end of the seventies. During the Hamburg meeting (1983), IAGA reasserted that *K* indices are to be hand scaled (Resolution 4, IAGA News 22 1984, p. 12), using if necessary print outs of the magnetograms obtained from digital data. Niblett et al. (1984) showed that a sampling resolution interval shorter than about 30 s is then necessary to avoid the underestimating of *K* indices.

The increasing number of digital and sometimes unmanned observatories and the creation of INTERMAGNET (see, e.g., Coles et al. 1990) put the question of computer production of *K* at the centre of the debate. Many teams proposed algorithms for computer derivation of *K* indices, and it was decided to organize a quantitative estimate of the relevance of the computer-derived indices, to find which algorithms would be recommended for future use (Menvielle 1991; Coles & Menvielle 1991). Four algorithms were thus selected during the Vienna meeting (1991), and endorsed by IAGA for computer production of *K* indices (IAGA News 32 1993, p. 27–28). The reader is referred to Menvielle et al. (1995) for a review.

Since that time, *K* indices can still be hand-scaled from magnetograms by an experienced observer, or computer-derived using one of the four algorithms that are acknowledged by IAGA.

[12] http://wdc.kugi.kyoto-u.ac.jp/ae_realtime

Table 8.3 Limits of classes for K indices at Niemegk observatory. The limits of range classes are expressed in nT

Range (nT)	0–5	5–10	10–20	20–40	40–70	70–120	120–200	200–330	330–500	>500
K value	0	1	2	3	4	5	6	7	8	9

8.4.2 Definition of the K Index

The K indices are based upon the range in the irregular variations, measured in the two horizontal geomagnetic components, after eliminating the so-called non-K variations; the vertical component Z is not considered because Z transient variations may be dominated by internal induction effects (see, e.g., Menvielle et al. 1982).

The difficulty in the scaling of K indices lies in the identification of the non-K variations using magnetograms from an individual observatory. When introducing K indices, Bartels et al. (1939) did not give any straightforward guidelines in order to estimate the non-K variations: they were defined in terms of a smooth curve to be expected for the considered element and the current day, during periods of magnetic quietness.

Mayaud (1967) established morphological rules as guidelines.[13] According to Mayaud, the non-K variations are defined as the simplest and least speculative smooth curve which corresponds to a possible S_R variation. In practice, the curve should be estimated from the quiet parts of the record, if any. This also includes the slow recovery of the ring current magnetic field which follows a magnetic storm, because this variation has a time constant of several hours so that it generally cannot be separated from the S_R. With this exception the resulting index is expected to be sensitive to the irregular variations only.

The main criticism of the K index has been the subjective nature of the S_R determination. It is at present accepted "that, when suitably trained, two independent observers scale the same magnitude of K indices most of the time, and rarely, if ever, does the difference exceed one unit" (Rangarajan 1989, p. 330). This is illustrated by results presented in, e.g., Mayaud and Menvielle (1980), Sucksdorff at al. (1991), and Menvielle et al. (1995).

[13] The morphological rules are also published in Menvielle et al. (1995)

A 3-h time interval was chosen for the derivation of the index since such intervals "seem to be long enough to give correct indications for such details as bays and other perturbations of only one hour or two in duration. At the same time, it is short enough not to affect too much of the day in cases where two successive intervals might be affected by a disturbance, such as the bay, occurring centred on their common point" (Bartels 1940, p. 28).

Ten classes of ranges were defined by Bartels at Niemegk observatory: proceeding in multiples of 2 up to 40 nT, then increasing more slowly to larger ranges (Table 8.3), this definition ensures a good description of both low and high levels of geomagnetic disturbances. At any station, the limits of the classes are proportional to those of Niemegk, and the grid is defined by its $K = 9$ lower limit L9 (see Section 8.4.3). An individual K index is an integer in the range 0 to 9 corresponding to a class that contains the largest range of geomagnetic disturbances in either of the two horizontal components during a 3 h UT interval.

8.4.3 The Irregular Activity as Described by the K Indices

The frequency distribution of K indices over 100 years has been analyzed by Mayaud (1976) for the stations used in the derivation of aa (see Section 8.5.2.4). He proved that the distribution is lognormal, which implies that geomagnetic activity is due to independent causes whose effects are multiplicative. Two components have been isolated: (1) a component of persistence, where a low value is more likely followed by a low one and a high value by a high one; (2) a component due to the 11-year cycle of solar activity.

The K grids used at the different stations should be defined so that K indices keep the same significance at all stations in spite of the variations of the observed geomagnetic perturbations with geomagnetic latitude. Mayaud (1968) showed that this is the case in the 45–55° *N* or *S* corrected geomagnetic latitude belts, hereafter called "subauroral latitudes", if the variation of L9 with corrected geomagnetic latitude[14] is identified with that of the irregular variations with corrected geomagnetic latitude (Resolution 4, IAGA Bulletin 19 1963, p. 359). This relation[15] is used to define the limits of the *K* grids by the International Service of Geomagnetic Indices.

Note that proportional grids lead to similar significance of *K* values if the indices are measured at stations where geomagnetic perturbations have similar local time and seasonal variations. This is the case at subauroral latitudes, where geomagnetic perturbations have midnight and equinoxial maxima (Berthelier 1979). The *K* standardization with respect to latitude is therefore the most effective at these latitudes.

Any *K* index is a code, and letters could as well have been used as numbers. It is thus obvious that although the comparison of individual *K* values is reliable, it is nonsense to make calculations, such as arithmetic averages, using the *K* values themselves. Some kind of standardization is then mandatory, and the reliability of the result clearly depends on the standardization process. The most relevant procedure is to return to the amplitude of the variations by means of *aK* equivalent amplitudes: IAGA recommends to use as the equivalent amplitude that amplitude at the middle of the corresponding *K* class, taken on the local grid for local purposes, and on the grid of Niemegk[16] when considering indices from different observatories located at different corrected geomagnetic latitudes (IAGA News 32 1993, p. 23–25).

8.4.4 Physical Meaning of K Indices

Menvielle (1979) showed that, for most 3 h intervals the ratio β/r takes values in a bounded domain, provided the typical time scale of the observed variations is less than the 3 h length of the time intervals for which *K* is measured; *r* is the range (in nT) from which the *K* index is derived and β is the root mean square (*rms*, in nT) over the 3 h interval of the irregular variations in the horizontal components. This is the case at subauroral latitudes, but it is not completely true at auroral and at low latitudes (see, e.g., Mayaud 1978).

The β/r values depend on the morphology of the irregular variations. On the basis of morphological considerations, Menvielle (1979) proposed:

$$0.3 \leq \beta/r \leq 0.8 \tag{6}$$

Since the effect of sources varies at random from one 3 h interval to the other (Mayaud 1976), this result can be interpreted in a statistical way by considering β/r as a random quantity; one may thus reasonably assume that when considering averages over time intervals, the mean value of β/r is its expected value, with an uncertainty which decreases with the number of intervals. Note that the same relation holds between β and the equivalent amplitudes a*K*, but with larger uncertainty [see Menvielle (1979) or Mayaud (1980) for a complete discussion].

Figure 8.7 presents a comparison between the rms β and the range *r* deduced from computer estimated *K* variations. It shows that the statistical relation

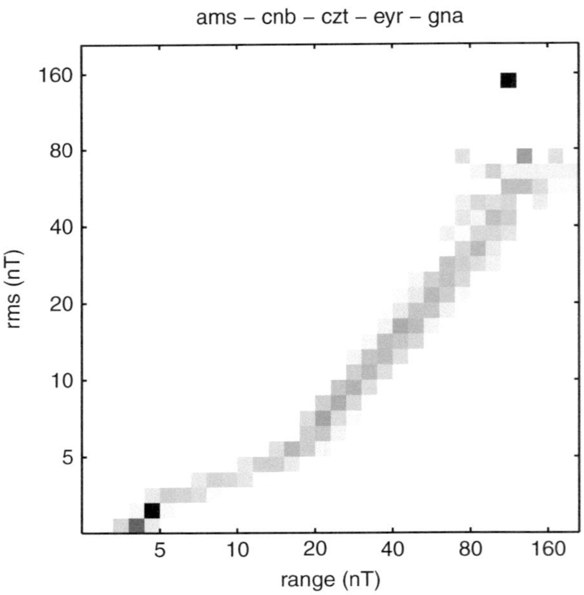

Fig. 8.7 Empirical histogram of the range vs rms distribution, for 2 years (1996–1997) and five am observatories (see Fig. 8.5 and Table 8.6). *Grey* levels correspond to percentages of range values for a given rms value: the darker the pixel, the higher the percentage

[14] Centred dipole coordinates are obtained by approximating the main field of the Earth by a centred dipole (that is, the degree one terms of a spherical analysis of this field); corrected geomagnetic coordinates differ from the former by taking higher-order spherical harmonic terms of the main field into account.

[15] For the sake of coherency with the Bartels' definition, one takes $L9 = 500$ nT at 50° of corrected geomagnetic latitude.

[16] Niemegk is a typical subauroral station

between rms and range established by Menvielle (1979) holds for range values smaller than ~5 nT, or larger than ~15 nT but with different expected values for the proportionality coefficient.

Because of the Poynting theorem, β^2 is proportional to the temporal average over the 3 h interval of the magnetic energy density related to the irregular geomagnetic variations. When measured at subauroral stations, *K* indices are therefore directly related to the magnetic energy density.

8.4.5 Use and Misuse of K Indices

The modern consensus is that $K = 0$–2 correspond to periods of magnetic quietness; $K = 3$–5 correspond to periods of moderate geomagnetic activity; $K =$ 6–9 correspond to periods of intense to very intense geomagnetic activity.

Before moving to the definition of planetary indices let us emphasize the three following points: (1) *K* indices are more relevant at subauroral latitudes; (2) if given in isolation, an individual *K* index gives only very poor information; and (3) one has to revert back to amplitudes before averaging the activity described by *K* indices.

8.5 K-Derived Geomagnetic Indices

Since their introduction by Bartels et al. (1939), the *K* indices have been regularly calculated at almost all the magnetic observatories, and they are used in the derivation of the IAGA planetary geomagnetic indices Kp (Section 8.5.2.1), am, an, as (Section 8.5.2.2), and aa (Section 8.5.2.4), and of longitude sector indices (Section 8.5.2.3).

The reader is referred to Mayaud (1980), Menvielle and Berthelier (1991), and Berthelier (1993) for detailed reviews of these indices.

8.5.1 History

The basic idea of using *K* indices from a network of observatories to derive a planetary index of geomagnetic activity was proposed by Bartels et al. (1939), in the same paper in which *K* indices were defined. The very first planetary index was then defined as the arithmetic average of *K* indices, using the eight observatories where *K* scalings were available at that time.

It was soon apparent that such a crude definition of a planetary activity index was not satisfactory; Bartels then introduced "standardized K-indices", the so-called Ks indices that are derived from *K* indices using conversion tables that aim at eliminating LT and seasonal features. The Kp index was later defined as the average of K_s indices from a network of 13 stations (for more details see Section 8.5.2.1); the calculation has remained the same since it was defined by Bartels (1949).

The next milestone in the history of *K*-derived planetary geomagnetic indices was the introduction by Mayaud (1968) of the "am" planetary, or "mondial" index and of the related "an" and "as" hemispheric indices, that take advantage of the large extension of the number of observatories operated over the world. The am, an, and as indices are weighted averages of the *aK* equivalent amplitudes derived from *K* indices from about 20 subauroral stations, evenly spaced in longitude in both hemispheres. In 2000, Menvielle and Paris (2001) proposed to use the am network for deriving longitude sector indices "aλ".

Mayaud (1971) also introduced the "aa" antipodal activity index in order to provide a very long series of geomagnetic activity indices. The aa index is a weighted average of the aK equivalent amplitudes from two almost antipodal stations, one in Western Europe, the other in Eastern Australia.

Quick-look values of *K*-derived geomagnetic indices are routinely computed at Niemegk observatory (Kp and ap) and LATMOS laboratory (am, an, as, aa, and longitude sector indices); they are made available on line via the Niemegk and ISGI web pages, respectively.[17]

8.5.2 Definition of the K-Derived Geomagnetic Indices

K indices are a physically meaningful measure of geomagnetic activity at subauroral latitudes, and they are not so useful at measuring activity at other latitudes

[17] K_p and ap: http://www.gfz-potsdam.de/pb2/pb23/Niemegk/en/index.html am, an, as, aa, and aλ: http://isgi.latmos.ipsl.fr

(see Section 8.4). Ideally, a K-derived planetary geomagnetic index is therefore based on observations made at subauroral stations evenly distributed in longitude in both hemispheres. Since the index is intended to monitor the activity level at corrected magnetic latitude of about 50°, the most straightforward solution is to use aK equivalent amplitudes from the original Niemegk scale and to express the index in nanoTesla.

In fact, an ideal network cannot be achieved: gaps will occur due to oceans (a rather severe limitation in the southern hemisphere), and to the non availability of observatories on land-masses.

8.5.2.1 Kp (ap) Indices

Kp was introduced by Bartels in 1949 (Bartels 1949). It has since been derived back to 1932 and the present data series is homogeneous and continuous from 1932 onwards. Kp is acknowledged by IAGA (Resolution 6, IATME Bulletin 14 1954, p. 368); Kp and ap are currently routinely produced by GeoForschung Zentrum (GFZ) Potsdam, Germany, as part of the International Service of Geomagnetic Indices (ISGI); they are made available electronically at the GFZ and ISGI Internet sites.[18]

The Kp (ap) Network

Because of the historical context when Kp was introduced, the Kp network is heavily weighted towards Western Europe and Northern America, where respectively seven and four out of the thirteen stations are located; only two stations are located in the Southern hemisphere and none in Eastern Europe or Asia (Table 8.4 and Fig. 8.5). In addition, the data for two European stations (Brorfelde and Uppsala), as well as those for the two stations from the Southern hemisphere (Eyrewell and Canberra), are combined so that their average enters into the final calculation; the Southern hemisphere thus contributes to the planetary index as one component out of eleven.

[18] Reference Kp and ap values are available on-line at http://www.gfz-potsdam.de/pb2/pb23/Niemegk/en/index.html and at http://isgi.latmos.ipsl.fr/

Although the Kp network was one of the best possible ones during the postwar year, it however clearly does not provide a true planetary description of the activity: for example, the global impact of a magnetic substorm will be overestimated when it occurs during local night at European or American longitudes.

The Kp Index Derivation

Each individual planetary Kp index is the average of the standardized Ks indices from the Kp observatories.

At each observatory, the Ks values are derived from the K indices by means of conversion tables that have been established through the following rather complicated procedure:

- a frequency distribution of reference (*fdr*) was defined separately for each three Lloyd seasons (four calendar months around summer and winter solstices, two times two months around the equinoxes). More precisely, the *fdr* is the frequency distribution of the K indices measured during the two intervals nearest to local midnight at all the eleven Kp observatories used at that time, for a selected set of days. The days were selected in the years 1943–1948 in order to have a good representation of low and high activity levels; they altogether correspond to 42 months;
- for the same set of days, for each 3 h interval, and for the three seasons, the frequency distributions of measured K are calculated for each station. Tables of conversion $K \Rightarrow Ks$ are prepared for each case, so that the Ks frequency distribution is as close as possible to the *fdr*. These tables are established by conceiving K and Ks as continuous variables between 0.0 and 9.0, then dividing each interval into thirds, for example 1.5–2.5 (say) is labelled as 2−, 2o, and 2+. The values of Ks are thus scaled as 0o, 0+, 1−, 1o, 1+ ... to 9o (0o and 9o correspond to the 0.0–0.166 and 8.833–9.0 intervals respectively), or expressed as $3Ks$, 0 to 27.

This standardization was introduced in order to avoid local time influences, which are different from season to season, even if "it obliterates also the possible universal time daily variation" in the worldwide index (Bartels et al. 1940). One may wonder whether such frequency distribution adjustment is significant

Table 8.4 The Kp network from 1932 onwards. The standardization tables have been computed by Bartels for the stations of the initial network (indicated by a "Y" in the "initial network" column). The stations of the network in 2010 are indicated by a "Y" in the "present" network column. In case of change of site, the previous station is indicated below the current one. For each station, the dates indicated in brackets correspond to the period during which the station is part of the network. Two stations were introduced after the network definition: Toolangui (replaced in 1981 by Canberra) and Lovö (replaced in 2004 by Uppsala), The *Ks* data for the two stations Brorfelde and Uppsala, as well as for Eyrewell and Canberra, are combined so that their average enters into the final calculation. Coordinates are taken from WDC for Geomagnetism, Kyoto, Data Catalogue No.28, April 2008 (from McCreadie et al. 2010)

	Observatory	Code	Geographic Lat. (°N)	Geographic Long. (°E)	Mag. Lat. (°N)	Init. netw.	Pres. netw.
(1)	Meannook, Canada (1932–. . .)	MEA	54.62	246.66	61.57	Y	Y
(2)	Sitka, USA (1932–. . .)	SIT	57.06	224.67	60.34	Y	Y
(3)	Lerwick, UK (1932–. . .)	LER	60.13	358.82	61.98	Y	Y
(4)	Ottawa, Canada (1969–. . .)	OTT	45.40	284.45	55.63		Y
	Agincourt, Canada (1932–1969)	AGN	43.78	280.73	53.93	Y	
(5)	Eskdalemuir, UK (1932–. . .)	ESK	55.32	356.80	57.80	Y	Y
(6)	Brorfelde, Denm. (1984–. . .)	BFE	55.63	11.67	55.45		Y
	Rude-Skov, Denm. (1932–1984)	RSV	55.48	12.46	55.18	Y	
	Uppsala, Sweden (2004–. . .)	UPS	59.90	17.35	58.51		Y
	Lovö, Sweden (1954–2004)	LOV	59.34	17.82	57.90		
(7)	Fredericksburg, USA (1957–. . .)	FRD	38.20	282.63	48.40		Y
	Cheltenham , USA (1932–1957)	CLH	38.70	283.20	48.91	Y	
(8)	Wingst, Germany (1932–. . .)	WNG	53.74	9.07	54.12	Y	Y
(9)	Niemegk, Germany (1988–. . .)	NGK	52.07	12.68	51.88		Y
	Witteveen, Netherl. (1932–1988)	WIT	52.81	6.67	53.66	Y	
(10)	Hartland, UK (1957–. . .)	HAD	51.00	355.52	53.90		Y
	Abinger, UK (1932–1957)	ABN	51.19	359.61	53.35	Y	
(11)	Eyrewell, N; Zeal. (1978–. . .)	EYR	−43.41	172.35	−47.11		Y
	Amberley, N; Zeal. (1932–1978)	AML	−43.15	172.72	−46.80	Y	
	Canberra, Austral. (1981–. . .)	CNB	−35.32	149.36	−42.71		Y
	Toolangui, Austral. (1972–1981)	TOO	−37.53	145.47	−45.38		

for auroral Kp stations (Meanook, Sitka) where local time and seasonal variation of activity are *a priori* different from those at the other Kp stations.

Figure 8.8 summarizes the Kp index derivation procedure.

The Kp index clearly depends on the set of days used to establish the tables of conversion. Mayaud (1980) indeed showed that a different set of day would have led to different conversion tables, and then to a different definition of the Kp index. The conversion tables calculated by Bartels (1951) are still used in deriving Kp; in particular, no correction has been made when any of the sites were changed.

ap Index

Kp behaves as *K*, and it is therefore not linearly related to the activity. Soon after its introduction, it became clear that computing averages of the activity required an index that is linearly related to the activity. The ap index was therefore introduced few years after (Bartels and Veldkamp 1954). It is expressed in "ap units": 1 ap unit ~2 nT. Any ap index is deduced from the corresponding Kp through a one to one correspondence table (Table 8.5): there is thus only 28 possible ap values. The well known, but generally not properly known, Ap index is the daily average of ap; it is expressed also in ap units.

8.5.2.2 am, an, and as Indices

am, an, and as were introduced by Mayaud in 1968 (Mayaud 1968). They since have been computed back to 1959 and the present data series is continuous and homogeneous from 1959 onwards. am, an, and as are acknowledged by IAGA (Resolution 2, IAGA Bulletin 27 1969, p. 123); they are currently routinely produced by LATMOS, Guyancourt, France, as part of the International Service of Geomagnetic Indices (ISGI);

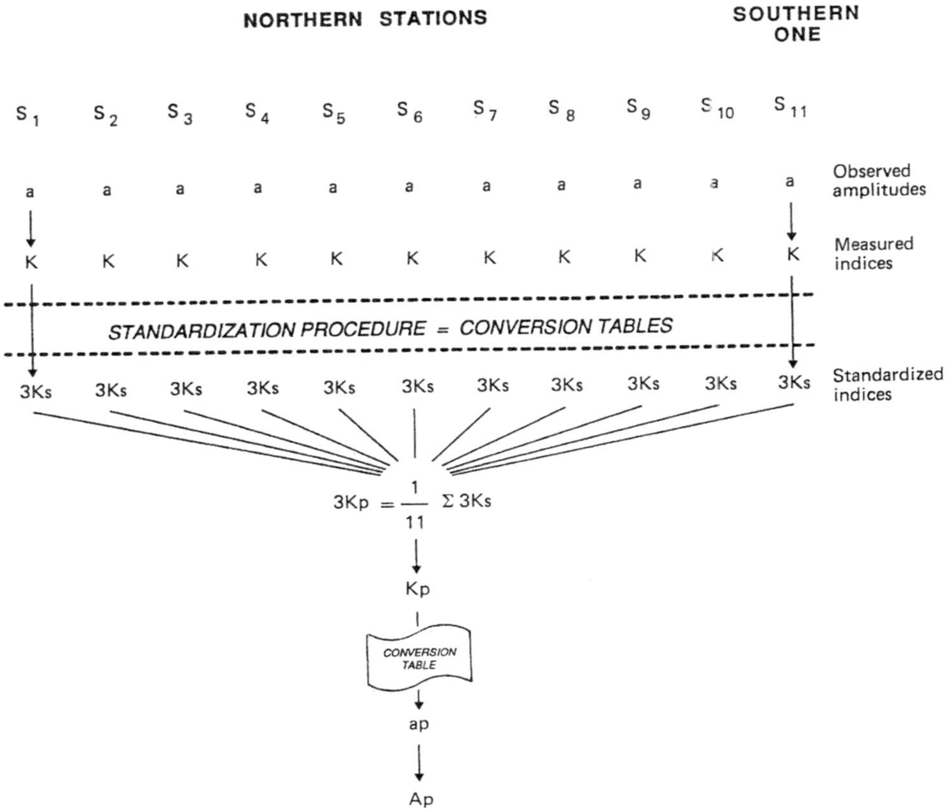

Fig. 8.8 Derivation scheme of 3 h Kp and ap indices. Ap is the daily mean value of ap (both expressed in 'ap units'; 1 'ap unit' ~2 nT)

Table 8.5 Converting from Kp to ap

Kp	ap	Kp	ap	Kp	ap
		0o	0	0+	2
1−	3	1o	4	1+	5
2−	6	2o	7	2+	9
3−	12	3o	15	3+	18
4−	22	4o	27	4+	32
5−	39	5o	48	5+	56
6−	67	6o	80	6+	94
7−	111	7o	132	7+	154
8−	179	8o	207	8+	236
9−	300	9o	400		

they are made available electronically at the ISGI Internet site.[19]

[19] Reference am, an, and as values are available on-line at http://isgi.latmos.ipsl.fr

The am Network

After the International Geophysical Year (IGY) and the end of the cold war, it became possible to build a network made up of mainly subauroral stations representing all longitudes in both hemispheres: the corrected geomagnetic latitude of the am stations therefore have an average value of ≈50° (ranging from 37° to 59°, one station is at 28°; see Table 8.6).

Stations are divided into groups according to their longitude (see Table 8.6), with five longitude sectors in the Northern hemisphere and four in the Southern hemisphere (there were only three before 1979). The number of sectors in each hemisphere is therefore large enough to get a proper description of the LT/longitude dependence of the irregular geomagnetic activity, as illustrated by Fig. 8.9.

Table 8.6 The am and aλ network from 1959 onwards. The stations are arranged in groups (G1 to G9), each group representing a longitude sector in one of the hemisphere. The stations of the network in 2010 are indicated in bold character. In case of change of site, the previous station is indicated in normal character. For each station, the dates indicated in brackets correspond to the period during which the station is part of the network. Coordinates are taken from WDC for Geomagnetism, Kyoto, Data Catalogue No.28, April 2008 (from McCreadie et al. 2010)

	Observatory	Code		Geographic Lat. (°N)	Long. (°E)	Mag. Lat. (°N)
(G1)	Magadan, Russia	MGD	(1967–...)	60.12	151.02	52.01
	Petropavlovsk, Russia	PET	(1969–...)	53.10	158.63	45.95
	Memambetsu, Japan	MMB	(1959–...)	43.91	144.19	35.35
(G2)	Arti (Sverdlovsk), Russia	ARS	(1959–...)	56.43	58.57	49.13
	Novosibirsk, Russia	NVS	(2002–...)	55.03	82.90	44.92
	Podkammenaya T., Russia	POD	(1973–2001)	61.40	90.00	51.54
	Tomsk, Russia	TMK	(1959–1970)	56.47	84.93	46.88
(G3)	Hartland, UK	HAD	(1959–...)	51.00	355.52	53.90
	Niemegk, Germany	NGK	(1959–...)	52.07	12.68	51.88
	Chambon-la-Forêt, France	CLF	(1996–...)	48.03	2.26	49.84
	Witteveen, Netherland	WIT	(1959–02/1988)	52.81	6.67	53.66
(G4)	Ottawa, Canada	OTT	(1975–...)	45.40	284.45	55.63
	Fredericksburg, USA	FRD	(1959–...)	38.20	282.63	48.40
(G5)	Newport, USA	NEW	(1975–...)	48.27	242.88	54.85
	Victoria, Canada	VIC	(1959–...)	48.52	236.58	54.14
	Tucson, USA	TUC	(1959–...)	32.17	249.27	39.88
(G6)	Canberra, Australia	CNB	(1986–...)	−35.32	149.36	−42.71
	Eyrewell, New Zealand	EYR	(1978–...)	−43.41	172.35	−47.11
	Amberley, New Zealand	AML	(1959–1977)	−43.15	172.72	−46.80
	Lauder, New Zealand	LDR	(1979–1985)	−43.03	169.41	−49.18
(G7)	Gnangara, Australia	GNA	(1959–...)	−31.78	115.95	−41.93
	Martin de Vivies, France	AMS	(1986–...)	−37.80	77.57	−46.39
	Toolangi, Australia	TOO	(1959–1984)	−37.53	145.47	−45.38
	Canberra, Australia	CNB	(1979–1985)	−35.32	149.36	−42.71
(G8)	Kerguelen Is., France	PAF	(1959–...)	−49.35	70.26	−56.94
	Crozet Is., France	CZT	(1973–...)	−46.43	51.86	−51.35
	Hermanus, South Africa	HER	(1959–...)	−34.43	19.23	−33.98
(G9)	Argentine Is., Ukraine	AIA	(1959–...)	−65.25	295.73	−55.06
	Trelew, Argentina	TRW	(1973–...)	−43.25	294.69	−33.05
	South Georgia, UK	SGG	(1975–03/1982)	−54.28	323.52	−45.57

The am Index Derivation

At each observatory of the am network, the grids used to measure K indices have been designed according to the procedure described in Section 8.4.3 above. Note that before 1979 the grids in use at some stations were slightly in error, so that it became necessary to adjust K values at these stations. An extra step was therefore added at that time to the derivation scheme (see Mayaud 1980, for further details).

For a given 3 h interval a unique K_i value is calculated for each sector of longitude by averaging the K values measured in the two or three stations belonging to the sector. The use of several observatories in each sector improves the quality of the derived am index as (1) it might compensate for differences in induced fields from one observatory to another, (2) a change in the site of a given observatory would have a smaller effect on the final result, and (3) small differences in K scalings are reduced within each sector (Mayaud and Menvielle 1980).

K_i is converted back to amplitude a_i standardized for 50° corrected geomagnetic latitudes,[20] and

[20] The conversion tables are based on the Niemegk scale: "each class is divided into ten equal parts, the tenth subclass being

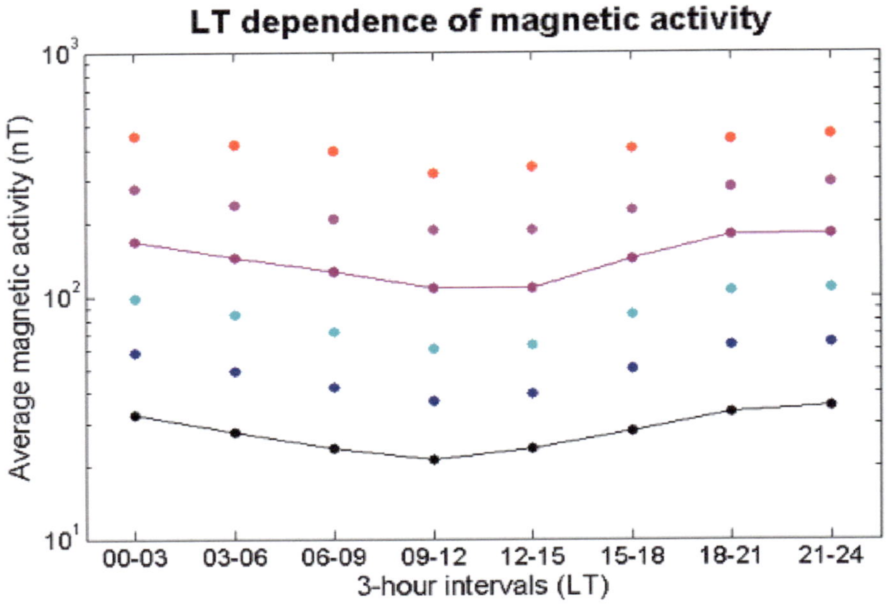

Fig. 8.9 Mean LT dependence of the regional geomagnetic activity for different levels of the planetary magnetic activity calculated over 1985–2005 using aλ indices. Each color corresponds to a different activity level from Km = 3 to Km = 8 in steps of 1 (see Section Km, Kn, Ks indices for Km definition). For the sake of clarity, values corresponding to Km = 3 have been linked by a black line and values corresponding to Km = 6 by a grey line. (from Lathuillère et al. 2008)

then multiplied by a weighting factor to balance the different ranges in longitude of the different sectors. For a given 3 h interval, an is then the weighted average value of a_i in the northern hemisphere, as is the weighted average of a_i in the Southern hemisphere, and am = (an + as)/2. An, As, and Am are mean daily values of an, as, and am, respectively. All of them are expressed in nanoTeslas.

Figure 8.10 summarizes the am, an, and as indices derivation procedure.

Km, Kn, Ks Indices

For the sake of tradition and convenience, Kn, Ks, and Km equivalent values are made available by means of a conversion table introduced by Mayaud, and expressed, as usual, by values from 0o, 0+, . . . to 9o (see Mayaud 1980 or Menvielle and Berthelier 1991 for further details).

Kpn, Kps, and Kpm equivalent values are also made available. They are expressed in Kp units, and derived using a conversion table deduced from that relating ap to Kp.

on either side of each limit between subsequent classes. (. . .) Special treatment is applied for Kj = 0.0 to 1.5: both classes K = 0 and 1 are considered as a single class, which is divided in 15 equal parts." (Mayaud, 1980, p. 56).

8.5.2.3 aλ Longitude Sector Index

In 2001, Menvielle and Paris (2001) proposed to use the am network for deriving longitude sector indices.

The longitude sectors are the same as those defined for the am derivation. In each longitude sector, the activity is characterized by the average of the *K* measured at the observatories of the sector. They are converted back into amplitudes using the same conversion tables as those used for am; the aλ longitude sector indices are expressed in nT.

This definition allows the derivation of homogeneous aλ indices series since the beginning of the am index series, in 1959. The aλ indices are currently routinely produced by LATMOS, Guyancourt, France, and made available electronically at the ISGI Internet sites.[21]

8.5.2.4 aa Index

aa was introduced by Mayaud (Mayaud 1971) as a simple means of monitoring global geomagnetic activity continuously back to 1868. In 1975, IAGA

[21] Reference aλ values are available on-line at http://isgi.latmos.ipsl.fr

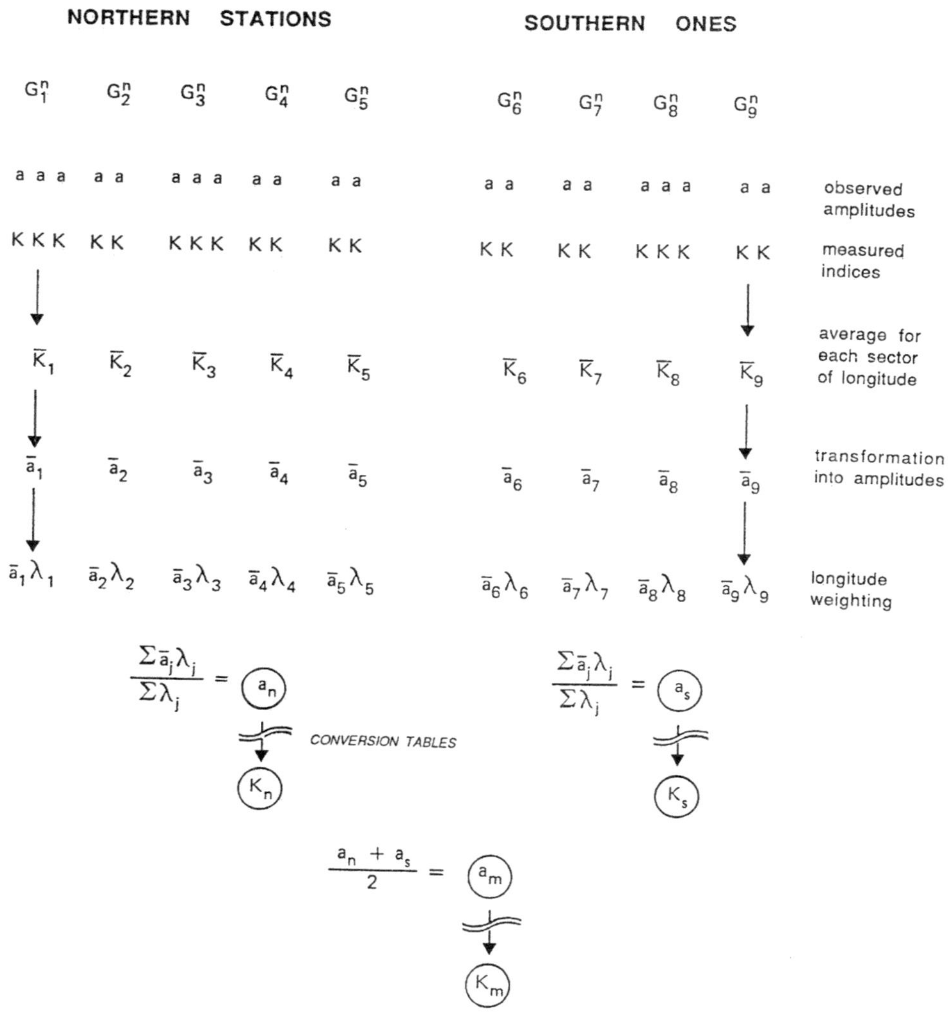

Fig. 8.10 Derivation scheme of 3 h an (northern hemisphere), as (southern hemisphere) and am indices, expressed in nT

recommended the replacement of the traditional Ci[22] index by aa, which can be determined in a more objective way than Ci (see Mayaud 1980). Mayaud measured aa for the period 1868–1968, then for the period 1969–1975 (IAGA Bulletin 39). The present data series is continuous and homogeneous from 1868 onwards. aa is acknowledged by IAGA (Resolution 3, IAGA Bulletin 37 1975, p. 128); it is currently routinely produced by LATMOS, Guyancourt, France, as part of the International Service of Geomagnetic Indices (ISGI); they are made available electronically at the ISGI Internet sites.[23]

The aa Network

aa is produced from the *K* indices of two nearly antipodal magnetic observatories in England and Australia (see Table 8.7).

[22] See note 1.

[23] Reference aa values are available on-line at http://isgi.latmos.ipsl.fr

Table 8.7 The aa network, from 1868 to the present: (**a**) Northern observatory; (**b**) Southern observatory. For each observatory, the weighting coefficient is given in brackets. All corrected geomagnetic latitudes are computed for the same period (the beginning of the 1970s) from the same geomagnetic field model. This result in a negligible effect (few %) on the aa secular variation

	Observatory	Corr. Geom. Lat.
(a) Northern hemisphere		
1868–1925	Greenwich (1.007)	50.1°
1926–1956	Abinger (0.934)	49.8°
1957–	Hartland (1.059)	50.0°
(b) Southern hemisphere		
1868–1919	Melbourne (0.967)	48.9°
1920–1979	Toolangui (1.033)	48.0°
1980	Canberra (1.084)	45.2°

The aa Index Derivation

For each three hour interval, *K* indices are measured at the two stations and converted back to amplitude by using the mid-class amplitudes of the original Niemegk grid: aa is thus standardized for 50° of corrected geomagnetic latitude. aa is computed as the weighted average of the northern and southern values; the weighting coefficients insure that the ratio between northern and southern activity remains the same despite the changes in the network. For the northern observatories, the normalization was carried out using the current French geomagnetic observatory as reference station (Greenwich to Abinger: Val Joyeux; Abinger to Hartland: Chambon-la-Forêt). In absence of a neighbouring observatory, the normalization between Melbourne and Toolangui was carried out with respect to the 100 years (1868–1967) of the normalized northern series. The normalization between Toolangui and Canberra was made by direct comparison of the activity recorded at the two observatories for the period January 1980 to September 1984. The reader is referred to Mayaud (1973; 1980) or Menvielle et al. (2010) for further details.

The aa index is expressed in nanoTesla (nT).

The aa Long Term Homogeneity

The long term homogeneity of the aa indices series was challenged during the early years of the twenty-first century. By comparing the 'official' aa index with reconstructed aa indices using independent long-running stations to provide data for the northern component of the index, Clilverd et al. (2005) demonstrated that the magnitude of the likely effect of any system changes on aa is ~2 nT. They also observed in 1957 a step change of the order of 1 nT in the deviation of the reconstructed series from the official aa series. Using a new daily index of geomagnetic activity, the Inter-Hourly Variability index (IHV, see Section 8.7.1), Svalgaard and Cliver (2007) also noted such a step change in 1957, and estimated it to be ~3 nT.

Using new *K* measurements on original 1956–1957 magnetograms from the Niemegk, Hartland, and Abinger observatories, as well as available *K* indices from the Abinger/Hartland, Eskdalemuir, Niemegk and Wingst observatories for the period 1947–1967, Menvielle et al. (2010) demonstrated that the jump evidenced in 1957 resulted from an underestimate of the activity level at the Hartland observatory: they proposed to take for this observatory a weighting coefficient very close to 1.0 instead of the previous factor of 1.059. When using monthly mean values, a very good approximation of the corrected aa is obtained by taking 94% of the aa computed using the Mayaud's coefficient.

Note that this correction only concerns the aa values for the period from 1957 onwards.

8.5.3 Comparison Between ap, am, and aa

Figures 8.11a and b illustrate the correlations between the am, ap, and aa indices.

The limited correlation between 3 h am and ap indices (Fig. 8.11a) results from the facts that (i) ap has only 28 possible values, and (ii) the ap and am daily variations are not the same. In fact, the Kp standardization tables aim at cancelling out both LT and UT daily variations and, on the contrary, the am derivation scheme naturally averages out the LT daily variation because of the even representation of all longitudes (remaining part estimated to be less than 2%, see Mayaud 1980), while it preserves the UT daily variation. Figure 8.11a also illustrates that, using ap instead of am indices to select time intervals according to the activity level may result in bias.

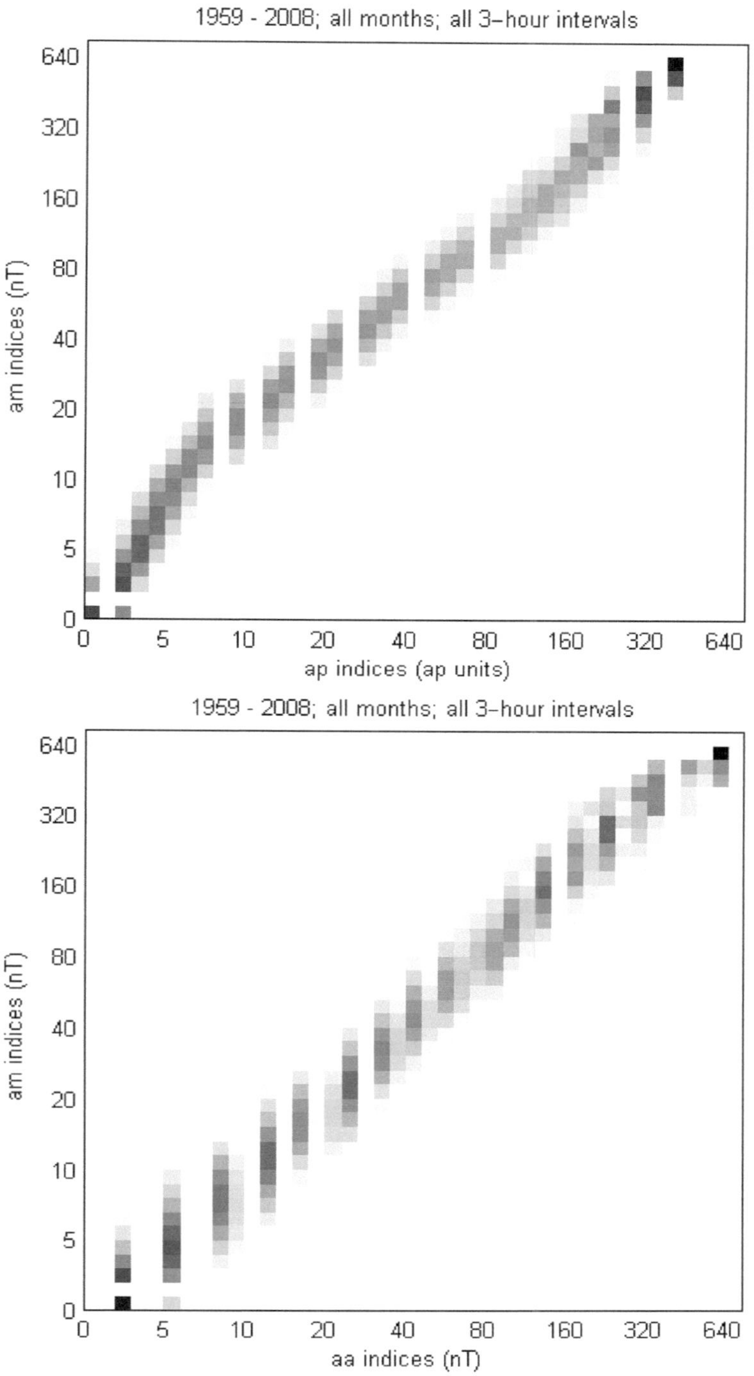

Fig. 8.11 (**a**) Empirical histogram of am vs ap indices distribution, for 50 years (1959–2008). Grey levels correspond to percentages of am values for a given ap value: the darker the pixel, the higher the percentage. (**b**) Empirical histogram of am vs aa indices distribution, for 50 years (1959–2008). Grey levels correspond to percentages of am values for a given aa value: the darker the pixel, the higher the percentage

On the contrary, 3 h am and aa indices are linearly correlated (Fig. 8.11b). The observed dispersion results from the fact that aa is based on only two observatories. One should however note that the distributions of am values for a given aa value are symmetrical, with a maximum for am = aa. The difference between aa and am mean values computed for the same set of 3 h intervals is thus expected to decrease with increasing number of intervals. Mayaud (1973) showed that the half-daily and daily mean values of aa are very close to the corresponding values of the am indices.

8.5.4 Classification of Days

Two activity classifications of days are currently used.

The first one was proposed by Johnston (1943). It is based upon Kp indices, and aims at identifying the *quietest* and *most disturbed* days of the month. One should recognize that it may happen that one of the selected quietest days is definitely not a quiet day, or that one of the selected most disturbed days is in fact a quiet day.

The second classification was introduced by Mayaud (1973). This is based upon aa indices, and aims at selecting days that are magnetically quiet or very quiet.

The following descriptions are based on IAGA Bulletin 32z. For further details, refer to Mayaud (1980).

8.5.4.1 Classification of Days as Deduced from Kp Indices

The identification of the quietest and most disturbed days of each month is made on the basis of three criteria:

- the sum of the eight values of Kp;
- the sum of the squares of these values;
- the greatest of the eight values of Kp.

According to each of these criteria, a relative "order number" is assigned to each day of a month, the three order numbers are averaged and the days with the lowest and the highest mean order numbers are selected as the five quietest, the ten quietest (Q-) and the five most disturbed (D-) days.

As already mentioned, it should be noted that these selection criteria give only a relative indication of the character of the selected days with respect to the other days of the same month. As the general disturbance level may be quite different for different years and also for different months of the same year, the selected quietest days of a month may sometimes be rather disturbed or vice versa. In order to indicate such a situation, selected days which do not satisfy certain absolute criteria are marked as follows:

- a selected "quiet day" is considered "not really quiet" and is marked by the letter *A* if for that day: Ap > 6, or marked by the letter K, if Ap < 6, with one Kp value greater than 3 or two Kp values greater than 2+.
- a selected "disturbed day" is considered "not really disturbed" and marked by an asterisk if Ap is lower than 20.

8.5.4.2 Classification of Days as Deduced from aa Indices

The identification of the quiet 24 h intervals is made firstly on the basis of the mean value of aa which must be lower than the fixed value 13 nT. Then, each individual aa value of the day is represented by a weight p (Table 8.8). A day with a mean value of aa < 13 nT and for which the sum of weights p, Σp is higher than, or equal to 4 is a quiet day; if Σp is lower than 4, the day is a really quiet day.

Table 8.8 Weights p attributed to the aa indices for quiet and very quiet day determination

p	0	1	2	4	6
aa	<17	17 < aa < 21	21 < aa < 28	28 < aa < 32	>32

The same rules are applied to select the 48 h quiet or really quiet intervals, with the same limit for the aa mean value (13 nT) and a limit for Σp equal to 6. One must note that in these intervals every local day (0 h to 24 h in local time) is really quiet, at any longitude.

8.5.5 Relation with Solar Wind Parameters

In situ continuous observations of solar wind parameters and the IMF began in November 1963. Since then, many authors have investigated the dependence of geomagnetic activity on solar wind and IMF parameters. The reader is referred to, e.g., Rangarajan (1989) for reviews of such works published before the end of the 1980s.

For example, Svalgaard (1977) performed an extensive regression analysis of am and solar wind data, to investigate any dependence. He used the am index because this index can be considered as a truly global one. He established that:

$$am \approx 6,6 \left\{ \frac{nV_0^2}{105} \right\}^{1/3} \left\{ \frac{BV_0 q\,(f,\alpha)}{21} \right\} \frac{1,157}{\left(1 + 3\cos^2\psi\right)^{2/3}} \quad (8.9)$$

where $V_0 = Vsw/100$. n is the number of protons per cm^3, and nV^2 is the solar wind dynamic pressure, $BV_0q(f,\alpha)$ is the influx of merging interplanetary field lines (where $q(f,\alpha)$ is a geometric factor, α the angle between the direction of the IMF and geomagnetic field at the magnetopause subsolar point, and f the IMF variability), and Ψ the angle between the solar wind velocity and the Earth dipole axis. The q function and the coefficients were empirically identified from the data. The reconstructed am indices from the above semi-empirical formula are in good agreement with the observed ones.

This result indicates that any am index is somehow a measure of the energy state of the magnetosphere during the corresponding 3 h interval. This supports the use of K-derived planetary indices as indicators of overall magnetosphere state. However, because their 3 h time resolution is not appropriate for the dynamics of most of magnetosphere processes, their contribution in case studies remains limited.

Legrand and Simon (1991 and references therein) studied solar activity by using aa and storm sudden commencements (ssc—see Section 8.6.3) in the long series of geomagnetic data beginning in 1868, and the series of solar wind data starting in 1963. They identified four classes of geomagnetic activity, corresponding to different sources.

They defined as quiet magnetic days those days for which Aa < 20 nT. Geomagnetic quietness is the most frequent state (67% of the time). It corresponds to periods during which the Earth is in regions of low solar wind velocity (<450 km s^{-1} typically). The number of magnetic quiet days per year varies from ~100 to ~300. It varies from one solar cycle to the other, depending on the maximum sunspot number for the cycle. It also depends on the phase of the cycle: it is larger when the solar dipole axis is close to the solar rotation one, during the minimum and ascending phases of the cycle.

During transient solar events which disturb the solar wind, Coronal Mass Ejections generating magnetic clouds with high solar wind pressure result in the most intense storms. This "shock activity" occurs only 8.5% of the time: there is no clear link between these storms and the number of sunspots, although their occurrence is linked with strong chromospheric eruptions within sunspot groups.

Legrand and Simon identified two other classes of activity:

- the recurrent activity that is related to high velocity solar wind streams that remain active during at least four Bartels rotations; they originate at high solar magnetic latitude and co-rotate with the Sun. The recurrent activity arises during the descending phase of the sunspot cycle, and it is responsible for the aa peak during this phase of the solar cycle; it occurs 7% of the time. It has a seasonal effect related to the inclination of the solar axis with respect to the ecliptic plane.
- the fluctuating activity that is observed during the rest of the time (17.5%). This mostly arises during the reversal phase of the coronal dipole, when the topology of the coronal field is rapidly evolving.

Later, Richardson et al. (2000) achieved a similar classification of geomagnetic activity, on the basis of a detailed analysis of the aa indices series and of the solar wind structure.

The results presented in this section make it clear that long term geomagnetic data series enable one to trace the evolution of solar activity and of solar wind velocity back to 1868 (see, e.g., Lockwood et al. 1999; Echer et al. 2004; Clilverd et al. 2005; Svalgaard and Cliver 2007; Ouattara et al. 2009), and to infer its behaviour back to the beginning of sunspot observations, in the early 1600s (e.g., Cliver et al. 1998).

8.5.6 Secular Variation of Geomagnetic Activity

The long term variations of corrected aa indices (see Section "The aa Long Term Homogeneity") and sunspot numbers is shown on Fig. 8.12. The solar

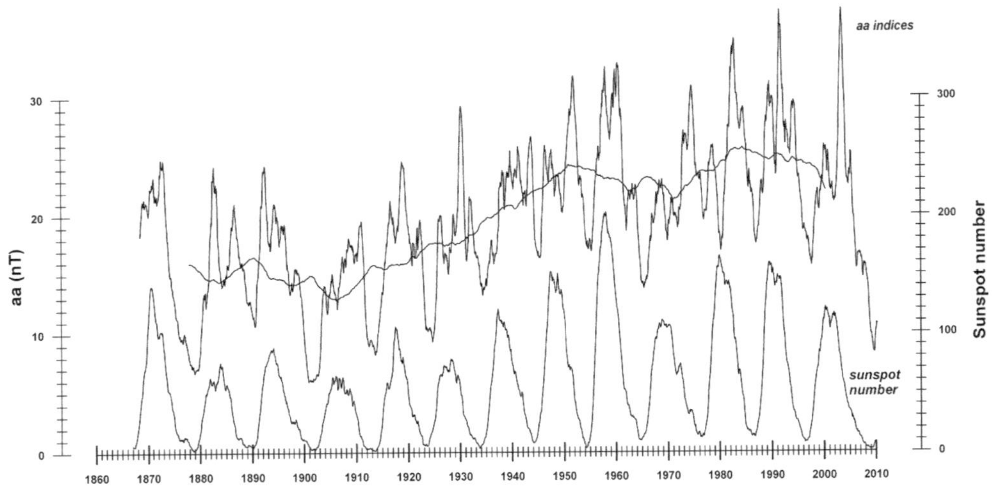

Fig. 8.12 Long-term variations of aa indices (12-month and 20-year running averages; *scale on the left*) and of sunspot numbers (12-month running averages; *scale on the right*) from 1868 until now

cycles are clearly marked by a 11-year periodicity in sunspot numbers. aa shows an increasing trend in both minima and maxima values of the cycle, with a dual-peak structure, with one peak close to the solar cycle maximum and the other one in the descending phase.

The 20-year running average of aa indices show that the geomagnetic activity level remains low before 1920 and high after 1950, with a regular increase between these two dates. Note that the level of geomagnetic activity during the current solar minimum (between cycle 23 and 24) is on the same order as that observed during the solar minimum of the 1910s.

Nevanlinna et al. (1993) used declination instantaneous visual readings carried out at the Helsinki observatory (once each other 10-min. from 1844 to 1856, and once per hour from 1857 to 1897) to extend the aa data series backwards in time to 1844. They showed that the magnetic activity during 1844–1856 was, on the average, about the same level as the activity measured at Nurmijärvi during the 1953–1992 period.

8.5.7 Annual and Diurnal Modulations

Annual and UT diurnal variations of the planetary geomagnetic activity were recognized a long time ago (e.g., Sabine 1856), and studied by many authors since (see a review in, e.g., de La Sayette and Berthelier 1996). Since its derivation scheme preserves annual and UT diurnal variations of planetary geomagnetic activity, analyses of the am data series has allowed researchers to differentiate and characterize various contributions in the observed annual/diurnal variation of the geomagnetic activity.

The first contribution to be identified was the McIntosh effect (McIntosh 1959), the incidence of which is maximum when the angle Ψ_m between the Earth-Sun line and the geomagnetic dipole axis is at 90°. The McIntosh effect is thus responsible for semi-annual and UT diurnal modulations. Boller and Stolov (1970) proposed that it may be related to the enhancement of Kelvin-Helmoltz instabilities at the magnetopause. The McIntosh effect is usually described in terms of a $\cos^2\Psi_m$ modulation.

Russell and McPherron (1973) investigated the significance of IMF polarity through the influence of Bz south[24] on geomagnetic activity. They proposed the so-called Russell-McPherron (RM) model as a possible source of the equinoctial maxima. In the RM model it is assumed (*i*) that in the Geocentric Solar Equatorial (GSEQ) coordinate system, the IMF is ordered following the Parker spiral model of the solar wind radial flow, and (*ii*) that energy can be transferred from the solar wind to the magnetosphere only when the IMF Bz component in geomagnetic solar

[24] see Section 8.2.

magnetosphere (GSM) coordinates[25] points southward. The first hypothesis implies that the GSEQ Bz component is considered to be nil; the RM effect thus only depends on the GSEQ By component (the IMF polarity) through its projection on the GSM North-South axis. It results in two annual maximum (April 5 and October 6) according to the IMF polarity, and has a characteristic UT variation.

However de La Sayette and Berthelier (1996) noticed that the GSEQ Bz component can be different from zero. One has therefore to consider another source for the effective GSM Bz component, namely the projection of the GSEQ Bz component on the GSM North-South axis. A careful analysis of 30-year long (1959–1988) am and IMF data series enabled them to demonstrate that the combination of the McIntosh effect and of this new "La Sayette-Berthelier" effect enable one to account fairly well for the part of the annual and diurnal variations of the geomagnetic activity that does not depend on the IMF polarity.

8.5.8 Use and Misuse of K-Derived Planetary Indices

The K-derived planetary indices are unique tools for selecting events, or time periods of, e.g., magnetic quietness (see Section 8.5.4) or storminess (Ap* and Aa* indices, extreme am values, ...).

When mapping the internal magnetic field, at any geographic scale (local to planetary), the variations of external origin are noise that may affect map accuracy, and separating fields of external origin is a key issue. One of the most common ways to achieve this objective is to use geomagnetic indices for selecting periods during which the unmodelled part due to external noise is below an a priori defined threshold. On the basis of numerical simulations, Mareschal and Menvielle (1986) showed that *K* indices defined from ground-based magnetic variations give reliable upper bounds to the mid-latitude, external transient fields recorded by Magsat (350 km altitude). In addition, Thomson & Lesur (2007) showed that using aλ longitude sector indices, instead of planetary indices, improves modelling accuracy.

K-derived geomagnetic indices are also widely used as inputs in models of the ionized environment to characterize the current overall state of the ionosphere and magnetosphere, or proxies to monitor the influence of the solar wind-magnetosphere-ionosphere coupling (see Lathuillère et al. (2002) an reference therein). Traditionally, Kp is used most, while am is a better proxy (Lathuillère et al. 2008; Müller et al. 2009).

It should however be noticed that the 3 h granularity of *K*-derived planetary indices results in some cases (e.g., solar wind/magnetosphere event analysis) in limited usefulness.

8.5.9 What Next?

As previously noted (Section 8.4.4), the root mean square (*rms*) of the irregular variations in the horizontal components of the magnetic field is proportional to the magnetic energy related to geomagnetic activity. Menvielle (2003) proposed new geomagnetic activity indices, based on the *rms* of the irregular variations in the magnetic horizontal components. Using such proxy does not put constraints on the length of the time interval over which the indices are derived.

rms indices can be computed at any observatory over time intervals significantly shorter than 3 h (typically few tens of minutes). Menvielle (2003) proposed to use such local *rms* indices to derive αm, αλ, and αa *rms* planetary indices by means of algorithms similar to those used to derive am, aλ, and aa planetary geomagnetic indices respectively.

Planetary and longitude sector αm and αλ indices have been computed for the years 2002–2005. Lathuillère and Menvielle (2010) used αm indices over 30-min intervals to study the time delay between low- to mid-latitude global thermosphere disturbance and magnetic activity, for very disturbed conditions. They pointed out a 2-h time delay between daytime and night-time responses, a difference that can hardly be resolved using indices with a 3 h time resolution.

It is foreseen that these indices will be routinely computed by LATMOS, Guyancourt, France, and made available on-line from the ISGI web page.

[25] The GSM coordinates are deduced from the GSEQ ones by a simple rotation around their common x axis, which points from Earth to the Sun.

8.6 Storm Indices

Studies of geomagnetic storms have shown that, at equatorial and mid latitudes, the decrease in the horizontal magnetic field (H) during a magnetic storm can approximately be represented by two components, i.e., a uniform magnetic field parallel to the geomagnetic dipole axis (the axially symmetric component) which is directed toward the south and a longitudinally non-uniform field (the asymmetric component). The onset of a magnetic storm is often characterized by a global sudden increase in H, which is referred to as the storm sudden commencement (*ssc*). In this section, the indices that measure the magnitude of the axially symmetric component, Dst and SYM, that of the asymmetric component, ASY, and the list of *ssc*s are introduced.

8.6.1 Dst Index

An equatorial ring current in the magnetosphere is generally assumed as the physical source of the axially symmetric component of storm disturbance fields. The Dst index was developed by Sugiura and colleagues to measure the magnitude of the current which produces this symmetric disturbance field (e.g., Sugiura and Kamei 1991 and references therein).

There are three classes of Dst index services from the World Data Center for Geomagnetism, Kyoto: the Quick Look Dst, Provisional Dst and (final) Dst, defined according to the stage of data processing provided by the observatories. For the (final) Dst index, four magnetic observatories, Hermanus, Kakioka, Honolulu, and San Juan are used. These observatories were originally chosen on the basis of the quality of observation and because they are located sufficiently distant from the auroral and equatorial electrojets and are fairly evenly distributed in longitude. For the Quick Look Dst and Provisional Dst, data from the Alibag observatory are also used. The coordinates of the observatories are given in Table 8.9, and a map of the network is given in Fig. 8.5. In the following section, the method of derivation for the (final) Dst is given. The methods for the Quick Look and provisional Dst are essentially the same as that for the (final) Dst, except for the method of the base line determination.

Dst index is acknowledged by IAGA (Resolution 2, IAGA Bulletin 27 1969, p. 123); it is currently routinely produced by WDC for Geomagnetism, Kyoto, Japan, as part of the International Service of Geomagnetic Indices (ISGI); it is made available electronically at the WDC-Kyoto and ISGI Internet sites.[26]

8.6.1.1 Definition and Method of Derivation of the Dst Index

The following descriptions are based on Sugiura and Kamei (1991). For further technical details, please refer to their paper.

The Dst index at a time 't', Dst(t), is defined as the average of the disturbance variation of the H component, $D_i(t)$, at the four observatories ($i = 1 - 4$) divided by the average of the cosines of the dipole latitudes at the observatories for normalization to the dipole equator;

$$\mathrm{Dst}\,(t) = \sum\nolimits_{i=1}^{4} D_i\,(t) / \sum\nolimits_{i=1}^{4} \cos\,(\lambda_i), \qquad (8.10)$$

where λ_i is the magnetic dipole latitude of the i-th observatory. Note that the normalization is by the average of the four cosines and NOT for each $\mathrm{D_i(t)}$ by $\cos(\lambda_i)$. According to Sugiura and Kamei (1991), this normalization procedure has been found to minimize undesired effects from missing hourly values. There is another reason why this normalization procedure is used, which will be discussed in a later section on the ASY/SYM indices.

Each $D_i(t)$ is calculated from the observed $H(t)$ by subtracting the base line value of the geomagnetic main field, $H_{\mathrm{base}}(t)$, and Sq (solar quiet) variation, Sq(t). The base line value is defined for each observatory in a manner that takes into account the secular variation of the annual mean values of H, calculated from the internationally selected five quietest days (five Q-days per month, see Section 8.5.4). The baseline is expressed by a power series in time and the coefficients for terms up to the quadratic are determined by the method of least squares fitting, using the annual means for the current year and the four preceding years;

[26] Reference Dst values are available on-line at http://wdc.kugi.kyoto-u.ac.jp/ and at http://isgi.latmos.ipsl.fr

Table 8.9 The Dst network in 2010. Coordinates are taken from WDC for Geomagnetism, Kyoto, Data Catalogue No. 28, April 2008

			Geographic		Geomagnetic	
Observatory	Country	Code	Lat (°N)	Lon (°E)	Lat (°N)	Lon (°E)
Kakioka	Japan	KAK	36.23	140.19	27.37	208.75
San Juan	Puerto Rico	SJG	18.11	293.85	28.31	6.08
Honolulu	Hawaii, USA	HON	21.32	202.00	21.64	269.74
Hermanus	South Africa	HER	−34.43	19.23	−33.98	84.02

$$H_{\text{base}}(\tau) = A + B\tau + C\tau^2 \tag{8.11}$$

where τ is time in years measured from a reference epoch.

The solar quiet daily variation, Sq, is derived for each observatory as follows. The average Sq variation for each month is determined from the values of $H(t)$ for the five Q-days of the month. These quietest days are determined in UT. In order to define an average Sq variation of each observatory, the five local days that have the maximum overlap with the five Q-days are used. Using hourly values immediately before and immediately after the local days thus selected, the linear change is subtracted from the Sq variation, and they are averaged at each local time.

The 12 sets of the monthly average Sq so determined for the year are expanded in a double Fourier series with local time, T, and month number, M, as two variables:

$$\text{Sq}\,(\text{T,M}) = \sum_{j}\sum_{k} \text{A}_{\text{jk}}\cos\left(\text{jT} + \alpha_{\text{j}}\right)\,\cos\left(\text{kM} + \beta_{\text{k}}\right) \tag{8.12}$$

From this representation, we calculate Sq(t) at any UT hour, t, of the year for each observatory.

The $\text{H}_{\text{base}}(\text{t})$ and Sq(t) determined in this way are subtracted from the observed H(t) to obtain the disturbance variation $\text{D}_{\text{i}}(\text{t})$ for each observatory.

8.6.1.2 Basic Characteristics of the Dst Index

The Dst index represents the axially symmetric disturbance magnetic field at the dipole equator on the Earth's surface. Major disturbances in Dst are negative; in other words, the disturbance field is southward. The statistical characteristics of the Dst index, such as seasonal and UT variations, have been examined by many authors (see, for example, Cliver et al. 2000; Takalo and Mursula 2001 and references therein).

8.6.1.3 Relationship with Magnetospheric, Ionospheric and Induced Currents

The southward disturbance fields in middle and low latitudes during a storm are produced by various currents in the magnetosphere and the ionosphere and by induced currents inside the Earth. It is widely believed that the axially symmetric part indicated by the Dst index is produced mainly by the equatorial current system in the magnetosphere, usually referred to as the ring current. The neutral sheet current flowing across the magnetospheric tail also contributes to the field decreases near the Earth. The compression of the magnetosphere from solar wind pressure increases also contributes to the symmetric variation which appears as a positive variation in the Dst index. A current is induced inside the Earth by the magnetic field variation of magnetospheric and/or ionospheric origin. The strength of the associated induced magnetic field is reported to be about 20–30% of the Dst index depending on the phase of magnetic storms (e.g., Hakkinen et al. 2002).

The effects from field-aligned currents could remain because of insufficient cancellation by the averaging process of the four observatory data, $D_i (i = 1 - 4)$. However, the effect should be smeared out when conducting a statistical analysis with a sufficiently long time series of the Dst index.

The reference level for Dst is set such as on the five Q-days the Dst index is zero on the average. However, even on these quietest days, a southward directed magnetic field produced by the equatorial current system in the magnetosphere, which is often referred to as the quiet time ring current, may exist. The magnitude has been estimated by various satellite observations such as the OGO 3 and 5 satellites (e.g., Sugiura and Poros 1973), Magsat (Langel et al. 1980), Oersted (Olsen et al. 2000) and Champ (Maus et al. 2005). For example, Langel et al. (1980), Olsen et al. (2000) and Maus et al. (2005) estimated it to be −25 nT, −20 nT, and −13 nT, respectively. Although all these values seem to be reasonable, there is a considerable difference among them, and therefore the absolute reference level for the Dst variation remains to be studied in the future because it may, for example, vary with the solar cycle.

8.6.1.4 Use and Misuse of the Dst Index

The Dst index was originally developed to monitor the axially symmetric ring current intensity during magnetic storms. However even during quiet periods, there are always currents in the magnetosphere contributing to the Dst index, such as the ring current, the magnetotail current and the magnetopause current, therefore this index is widely used to quantitatively monitor the currents.

However, we should note that we cannot separate the effects of these different current systems without the help of independent information such as the solar wind pressure, the southward interplanetary magnetic field, etc. For example, an increase of the Dst does not necessarily mean that a decay of the ring current has occurred; the increase may be due to the effect of solar wind pressure increase or a decay of the magnetotail current system. It should also be noted that the equatorial current is not a symmetric ring current, particularly not in the developing phase of a magnetic storm (e.g., Jordanova 2007 and references therein).

The Dst and SYM-H (see Section 8.6.2) indices are often used to estimate the total amount of kinetic energy of the charged particles that form the ring current by using the Dessler-Parker-Sckopke (D-P-S) equation (Dessler and Parker 1959; Sckopke 1966):

$$\Delta H_z/H_0 = 2E/3E_m \tag{8.13}$$

where $\Delta H_z(0)$ is the magnetic field caused by the trapped particles and E the total energy of the particles; H_o and E_m are constants denoting the horizontal component of the main field at the Earth's equator and the total energy of the main field external to the Earth, respectively.

However, it should be noted that the equation assumes that the magnetic field, in which the charged particles drift and form the ring current, is not modified by the ring current itself. For a more quantitatively accurate discussion, the generalized equation developed by several authors (Carovillano and Maguire 1968; Olbert et al. 1968; Siscoe 1970) should be used although we need additional information on the structure of the current system to apply the generalized equation. The D-P-S equation tends to overestimate the total kinetic energy and the difference of estimation could be more than 30–40% (Iyemori 1990). If the effect of the induced current inside the Earth is taken into account, the difference may be even larger.

The Dst index is frequently used to estimate the base-line value with 1 nT accuracy when one makes geomagnetic absolute measurements. In such case, it is necessary to be careful about the accuracy of the base value, i.e., Dst = 0 nT. As discussed in the previous section, there are causes of uncertainty in the base-line determination even if the measurement of H itself is perfect. One uncertainty is the so called "quiet time ring current". Another is the "day-to-day" variability of the Sq field (e.g., Takeda 1999). For the Quick Look Dst (QL-Dst) and provisional Dst, the absolute value of the H provided from each observatory is tentative. There is often a more than 10 nT difference of the level among QL-Dst, provisional Dst and final Dst. Therefore, we should not expect the absolute accuracy to in general be better than 10 nT for QL-Dst and Provisional Dst, and 5nT for (final) Dst (private communication with Sugiura). These values of accuracy themselves, i.e., 10 nT or 5 nT, remain to be studied further.

8.6.2 ASY and SYM Indices

It is known that the disturbance field in mid- and low-latitudes is generally not axially symmetric. In particular, in the developing phase of a magnetic storm, the asymmetric disturbance field can be even greater than the symmetric part (see, e.g., Sugiura and Chapman 1960; Akasofu and Chapman 1964).

To describe the asymmetric and symmetric disturbance fields in mid-latitudes with high-time (i.e., 1 min) resolution, longitudinally asymmetric (ASY) and symmetric (SYM) disturbance indices were introduced and derived for both H and D components, i.e., for the components in the horizontal direction *H* (directed positive towards the geomagnetic dipole pole in the Northern hemisphere) (ASY-H, SYM-H) and in the orthogonal (East-West) direction *D* (ASY-D, SYM-D). The symmetric disturbance field in H, SYM-H, is essentially the same as the hourly Dst index described in the previous section, although 1 minute values from different sets of stations and a slightly different coordinate system are used. Similarly, the asymmetric disturbance component in H, ASY-H, is close to the asymmetric indices proposed by Kawasaki and Akasofu (1971), Crooker and Siscoe (1971), or Clauer et al. (1983). The ASY-D was introduced and discussed in Iyemori et al. (1990; 1996).

ASY and SYM are currently routinely produced by WDC-C2, Kyoto, Japan; they are made available electronically at the WDC-C2 Internet site.[27]

8.6.2.1 Definition and Method of Derivation of the ASY and SYM Indices

The geomagnetic disturbance field at each observatory is divided into two parts, symmetric and asymmetric parts, for both the *H* (horizontal) and *D* (declination) components, δH^{SYM}, δH^{ASY}, δD^{SYM} and δD^{ASY}, respectively. For the *D* component, the disturbance in angle, δD, is converted to the value in nT unit by the equation $H_0 \tan(\delta D)$, where H_0 is the intensity of the horizontal component of the geomagnetic main field at a quiet time.

The asymmetric components, δH^{ASY} and δD^{ASY}, are obtained by subtracting the symmetric parts, δH^{SYM} and δD^{SYM} from each disturbance field.

Shown in Table 8.10 and Fig. 8.5 are the lists of the geomagnetic observatories used for the derivation and their distribution in the geomagnetic dipole coordinate system, respectively. Only six of the stations are used throughout a month, and a station can be replaced by another one (i.e., one station from the station pair linked by the lines in Fig. 8.5) for other months depending on the availability and quality of the data for the month at the station. The data are processed in units of 1 month. Therefore, some amount of jump in magnitude is seen between successive months.

The derivation procedure essentially consists of the following four steps:

1. *Subtraction of the geomagnetic main field and the Sq field*:
   To calculate the base value including the Sq field, the data of the five Q-days are used. That is, the original data of the five quietest days of the month that include the Sq field as well as the geomagnetic main field are averaged every UT minute and fitted by B-spline functions of 'rank'=4 ('order'=3). The fitted B-spline values are subtracted from the observed values to obtain the disturbance fields.
2. *Coordinate transformation to a dipole coordinate system*:
   As the ring current flows in the magnetosphere at a distance of several R_E and mainly in the magnetosphere equatorial plane which should be close to the equatorial plane of the geomagnetic dipole, its magnetic effect on the ground should be nearly parallel to the dipole axis. On the other hand, the *H* direction at each observatory is generally different from the dipole pole direction because of the non-dipole component of the geomagnetic main field and/or local geomagnetic anomalies. Therefore, the ring current effect is mixed into the *D* component. To minimize the ring current effect in the D-component, the data are transformed to the dipole coordinate system at each station. The differences between the dipole pole position and local geomagnetic direction are shown in Table 8.10.
3. *Calculation of the longitudinally symmetric indices SYM-H and SYM-D*:
   The longitudinally symmetric component is calculated by averaging the disturbance component at each minute for the 6 stations. For the *H* component, a latitudinal correction is made on the averaged value to get the value (SYM-H) which corresponds to the hourly Dst index. That is, the averaged value is divided by the 6-station average of cos(λ), where λ is the dipole latitude of each station. On the symmetric component of *D* (SYM-D), we make no latitudinal correction at all because we have, so far, no clear indication or theoretical idea of the current system which generates non-zero SYM-D.

[27] Reference SYM and ASY values are available on-line at http://swdcwww.kugi.kyoto-u.ac.jp/

Table 8.10 SYM and ASY network in 2010. The network consists in 6 stations (bold character) evenly distributed in longitude. Four of them have a backup station: the station and its backup station are replaced by each other in the index computation, depending on the availability and the condition of the data of the month. The rotation angle is the difference between the the dipole pole direction and the local geomagnetic North direction. Coordinates are taken from WDC for Geomagnetism, Kyoto, Data Catalogue No.28, April 2008

			Geographic		Magnetic		
Observatory	Country	Code	Lat. (°N)	Long. (°E)	Lat. (°N)	Long. (°E)	Rotation angle (°)
Fredericksburg	**USA**	**FRD**	**38.20**	**282.63**	**48.40**	**353.38**	**0.4**
San Juan	Puerto Rico	SJG	18.11	293.85	28.31	6.08	−8.9
Tucson	**USA**	**TUC**	**32.17**	**249.27**	**39.88**	**316.11**	**2.7**
Boulder	USA	BOU	40.13	254.76	48.40	320.59	2.5
Honolulu	**USA**	**HON**	**21.32**	**202.00**	**21.64**	**269.74**	**0.5**
Memambetsu	**Japan**	**MMB**	**43.91**	**144.19**	**35.35**	**211.26**	**−16.1**
Urumqi	**China**	**WMQ**	**43.80**	**87.70**	**34.34**	**162.53**	**7.7**
Alibag	India	ABG	18.64	72.87	10.19	146.16	6.8
Martin de Vivies	France	AMS	−37.80	77.57	−46.39	144.27	−32.4
Hermanus	**S. Africa**	**HER**	**−34.43**	**19.23**	**−33.98**	**84.02**	**−10.1**
Chambon-la-Forêt	France	CLF	48.03	2.26	49.84	85.68	13.6

4. *Derivation of the asymmetric indices ASY-H and ASY-D*:
   The asymmetric component at each station is obtained by subtracting the symmetric component from each disturbance field. For the *H* component, the symmetric part is subtracted after making a latitudinal correction assuming that the SYM-H represents the magnitude of the uniform field parallel to the dipole axis generated by the ring current. After subtracting the symmetric part, a latitudinal correction is made through a multiplication by a normalization coefficient for each station. The coefficient is determined empirically such as the standard deviations of the asymmetric variation for the 6 stations become equal. The ASY-H and ASY-D indices are calculated by taking the range between the maximum and minimum deviation for the *H* and *D* components, respectively.

8.6.2.2 Basic Characteristics of the ASY and SYM Indices

It has been shown that the variation of the asymmetric *H* component correlates well with the AE index (e.g., Crooker 1972; Clauer and McPherron 1980). Not only the ASY-H but also the ASY-D correlates well with the AE index, probably because the field-aligned currents, which closely relate to the auroral electrojets, generate magnetic disturbances even at mid-latitudes (e.g., Fukushima 1976; Nakano and Iyemori 2005).

The main difference between the SYM-H and hourly Dst index is in the time resolution, and the effects of rapid variation of the solar wind parameters such as dynamic pressure, southward component of the interplanetary magnetic field, or the effect of substorm onset are more clearly seen in the SYM-H than in the hourly Dst index.

8.6.2.3 Relationship of the ASY Indices with Magnetospheric and Ionospheric Currents

The asymmetric disturbance field has usually been attributed to a partial ring current (Akasofu and Chapman 1964; Cahill 1966; Frank 1970; Fukushima and Kamide 1973). However, it has also been suggested that the asymmetric disturbance field may be produced by a net field-aligned current system flowing into the ionosphere near noon and flowing out near midnight (Crooker and Siscoe 1981; Suzuki and Fukushima 1984; Nakano and Iyemori 2003; 2005).

Recent high-altitude satellites such as AMPTE and DE-1 and numerical simulation studies have shown that the equatorial current is largely asymmetric centred on the strongest part at midnight or pre-midnight (Iijima et al. 1990; Nakabe et al. 1997; Jordanova 2007 and references therein). On the ground and at low-altitude satellite orbit, on the other hand, the magnetic disturbance field is more positive (i.e., northward) on the dawn side than on the dusk side. This 90 degree difference compared with the high-altitude satellite observations, i.e., noon-midnight asymmetry or dawn-dusk asymmetry, indicates that, the field-aligned currents and not the asymmetric equatorial current dominate the asymmetric disturbance fields on the ground.

The effects from the field-aligned currents are larger in the higher latitudes than in the lower ones, and the effects are sometimes greater than the effects from the ring current during the main phase of geomagnetic storms. In such a case, the latitude correction for each observatory $Di(t)$ by $\cos(\lambda_i)$ does not work as a correction, but instead enlarges the effect of high latitude currents (i.e., field-aligned currents). This may be another reason why Sugiura adopted the normalization method described in Section 8.6.1.1.

8.6.2.4 Use and Misuse of ASY and SYM Indices

The main qualitative differences between the ASY and AE indices are in the spectral characteristics and sensitivity to minor substorm activities. The ASY indices vary more smoothly, and it is thus rather difficult for them to reflect minor substorm activity in the polar region.

Users need to keep in mind that the ASY indices should not be used as substitutes for the AE indices, although there are some similarities between them.

The ASY-H and ASY-D do not necessarily indicate the asymmetry of the ring current but the effects of the field-aligned currents (Suzuki and Fukushima 1984; Nakano and Iyemori 2005). The ASY indices can monitor the positive bays at substorm onsets. However, it is difficult to monitor the positive bays for small substorms, because of the error in Sq subtraction, which is mainly caused by the day-to-day variability of the Sq field.

8.6.3 Storm Sudden Commencements (ssc)

Storm Sudden Commencements (*ssc*) are defined by an abrupt increase or decrease in the horizontal component of the geomagnetic field, which marks the beginning of a geomagnetic storm or an increase in activity lasting at least one hour.

In the definition introduced by Mayaud (1973), the decisive factors are (i) the suddenness of the event's beginning and (ii) the change of rhythm in the magnetic activity before and after the sudden move. The latter point implies that some *ssc*s included in the list are not followed by a real magnetic storm.

This definition was modified during the 2009 IAGA Assembly in Sopron (Hungary) to take into account the present level of knowledge on rapid variations, and to open the way for computer detection:

- a threshold has been added for the rate of change in dX/dt;
- a quantitative criterion, based on am and Dst indices, is used for deciding whether the event is a *ssc*, when the impulse is followed by a storm, or a sudden impulse (*si*) otherwise.

The reader is referred to Curto et al. (2007a; 2007b) for further details on the method of listing *ssc*s as well as the history of the *ssc* lists. The first series of *ssc* lists covering 100 years (1868–1967) were obtained by Mayaud (1973). Lists of *ssc* were afterward established for a second period, 1968–1994, a third one, 1995–2005, and a fourth one beginning in 2006. All these data are available from the home page of the Ebro observatory or from the ISGI home page.

Lists of *ssc*s are currently routinely established at the Service of Rapid Variation, Observatori de l'Ebre, Roquetes, Spain, as part of the International Service of Geomagnetic Indices (ISGI); lists of *ssc*s are made available electronically at the Observatori de l'Ebre and ISGI Internet sites.[28]

8.6.4 What Next?

The Dst and SYM-H indices are robust indices in the sense that they do not change much depending on the stations and quality of the data used, probably because of the averaging process in the derivation.

[28] Reference lists of *ssc*s are available on-line at http://www.obsebre.es and at http://isgi.latmos.ipsl.fr

They also have a basis in a physical principle (Siscoe 1971). Therefore, it is important and meaningful to derive them, in particular the Dst index before IGY, to study long term variation of solar-terrestrial activity: some trials have already been done (e.g., Karinen and Mursula 2005) .

The ASY and SYM indices have been published as provisional indices because reliable digital data were not available in the 1980s and early 1990s at some stations used in the derivation, and it was necessary to change the station month by month depending on the quality or availability of the data in a particular month. However, through the efforts led by IAGA, INTERMAGNET and by each magnetic observatory, the quality has been greatly improved, and it may thus be possible to establish which observatories will provide data as well as the method of derivation to derive (final/definitive) ASY and SYM indices in a near real time basis.

8.7 Some Other Indices

We give in this section only some examples of newly proposed quantities that open new perspectives.

8.7.1 IHV and IDV Indices

Svalgaard et al. (2004) proposed the Inter Hourly Variability (IHV) index of geomagnetic activity. This index can be computed at any observatory: it is derived from hourly values or means of the H component as the sum of the unsigned differences between adjacent hours over a 7 h interval centred on local midnight. Svalgaard and Cliver (2007) derived IHV indices separately for stations in both hemispheres within six longitude sectors spanning the Earth using only local night hours. IHV is intended as a long-term index and available data allows derivation of the index back well into the nineteenth century. On a timescale of a 27-day Bartels rotation, IHV averages for stations with corrected geomagnetic latitude less than 55° are strongly correlated with mid latitude range indices, and also with the $BVsw^2$ solar wind quantity.

Svalgaard and Cliver (2005) devised the Inter Diurnal Variability (IDV) index, defined as the unsigned difference between two consecutive days of the 00–01 LT average value of H; the difference is assigned to the first day. The individual daily values are then averaged over longer intervals so as to minimize various geometric and seasonal effects. The IDV index has the interesting and useful property of being highly correlated with the strength B of the interplanetary magnetic field and essentially unaffected by the solar wind speed Vsw. This enables these authors to get a new estimate of the variation of B from 1872 to the present.

The IHV and IDV indices are very interesting, because they demonstrate the possibility of using the magnetic observatory data archives to derive quantities enabling one to get insights on the evolution of the geomagnetic activity—and therefore the Sun—since the end of the nineteenth century. They also illustrate the fact that different quantities are likely to contain different and complementary information, as it is the case for images of the same object taken using different filters and from different standpoints.

8.7.2 Pulsations Indices

There have been many attempts to derive geomagnetic indices from pulsations. A complete review of these indices is out the scope of this chapter, and we consider hereafter only some examples.

8.7.2.1 The Wp Index

A new substorm index, the Wp (Wave and planetary) index, which is related to the wave power of low-latitude Pi2 pulsations,[29] was recently proposed by Nosé et al. (2009). This index is derived from high time resolution geomagnetic field data at about 10 low-latitude stations (i.e., Kakioka, Urumqi, Iznik, Tihany, Fürstenfeldbruck, Ebro, Tristan da Cunha, San Juan, Teoloyucan, Tucson, and Honolulu). These stations are distributed over the globe in a longitudinal direction with maximum separation between neighbouring

[29] Geomagnetic pulsations are quasi-periodic variations of the Earth's magnetic field which are classified by their structure and frequency

stations of ~120°. Thus at least one station is located on the night side where Pi2 pulsations have dominant power. Thus the Wp index can be considered as "a global index", which means an index reflecting Pi2 wave power during local night-time at any given UT time. Plots and digital data of the Wp index are available on line.[30]

8.7.2.2 ULF Indices

Evidence exists that intense ULF waves are associated with magnetic storms (Pilipenko et al. 2001). In order to quantify the level of low-frequency turbulence/variability of the geomagnetic field, IMF, and solar wind plasma, and to facilitate the use of ULF wave power as an aid for storm data analysis a set of magnetic ULF indices was recently devised (Kozyreva et al. 2007). These simple hourly indices are based on the integrated spectral power in the band 2–7 mHz or wavelet power with time scales ~10–100 min. A ground wave index has been produced from the data of global magnetometer arrays in the northern hemisphere. Interplanetary and geostationary wave indices have been calculated using magnetometer and plasma data from interplanetary and geosynchronous satellites.[31]

8.7.2.3 Localised PC3 Indices

Geomagnetic pulsations constitute a definite source of noise in, e.g., petroleum exploration surveys, that require that the ground magnetic deviation does not exceed 2 nT over any 2 min interval, in order for the survey data to be acceptable. Such amplitudes and periods are typical of geomagnetic pulsations in the Pc3/4 range. Pc3 indices from a network of stations can then be used by space weather agencies (e.g., the Australian Space Weather Agency) to produce in near real-time regional contour maps of pulsation activity and alert exploration survey groups, using geomagnetic methods, of data quality.

[30] http://s-cubed.info

[31] Digital data of the ULF index are available from: ftp://space.augsburg.edu/MACCS//ULF_Index/

8.8 Concluding Remarks

Since the publication of the very first geomagnetic indices at the end of the nineteenth century, geomagnetic indices producers had to adapt their products and procedures to the evolution of the user needs and to that of the technological environment. The review presented in this chapter suggests that future activities in the field of geomagnetic indices could be developed along the following axes:

- derivation of indices based upon magnetic variations (e.g., pulsations) the precise recording and analysis of which become possible thanks to the technological progress in digital magnetograms acquisition and dissemination;
- derivation of long homogeneous data series of new indices, to take advantage of the geomagnetic observatory historical data basis and improve our knowledge and understanding of the long term evolution of the Sun;
- derivation of regional indices, to improve the resolution in longitude of existing planetary indices whenever necessary or to fit with Space Weather related requirements such as, for example, indices dedicated to GIC currents monitoring (see, e.g., Menvielle and Marchaudon 2007)

During the last decades, producers of geomagnetic indices had to cope with crucial changes that occurred in geomagnetic observatory practice and data dissemination: digital magnetometers replaced analogue magnetometers, Internet and computer developments resulted in a revolution in data handling, and—as a consequence of Internet facilities—users strongly required to have preliminary values of geomagnetic indices available on line within delays as short as possible. This challenged the quality stamping data policy, and even the possibility to continue the production of high quality long term data series. The resulting current policy for geomagnetic indices derivation and dissemination succeed both in fitting with this new environment and in preserving the historical heritage, namely the high quality, homogeneous long term data series.

The evolution in geomagnetic indices activity management during this transition period evidenced the importance of IAGA as a reference institution for the policy in matter of indices, that of the International Service of Geomagnetic Indices and of its Collaborating Institutes in the quality data stamping, and that of close links between observatory and research activities. This is strikingly illustrated by on-line dissemination of preliminary (or quick-look) values of indices acknowledged by IAGA: ISGI Collaborating Institutes decided to routinely make available on-line state of the art preliminary values (few hours to two days delay: in 1996 for am, aa, Dst, and AE indices, and few years after for Kp index; 30 min delay: in 2004 for aa), and IAGA urged the *"producers of the estimated indices to clearly label them"* (Resolution 5, IAGA News 38 1998, p. 42); this was made possible because of the deep involvement of the geomagnetic activity producers in Solar Terrestrial physics and Space Weather research activities. One of the major challenges is to keep this ability to find out solutions enabling the satisfaction of new requests from users without loosing the present quality of geomagnetic indices and the homogeneity of long term data series.

To conclude, let emphasize on the fact that the current status of geomagnetic indices relies on the contribution of those who have worked in geomagnetism since more than 150 years. Among them, there are outstanding figures of the history of the geomagnetism, and their contributions are referred to in this review. But one should never forget that present geomagnetic indices data series do rely on the daily conscientious work of those who took care, every day of the year, of the observatories and provided the community with the high quality observations on which geomagnetic indices are based, although the contributions of these observers are almost never explicitly quoted. Let dedicate this review to them.

Acknowledgements The authors thank A.W.P. Thomson for his constructive review that helped us to improve the manuscript.

References

Ahn B-H, Kroehl HW, Kamide Y, Kihn EA (2000a) Universal time variations of the auroral electrojet indices. J Geophys Res 267–275

Ahn B-H, Kroehl HW, Kamide Y, Kihn EA (2000b) Seasonal and solar cycle variations of the auroral electrojet indices. J Atmos Solar-Terr Phys 62:1301–1310

Akasofu S-I, Chapman S (1964) On the asymmetric development of magnetic storm field in low and middle latitudes. Planet Space Sci 12:607

Akasofu S-I, Chapman S (1972) Solar terrestrial physics. Oxford University Press, Oxford

Allen J.H, Kroehl HW (1975) Spatial and temporal distributions of magnetic effects of auroral electrojets as derived from AE indices. J Geophys Res 80:3667–3677

Bartels J (1940) Report on the numerical characterization of days, IATME Bull. I1, p. 27, International Union of Geodesy and Geophysics, Publ. Off., Paris.

Bartels J (1949) The standardized index, Ks, and the planetary index, Kp, IATME Bulletin 12b, 97.

Bartels J (1951) An attempt to standardize the daily international magnetic character figure, IATME Bull. 12e, p. 109, International Union of Geodesy. and Geophysics, Publ. Off., Paris.

Bartels J, Veldkamp J (1954) International data on magnetic disturbances, fourth quarter, 1953. J Geophys Res 59:295

Bartels J, Heck NH, Johnston HF (1939) The three-hour-range index measuring geomagnetic activity. J Geophys Res 44:411

Bartels J, Heck NH, Johnston HF (1940) Geomagnetic three-hour-range indices for the years 1938 and 1939. J Geophys Res 45:309

Baumjohann W, Kamide Y (1984) Hemispherical Joule heating and the AE indices. J Geophys Res 89 :383–388

Berthelier A (1979) Étude des influences du vent solaire sur l'activité magn6tique terrestre, particulièrement aux hautes latitudes, doctorat d'etat, Univ. Pierre et Marie Curie

Berthelier A (1993) The K-derived planetary indices: derivation, meaning and uses in solar terrestrial physics. In: Hruska J et al (eds) STPW-IV proceedings, vol 3. US Government Publications Office, Boulder, CO, pp. 3–20

Basu S (1975) Universal time seasonal variations of auroral zone magnetic activity and VHF scintillations. J Geophys Res 80:4725–4728

Boller BR, Stolov HL (1970) Kelvin-Helmholtz instability and the semiannual variation of geomagnetic activity. J Geophys Res 75:6073

Burton RK, McPherron RL, Russell CT (1975) An empirical relationship between interplanetary conditions and Dst. J Geophys Res 80:4204

Cahill LJ Jr (1966) Inflation of the inner magnetosphere during a magneticstorm. J Geophys Res 71:4505

Carovillano, RL, Maguire JJ (1968) Magnetic energy relationships in the magnetosphere, In: Carovillano RL and McClay JF (eds) Physics of the magnetosphere, Astrophysics and Space Science Library, Reidel, Dordrecht, The Netherlands, pp 270–300

Chun FK, Knipp DJ, McHarg MG, Lu G, Emery BA, Troshichev OA (1999) Polar cap index as a proxy for hemispheric Joule heating. Geophys Res Lett 26:1101

Clauer CR, McPherron RL (1980) The relative importance of the interplanetary electric field and magnetospheric substorms on the partial ring current development. J Geophys Res 85:6747–6759

Clauer CR, McPherron RL, Searls C, Kivelson MG (1981) Solar wind control of auroral zone geomagnetic activity. Geophys Res Lett 8:915–918

Clilverd MA, Clark T, Clarke E, Rishbeth H, Ulich T (2002) The causes of long-term change in the aa index. J Geophys Res 107:A12441. doi:10.1029/2001JA000501

Clilverd MA, Clarke E, Ulich T, Linthe J, Rishbeth H (2005) Reconstructing the long-term aa index. J Geophys Res 110:A07205. doi:10.1029/2004JA010762

Cliver EW, Boriakoff V, Bounar KH (1998) Geomagnetic activity and the solar wind during the Maunder Minimum. Geophys Res Lett 25:897

Cliver EW, Kamide Y, Ling AG (2000) Mountains versus valleys: Semiannual variation of geomagnetic activity. J Geophys Res 105:2413–2424

Clauer CR, McPherron RL, Searls C (1983) Solar wind control of the low- latitude asymmetric magnetic disturbance field. J Geophys Res 88:2123–2130

Coles R, Menvielle M (1991) Some thoughts concerning new digital magnetic indices. Geophys Trans 36:303–312

Cowley SWH (1982) The causes of convection in the Earth's magnetosphere: a review of developments during the IMS. Rev Geophys Space Phys 20:531–565

Crooker NC, Siscoe GL (1971) A study of the geomagnetic disturbance field asymmetry. Radiol Sci 6:495–501

Crooker NC (1972) High-time resolution of the low-latitude asymmetric disturbance in the geomagnetic field. J. Geophys Res 77:773–775

Crooker NU, Siscoe GL (1981) Birkeland currents as the cause of thelow-latitude asymmetric disturbance field. J Geophys Res 86:11201

Curto J-J, Araki T, Alberca LF (2007a) Evolution of the concept of Sudden Storm Commencements and their operative identification. Earth Planet Space 59:I–XII

Curto J-J, Cardùs JO, Alberca LF, Blanch E (2007b) Milestones of the IAGA International Service of Rapid Magnetic Variations and its contribution to geomagnetic field knowledge. Earth Planets Space 59:463–471

Davis TN, Sugiura M (1966) Auroral electroject activity index AE and its universal time variations. J Geophys Res 71:785–801

de La Sayette P, Berthelier A (1996) The am annual-diurnal variations 1959–1988: A 30 year evaluation. J Geophys Res 101:10,653

Dessler AJ, Parker EN (1959) Hydromagnetic theory of geomagnetic storms. J Geophys Res 64:2239–2252.

Echer E, Gonzalez WD, Gonzalez ALC, Prestes A, Vieira LEA, Dal Lago A, Guarnieri FL, Schuch NJ. (2004) Long-term correlation between solar and geomagnetic activity. J Atmos Solar-Terr Phys 66:1019–1025

Fiori RAD, Koustov AV, Boteler D, Makarevich RA (2009) PCN magnetic index and average convection velocity in the polar cap inferred from SuperDARN radar measurements. J Geophys Res 114:A07225. doi:10.1029/2008JA013964

Frank LA (1970) Direct detection of asymmetric increases of extraterrestrial ring proton intensities in the outer radiation zone. J Geophys Res 75:1263

Fukushima N (1976) Generalized theorem for no ground magnetic effect of vertical currents connected with Pedersen currents in the uniform-conductivity ionosphere. Rep Ionos Res Japan 30:35–40

Fukushima N, Kamide Y (1973) Partial ring current models for world geomagnetic disturbances. Rev Geophys Space Phys 11:795

Hakkinen LVT, Pulkkinen TI, Nevanlinna H, Pirjola RJ, Tankanen EI (2002) Effects of induced currents on Dst and on magnetic variations at midlatitude stations. J Geophys Res 107:1014–1021. doi:10.1029/2001JA900130

Huang C-S (2005) Variations of polar cap index in response to solar wind changes and magnetospheric substorms. J Geophys Res 110:A01203. doi:10.1029/2004JA010616

Iijima T, Potemra TA, Zanetti LJ (1990) Large-scale characteristics of magnetospheric equatorial currents. J Geophys Res 95:991

Iyemori T (1990) Storm-time magnetos-pheric currents inferred from mid-latitude geomagnetic field variations. J Geomagnetics Geoelectrics 42:1249–1265

Iyemori T, Rao DRK (1996) Decay of the Dst component of geomagnetic disturbance after substorm onset and its implication to storm substorm relation. Ann Geophys 14:608–618

Iyemori T, Maeda H, Kamei T (1979) Impulse response of geomagnetic indices to interplanetary magnetic field. J Geomagnetics Geoelectrics 31:1–9

Janzhura A, Troshichev O, Stauning P (2007) Unified PC indices: Relation to the isolated magnetic substorms. J Geophys Res 112:A09207. doi:10.1029/2006JA012132

Johnston HF (1943) Mean K-indices from twenty one magnetic observatories and five quiet and five disturbed days for 1942. Terr Magn Atmos Elec 47:219

Jordanova VK (2007) Modeling geoma-gnetic storm dynamics: New results and chal-lenges. J Atmosp Solar-Terr Phy 69:56–66

Karinen A, Mursula K (2005) A new reconstruction of the Dst index for 1932–2002. Ann Geophys 23:475–485

Kawasaki K, Akasofu S-I (1971) Low-latitude DS component of geomagnetic storm field. J Geophys Res 76:2396–2405

Kitamura K, Shimazu H, Fujita S, Watari S, Kunitake M, Shinagawa H, Tanaka T (2008) Properties of AE indices derived from real-time global simulation and their implications for solar wind-magnetosphere coupling. J Geophys Res 113:A03S10. doi:10.1029/2007JA012514

Kozyreva OV, Pilipenko VA, Engebretson MJ, Yumoto K, Watermann J, Romanova N (2007) In search of a new ULF wave index: Comparison of Pc5 power with dynamics of geostationary relativistic electrons. Planet Space Sci 55:755–769. doi:10.1016/j.pss.2006.03.013

Langel RA, Estes RH, Mead GD, Fabiano EB, Lancaster ER (1980) Initial geomagnetic field model from Magsat vector data. Geophys Res Lett 7:793

Lathuillère C, Menvielle M (2010) Comparison of the observed and modeled low- to mid-latitude thermosphere response to magnetic activity: Effects of solar cycle and disturbance time delay. J Adv Space Res 45:1093–1100. doi:10.1016/j.asr.2009.08.016

Lathuillère C, Menvielle M, Lilensten J, Amari T, Radicella SM (2002) From the Sun's atmosphere to the Earth's atmosphere: an overview of scientific models available for space weather developments. Ann Geophys 20:1081–1104

Lathuillère C, Menvielle M, Marchaudon A, Bruinsma S (2008) A statistical study of the observed and modeled global thermosphere response to magnetic activity at middle and low latitudes. J Geophys Res 113:A07311. doi:10.1029/2007JA012991

Legrand JP, Simon P (1991) A two components solar cycle. Sol Phys 131:187

Li X, Oh KS, Temerin M (2007) Prediction of the AL index using solar wind parameters. J Geophys Res 112:A06224. doi:10.1029/2006JA011918

Liou K, Carbary JF, Newell PT, Meng C-I, Rasmussen O (2003) Correlation of auroral power with the polar cap index. J Geophys Res 108(A3):1108. doi:10.1029/2002JA009556

Lockwood M, Stamper R, Wild MN (1999) A Doubling of the Sun's Coronal Magnetic Field during the Last 100 Years. Nature 399:437–439

Lukianova R (2003) Magnetospheric response to sudden changes in solar wind dynamic pressure inferred from polar cap index. J Geophys Res 108(A12):1428. doi:10.1029/2002JA009790

Lukianova R (2007) Comment on Unified PCN and PCS indices: method of calculation, physical sense, dependence on the IMF azimuthal and northward components In: Troshichev O, Janzhura A, Stauning P(eds). J Geophys Res 112:A07204. doi:10.1029/2006JA011950

Lukianova R, Troshichev OA, Lu G (2002) The polar cap magnetic activity indices in the southern (PCS) and northern (PCN) polar cap: Consistency and discrepancy. Geophys Res Lett 29:1879. doi:10.1029/2002GL015179

Lyatsky W, Lyatskaya S, Tan A (2007) A coupling function for solar wind effect on geomagnetic activity. Geophys Res Lett 34:L02107. doi:10.1029/2006GL027666

Lyatskaya S, Lyatsky W, Khazanov GV (2008) Relationship between substorm activity and magnetic disturbances in two polar caps. Geophys Res Lett 35:L20104. doi:10.1029/2008GL035187

Mansurov SM (1969) New evidence of a relationship between magnetic fields in space and on Earth. Geomagn Aeron 9:622

Mareschal M, Menvielle M (1986) On the use of K indices to define maximum external contributions to Magsat data at midlatitudes. Phys Earth Planet Inter 43:799–204

Maus S, McLean S, Dater D, Lühr H, Rother M, Mai W, Choi S (2005) NGDC/GFZ candidate models for the 10th generation International Geomagnetic Reference Field. Earth Planets Space 57:1151–1156

Mayaud, PN (1967) Atlas des indices K, IAGA Bull. 21, International Union of Geodesy and Geophysics, Paris

Mayaud PN (1968) Indices Kn, Ks, Km, 1964–1967, Centre National de la Recherche Scientifique, Paris, 156p

Mayaud PN (1971) Une mesure planétaire d'activité magnétique basée sur deux observatoires antipodaux. Ann Geophys 27:67

Mayaud PN (1973) A hundred year series of geomagnetic data, 1868–1967: indices aa, storm sudden commencements, IUGG Publication Office, Paris, 256p

Mayaud PN (1976) Analyse d'une série centenaire d'indices d'activité magnétique, III, la distribution de fréquence est-elle logarithmo-normale ? Ann Geophys 32:443

Mayaud PN (1978) Morphology of the transient irregular variations of the terrestrial magnetic field, and their main statistical laws. Ann Geophys 34:243

Mayaud PN (1980) Derivation, meaning, and use of geomagnetic indices, Geophys. Monogr. Ser., vol 22. AGU, Washington, DC

Mayaud PN, Menvielle M (1980) A report on Km observatories visit, in IAGA Bull. 32i, International Union of Geodesy and Geophysics Publication Office, Paris, p 113

McCreadie H, Menvielle M (2010) The PC Index: Review of methods. Ann Geophys (in press)

McCreadie H, Menvielle M, Barton C (2010) A guide to geomagnetic indices derived from Earth surface data. Ann Geophys submitted.

McIntosh DH (1959) On the annual variation of magnetic disturbance. Phil Trans R Soc London A 251:525

McPherron RL (1995) Magnetospheric dynamics, In: Kivelson MG, Russel CT (eds) Introduction to space physics. Cambridge University Press, New York, USA pp 400–458

Menvielle M. (1979) A possible geophysical meaning of K indices. Ann Géophys 35:189–196

Menvielle M (1991) Evaluation of aigorithms for computer production of K indices. Geophys Trans 36:313–320

Menvielle M (2003) On the possibility to monitor the planetary activity with a time resolution better than 3 hours, In: Loubser L (ed) Proceedings of the Xth IAGA Workshop on Geomagnetic Instruments Data Acquisition and Processing, HMO publication, pp 246–250

Menvielle M, Berthelier A (1991) The K-derived planetary indices: description and availability. Rev Geophys Space Phys Hermanus, Republic of South Africa 29:415–432; erratum: 30:91 1992

Menvielle M, Marchaudon A (2007) Geomagnetic indices. In: Lilensten J (ed) Solar-Terrestrial Physics and Space Weather, Space Weather, Springer, Dordrecht, The Netherlands pp 277–288

Menvielle M, Paris J (2001) The aλ longitude sector geomagnetic indices. Contrib Geophys Geod 31:315–322

Menvielle M, Clarke E, Thomson A (2010) The aa data series revisited: the Abinger to Hartland normalization, XIVth Workshop on Geomagnetic Instruments, Data Acquisition and Processing, Changchun, China, Oral communication

Menvielle M, Papitashvili NE, Häkkinen L, Sucksdorff C (1995) Computer production of K indices: review and comparison of methods. Geophys J Int 123:866–886

Menvielle M, Rossignol J-C, Tarits P (1982) The coast effect in terms of deviated electric currents: a numerical study. Phys Earth Planet Inter 28:118–128

Müller S, Lühr H, Rentz S (2009) Solar and magnetospheric forcing of the low latitude thermospheric mass density as observed by CHAMP. Ann Geophys 27:2087–2099

Nakabe S, Iyemori T, Sugiura M, Slavin JA (1997) A statistical study of the Magnetic field structure in the inner magnetosphere. J Geophys Res 102:17571–17582

Nakano S, Iyemori T (2003) Local-time distribution of net field-aligned currents derived from high-altitude satellite data. J Geophys Res 108(A8):1314. doi:10.1029/2002JA009519

Nakano S, Iyemori T (2005) Storm-time field-aligned currents on the nightside inferred from ground-based magnetic data at midlatitudes: Relationships with the interplanetary magnetic field and substorms. J Geopys Res 110:A07216. doi:10.1029/2004JA010737

Nevanlinna H, Ketola A, Häkiinen L, Viljanen A, Ivory K (1993) Geomagnetic activity during solar cycle 9 (1844–1856). Geophys Res Lett 20:743–746

Niblett ER, Loomer EI, Coles RL, Jansen G Van Beek (1984) Derivation of K indices using magnetograms constmcted from digital data. Geophys Surv 6:431

Nosé M, Iyemori T, Takeda M, Toh H, Ookawa T, Cifuentes G-Nava, Matzka J, Love JJ, McCreadie H, MK, Tunçer,

Curto JJ (2009) New substorm index derived from high-resolution geomagnetic field data at low latitude and its comparison with AE and ASY indices, In: Love JJ (ed) Proceedings of XIIIth IAGA Workshop on Geomagnetic Observatory Instruments, Data Acquisition, and Processing, U.S. Geological Survey Open-File Report 2009–1226, pp 202–207

Østgaard N, Vondrak RR, Gjerloev JW, Germany G (2002) A relation between the energy deposition by electron precipitation and geomagnetic indices during substorms. J Geophys Res 107:1246. doi:10.1029/2001JA002003

Olbert S, Siscoe GL, Vasyliunas VM (1968) A simple derivation of the Dessler-Parker-Sckopke relation. J Geophys Res 73:1115–1116

Ouattara F, Amory-Mazaudier C, Menvielle M, Simon P, Legrand J-P (2009) On the long term change in the geomagnetic activity during the 20th century. Ann Geophys 27:2045–2051

Papitashvili VO, Gromova LI, Popov VA, Rasmussen O (2001) Northern Polar Cap magnetic activity index PCN: Effective area, universal time and solar cycle variations, Scientific Report 01-01, Danish Meteorological Institute, Copenhagen, Denmark, 57 pp

Perreault P, Akasofu S-I (1978) A study of geomagnetic storms. Geophys J R Astr Soc 54:547–573

Pilipenko V, Kleimenova N, Kozyreva O, Engebretson M, Rasmussen O (2001) Global ULF wave activity during the May 15, 1997 magnetic storm. J Atmos Sol Terr Phys 63:489

Rangarajan GK (1989) Indices of geoma-gnetic activity, in Geomagnetism, edited by Jacobs JA, Academic, San Diego, California, p 323

Richardson G, Cliver EW, Cane HV (2000) Sources of geomagnetic activity over the solar cycle: Relative importance of coronal mass ejections, high-speed streams, and slow solar wind. J Geophys Res 105:18203–18213

Rostoker G. (1972) Geomagnetic indices. Rev Geophys Space Phys 10:935–950

Russell CT, McPherron RL (1973) Semiannual variation of geomagnetic activity. J Geophys Res 78:92

Sabine E (1856) On periodical laws discoverable in the mean effects of the larger magnetic disturbances. Phil Trans R Soc London A, 146:357

Sckopke N (1966) A general relation between the energy of trapped particles and the disturbance field near the Earth. J Geophys Res 71:3125–3130

Shue J-H, Kamide Y (2001) Effects of solar wind density on auroral electrojets. Geophys Res Lett 28:2181–2184

Siscoe GL (1970) The virial theorem applied to magnetospheric dynamics. J Geophys Res 75:5340–5350

Stauning P (2007) A new index for the interplanetary merging electric field and geomagnetic activity: Application of the unified polar cap indices. Space Weather 5:S09001. doi:10.1029/2007SW000311

Stauning P, Troshichev OA (2008a) Polar cap convection and PC index during sudden changes in solar wind dynamic pressure. J Geophys Res 113:A08227. doi:10.1029/2007JA012783

Stauning P, Troshichev OA, Janzhura, AS: Polar Cap (PC) Index. Unified PC-N (North) index procedures and quality, Scientific Report 06–04, Danish Meteorological Institute, Copenhagen, Denmark, 2006

Stauning P, Troshichev OA, Janzhura A (2008b) The Polar Cap (PC) indices: Relations to solar wind parameters and global magnetic activity. J Atmos Terr Phys 70:2246–2261

Sucksdorff C, Pirjola R, Hàkkinen L (1991) Computer production of K-values based on linear elimination. Geophys Trans 36:333–345

Sugiura M. (1964) Hourly values of equatorial Dst for the IGY., Annals of International Geophysics Year, vol 35, Chapter 9. Pergamon Press, Oxford

Sugiura M. (1973) Quiet time magneto-spheric field depression at 2.3–3.6 R_E. J Geophys Res 78:3182

Sugiura M, Chapman S (1960) The average morphology of geomagnetic storms with sudden commencement, Abandl. Akad. Wiss. Getingen Math. Phys. Kl., Sondernheft Nr.4, Götingen

Sugiura M, Hendricks S (1967) Provisional hourly values of equatorial Dst for 1961, 1962 and 1963, NASA Tech. note D-4047

Sugiura M, Kamei T (1991) Equatorial Dst index 1957–1986, IAGA Bulletin No. 40

Sugiura M, Poros DJ (1973) A magnetospheric field model incorporating the OGO-3 and -5 magnetic field observations. Planet Space Sci 21:1763

Suzuki A, Fukushima N (1984) Anti-sunward current below the MAGSAT level during magnetic storms. J Geomagnetics Geoelectrics 36:493–506

Svalgaard L (1968) Sector Structure of the Interplanetary Magnetic Field and Daily Variation of the Geomagnetic Field at High Latitudes, Geophysical papers R-6, Danish Meteorogical Institute

Svalgaard L (1977) Geomagnetic activity: Dependence on solar wind parameters, In: Zirker JB (ed) Skylab workshop monograph on coronal holes, Chapter 9. Columbia University Press, New York, NY, p 371

Svalgaard L, Cliver EW (2005) The IDV index: its derivation and use in inferring long-term variations of the interplanetary magnetic field strength. J Geophys Res 110:A12103. doi:10.1029/2005JA011203

Svalgaard L, Cliver EW (2007) Interhourly variability index of geomagnetic activity and its use in deriving the long-term variation of solar wind speed. J Geophys Res 112:A10111. doi:10.1029/2007JA012437

Svalgaard L, Cliver EW, Le Sager P (2004) IHV: a new long-term geomagnetic index. Adv Space Res 34, 436–439.

Takahashi K, Meng C, Kamei T, Kikuchi T, Kunitake M (2004) Near-real-time Auroral Electrojet index: An international collaboration makes rapid delivery of Auroral Electrojet index. Space Weather 2:S11003. doi:10.1029/2004SW000116

Takalo J. Mursula K (2001) A model for the diurnal universal time variation of the Dst index. J Geophys Res 106:10905–10921

Takeda M (1999) Time variation of geomagnetic Sq field in 1964 and 1980. JASTP 61:765–774

Thomson AWP, Lesur V (2007) An improved geomagnetic data selection algorithm for global geomagnetic field modeling. Geophys J Int 169:951–963. doi:10.1111/j.1365-246X.2007.03354

Troshichev OA, Andrezen VG (1985) The relationship between interplanetary quantities and magnetic activity in the southern polar cap. Planet Space Sci 33:415–419

Troshichev OA, Andrezen VG, Vennerstrøm S, Friis-Christensen E (1988) Magnetic activity in the polar cap: Anew index. Planet Space Sci 36:1095–1102

Troshichev OA, Dmitrieva NP, Kuznetsov BM (1979) Polar cap magnetic activity as a signature of substorm development. Planet Space Sci 27:217–221

Troshichev OA, Hayakawa H, Matsuoka A, Mukai T, Tsuruda K (1996) Cross polar cap diameter and voltage as a function of PC index and interplanetary quantities. J Geophys Res 101:13,429

Troshichev OA, Janzhura A, Stauning P (2006) Unified PCN and PCS indices: Method of calculation, physical sense, and dependence on the IMF azimuthal and northward components. J Geophys Res 111:A05208. doi:10.1029/2005JA011402

Troshichev OA, Janzhura A, Stauning P (2007a) Magnetic activity in the polar caps: Relation to sudden changes in the solar wind dynamic pressure. J Geophys Res 112:A11202. doi:10.1029/2007JA012369

Troshichev OA, Janzhura A, Stauning P (2007b) Reply to Comment of Lukianova R, R. on paper In: Troshichev OA, Janzhura A, Stauning P (eds) The unified PCN and PCS indices: method of calculation, physical sense, dependence on the IMF azimuthal and northward components J Geophys Res 112:A07205. doi:10.1029/2006JA012029

Troshichev OA, Lukianova RY, Papitashivili VO, Rich FJ, Rasmussen O (2000) Polar cap index (PC) as a proxy for ionospheric electric field in the near-pole region. Geophys Res Lett 27:3809

Vennerstrøm S, Friis-Christensen E, Troshichev OA, Andrezen VG (1991) Comparison between the polar cap index, PC, and the auroral electrojet indices AE, AL, and AU. J Geophys Res 96:101

Vennerstrøm S, Friis-Christensen E, Troshichev OA, Andrezen VG (1994) Geomagnetic Polar Cap (PC) Index 1975–1993, Report UAG-103, WDC-A for STP, NGDC, Boulder. (Cited from Papitashvili et al. 2001)

Weigel RS (2007) Solar wind time history contribution to the day-of-year variation in geomagnetic activity. J Geophys Res 112:A10207. doi:10.1029/2007JA012324

Weygand JM, Zesta E (2008) Comparison of auroral electrojet indices in the northern and southern hemispheres. J Geophys Res 113:A08202. doi:10.1029/2008JA013055

Chapter 9

Modelling the Earth's Magnetic Field from Global to Regional Scales

Jean-Jacques Schott and Erwan Thébault

Abstract In the recent years, a large amount of magnetic vector and scalar data have been measured or made available to scientists. They cover different ranges of altitudes from ground to satellite levels and have high horizontal densities over some geographical areas. Processing these potential field data may require alternatives to the widely used Spherical Harmonics. During the past decades, new techniques have been proposed to model regionally the magnetic measurements. They complement the set of older approaches that were revived and sometimes revised in the meantime. The amount of available techniques is intimidating and one often wonders which method is the most appropriate for what purpose. In this paper, we review several modelling strategies. Starting from the Spherical Harmonics, we discuss methods with global support (wavelets, multi-scale, Slepian functions,...) and then bring the focus on regional methods with local support (Rectangular Harmonic Analysis, Cylindrical Harmonic Analysis, Spherical Caps,...). We briefly examine the theoretical aspects and properties of each approach. We compare them with the help of a unique set of perfect synthetic data that mimic an ideal spatial distribution at a fixed surface. This helps us to better emphasize the theoretical characteristics of each approach and suggest, when relevant, improvements that would be useful for future practical applications.

J.-J. Schott (✉)
Ecole et Observatoires des Sciences de la Terre, Université de Strasbourg, F-67084 Strasbourg, France
e-mail: jj.schott@eost.u-strasbg.fr

9.1 Introduction

During the last assembly of the International Association of Geomagnetism and Aeronomy (IAGA) that took place in Sopron, Hungary, and within the division V ("Geomagnetic Observatories, Surveys and Analyses"), a significant number of contributions were related to global and regional modelling of the Earth's magnetic field. The ubiquity of this topic through sessions is a nice tribute paid to recent successful and future satellite missions (Friis-Christensen et al., 2006) and to continuous efforts made by the geomagnetic community towards the acquisition, maintenance, compilation, and fast online availability of magnetic data.

We noticed different modelling strategies among the variety of presentations. A first philosophy relied on the properties of Spherical Harmonics (SH) either by modelling all available data in a grand inversion or by modelling sources separately. The former approach is often referred to as a comprehensive inversion, was initiated decades ago (Sabaka and Baldwin, 1993; Langel et al., 1996), and was later pursued until today by including long series of magnetic observatory and recent satellite measurements (Sabaka et al., 2004). This approach can deal with the coupling between the ionosphere and the magnetosphere (see Sabaka et al., 2009; for instance) and the field does not need to be potential. It is based on the Mie representation well described in Backus et al., (1996; Chapter 5) that is not the scope of the present paper as no regional technique currently consider non potential fields. Other models are comprehensive-like but are based on potential field theory (see Gillet et al., 2010 for a review). Such models like, for instance, CHAOS (Olsen et al.,

M. Mandea, M. Korte (eds.), *Geomagnetic Observations and Models*, IAGA Special Sopron Book Series 5,
DOI 10.1007/978-90-481-9858-0_9,

2006), CHAOS-2 models (Olsen et al., 2009), and GRIMM (Lesur et al., 2008), are all exploiting a selection of recent Ørsted (Olsen et al., 2000) and CHAMP (Reigber et al., 2002) satellite data. They mostly focus on the Earth's core field and have therefore low spatial resolution. Improving the resolution may be achieved by modelling the satellite data sequentially. This approach, traditionally dedicated to the modelling of the lithospheric field, uses stringent data selection and correction (Maus et al., 2008), and is therefore more subjective (see Sabaka and Olsen., 2006 for a formal discussion and Thébault et al., 2010 for some practical implications). Whatever the selected SH modelling approach (comprehensive or sequential), the hundreds of kilometer distance between Low Earth Orbiting satellite and the crustal sources introduce a blurring effect. Thus, an horizontal spatial resolution of about 350 km (about SH degree 130) is probably the maximum achievable with data measured at 350 km altitude by a single satellite. Dense near-surface measurements, on the contrary, are closer to the crustal sources and have kilometric spatial resolution (see the World Digital Magnetic Anomaly Map project—WDMAM, Korhonen et al., 2007). Unfortunately, they are also so unevenly distributed at the Earth's scale that the internal SH Gauss coefficients cannot be estimated readily without data interpolation (Hamoudi et al., 2007; Maus et al., 2007b; Maus et al., 2009). This can only be done at the cost of manufacturing synthetic data and thus, possible wrong wavelengths and artefacts. Would the data be uniformly distributed, the number of required Gauss coefficients to represent the data to their intrinsic resolution would be anyway daunting.

The concept of regional modelling is precisely devoted to process dense sets of data available at different altitudes and to adjust the model to the data resolution; the ulterior motive being often to perform spectral analyses. Some methods were proposed in the past but only since the 1980's with the availability of MAGSAT vector satellite measurements (Langel et al., 1980) are they obeying Laplace equation (Alldredge, 1981). Among them, the first family uses functions with global support on the sphere; they are based on spherical splines (Shure et al., 1982), wavelets (e.g., Holschneider et al., 2003) or other types of localized spherical functions (e.g., Lesur, 2006; Simons et al., 2006). The second family relies on functions with local support. They may rely on a flat Earth approximation like the Rectangular Harmonic Analysis (Alldredge, 1981) and the Cylindrical Harmonic Analysis (Alldredge, 1982) or may consider the spherical curvature of the Earth like the Spherical Cap Harmonics Analysis (SCHA, Haines, 1985a) and its revision (Revised-SCHA, Thébault et al., 2004, for instance). This diversity of techniques is confusing and one often wonders what method is the most appropriate for his purpose. All techniques are obviously not equivalent in practice. They are founded on different theoretical arguments and were often originally derived in a framework far from geomagnetism. They address problems using assumptions with which any new application in geomagnetism must be consistent. We easily understand that modelling data in wide areas does not always bear the flat Earth approximation. Likewise, processing multi-level data with a technique not initially designed to allow upward and downward continuation makes little sense, even though it might give some numerical results. Most of the techniques presented here are in a development stage in the framework of geomagnetism. They currently allow mapping the data with more or less success. We consider that a better knowledge of their mathematical foundations will certainly help developing them towards more geophysical applications related, for instance, to spectral analysis, internal/external field separation and source characterization.

In this paper, we focus on potential field modelling techniques and outline the general theoretical properties of each approach by recalling some of their fundamentals. We begin with some generalities deduced from the Spherical Harmonics and proceed from global to regional scales keeping the same conventions when possible. We emphasize the orthogonality and completeness properties of the methods developed and discuss, when applicable, their relationship with Spherical Harmonics. We provide an example of inverse problem using a set of synthetic data distributed equally over a region at a unique altitude. This helps us to discuss the practical feasibility of the techniques regarding rates of convergence of the solutions and edge effects. However, we should keep in mind that real inverse problems often necessitate subtleties and, sometimes, *had hoc* procedures, regularization, or other kinds of *a priori* information. This requires specific 'know-how' acquired by experience. The examples are thus for illustrative purpose and by no means aimed at demonstrating the performance of one particular technique. In some examples, the setting of the inverse problem is purposefully designed to enhance a specific weakness or strength and therefore precludes direct figure comparisons between techniques.

9.2 Global Modelling With Spherical Harmonics in a Shell

The Spherical Harmonic (SH) expansion is well known and explained in many papers and books (in particular, we refer to Backus et al., 1996). However, for forthcoming discussions and comparisons with regional modelling methods, we provide a general solution of the Laplace equation in a shell and recap some important properties of SH.

The shell $S(b, c)$ is the open bounded set of $\mathbb{R}^3$ defined by $S(b, c) = \{\mathbf{r} \in \mathbb{R}^3 \mid b < \mid \mathbf{r} \mid < c\}$. Due to the spherical symmetry of the problem and to the shape of the boundaries of the domain, the most appropriate solutions are expressed in spherical coordinates (r, θ, φ). Laplace equation then writes

$$\nabla^2(V) = \frac{1}{r^2}\partial_r\left(r^2\partial_r V\right) + \frac{1}{r^2\sin\theta}\partial_\theta\left(\sin\theta\,\partial_\theta V\right) + \frac{1}{r^2\sin^2\theta}\partial^2_\varphi V, \quad (9.1)$$

$$= \frac{1}{r^2}\partial_r\left(r^2\partial_r V\right) + \frac{1}{r^2}\nabla^2_S(V) = 0, \quad (9.2)$$

where

$$\nabla^2_S(V) = \frac{1}{\sin\theta}\partial_\theta\left(\sin\theta\,\partial_\theta V\right) + \frac{1}{\sin^2\theta}\partial^2_{\varphi^2} V, \quad (9.3)$$

is the Beltrami-Laplace operator. The spectral properties of this operator are essential for the functions belonging to the Hilbert space defined on the unit sphere $S(1)$, hence for the solid SH. The solutions of Eq. (9.1) in a ball $0 < \mid r \mid < c$ may be expressed in terms of harmonic homogeneous polynomials, an approach adopted by Backus et al. (1996, Section 3.1). This approach has close connections with rotational symmetries on the sphere and with commutativity properties of a class of differential operators on the sphere. However, we cannot generalize this way of doing to domains with geometry breaking up the rotational symmetry. We therefore prefer to deal with the problem from another viewpoint, which yet remains a standard one (Hobson, 1965).

9.2.1 Resolution of Laplace Equation by the Fourier Decomposition Method

In the geocentric reference frame, the Fourier method provides solutions of Eq. (9.1) in terms of products of separate functions of r, θ and φ. This requires setting two Sturm-Liouville problems and one ordinary differential equation. Writing

$$V(r, \theta, \varphi) = R(r)P(\theta)F(\varphi), \quad (9.4)$$

we obtain the following equations

$$d_r\left(r^2 d_r R\right) = \nu(\nu+1)R, \quad (9.5)$$

$$d^2_{\varphi^2}F = -\kappa F, \quad (9.6)$$

Eq. (9.6) being associated to the boundary conditions

$$F(0) = F(2\pi);\ \left(d_\varphi F\right)_0 = \left(d_\varphi F\right)_{2\pi}, \quad (9.7)$$

$$d_u\left[\left(1-u^2\right)d_u P\right] + \left[\nu(\nu+1) - \frac{\kappa}{1-u^2}\right]P = 0, \quad (9.8)$$

and Eq. (9.8) associated to the boundary conditions

$$P \text{ and } d_u P \text{ finite } at \text{ } 0 \text{ and } \pi, \quad (9.9)$$

with $u = \cos\theta$. Equation (9.5) is an Euler equation of degree 2 without boundary condition. The constant is written $\nu(\nu+1)$ for well-known convenience. At this stage, ν could be real or complex. The differential equation has two independent solutions admissible in the range $[b, c]$

$$R(r) = r^\nu;\ R(r) = r^{-\nu-1}. \quad (9.10)$$

Equation (9.6) is a regular Sturm-Liouville problem with periodic boundary conditions (Eq. 9.7), which dictate the range of values κ and impose them to be of the form $\kappa = m^2, m \in \mathbb{Z}$. One may take m as a positive or null integer without loss of generality. Thus, Eq. (9.6) has two independent solutions

$$F(\varphi) = e^{im\varphi};\ F(\varphi) = e^{-im\varphi}. \tag{9.11}$$

Note that $F(\varphi)$ is an eigenvector of the operator $-d^2_{\varphi^2}F$ applied to functions belonging to the space $C^2([0,2\pi]) \cap L^2([0,2\pi])$, $L^2([0,2\pi])$ being endowed with the inner product

$$\langle F, G\rangle = \int_0^{2\pi} F(\varphi)\,\overline{G(\varphi)}\,d\varphi. \tag{9.12}$$

On the subspace of the functions verifying Eq. (9.7), $-d^2_{\varphi^2}F$ is self-adjoint. Hence, the eigenvectors $F(\varphi)$ given by Eq. (9.11) are orthogonal. The Sturm-Liouville problem defined by Eqs. (9.8) and (9.9) is termed 'singular' due to the vanishing of the coefficient of the highest derivative order occurring at both ends of the interval. Replacing κ with m^2, and making the successive changes $P(u) = \left(1-u^2\right)^{m/2} T(u)$, $s = (1-u)/2$, the differential Eq. (9.8) is reshaped into an hypergeometric equation

$$\begin{aligned} &s(1-s)\,d^2_{s^2}T + (m+1)(1-2s)\,d_sT \\ &\quad + (\nu - m)(\nu + m + 1)\,T = 0, \end{aligned} \tag{9.13}$$

which solutions have properties described for instance in Morse and Feshbach (1953, Section 5.2). There is only one analytical solution in the vicinity of each singular point ($s = 0$ or 1) and the solution is analytical at both ends if and only if ν is an integer l. This implies that $-l \leqslant m \leqslant l$ and $T(s)$ being a polynomial of degree $l-m$.

Turning back to $P(u)$, it may be shown (for instance, Olver, 1997, p.180) that if l and m are integers, $P(u)$ takes the familiar form derived from Rodrigues's formula and is called associated Legendre functions $P_l^m(\cos\theta)$. Eq. (9.8) may now be written

$$-d_u\left[\left(1-u^2\right)d_uP_l^m\right] + \frac{m^2}{1-u^2}P_l^m = l(l+1)P_l^m, \tag{9.14}$$

which shows that $P_l^m(u)$ is an eigenvector of the operator

$$D_m = -d_u\left[\left(1-u^2\right)d_u\right] + \frac{m^2}{1-u^2}I, \tag{9.15}$$

where I is the identity operator. D_m is self-adjoint on the subspace of the functions belonging to the space $C^2([-1,1]) \cap L^2([-1,1])$ and taking finite values at $|u| = 1$, the Hilbert space $L^2([-1,1])$ being equipped with the inner product $\int f, g\rangle = \int_{-1}^{1} f(u)g(u)du$, with respect to which the Legendre associated functions are orthogonal. Together with the orthogonality properties of the functions $F(\varphi)$ (Eq. 9.11), the orthogonality of P_l^m is a fundamental property of the spherical harmonic expansions. Consider now the space $L^2(S(\rho))$ of the functions defined on the sphere $S(\rho)$ centered on the origin, with radius ρ. $L^2(S(\rho))$ is a Hilbert space for the inner product

$$\langle f, g\rangle = \frac{1}{4\pi}\int_0^{\pi}\int_0^{2\pi} f(\rho,\theta,\varphi)\,\overline{g(\rho,\theta,\varphi)}\,\sin\theta d\theta d\varphi. \tag{9.16}$$

The operator ∇_S^2 (Eq. 9.3) is self-adjoint on the subspace $C^2(S(1)) \cap L^2(S(1))$ of the functions taking finite values at $\theta = 0$ and $\theta = \pi$. From the properties of $F(\varphi)$ and $P_l^m(\cos\theta)$ we derive readily that

$$\beta_l^m(\theta,\varphi) = P_l^m(\cos\theta)e^{im\varphi}\ (m = -l,\ldots,l), \tag{9.17}$$

are eigenfunctions of $-\nabla_S^2$ associated to the eigenvalues $l(l+1)$. Thus, to the eigenvalue $l(l+1)$ is associated an eigensubspace of dimension $2l+1$. The functions $\beta_l^m(\theta,\varphi)$ are orthogonal with respect to the inner product defined on $L^2(S(1))$. This property is a straightforward consequence of the orthogonality properties of $F(\varphi)$ and P_l^m. In Geomagnetism, the common convention is to use the Schmidt functions written p_l^m (see Langel, 1987, p. 254 for a definition). However, the norm of the SH $\|\beta_l^m\|_{L^2(S(1))}$ may take various expressions (see Langel, 1987, p. 255 for the most common ones). The final solutions of the Laplace equation are, according to the Fourier decomposition

$$\psi_{i,\,l}^m(r,\theta,\varphi) = R_E\left(\frac{R_E}{r}\right)^{l+1}\beta_l^m, \tag{9.18a}$$

$$\psi_{e,\,l}^{m}(r,\theta,\varphi) = R_E \left(\frac{r}{R_E}\right)^l \beta_l^m, \tag{9.18b}$$

where R_E is the mean earths' radius. Its incorporation in expressions (9.18) is common in earth's magnetic field modelling and traces back at least to Chapman and Bartels (1940). The subscripts 'i' (inner) and 'e' (external) are self-explanatory for readers familiar with SH.

9.2.2 Orthogonality and Completeness Properties

The Laplace equation is the most famous example of second-order partial differential equations. In modern studies of second-order PDE in an open set Ω, an extensive use is made of the Sobolev space $H^1(\Omega)$. It is the space of functions belonging to $L^2(\Omega)$ as well as their first derivatives. $H^1(\Omega)$ is a Hilbert space for the inner product $\langle f,\, g\rangle_{H^1} = \int_\Omega \left[f(\mathbf{r})\overline{g(\mathbf{r})} + \vec{\nabla}(f(\mathbf{r}))\cdot\vec{\nabla}(\overline{g(\mathbf{r})})\right] d\tau$ (Reddy, 1998, p. 227). Let be $\Omega = S(b,c)$, and $\partial\Omega$ its boundary: $\partial\Omega = S(b) \cup S(c)$. It may be shown that the functions $\psi_{i,l}^{m}$ and $\psi_{e,l'}^{m'}$,which belong to $H^1(\Omega)$) are orthogonal with respect to the inner product $\langle .,.\rangle_{H^1}$ unless $l = l'$ and $m = m'$. However, within the frame of the earth's magnetic field modelling, $H^1(\Omega)$ is not the most relevant space because the measured data is the gradient of the potential, not the potential itself. Beside $H^1(\Omega)$, $H_0^1(\Omega)$ which is the subspace of $H^1(\Omega)$ of the functions taking the value 0 on the boundary $\partial\Omega$, is another Sobolev space that plays a prominent role. $H_0^1(\Omega)$ is a closed subspace of $H^1(\Omega)$ with respect to the inner product

$$\langle f,g\rangle_{H_0^1} = \int_\Omega \vec{\nabla}(f(\mathbf{r}))\cdot\vec{\nabla}(\overline{g(\mathbf{r})})d\tau. \tag{9.19}$$

This inner product defines a true norm on $H_0^1(\Omega)$ because $\|f\|_{H_0^1} = 0$ implies $f = 0$ due to the boundary condition. However, $H_0^1(\Omega)$ is still unsuitable in the case of harmonic functions because $\nabla^2(f) = 0$, associated with the condition $f = 0$ on $\partial\Omega$, is an homogeneous Dirichlet problem which unique solution is $f = 0$. It is important, however, regarding the uniqueness of the inverse problem, to find a subspace where Eq. (9.19) provides a true norm. Backus (1986), showed that the scalar magnetic potential V could be chosen such that $\langle V\rangle_r = 0$ without loss of generality, where $\langle V\rangle_r$ stands for the mean value of V on any sphere of radius r ($b < r < c$). Let thus $U(\Omega)$ be the subset of $H^1(\Omega)$ of the functions verifying the property $\langle f\rangle_r = 0$. $U(\Omega)$ is evidently a subspace of $H^1(\Omega)$. In order to avoid confusions, we will note $\langle f,g\rangle_U$ the inner product defined by Eq. (9.19) when it applies to functions belonging to $U(\Omega)$. This inner product defines a true norm on $U(\Omega)$ because if $\|f\|_U = 0$ then $f =$ constant on $U(\Omega)$ but since $\langle f\rangle_r = 0$, then $f = 0$. Backus et al. (1996, p. 125) showed that $\psi_{i,l}^{m}(r,\theta,\varphi)$ and $\psi_{e,l}^{m}(r,\theta,\varphi)$ belong to $U(\Omega)$ except for $l = 0$, though for two different reasons. These basis functions are therefore excluded hereafter.

Furthermore, it may be shown that $\psi_{i,l}^{m}(r,\theta,\varphi)$ and $\psi_{e,l}^{m}(r,\theta,\varphi)$ are orthogonal with respect to the inner product $\langle .,.\rangle_U$, which means that

$$\left\langle \psi_{i,l}^{m}, \psi_{e,l'}^{m'}\right\rangle_U = \int_\Omega \vec{\nabla}(\psi_{i,l}^{m}(\mathbf{r}))\cdot\vec{\nabla}(\overline{\psi_{e,l'}^{m'}(\mathbf{r})})d\tau = 0, \tag{9.20}$$

and

$$\left\langle \psi_{i,l}^{m}, \psi_{i,l'}^{m'}\right\rangle_U \text{ or } \left\langle \psi_{e,l}^{m}, \psi_{e,l'}^{m'}\right\rangle_U = 0 \text{ if } l \neq l' \text{ or } m \neq m'. \tag{9.21}$$

Then, using definitions (9.18)

$$\|\psi_{i,l}^{m}\|_U^2 = \int_\Omega \left|\vec{\nabla}\psi_{i,l}^{m}\right|^2 d\tau = 4\pi R_E^3\,(l+1) \left[\left(\frac{R_E}{b}\right)^{2l+1} - \left(\frac{R_E}{c}\right)^{2l+1}\right] \|\beta_l^m\|_{S(1)}^2, \tag{9.22}$$

and

$$\|\psi_{e,l}^{m}\|_U^2 = \int_\Omega \left|\vec{\nabla}\psi_{e,l}^{m}\right|^2 d\tau = 4\pi R_E^3 l \left[\left(\frac{c}{R_E}\right)^{2l+1} - \left(\frac{b}{R_E}\right)^{2l+1}\right] \|\beta_l^m\|_{S(1)}^2, \tag{9.23}$$

where $\|.\|_{S(1)}$ is the norm defined previously on the Hilbert space $L^2(S(1))$. At last, the question arises of this set being a base on $U(\Omega)$. If so, only the null function is orthogonal to $\psi_{i,l}^m$ or $\psi_{e,l'}^{m'}$. An elementary proof of this property may be given thanks to the following Green identity (for instance, Reddy, 1998, p. 219)

$$\int_{\Omega} f\nabla^2\overline{h}d\tau = \int_{\partial\Omega} f\frac{\partial\overline{h}}{\partial n}ds - \int_{\Omega}\overrightarrow{\nabla}f\cdot\overrightarrow{\nabla}\overline{h}d\tau = \int_{\partial\Omega} f\frac{\partial\overline{h}}{\partial n}ds - \langle f,h\rangle_U, \tag{9.24}$$

where n is the outward unit vector, orthogonal to the boundary. Let f be an harmonic function belonging to $U(\Omega)$. With $h=\psi_{i,l}^m$, Eq. (9.24) becomes

$$\langle f,\psi_{i,l}^m\rangle_U = (l+1)\left[b^2\left(\frac{R_E}{b}\right)^{l+2}\int_{S(1)} f(b,\theta,\varphi)\overline{\beta}_l^m d\sigma - c^2\left(\frac{R_E}{c}\right)^{l+2}\int_{S(1)} f(c,\theta,\varphi)\overline{\beta}_l^m d\sigma\right], \tag{9.25}$$

and with $h=\psi_{e,l}^m$

$$\langle f,\psi_{e,l}^m\rangle_U = l\left[c^2\left(\frac{c}{R_E}\right)^{l-1}\int_{S(1)} f(c,\theta,\varphi)\overline{\beta}_l^m d\sigma - b^2\left(\frac{b}{R_E}\right)^{n-1}\int_{S(1)} f(b,\theta,\varphi)\overline{\beta}_l^m d\sigma\right]. \tag{9.26}$$

The unique solution to the system of equations $< f,\psi_{i,l}^m >_U = < f,\psi_{e,l}^m >_U = 0$ is

$$\int_{S(1)} f(b,\theta,\varphi)\overline{\beta}_l^m d\sigma = \int_{S(1)} f(c,\theta,\varphi)\overline{\beta}_l^m d\sigma = 0. \tag{9.27}$$

Since $\{\beta_l^m\}$ is an orthonormal base on $L^2(S(1))$ (Backus et al, 1996), f is null on the boundary $\partial\Omega$ and since the function f is the solution of the following Dirichlet problem: $\nabla^2(f)=0$ on $\Omega, f=0$ on $\partial\Omega$, the unique solution is $f=0$ on Ω.

9.2.3 Spherical Harmonic Expansion and Convergence Properties

Consider now a potential V belonging to $U(\Omega)$. Its SH expansion, S_V, on the basis $\{\psi_{i,l}^m,\psi_{e,l}^m\}$ is the double series

$$S_V = \sum_{l=1}^{\infty}\sum_{m=-n}^{n}\left(g_n^m\psi_{i,l}^m + q_n^m\psi_{e,l}^m\right). \tag{9.28}$$

The internal and external Gauss coefficients g_l^m and q_l^m are respectively given by the relations

$$g_l^m = \frac{1}{\left\|\psi_{i,l}^m\right\|_U^2} < V,\psi_{i,l}^m >_U, \tag{9.29}$$

$$q_l^m = \frac{1}{\left\|\psi_{e,l}^m\right\|_U^2} < V,\psi_{e,l}^m >_U. \tag{9.30}$$

The Green's identity (9.24) may be used to compute the Gauss coefficients in two other ways, which are equivalent to solving a Dirichlet or a Neumann boundary value problem, but give nevertheless the same expression of Gauss coefficients. In geomagnetism and potential theory, V is an harmonic function on Ω, which gradient $\overrightarrow{B} = -\overrightarrow{\nabla}V$ is known on Ω. A standard way of solving this problem is to search for the Gauss coefficients of the SH expansion S_V of V, which minimize the functional

$$d^2 = \int_{\Omega}\left|\overrightarrow{\nabla}S_V - \overrightarrow{B}\right|^2 d\tau = \langle S_V, V\rangle_U. \tag{9.31}$$

This problem is closely connected to the inverse problem based on the least squares method, which is widely used in geomagnetic field modelling. The functional d^2 is a quadratic form in the Gauss coefficients and it turns out that the coefficients which minimize d^2 are given by Eq. (9.29) and (9.30). Thus, the mean-square solution is the same as the solution of the Dirichlet problem and the Neumann problem. The equivalence between Dirichlet and Neumann problems is specific to the space $U(\Omega)$ but the equivalence between the Dirichlet problem and the minimization of the functional d^2 (Eq. 9.31) is a general property in $H^1(\Omega)$ (Dautray and Lions, 1987, vol. II, p. 632).

We thus now examine the convergence properties of expansion (9.28). From the property of the set $\{\psi_{i,l}^{m}, \psi_{e,l'}^{m'}\}$ (Eq. 9.18) being a basis of the space $U(\Omega)$, the SH expansion (9.28) converges towards V with respect to the norm $\|\cdot\|_U$. Such convergence is consistent with the least-squares minimization problem set in Eq. 9.31 but does not preclude the Gibbs phenomenon. This typical well-known approximation error occurs in Fourier-like expansions and its quantitative description refers to uniform convergence, which is the convergence associated to the infinity or Chebyshev norm $\|f\|_\infty = \sup|f|$or $\|f\|_\infty = \sup\left|\vec{\nabla} f\right|$ on Ω. These norms (or semi-norm regarding the second expression) are typical for spaces of continuous or continuously differentiable functions. The relationship between these spaces and $H^1(\Omega)$ or $U(\Omega)$, and hence between $\|\cdot\|_\infty$ and $\|\cdot\|_U$, is not obvious. Harnack's first theorem on uniform convergence (Kellog, 1929, p. 248) enunciates a condition relevant for the earth's magnetic field modelling, which states that the infinite series S_V converges uniformly towards V if $\vec{\nabla}_s V$ (the surface gradient, see, for instance, Backus et al., 1996 p. 324) is continuous on the sphere or, in the present case, on the set of two concentric spheres. Thus, in practice, the SH expansion is uniformly convergent if we exclude singular sources on the boundary of the domain of interest, which explains why, to our knowledge, Gibbs phenomenon has not been reported in SH but in very few cases for which a small number of outliers precisely behaved like singular sources on the sphere (e.g., Hamoudi et al., 2007 Section 4.6).

For the following discussions, we construct a benchmark magnetic field over Western Europe. We use the SH models associated to published Gauss coefficients to synthesize a set of perfect data for X, Y and Z magnetic field components. In Fig. 9.1 we present the Z component that results from the superimposition of the main field at epoch 2010 (IGRF11 model, see http://www.ngdc.noaa.gov/IAGA/vmod/igrf.html for details) and an estimation of the crustal field up to SH degree 720 (i.e., a maximum spatial resolution of about 55 km; see Maus, 2010 for details). We call it Z_{all} in the following.

9.3 Other Modelling at a Global Scale

Spherical harmonic expansions remain the fundamental tool for modelling the Earth's magnetic field thanks to their completeness and convergence properties, be it through the Gauss or the Mie representation. Both rely strongly on Newtonian potentials, which verify the Laplace equation in any source-free domain. However, invoking concepts more familiar in the physics of wave propagation and signal processing, some authors argue that there is no possible balance between spectral and spatial localization with SH (e.g., Freeden and Michel, 2000; Lesur, 2006; Simons et al., 2006). SH are indeed perfectly localized in the frequency domain but not localized in space, their support being the whole sphere, and the necessary truncation of the expansion

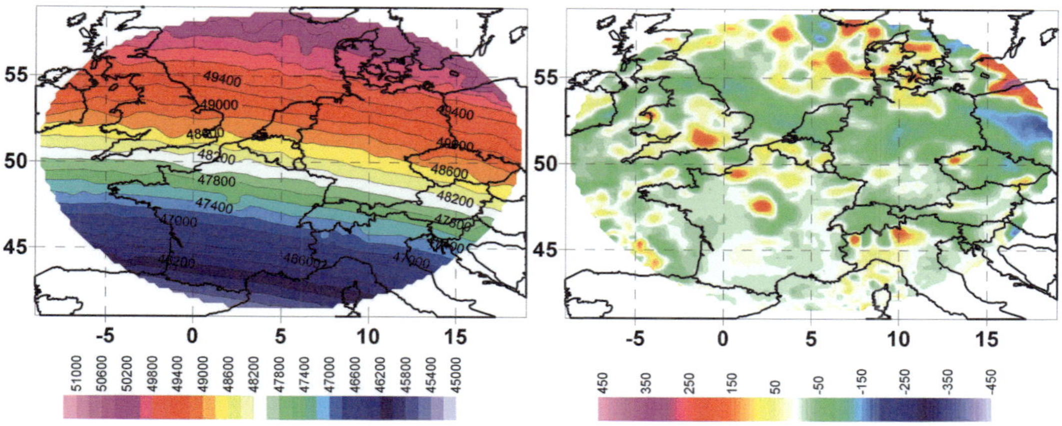

Fig. 9.1 Z component of the magnetic field at the Earth's reference radius within a Spherical Cap centred on Europe. *Left*: superimposition of the field at 2010.0 and an estimate of the crustal field (see text for details) and (*right*) the crustal field alone with about 50 km horizontal spatial resolution. Units are in nT

introduces some further level of subjectivity. Shure et al. (1982) tackled this last inconvenience by introducing the concept of harmonic splines. Concerning the localization in space, several proposals were made during the last decade, which amount to generate the Hilbert space of harmonic functions by other functions than the SH defined as β_n^m by Eq. (9.17) but still based upon Legendre polynomial expansions. Strictly speaking, these functions are only numerically localized, the counterpart being that their spectrum covers a more or less extended range of frequencies. Hereafter, we selected three representations that emerged recently in geomagnetism and we refer the reader to Shure et al. (1982) for a presentation of spherical splines (see also Langel, 1987; section 13.1).

9.3.1 Wavelets

We consider the following Dirichlet problem: to find the potential $V(\vec{r})$, which is harmonic in the infinite shell $S(R,\infty)$ and which is known on the sphere $S(R)$. Note that R does not need to be the Earth's mean radius R_E. We assume that the potential vanishes at infinity. Thus, the Gauss coefficients g_l^m are given by the limiting expression of Eq. (9.29) when c is put to infinity

$$g_l^m = \frac{1}{4\pi R \left\| \beta_l^m \right\|^2_{L^2(S(1))}} \int_{S(1)} V(R,\theta,\varphi)\beta_l^m(\theta,\varphi)\, d\sigma, \tag{9.32}$$

where we assume that β_l^m takes real values. We adopt this formulation in order to avoid unnecessary complications with complex functions and Hermitian inner products. For the sake of convenience, we continue to select the orders m in the range $[-l,+l]$, thus adopting the notations used by other authors (Lesur, 2006; Simons and Dahlen, 2006). We further consider the β_l^m as fully normalized with respect to the inner product defined by Eq. (9.16). Following the notation of Backus et al. (1996), we define $\vec{r} = r\hat{\mathbf{r}}$ so that Eq. (9.32) becomes

$$g_l^m = \frac{1}{4\pi R} \int_{S(1)} V(R\,\hat{\mathbf{s}})\beta_l^m(\hat{\mathbf{s}})\, d\sigma. \tag{9.33}$$

Assuming that interchanging the integration and summation makes sense, the expansion given by Eq. (9.28) may be written

$$V(r\hat{\mathbf{r}}) = \frac{1}{4\pi R}\sum_{l=1}^{\infty}\sum_{m=-l}^{l} \int_{S(1)} \psi_l^m(r\hat{\mathbf{r}})\beta_l^m(\hat{\mathbf{s}})\, V(R\,\hat{\mathbf{s}})d\sigma, \tag{9.34}$$

$$= \frac{1}{4\pi}\int_{S(1)} V(R\hat{\mathbf{s}})\sum_{l=1}^{\infty}\sum_{m=-n}^{n}\left(\frac{R}{r}\right)^{l+1} \beta_l^m(\hat{\mathbf{r}})\beta_l^m(\hat{\mathbf{s}})\, d\sigma(\hat{\mathbf{s}}), \tag{9.35}$$

where $d\sigma(\hat{\mathbf{s}})$ means that the integration on the unit sphere is performed with respect to the variable $\hat{\mathbf{s}}$. Eq. (9.35) is more concise using the spherical harmonic addition theorem (Backus et al., 1996, p. 62)

$$\sum_{m=-l}^{l} \beta_l^m(\hat{\mathbf{r}})\beta_l^m(\hat{\mathbf{s}}) = (2l+1)\, P_l(\hat{\mathbf{r}}\cdot\hat{\mathbf{s}}), \tag{9.36}$$

where P_l is the Legendre polynomial of degree l (which expression must be consistent with the norm chosen for β_l^m). We obtain finally

$$V(r\hat{\mathbf{r}}) = \frac{1}{4\pi}\int_{S(1)} K(r\hat{\mathbf{r}},\hat{\mathbf{s}})V(R\hat{\mathbf{s}})d\sigma(\hat{\mathbf{s}}), \tag{9.37}$$

with

$$K(r\hat{\mathbf{r}},\hat{\mathbf{s}}) = \sum_{l=1}^{\infty}(2l+1)\left(\frac{R}{r}\right)^{l+1} P_l(\hat{\mathbf{r}}\cdot\hat{\mathbf{s}}). \tag{9.38}$$

Equation (9.37) teaches us that the potential V can be computed at any point $r\hat{\mathbf{r}}$ within the infinite shell $S(R,\infty)$ by an integral transform based upon the kernel $K(r\hat{\mathbf{r}},\hat{\mathbf{s}})$, which maps the potential known on the boundary $S(R)$ to the potential at any point $r\hat{\mathbf{r}}$. Note that this formalism is defined for $r > R$,where R is the chosen reference surface (again not necessarily Earth's mean radius and it could be the core mantle boundary, see Constable et al., 1993, for instance). Eq. (9.37) and (9.38) are the departure point of many representation using the global support. Constable et al., (1993; Appendix) expanded Eq. (9.38) making use of the generating Legendre polynomials and called $K(r\hat{\mathbf{r}},\hat{\mathbf{s}})$ the Green's function. In a recent paper presented during the IAGA in Sopron, Stockmann et al. (2009) used this kernel and a spherical triangle tesselation to estimate from satellite data the lithospheric field near the Earth's surface. In fact, Eq. (9.37) and (9.38) may be obtained when one takes the Poisson integral kernel as a departure point (Kellog, 1929, p. 251), hence the name of Poisson wavelet generally given to this solution. Eq. (9.37) also writes

$$V(r\widehat{\mathbf{r}}) = \langle K(r\widehat{\mathbf{r}},\widehat{\mathbf{s}}), V(R\widehat{\mathbf{s}})\rangle_{L^2(S(1))}, \quad (9.39)$$

which is the definition of the convolution on the sphere (Holschneider et al., 2003). In addition, Eq. (9.37) looks like defining $K(r\widehat{\mathbf{r}},\widehat{\mathbf{s}})$ as being a reproducing kernel on the Hilbert space $L^2(S(R,\infty))$, the subtlety being that the equality (9.37) has to be interpreted in terms of the norm of $L^2(S(R,\infty))$ and not in terms of a pointwise equality between functions (see Backus et al., 1996, section 3.3 for a definition and properties of reproducing kernels). The kernel $K(r\widehat{\mathbf{r}},\widehat{\mathbf{s}})$ has another important property regarding the construction of spherical wavelets or scaling functions since it can be interpreted as the rotated function $K(r\widehat{\mathbf{e}}_z\cdot\widehat{\mathbf{s}})$ in the rotation $R_{\widehat{\mathbf{r}}}$ on the sphere such as $\widehat{\mathbf{r}} = R_{\widehat{\mathbf{r}}}(\widehat{\mathbf{e}}_z)$. Here, $\widehat{\mathbf{e}}_z$ is the unit vector carried by the axis Oz of the Cartesian reference frame to which the spherical coordinate θ is referred (see Backus et al., 1996, p.59, for further details about rotations on the sphere). Knowing that

$$K(r\widehat{\mathbf{e}}_z,\widehat{\mathbf{s}}) = \sum_{l=1}^{\infty}(2l+1)\left(\frac{R}{r}\right)^{l+1} P_l(\widehat{\mathbf{e}}_z\cdot\widehat{\mathbf{s}}), \quad (9.40)$$

Eq. (9.40) shows that $K(r\widehat{\mathbf{e}}_z,\widehat{\mathbf{s}})$, regarded as a function of $\widehat{\mathbf{s}}$, is the sum of zonal spherical harmonics in the usual sense, and, therefore, is itself a zonal function. Therefore, $K(r\widehat{\mathbf{r}},\widehat{\mathbf{s}})$ is a zonal function around the axis defined by the unit vector $\widehat{\mathbf{r}}$.

If $r = R$, the series (Eq. 9.40) no longer converges in a classical sense and cannot be used to define a wavelet transform (actually, $K(r\widehat{\mathbf{r}},\widehat{\mathbf{s}}) \to \delta(\widehat{\mathbf{r}},\widehat{\mathbf{s}})$, the Dirac distribution when $r \to R$, see Simons et al., 2006). This inconvenience is mitigated in spherical wavelets and scaling functions theory by a flexible modification of the kernel, which then writes

$$K(r\widehat{\mathbf{e}}_z,\widehat{\mathbf{s}}) = \sum_{l=1}^{\infty}(2l+1)\,\gamma(l)\left(\frac{R}{r}\right)^{l+1} P_l(\widehat{\mathbf{e}}_z\cdot\widehat{\mathbf{s}}), \quad (9.41)$$

where γ is an appropriate function defined on $\mathbb{N}$ which gives sense to the infinite sum for $r = R$. Furthermore, the well-known scaling in wavelet theory is introduced through a dilation generator D_a (Holschneider, 1995, p. 3, Freeden and Michel, 2000, p. 209), which define a dilated function $\gamma(an)$ by

$$D_a(\gamma(n)) = \gamma(an), \quad (9.42)$$

where a is a real positive number. Finally, the modified kernel writes

$$K_a(r\widehat{\mathbf{e}}_z,\widehat{\mathbf{s}}) = \sum_{l=1}^{\infty}(2l+1)\,\gamma(al)\left(\frac{R}{r}\right)^{l+1} P_l(\widehat{\mathbf{e}}_z\cdot\widehat{\mathbf{s}}). \quad (9.43)$$

Equation (9.43) is a relevant expression for introducing wavelets and scaling functions.

9.3.1.1 Poisson Wavelets

We focus on the family of Poisson wavelets, which are scalar wavelets that were proposed by Holschneider et al. (2003), because it has a simple and attractive interpretation in terms of multipolar potentials. The properties of the wavelets require so-called admissibility conditions on the function γ. According to the expression given in Panet et al. (2006), the Poisson wavelets write (not to be confused with g_n^m the SH Gauss coefficients)

$$g_a^n(r\widehat{\mathbf{r}}) = \frac{1}{R}\sum_{l=1}^{\infty}(2l+1)\,e^{-al}(al)^n\left(\frac{R}{r}\right)^{l+1} P_l(\widehat{\mathbf{e}}_z\cdot\widehat{\mathbf{r}}). \quad (9.44)$$

Between Eqs. (9.43) and (9.44), the function γ takes the particular expression $\gamma(t) = e^{-t}t^n$. We denote $g_{\widehat{\mathbf{s}},a}^n(r\widehat{\mathbf{r}})$ the scalar field derived from $g_a^n(r\widehat{\mathbf{r}})$ by a rotation $R_{\widehat{\mathbf{s}}}$ and write

$$g_{\widehat{\mathbf{s}},a}^n(r\widehat{\mathbf{r}}) = \frac{1}{R}\sum_{l=1}^{\infty}(2l+1)\,e^{-al}(al)^n\left(\frac{R}{r}\right)^{l+1} P_l(\widehat{\mathbf{s}}\cdot\widehat{\mathbf{r}}). \quad (9.45)$$

As mentioned above, the wavelet family g_a^n (Eq. 9.44) has an interesting interpretation in terms of multipoles. Consider the potential

$$\psi_n^\lambda(r\widehat{\mathbf{r}}) = \left[\lambda\partial_{z'}\circ\left(z'\partial_{z'}\right)\circ\ldots\circ\left(z'\partial_{z'}\right)\right]\frac{1}{|r\widehat{\mathbf{r}} - r'\widehat{\mathbf{r}}'|}, \quad (9.46)$$

where the derivatives are taken n times, $r\widehat{\mathbf{r}} = x\widehat{\mathbf{e}}_x + y\widehat{\mathbf{e}}_y + z\widehat{\mathbf{e}}_z$, $r'\widehat{\mathbf{r}}' = z'\widehat{\mathbf{e}}_z$, and λ is an arbitrary, dimensionless constant. Developing the derivations between brackets, Eq. (9.46) becomes

$$\psi_n^\lambda(r\widehat{\mathbf{r}}) = \sum_{k=1}^{n} C_n^k \lambda^k \left(\partial_{z'}^k \frac{1}{|r\widehat{\mathbf{r}} - r'\widehat{\mathbf{r}}'|}\right)_{z'=\lambda R}, \quad (9.47)$$

where the coefficients C_n^k are computed recursively using the recurrence relation $C_k^n = C_{k-1}^{n-1} + kC_k^{n-1}$, $k = 1, \ldots, n-1$, with the convention $C_0^{n-1} = C_n^{n-1} = 0$. Each term $\lambda^k \left(\partial_{z'}^k \frac{1}{|r\widehat{\mathbf{r}} - r'\widehat{\mathbf{r}}'|}\right)_{z'=\lambda R}$ is a zonal multipole of order k, having k identical axes along Oz and located at the point $(0,0,\lambda R)$. Thus, ψ_n^λ is the sum of n zonal multipoles of orders ranging from 1 (dipole) to n, all located on the Oz axis, at the point $(0,0,\lambda R)$. It may be shown, using the expansion of $\frac{1}{|r\widehat{\mathbf{r}} - r'\widehat{\mathbf{r}}'|}$ in terms of Legendre polynomials, that

$$\psi_n^\lambda(r\widehat{\mathbf{r}}) = \frac{1}{r} \sum_{l=1}^{\infty} \lambda^l l^n \left(\frac{R}{r}\right)^l P_l\left(\widehat{\mathbf{r}} \cdot \widehat{\mathbf{e}}_z\right). \quad (9.48)$$

Writing now $\lambda = e^{-a}$, the potential ψ_n^λ takes the form

$$\psi_n^\lambda(r\widehat{\mathbf{r}}) = \frac{a^{-n}}{R} \sum_{l=1}^{\infty} e^{-al} (al)^n \left(\frac{R}{r}\right)^{l+1} P_l\left(\widehat{\mathbf{r}} \cdot \widehat{\mathbf{e}}_z\right). \quad (9.49)$$

Comparing Eq. (9.49) with Eq. (9.44), we obtain

$$g_n^a(r\widehat{\mathbf{r}}) = a^n \left(2\psi_{n+1}^\lambda(r\widehat{\mathbf{r}}) + \psi_n^\lambda(r\widehat{\mathbf{r}})\right). \quad (9.50)$$

Thus, the wavelet $g_n^a(r\widehat{\mathbf{r}})$ is the sum of the potentials produced by a set of $(n+1)$ zonal multipoles, with orders ranging from 1 to $(n+1)$, having all their axes along Oz, and located at $(0,0,R\exp(-a))$. The rotational properties of $g_{\widehat{\mathbf{s}},a}^n$ (Eq. 9.45) are such that $g_{\widehat{\mathbf{s}},a}^n$ is the sum of the same multipoles located at point $R\exp(-a)\widehat{\mathbf{s}}$ with axes along the direction defined by the unit vector $\widehat{\mathbf{s}}$.

The set $\left\{g_{\widehat{\mathbf{s}},a}^n\right\}$ is a continuous family of wavelets, where $\widehat{\mathbf{s}}$ defines the radial axis $\Delta(\widehat{\mathbf{s}})$ carrying the set of $(n+1)$ multipoles as well as the direction of the axes of the multipoles, and a refers to the location of the multipoles on Δ. In practice, the number of data being finite, the family $\left\{g_{\widehat{\mathbf{s}},a}^n\right\}$ must be discretized. This operation leads to the concept of a frame in the Hilbert space H of the harmonic functions belonging to $L^2(S(R,\infty))$. A frame is a generating system which linear combinations are dense in the Hilbert space, the elements of the frame being neither linearly independent nor orthogonal to each other. Holschneider et al. (2003) provided some qualitative evidences about the completeness of the frame by comparing the dimensions of wavelet and spherical harmonic subspaces and by computing misfits between spherical harmonics and their approximation by a finite series of discrete wavelets.

Discretizing the dilation factor a (the depth of the $(n+1)$ multipoles) is straightforward but requires the definition of a reference radius R. Various spheres of geophysical importance may be used, for instance core-mantle boundary, which may offer more flexibility in the distribution of the depths of the multipoles (see Chambodut et al., 2005). The discretizing of the directions $\Delta(\widehat{\mathbf{s}})$ is, however, a more heavy task, connected to the long standing difficulty of defining a quasi-uniform distributions of a finite number of points on the sphere (Holschneider et al., 2003; Chambodut et al., 2005).

The inverse problem formally consists in approximating a potential $V(r\widehat{\mathbf{r}})$ with a linear combination of a given finite subspace of a frame of discrete wavelets. This writes

$$W_V(\vec{r}) = \sum_{j=1}^{J} \sum_{k=1}^{K} \alpha_{j,k} g_{\widehat{\mathbf{s}}(j),a(k)}^n(\vec{r}), \quad (9.51)$$

where the discrete family $\left\{g_{\widehat{\mathbf{s}}(j),a(k)}^n\right\}$ has been indexed according to a pair of indexes (j,k) for the sake of clarity. Actually, for inversion purposes, a single indexing was used by Holschneider et al. (2003). W_V is the approximation of V in the subspace spanned by the wavelets $g_{\widehat{\mathbf{s}}(k),a(k)}^n$. There is a fundamental difficulty raised by the redundancy of the wavelet frame. Whereas the Gauss coefficients are theoretically unique, the coefficients α_k are not. Therefore, the inverse problem is by essence ill-conditioned and requires some regularization. Fortunately, the wavelets have convenient properties with respect to the inner product on $L^2(S(1))$ that allows the quadratic term involved in the smoothness constraint to be written in a concise form. The reader is referred to Holschneider et al. (2003), Chambodut et al. (2005) and Panet et al. (2006) for applications in geomagnetism and geodesy.

9.3.1.2 Multi-scale Modelling

Mayer and Maier (2006) proposed a modelling of CHAMP satellite measurements based upon vector scaling functions as an alternative to the Mie representation. However, we discuss the expression for the scalar potential only that was elaborated by Maier (2003, Chapter 4), and was applied to the crustal field modelling by Maier and Mayer (2003). More specifically, they proposed a multi-scale method for downward continuation of the crustal field estimated at satellite altitude. A less sophisticated and older approach based on scalar data may also be found in Achache et al. (1987). The method may apply to vector data but hereafter we restrict ourselves to the radial component modelling. The problem is the following: how from the given radial component B_r known over the surface of a sphere of radius r can we express B_r over a lower spherical surface R. The solution in terms of spherical harmonics is of course well-known (for instance Maus et al. 2007a), but we review it because, first, it is interesting to see which advantages could be drawn from the flexibility of the wavelet representation, second, it is the heart of many problems in geomagnetism, and thus regional modelling. We start again with the expansion (9.28) and we assume internal fields only. The radial component then simply writes

$$B_r(r\widehat{\mathbf{r}}) = -\partial_r V = \sum_{l=1}^{\infty}\sum_{m=-l}^{l} g_l^m (l+1)\left(\frac{R}{r}\right)^{l+2} \beta_l^m(\widehat{\mathbf{r}}). \tag{9.52}$$

As in the previous section, β_l^m is a real, normalized, spherical harmonic function and $r \geq R$. We remark that rB_r is itself an harmonic function in $S(R,\infty)$. In particular for $r = R$

$$RB_R(R\widehat{\mathbf{r}}) = R\sum_{l=1}^{\infty}\sum_{m=-l}^{l} g_l^m (l+1)\,\beta_l^m(\widehat{\mathbf{r}}). \tag{9.53}$$

where B_R stands for B_r calculated on the sphere $S(R)$. The coefficients $Rg_l^m(l+1)$ are obtained straightforwardly by

$$Rg_l^m(l+1) = \frac{1}{4\pi}\int_{S(1)} \beta_l^m(\widehat{\mathbf{r}})\, RB_R(R\widehat{\mathbf{r}})\, d\sigma. \tag{9.54}$$

Introducing the expression of $g_l^m(l+1)$ into Eq. (9.52), we obtain a relationship between $rB_r(r\widehat{\mathbf{r}})$ and $RB_R(R\widehat{\mathbf{s}})$ similar to that given by Eq. (9.37)

$$rB_r(r\widehat{\mathbf{r}}) = \frac{1}{4\pi}\int_{S(1)} K(r\widehat{\mathbf{r}},\widehat{\mathbf{s}})\, RB_R(R\widehat{\mathbf{s}})\, d\sigma(\widehat{\mathbf{s}}), \tag{9.55}$$

with $K(r\widehat{\mathbf{r}},\widehat{\mathbf{s}})$ being explicitly written in Eq. (9.38). $K(r\widehat{\mathbf{r}},\widehat{\mathbf{s}})$ is the kernel of an operator designated by Λ_{AP}, according to Maier (2003, p. 99), which links rB_r to RB_R. Formally

$$\Lambda_{AP}(RB_R) = rB_r. \tag{9.56}$$

RB_R (respectively rB_r) is an element of the Hilbert space $L^2(S(R))$ (respectively $L^2(S(r))$), the inner product on $L^2(S(\rho))$ ($\rho = R$ or r) being defined by Eq. (9.16). With respect to this inner product, the functions

$$Y_{\rho,l}^m(\rho\widehat{\mathbf{r}}) = \beta_l^m(\widehat{\mathbf{r}}), \tag{9.57}$$

are still orthonormal (note that we use explicit notations to designate elements belonging to each of the spaces $L^2(S(R))$ and $L^2(S(r))$). Λ_{AP} is an operator mapping the Hilbert space $L^2(S(R))$ onto the Hilbert space $L^2(S(r))$ and defines the upward continuation operation. It may be shown that its adjoint operator Λ_{AP}^* is given by

$$\Lambda_{AP}^* rB_r(r\widehat{\mathbf{r}}) = \frac{1}{4\pi}\int_{S(1)} K(r\widehat{\mathbf{r}},\widehat{\mathbf{s}})\, rB_r(r\widehat{\mathbf{s}})\, d\sigma(\widehat{\mathbf{s}}), \tag{9.58}$$

and that $\psi_{R,l}^m$ (respectively $\psi_{r,l}^m$) is an eigenvector of Λ_{AP} (respectively Λ_{AP}^*) associated to the eigenvalue

$$\sigma_l = \left(\frac{R}{r}\right)^{l+1}. \tag{9.59}$$

Thus, the limit of σ_l, when l tends toward infinity, is 0 and there is no theoretical difficulty in the calculation of rB_r knowing RB_R (Eq. 9.56). In general, we also face the problem of calculating RB_R knowing rB_r because small scales are geometrically more enhanced than larger scales with downward continuation. The difficulty is more explicit if we write rB_r in terms of spherical harmonics

$$rB_r = \sum_{l=1}^{\infty}\sum_{m=-l}^{l} q_l^m Y_{r,l}^m \text{ with } q_l^m = \langle rB_r, Y_{r,l}^m\rangle_{L^2(S(r))}. \tag{9.60}$$

Using Eqs. (9.56, 9.60) and the above-mentioned properties of $Y_{R,l}^m$ and $Y_{r,l}^m$ with respect to the operators Λ_{AP} and Λ_{AP}^* respectively, we obtain

$$\Lambda_{AP}^* \circ \Lambda_{AP}(RB_R) = \sum_{l=1}^{\infty}\sum_{m=-l}^{l} \sigma_l q_l^m Y_{R,l}^m. \tag{9.61}$$

On the other hand, we are looking for the expansion of RB_R of the form

$$RB_R = \sum_{l=1}^{\infty}\sum_{m=-l}^{l} p_l^m Y_{R,l}^m. \tag{9.62}$$

Applying the operator $\Lambda_{AP}^* \circ \Lambda_{AP}$ to this expansion, we obtain

$$\Lambda_{AP}^* \circ \Lambda_{AP}(RB_R) = \sum_{l=1}^{\infty}\sum_{m=-l}^{l} \sigma_l^2 p_l^m Y_{R,l}^m. \tag{9.63}$$

Comparing it to expression (9.61) and using $\Lambda_{AP}\left(Y_{R,l}^m\right) = \sigma_l Y_{R,l}^m$, we obtain finally

$$RB_R = \sum_{l=1}^{\infty}\sum_{m=-l}^{l} \sigma_l^{-1}\langle rB_r, Y_{r,l}^m\rangle_{L^2(S(r))} Y_{R,l}^m. \tag{9.64}$$

Equation (9.64) makes the generalized, Moore-Penrose, inverse of Λ_{AP} explicit. Hereafter, we denote Λ_{AP}^+ this generalized inverse (hence, formally, $RB_R = \Lambda_{AP}^+(rB_r)$). Due to the behavior of σ_l^{-1}, the convergence of the double series is by no means ensured and some regularization method has to be invoked. It is precisely at this point that the multi-scaling approach can be involved. We assume for a while that Eq. (9.64) makes sense, and we split this expression into two successive operations following Freeden et al. (1999)

$$A^D(R\hat{\mathbf{r}}) = \sum_{l=1}^{\infty}\sum_{m=-l}^{l} \sigma_l^{-1/2}\langle rB_r, Y_{r,l}^m\rangle_{L^2(S(r))} Y_{r,l}^m(\hat{\mathbf{r}}), \tag{9.65}$$

and

$$A^R(R\hat{\mathbf{r}}) = \sum_{l=1}^{\infty}\sum_{m=-l}^{l} \sigma_l^{-1/2}\langle A^D, Y_{R,l}^m\rangle_{L^2(S(R))} Y_{R,l}^m(\hat{\mathbf{r}}). \tag{9.66}$$

It may be shown that $A^R(R\hat{\mathbf{r}}) = RB_R(R\hat{\mathbf{r}})$, at least formally. Now, Eq. (9.65) may be written as a mapping from $L^2(S(r))$ onto $L^2(S(R))$, which gives $A^D(R\hat{\mathbf{r}})$ knowing $rB_r(r\hat{\mathbf{s}})$. Likewise, Eq. (9.66) is an internal mapping on $L^2(S(R))$. Each of these mappings is expressed through an integral equation using a kernel Φ

$$A^D(R\hat{\mathbf{r}}) = \frac{1}{4\pi}\int_{S(1)} \Phi^D(\hat{\mathbf{r}},\hat{\mathbf{s}})\, rB_r(r\hat{\mathbf{s}})\, d\sigma(\hat{\mathbf{s}}), \tag{9.67}$$

with

$$\begin{aligned}\Phi^D(\hat{\mathbf{r}},\hat{\mathbf{s}}) &= \sum_{l=1}^{\infty}\sum_{m=-l}^{l} \sigma_l^{-1/2} Y_{R,l}^m(\hat{\mathbf{r}})\, Y_{r,l}^m(\hat{\mathbf{s}})\\ &= \sum_{l=1}^{\infty} \sigma_l^{-1/2}(2l+1)P_l(\hat{\mathbf{r}}\cdot\hat{\mathbf{s}}),\end{aligned} \tag{9.68}$$

and

$$A^R(R\hat{\mathbf{r}}) = RB_R(R\hat{\mathbf{r}}) = \frac{1}{4\pi}\int_{S(1)} \Phi^R(\hat{\mathbf{r}},\hat{\mathbf{s}})\, A^D(R\hat{\mathbf{s}})\, d\sigma(\hat{\mathbf{s}}), \tag{9.69}$$

with

$$\begin{aligned}\Phi^R(\hat{\mathbf{r}},\hat{\mathbf{s}}) &= \sum_{l=1}^{\infty}\sum_{m=-l}^{l} \sigma_l^{-1/2} Y_{R,l}^m(\hat{\mathbf{r}})\, Y_{R,l}^m(\hat{\mathbf{s}})\\ &= \sum_{l=1}^{\infty} \sigma_l^{-1/2}(2l+1)P_l(\hat{\mathbf{r}}\cdot\hat{\mathbf{s}}).\end{aligned} \tag{9.70}$$

On the right-hand sides of Eqs. (9.68) and (9.70), we have applied the addition theorem (Eq. 9.36) and we recognize, again, the expressions in terms of Legendre polynomials. Of course, $\Phi = \Phi^D = \Phi^R$ but their expressions are formally different for the sake of clarity, Φ^D being the kernel of an operator mapping $L^2(S(r))$ onto $L^2(S(R))$ and Φ^R being the kernel of an operator on $L^2(S(R))$. The right-hand sides of Eqs. (9.68) and (9.70) show that the series do not converge. In order to remedy this drawback,

Freeden et al. (1999) replaced the problematic coefficients $\sigma_l^{-1/2}$ by a family of coefficients $\{\gamma_j(l)\}$ called filters, j, being a positive or negative integer. The kernel $\Phi_j^D(\widehat{\mathbf{r}},\widehat{\mathbf{s}})$, for which $\sigma_l^{-1/2}$ is replaced by $\gamma_j(l)$, is called regularization decomposition kernel whereas $\Phi_j^R(\widehat{\mathbf{r}},\widehat{\mathbf{s}})$ is called regularization reconstruction kernel. We define $A_j^D(R\widehat{\mathbf{r}})$ and $A_j^R(R\widehat{\mathbf{r}})$ the functions obtained in Eqs. (9.65) and (9.66), with $\sigma_l^{-1/2}$ replaced by $\gamma_j(l)$. These functions are smoothed, and approximate, versions of the exact solutions A^D and A^R. If the families $\{\gamma_j(l)\}$ verify appropriate constraints (see Freeden et al., 1999, for the details and for some relevant functions $l \to \gamma_j(l)$) it may be shown that $\lim_{j\to\infty} \left\| A_j^R(R\widehat{\mathbf{r}}) - \Lambda_{AP}^{+}(rB_r) \right\|_{L^2(S(R))} = 0$, which is obviously a desired property of the regularization. The regularized solution finally writes

$$A_j^R(R\widehat{\mathbf{r}}) = P_j(rB_r) = \frac{1}{16\pi^2} \int_{S(1)} \int_{S(1)} \Phi_j^R(\widehat{\mathbf{r}},\widehat{\mathbf{s}})\, \Phi_j^D(\widehat{\mathbf{s}},\widehat{\mathbf{t}})\, rB_r(r\widehat{\mathbf{t}})\, d\sigma(\widehat{\mathbf{t}})\, d\sigma(\widehat{\mathbf{s}}). \tag{9.71}$$

P_j being defined by the right-hand side and being an approximation of Λ_{AP}^{+}. The functions rB_r that are upward continuations onto the sphere $S(r)$ of radial components known on the sphere $S(R)$, belong to the range $Image(\Lambda_{AP}) \subset L^2(S(r))$ of Λ_{AP}. This implies that $A_j^R = P_j(rB_r)$ belongs to the subspace $V_j = \{P_j(f) \mid f \in Image(\Lambda_{AP})\}$. It may be shown that $V_j \subset V_{j'}$ when $j < j'$ and that the closure of $\lim_{j\to\infty} V_j = L^2(S(R))$. Thus the solution of the generalized inverse problem may be approximated to arbitrary accuracy (in the sense of $\|\cdot\|_{L^2(S(R))}$) by increasing the scaling index j. However, every approximation $A_j^R(R\widehat{\mathbf{r}})$ has to be computed by means of a numerical surface integration. Freeden et al. (1999) suggest a possibly more efficient way. The decomposition $\Psi_j^D(\widehat{\mathbf{r}},\widehat{\mathbf{s}})$ and reconstruction $\Psi_j^R(\widehat{\mathbf{r}},\widehat{\mathbf{s}})$ wavelets take the same expressions as the corresponding decomposition and reconstruction kernels when the family of coefficients $\{\gamma_j(l)\}$ is replaced by the family $\{\varphi_j(l)\}$

$$\varphi_j(l) = \left[(\gamma_{j+1}(l))^2 - (\gamma_j(l))^2 \right]^{1/2}. \tag{9.72}$$

Using $\Psi_j^D(\widehat{\mathbf{r}},\widehat{\mathbf{s}})$ and $\Psi_j^R(\widehat{\mathbf{r}},\widehat{\mathbf{s}})$, Freeden et al. (1999) define a new operator R_j

$$R_j(rB_r) = \frac{1}{16\pi^2} \int_{S(1)} \int_{S(1)} \Psi_j^R(\widehat{\mathbf{r}},\widehat{\mathbf{s}})\, \Psi_j^D(\widehat{\mathbf{s}},\widehat{\mathbf{t}})\, rB_r(r\widehat{\mathbf{t}})\, d\sigma(\widehat{\mathbf{t}})\, d\sigma(\widehat{\mathbf{s}}), \tag{9.73}$$

and subspaces $W_j = \{R_j(f) \mid f \in Image(\Lambda_{AP})\}$. It may be shown that $P_J(rB_r) = P_0(rB_r) + \sum_{j=0}^{J-1} R_j(rB_r)$ and that $V_J = V_0 \oplus \sum_{j=0}^{J} W_j$ where the symbol $\oplus$ stands for the direct sum of the subspaces W_j. Thus, V_0 and $\{W_j\}_{j=0,\cdots J}$ are a partition of the approximation subspace V_J. Using this wavelet approach, the approximation gained at step $j+1$ is directly obtained by upgrading it from step j thanks to Eq. (9.73). However, as noticed by Maier (2003) there are some practical difficulties in the implementation. First, this method assumes data located on the sphere $S(r)$, thus neglecting altitude variations. Second, if the crustal field modelling is the target, an appropriate low-frequency global model has to be subtracted from the selected (and already processed) data. Third, since surface integrations have to be performed (Eq. 9.71 and 9.73), it is necessary to resample scattered data onto the nodal points of an appropriate grid and use integration algorithms (see Lesur and Gubbins, 1999, for a review). The multi-scale resolution was applied by Maier (2003, Chapter 4) on two spherical caps, one enclosing the Bangui anomaly and one enclosing the European continent. Due to the limited areas, Gibbs effects appeared on the boundaries that could be hidden using caps larger than the integration domain, themselves larger than the visualization caps. As we shall see in Section (9.4), this is reminiscent of a numerical 'trick' often employed in regional modelling that help artificially improving the convergence of the numerical solution by in fact implicitly imposing homogeneous conditions near the boundaries. The field is free to take any value and shape in regions with no data. This will, in turn, improve the fit in regions where data are available.

9.3.2 Localized Harmonic Functions

The localized functions proposed by Lesur (2006) are similar to the discretized Poisson wavelets described in Section (9.3.1.1) in the sense that they are linear combinations of zonal solid spherical harmonics. However, they do not conform to the wavelet concept (described so far) because they are not constructed using a dilated mother wavelet and are band-limited. From this last viewpoint, the localized functions are closer to the Slepian functions (Section 9.4.5). The unit vectors $\widehat{\mathbf{s}}_k$ which defines the symmetry axis of the zonal spherical harmonics $P_l\left(\widehat{\mathbf{s}}_k \cdot \widehat{\mathbf{r}}\right)$ are distributed on a grid according to the following scheme

$$\begin{aligned} \theta_i &= \arccos\left(u_i\right) \;\; i = 1, \ldots, (L+1)\,; \\ \varphi_j &= \frac{2\pi j}{2L+1} \;\; j = 1, \ldots, (2L+1)\,, \end{aligned} \tag{9.74}$$

where u_i is the ith zero of the Legendre polynomial P_{L+1}. This distribution addresses the issue of computing spherical integrals using quadrature methods. The grid defined by Eq. (9.74) is often referred to as a Gauss-Neumann grid (see Sneeuw, 1994 and references therein). Accordingly, hereafter, we will use the double index (i, j) instead of the single one k, although it would not be difficult to map the pair (i, j) to a single index. The Gauss coefficients of an expansion $V\left(R\widehat{\mathbf{r}}\right) = R \sum_{l=1}^{L} \sum_{m=-l}^{l} g_l^m \beta_l^m\left(\widehat{\mathbf{r}}\right)$, assuming that the β_l^m are real and Schmidt quasi-normalized, are given by the classical integral

$$g_l^m = \frac{2l+1}{4\pi R} \int_{S(1)} V\left(\widehat{\mathbf{r}}\right) \beta_l^m\left(\widehat{\mathbf{r}}\right) d\sigma. \tag{9.75}$$

Using the grid with the associated weight

$$w_i^{L+1} = \frac{2}{1-u_i^2} \partial_u \left(P_{L+1}\left(u_i\right)\right)^{-2} \;\; i = 1, \ldots, (L+1)\,, \tag{9.76}$$

the integral may be approximated to high accuracy, by the finite sum (Lesur, 2006)

$$g_l^m = \frac{2l+1}{2\left(2L+1\right)R} \sum_{i=1}^{L+1} w_i^{L+1} \sum_{j=1}^{2L+1} V\left(R\widehat{\mathbf{r}}_{ij}\right) \beta_l^m\left(\widehat{\mathbf{r}}_{ij}\right). \tag{9.77}$$

Now, the localized functions write

$$F_{ij}^L\left(r\widehat{\mathbf{r}}\right) = R \sum_{l=1}^{L} \sum_{m=-l}^{l} \left(\frac{R}{r}\right)^{l+1} f_l \beta_l^m\left(\widehat{\mathbf{s}}_{ij}\right) \beta_l^m\left(\widehat{\mathbf{r}}\right). \tag{9.78}$$

where the coefficient f_l is a tuning factor allowing to tighten more or less the functions $f_l \beta_l^m\left(\widehat{\mathbf{s}}_{ij}\right) \beta_l^m\left(\widehat{\mathbf{r}}\right)$ around the point $\widehat{\mathbf{s}}_{ij}$. Note that as before the functions $F_{ij}^L\left(r\widehat{\mathbf{r}}\right)$ could be again expressed in terms of Legendre polynomials $\left(\frac{R}{r}\right)^{l+1} P_l\left(\widehat{\mathbf{s}} \cdot \widehat{\mathbf{r}}\right)$ using the addition theorem. It may be shown that, like the basis functions $\left(\frac{R}{r}\right)^{l+1} \beta_l^m$, they span the space H_L of the harmonic functions in the domain $S(R, \infty)$, of maximum degree L. The dimension of H_L being $L\,(L+2)$, they are not linearly independent (this property is similar to the concept of frame in Section 9.3.1.1) and they do not necessarily form a basis. Thus, the coefficients γ_{ij} of the expansion

$$V\left(r\widehat{\mathbf{r}}\right) = \sum_{i=1}^{L+1} \sum_{j=1}^{2L+1} \gamma_{ij} F_{ij}^L\left(r\widehat{\mathbf{r}}\right) \tag{9.79}$$

are not unique. However, thanks to the orthogonality properties of the spherical harmonics β_l^m with respect to the quadrature expressed by Eq. (9.77), Lesur (2006) gave an elegant expression of the γ_{ij} in terms of the Gauss coefficients. In the framework of the inverse problem, Lesur (2006) discussed the choice of f_l in connection with the weight functions $w_L(\theta)$ and with the decrease rate of the gradient away from $\widehat{\mathbf{s}}_{ij}$. The inverse problem amounts to find the coefficients γ_{ij} which parameterize the model

$$\vec{B} = -\vec{\nabla} \left\{ \sum_{i=1}^{L+1} \sum_{j=1}^{2L+1} \gamma_{ij} F_{ij}^L\left(r\widehat{\mathbf{r}}\right) \right\}. \tag{9.80}$$

Due to the non-uniqueness of this expansion, a smoothness constraint built *via* a damping matrix may be introduced. Localized harmonic functions are used in Lesur and Maus (2006) model globally the lithospheric field with reduced spatial resolution at high latitudes. According to Lesur and Maus (2006), this flexibility allowed reducing the spurious effects visible in the polar regions with model MF4 (Maus et al., 2006) and dealing with multi-level data.

Figure 9.2 shows a reconstruction of the crustal part of the synthetic Z_{all} data (see Fig. 9.1-right) using Eq. (9.80) with Eq. (9.78) up to $L = 400$ (about 100-km wavelength). We recall that the synthetic data are calculated to SH 720 ($L = 720$). This difference introduce some spatial aliasing in the modelling that may explain part of the observed tiny wiggles both in the

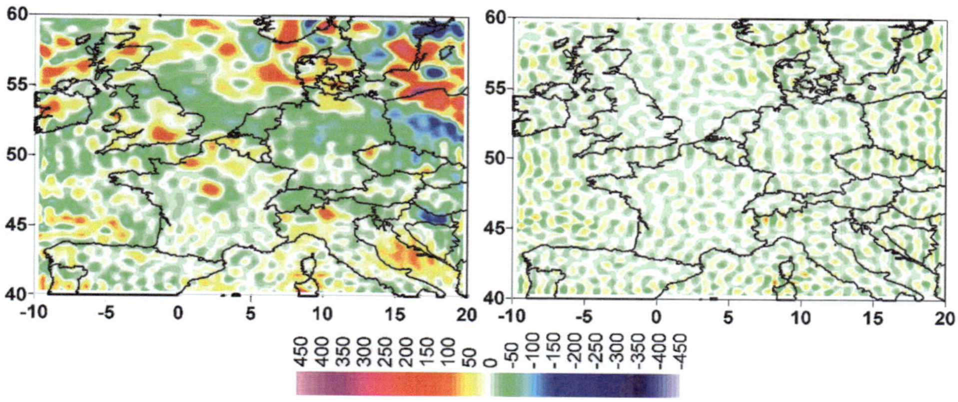

Fig. 9.2 Example of radial magnetic field reconstruction using the philosophy of band-limited functions (*left*) and the residuals (*right*) between this approximation and the original data shown in Fig. 9.1

model and the residual maps. However, the setting of the tuning factor f_l plays also a key role in the apparent stability of the modelling. Increasing the expansion of the series (9.78) and selecting a more appropriate f_l factor, for instance, would provide an almost perfect residual mean squares fit.

9.4 Modelling the Field Regionally

SH basis functions are neither well suited for modelling unevenly distributed data nor for crustal field modelling because their sensitivity at the global scale is in poor agreement with the local nature of the geological sources. We saw that only combinations of band limited and weighted SH harmonics could help circumventing this difficulty. Another philosophy, however, is to perform data fitting at a regional scale using functions with local support. Such an approach has a long history (e.g., Howarth, 2001) and we focus here on regional modelling methods based upon the resolution of the Laplace equation in a bounded domain Ω leading to Fourier-like expansions. Before proceeding further, it is important to keep in mind that the concept of internal and external field at regional scale is complicated by the existence of a lateral boundary, be it a square or a circle (or any other type of boundary). We thus assert without formally demonstrating it (but we will give some arguments below in Section 9.4.2.2) that regional basis functions are not able to distinguish between magnetic fields generated below or above the Earth's surface. Thus, if one wants to study and interpret the modelled magnetic field source, specific data pre-processing are required in order to remove unwanted contributions. We do not take much risk by further asserting that this difficulty arises also with global modelling techniques as long as they are used over a small portion of the Earth only, even though they are based on functions with global support. For this reason it is advisable to filter out the undesired magnetic field contributions, generally of external origin, before performing the regional modelling. At present, no comprehensive modelling of the magnetic field was undertaken. Some recent general reviews regarding other methods of local modelling, the availability of magnetic data at the regional scale, and applications may be found in several papers or books (Langel and Hinze, 1998; Mandea and Purucker, 2005; Purucker and Whaler, 2007 and Thébault et al. 2010, for instance).

9.4.1 *Review of Modelling in the Flat Earth Approximation*

Every method leading to a Fourier series expansion could be presented in a way similar to the SH formalism that is, via the resolution of a boundary value problem for the Laplace equation in the domain Ω using the method of variable separation. For some reason, this way of doing has not been systematically applied to the flat earth approximation methods as is outlined below. In these methods, the earth is locally approximated by its tangent plane and the domain of interest is built upon this plane (see Langel and

Hinze, 1998, p. 134 for a qualitative discussion about the validity of this approximation). The main advantage of this assumption is that the involved functions are much simpler to compute than in the spherical geometry.

9.4.1.1 Rectangular Harmonic Analysis

Rectangular harmonic analysis (RHA) refers to a local domain consisting in a rectangular box. Alldredge (1981, 1982, 1983) applied RHA to surface data whereas Nakagawa and Yukutake (1985) and Nakagawa et al. (1985) extended its use to the analysis of satellite data but at the expense of using an *had-hoc* weighting to minimize edge effects. Haines (1990) made a thorough analysis on RHA which led him to suggest basis functions provided by various boundary value problems.

Let us start however with the most frequently used expansion, written in terms of periodic functions. Following the notations of Langel and Hinze (1998, p. 132), the expression of the expansion S_V of a potential V can be conveniently expressed in complex form

$$S_V(x,y,z) = X_0 x + Y_0 y + Z_0 z + \sum_{k=-K}^{K} \sum_{l=-L}^{L} \chi_{kl} \exp \left[-2\pi i \left(\frac{kx}{L_X} + \frac{ly}{L_Y} \right) \right] \exp\left(D_{kl} z\right), \tag{9.81}$$

with

$$D_{kl} = 2\pi \left(\frac{k^2}{L_X^2} + \frac{l^2}{L_Y^2} \right)^{1/2}, \tag{9.82}$$

which, apart from the linear term, is valid in the unbounded domain $]0, L_X[$ x $]0, L_Y[$ x $]0, \infty[$ with the z axis oriented positively downwards (Fig. 9.3). The potential is essentially a L_X, L_Y periodic function in an horizontal plane, and vanishes when z tends towards minus infinity. The expansion is complemented with linear terms which are intended to reduce boundary effects (Note that Eq. 9.82 has been corrected for the error in Eq. 9.5) of Alldredge, 1981, as was underlined by Malin et al., 1996). Nakagawa and Yukutake (1985) and Nakagawa et al. (1985) worked on an area with square section ($L_X = L_Y$) and isotropic expansions ($K = L = 3$) whereas Alldredge used a domain with a rectangular section but restricted the sums in (9.81) by the relationship $k + l = N_{\max} + 1$. As the function to be modelled is not periodic at all, Gibbs effects are to be expected. They are all the more serious as the values at opposite boundaries are different. The linear terms, which are obviously harmonic, are intended to minimize the ringing effects and some authors (e.g., Nakagawa et al., 1985) further weighted the data in an area along the edges with a cosine taper function or even added some more terms solving Laplace equation (Malin et al., 1996).

Haines (1990) made a thorough analysis of RHA. To our knowledge, he was the first to spot the fundamental drawbacks of the original RHA expansion, the one based on periodic basis functions. Noting that these functions solve a particular boundary value problem, he suggested applying other boundary conditions that would be consistent with the properties of the function to be modelled. Haines focused his discussion on the uniform convergence properties of generalized Fourier expansions S_V. In the most general case of a regular Sturm-Liouville problem, the expansions are the solutions of the ordinary second-order differential equation on the interval $]a, b[$

$$-d_x\left(p(x) d_x f\right) + \left(q(x) - \lambda g(x)\right) = 0, \tag{9.83}$$

subject to the general mixed boundary conditions

$$\alpha_1 f(a) - \beta_1 \left(d_x f\right)_a = 0, \tag{9.84a}$$

$$\alpha_2 f(b) + \beta_2 \left(d_x f\right)_b = 0, \tag{9.84b}$$

where $p(x)$ is positive, continuously differentiable on $[a, b]$, $g(x)$ is positive and continuous, $q(x)$ is continuous. Note that $-(d_x f)_a$ and $(d_x f)_b$ are one-dimensional expressions of the normal derivative to the boundary of the domain. Setting $p(x) = g(x) = 1$ and $q(x) = 0$, and periodic boundary conditions on $f(x)$ and its derivative, we obtain the familiar expansion given by Eq. (9.81). The Dirichlet boundary value problem is defined by setting $\beta_1 = \beta_2 = 0$, whereas $\alpha_1 = \alpha_2 = 0$, define a boundary value problem of Neumann type. Note that the boundary value problem is incomplete as no condition is set neither on the lower nor the upper surface so that the solution with altitude is not a basis and thus does not necessarily agree with the

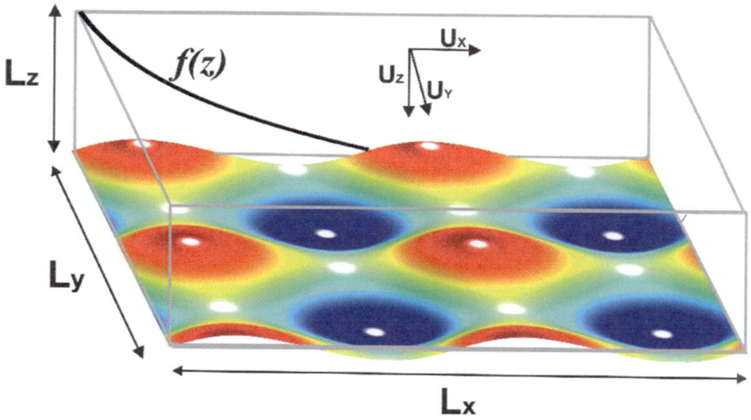

Fig. 9.3 Schematic representation of the domain of validity of the Rectangular Harmonic Analysis (RHA). The colour surface represents the rectangular harmonics for l = 3 and k = 3 (see Eq. 9.81). f(z) is the exponential radial field dependence with altitude

behavior of Newtonian potential fields with altitude. The discussions made by Haines (1990) concerning the choice of the most appropriate lateral boundary conditions rely on the Sturm-Liouville theorem (Gonzalez-Velasco, 1995, Section 4.4), which simplest expression is "*if f is continuous and satisfies the boundary conditions in Eqs. (9.84a), (9.84b) and f′ piecewise continuous on* $[a, b]$*, the generalized Fourier series* S_f *converges absolutely and uniformly towards* f *on* $[a, b]$".

In the absence of magnetic sources close to the boundary of the domain, the regularity conditions are fulfilled by the magnetic potential under consideration. However, the magnetic potential in general verifies neither Neumann nor Dirichlet nor mixed boundary conditions, which is particularly troublesome. In this case, the problem is no longer self-adjoint (i.e., the basis is not orthonormal) and becomes by far more difficult (Coddington, 1955, Chapter 12). The simplest way to overcome the difficulty, as advocated by Haines (1990), is to mix up basis functions of self-adjoint problems. This solution should preserve uniform convergence but at the expense of introducing non-orthogonal basis functions. Another way of circumventing the difficulty is to deal with potentials having wavelengths shorter than the dimension of the domain, thus reduced values on the boundary, closer to Dirichlet or Neumann conditions. These arguments should be kept in mind as they are particularly important to understand some of the properties of Spherical Cap Harmonic Analysis discussed in Section (9.4.2.2).

Figure 9.4 illustrates this previous discussion. The model is obtained by inverting the synthetic data Z_{all} (see Section 9.2.3) with the basis functions defined by $Z = -\partial_z S_V(x, y, z)$ using Eq. (9.81); thus without setting specific boundary conditions. The maximum series expansion defines a minimum wavelengths of about 100 km. As can be verified, the RHA does quite well in modelling single surface data and is able to represent both large (core) and small (crustal) wavelengths up to the required resolution. However, the residual map exhibits long oscillation, spreading from the edges to the center of the rectangle, that is symptomatic of Gibbs effect. The slight curvature in these large residuals also show the consequence of the flat Earth approximation. Whether or not the shape and magnitude of residuals are significant is a matter of judgement left to the reader as, in practice, it depends on the purpose for which the model is derived. The non-orthogonality of the basis functions is another property that forbids us to carry out spectral analyses and restricts ourselves to relatively low series expansion since expanding the series further keeps degrading the conditioning of the inverse matrix. Note that such a spectrum would anyway be difficult to interpret because of spatial aliasing unless some detrending is carried out *prior* to the inversion. At last, introducing data measured at different altitudes does not provide satisfactory solutions because the radial functions are

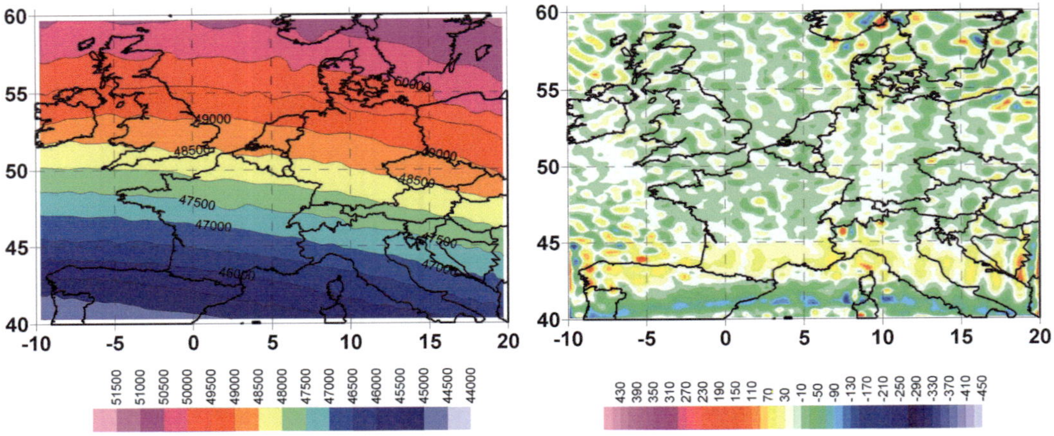

Fig. 9.4 Example of Rectangular Harmonic Analysis using the Z vector component only. The obtained RHA model (*left*) and the residuals between the model and data (*right*) illustrate some of the properties discussed in the text. Units are in nT

not designed for it. Setting appropriate boundary conditions on each surface of the whole domain (including upper and lower surfaces) would likely alleviate part of these practical difficulties.

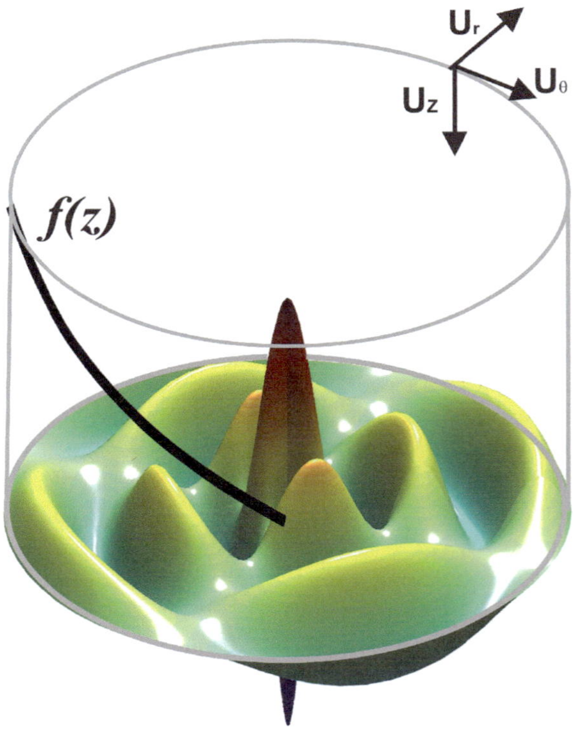

Fig. 9.5 Schematic representation of the domain of validity of the Cylindrical Harmonic Analysis (CHA). The colour surface represents the cylindrical harmonics for $m = 2$ and $k = 2$ (see Eq. 9.92). f(z) represents the exponential radial field dependence with altitude

9.4.1.2 Cylindrical Harmonic Analysis

Alldredge (1982) also studied the solutions of the Laplace equation in a circular cylindrical region. The vertical axis of the area is its axis of symmetry and its lateral boundary a cylinder of radius ρ (Fig. 9.5). In cylindrical coordinates (r, θ, z), the Laplace equation writes

$$\frac{1}{r}\partial_r\left(r\partial_r V\right) + \frac{1}{r^2}\partial^2_{\theta^2} V + \partial^2_{z^2} V = 0. \tag{9.85}$$

The method of variable separation, with $V(r,\theta,z) = R(r)T(\theta)Z(z)$ leads to the following set of ordinary differential equations

$$d^2_{z^2} Z = \mu^2 Z, \tag{9.86a}$$

$$r^2 d^2_{r^2} R + r d_r R + \left(\mu^2 r^2 - \lambda^2\right) R = 0, \tag{9.86b}$$

$$d^2_{\theta^2} T = -\lambda^2 T, \tag{9.86c}$$

where μ^2 and λ^2 are *a priori* complex constants. Equation (9.86c) associated to 2π-periodic conditions for the function and its first derivative, leads to the condition $\lambda = m$, positive integer and to the familiar solution

$$T(\theta) = A_m \mathrm{e}^{im\theta} + B_m \mathrm{e}^{-im\theta}. \tag{9.87}$$

With the change of variable $\mu r = s$, and the change of function $S(s) = R(s/\mu)$, Eq. (9.86b) is reshaped into the Bessel differential equation

$$s^2 d_{s^2}^2 S + s d_s S + \left(s^2 - m^2\right) S = 0, \tag{9.88}$$

with s also *a priori* a complex variable. The appropriate form of the solution and the values taken by μ are found when Eq. (9.88) is associated with boundary conditions at $s = 0$ and $s = \mu\rho$. This leads to a singular Sturm-Liouville problem because the coefficient of the second derivative vanishes at $s = 0$. Eq. (9.89) has two linearly independent solutions, the Bessel function of the first kind and integer order $J_m(s)$ and the Neumann function $N_m(s)$. However, the Neumann functions have to be discarded because they tend towards infinity when s tends towards 0, thus

$$R(r) = J_m(\mu r). \tag{9.89}$$

The factor μ can be specified by setting the boundary condition at $s = \mu\rho$. This writes

$$\alpha J_m(\mu\rho) + \beta J_m^{'}(\mu\rho) = 0. \tag{9.90}$$

The zeros of J_m and $J_m^{'}$ are real (Abramowitz and Stegun, 1965, section 9.5). In addition, if $\alpha = 0$ or $\beta = 0$, μ must be real but it may be shown that this remains true (in the general case) for any real value of α and β. Therefore, the variable s is real and μ is a root of the function $\alpha J_m(\mu\rho) + \beta J_m^{'}(\mu\rho)$. There are infinitely many values of μ which verify Eq. (9.90). They build up a countable subset of $\mathbb{R}$, depending on m and which can therefore be indexed by the pair (m, k) with $k \in \mathbb{N}^*$. Formally, the solution should write

$$\begin{aligned} V(r,\theta,z) =& \sum_{m=0}^{M}\sum_{k=0}^{K} J_m(\mu_{mk} r)\,(D_{mk}\cos m\theta \\ &+ E_{mk}\sin m\theta)\exp(\mu_{mk} z), \end{aligned} \tag{9.91}$$

and be a complete basis. In spite of these considerations, (Alldredge, 1982) adopted another form, with no definite boundary condition on the boundary $r = \rho$

$$\begin{aligned} V(r,\theta,z) =& Az + \sum_{m=0}^{M}\sum_{k=0}^{K} J_m(k\nu r)\,(D_{mk}\cos m\theta \\ &+ E_{mk}\sin m\theta)\exp(kz), \end{aligned} \tag{9.92}$$

with ν a scaling factor that is tuned manually and empirically by trials and errors.

Equation (9.92) is valid inside the cylinder, half-infinite towards negative z, apart from the linear term. Indices m are integers, as expected. In the formalism of Alldredge (1982), the choice of μ $(=k\nu)$ is not based upon boundary condition but on scale considerations. This raises some important practical difficulties illustrated by Fig. 9.6 that shows the CHA model obtained from the set of synthetic data Z_{all} using expression (Eq. 9.92) for the potential. After several tries, we could find a scaling parameter ν that allowed an apparent satisfying fit of the large scales of the magnetic field; there are certainly an infinite number of ν that would give comparable result. However, the same value of ν cannot represent both large and small scales and all crustal field contributions end up in the residual map. For some applications related to regional main field modelling, this low-pass property appears interesting as it seems to filter out crustal field contamination. This result is however misleading because the manual choice of ν act as a filter that has no real significance. By no means can we assert that the main field has been correctly represented because the set of functions do not form a complete basis; the residuals illustrate this incompleteness not a resolution problem imposed by the series truncation. The functions being not orthogonal, spectral analysis are not permitted and introducing multi-altitude data would have introduced other difficulties. As it stands, the CHA modelling is flawed. The mathematics would be correct after setting boundary conditions, at least on the lateral surface. They would define not one value of ν in Eq. (9.92) but a discrete set of μ_{mk} (Eq. 9.91) varying in m and k thus defining a complete basis function allowing the representation of any contribution of magnetic field (core and crustal) in the horizontal plane (dealing with multilevel data would require boundary conditions on the lower and upper surfaces).

9.4.2 SCHA and R-SCHA

Spherical Cap Harmonic Analysis and Revised Spherical Cap Harmonic Analysis are, in regional modelling, the closest relatives to SHA. SCHA was designed by Haines (1985a) to provide a reference field for Canada (Haines, 1985b). Since then, SCHA

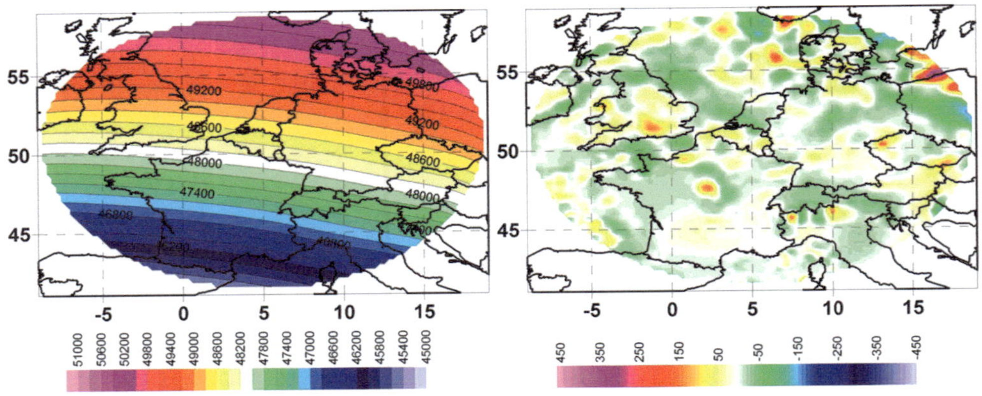

Fig. 9.6 Example of Cylindrical Harmonic Analysis using the Z vector component only. The obtained CHA model (*left*) and the residuals between the model and data (*right*) illustrate some of the properties discussed in the text. Units are in nT

has been widely used in a variety of regional models, including reference field models, secular variation, crustal field, external field, and even outside geomagnetism making it probably the most popular regional modelling method (see Torta et al., 2006, for a review).

9.4.2.1 Definition of the Domain

The domain of interest is shown on Fig. 9.7. It is the bounded volume Ω delimited by the intersection of a spherical shell $S(b, c)$ defined in Section (9.2), with a circular cone having its summit at the center of the Earth and aperture angle θ_0. The location of the cone axis and the half-angle θ_0 on the Earth depend of course on the area of interest. Generally, the radius of the inner sphere is the earth's mean radius (i.e., $b = R_E$ according to previous notations. We now set $R_E = a$ to avoid confusion with the radial function). The closed boundary $\partial\Omega$ of Ω consists in three pieces of geometrically simple boundaries: $\partial_{\theta_0}\Omega$ denotes the lateral portion of the cone $\theta = \theta_0$, $\partial_a\Omega$ and $\partial_c\Omega$ stand for the lower and upper cap at radii a and c respectively. Thus, the boundary $\partial\Omega = \partial_{\theta_0}\Omega \cup \partial_a\Omega \cup \partial_c\Omega$ is substantially more complicated than the boundary of the spherical shell.

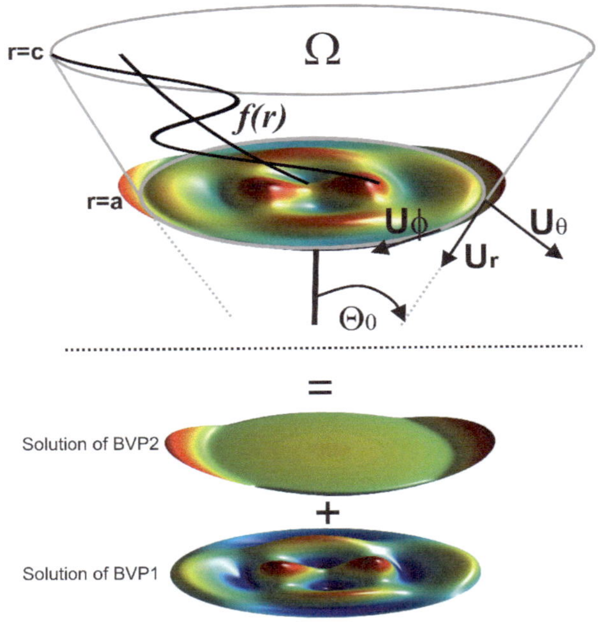

Fig. 9.7 Schematic representation of the spherical cone considered when solving a R-SCHA problem. The upper color surface (at $r = a$) represents the superimposition of the two independent solutions found when splitting the original BVP (Eq. 9.99) into two independent BVP's (Eq. 9.100 and 9.101). $f(r)$ represents the respective radial solution of BVP1 and BVP2 (see text for details)

9.4.2.2 Resolution of Laplace Equation in SCHA by the Fourier Decomposition Method

The resolution follows closely the pattern of Section (9.2.1). The only difference resides in the boundary conditions on Eq. (9.9). In SCHA, they are

$$P \text{ and } P' \text{ finite at } \theta = 0, \tag{9.93a}$$

$$\alpha P(\theta_0) + \beta P'(\theta_0) = 0. \tag{9.93b}$$

Instead of the mixed boundary condition defined in Eq. (9.93b), Haines (1985a) uses two separate

Dirichlet (with $\beta = 0$) and Neumann conditions (with $\alpha = 0$) and adds both sets of solutions. As already said above this non-orthodox procedure is applied with the hope to define a series expansion that convergences uniformly towards the solution. Here, for discussion purposes, we call SCHA the well-known solution of the Sturm-Liouville problem defined by Eq. (9.8) together with boundary conditions (9.93a) and (9.93b). We solve

$$\nabla^2 V = 0 \text{ on } \Omega, \tag{9.94a}$$

$$\alpha\,(V)_{\partial\theta_0} + \beta\left(\frac{\partial V}{\partial n}\right)_{\partial\theta_0} = 0. \tag{9.94b}$$

Note that in SCHA the Boundary Value Problem (BVP) is again incomplete as no condition is put on the boundary $\partial_a\Omega \cup \partial_c\Omega$. As was the case for RHA and CHA, there is no sufficient constraint on the radial function to ascertain that the solution will behave correctly with altitude. This, in turn, prevents us from dealing with multi-level data.

The solutions of Eq. (9.8), where the constant ν is *a priori* arbitrary, are the generalized Legendre functions (Hobson, 1965, Chapter V; Robin, 1958, Vol. II) of first (P_ν^m) and second (Q_ν^m) kind. As in SH, the condition expressed by (93a) excludes the second kind. The function P_ν^m is the eigenvector of the operator defined in Eq. (9.15). Thanks to the boundary conditions (9.93b), this operator is self-adjoint on the space $D = \{P \in L^2\,]u_0, 1[\, \cap C^2\,[u_0, 1],\ P$ fulfilling the boundary condition Eq. (9.93b)$\}$, where $u_0 = \cos\theta_0$. Therefore, the eigenvalues $\nu\,(\nu+1)$ are real. If, in addition, the constants α and β have the same sign (a condition obviously fulfilled with Dirichlet or Neumann conditions), the operator in Eq. (9.13) is positive (Reddy, 1998, section 6.5), which in turn implies that the eigenvalues $\nu\,(\nu+1)$ are real positive. The detailed resolution of the hypergeometric equation defined by Eq. (9.13) shows that $\nu > m$ if the boundary condition (93b) is also to be fulfilled. There is no loss of generality if we take ν real positive or null (with Neumann condition) since $P_\nu^m = P_{-\nu-1}^m$ (Robin, 1958, Vol. II, p.52). As in SH, the functions

$$\beta_k^m(\theta,\varphi) = P_{\nu(k,m)}^m(\cos\theta)e^{im\varphi}\ m \in \mathbb{Z}, \tag{9.95}$$

are eigenfunctions of $-\nabla_S^2$ associated to the eigenvalues $\nu\,(\nu+1)$ (in order to avoid unnecessary complications, we will hereafter discard the complex form because P_ν^m and P_ν^{-m} are connected to each other by a factor involving the gamma function — see Robin, 1958, Vol. II, p. 58). For simplicity we again keep the notation appropriate to the complex form. The constant function $\beta_0^0(\theta,\varphi)$ may or not be included into the set of basis functions, depending on the values taken by the coefficients α and β: if $\alpha = 0$ (Neumann boundary condition), $\beta_0^0(\theta,\varphi)$ fulfills the boundary condition and is therefore acceptable, but has to be discarded in all other cases.

The values of ν are the roots of the function $\alpha P_\nu^m(\theta_0) + \beta P_\nu'^m(\theta_0)$ and depend on m. The integer k indexes these roots for fixed m. Haines (1985a) showed how to compute them in the case of the Dirichlet or Neumann boundary conditions. The method applies likewise to the mixed boundary condition with some more numerical complexity. The general expressions of the basis functions are

$$\psi_{i,k}^m(r,\theta,\varphi) = a\left(\frac{a}{r}\right)^{\nu(m,k)+1}\beta_k^m, \tag{9.96a}$$

$$\psi_{e,k}^m(r,\theta,\varphi) = a\left(\frac{r}{a}\right)^{\nu(m,k)}\beta_k^m, \tag{9.96b}$$

and the potential simply writes

$$V(r,\theta,\varphi) = \sum_{k=0}^{\infty}\sum_{m=0}^{\infty} G_k^{i,m}\psi_{i,k}^m(r,\theta,\varphi) + G_k^{e,m}\psi_{e,k}^m(r,\theta,\varphi). \tag{9.97}$$

Such expressions look very similar to SH expansion (Eq. 9.18), which is misguiding. The degrees ν form a discrete set of real values depending on the order m. Therefore, the interpretation of the subscripts i and e in terms of truly inner and external field sources with respect to the sphere $S(a)$ is not as straightforward as in SH.

Despite its popularity and its apparently close relationship with SH, it was noticed by several authors (e.g., De Santis and Falcone, 1995) that it was difficult to model correctly the radial dependency, particularly when considering cones of small aperture (De Santis, 1991). This led some authors (Torta et al., 2006 for a review) to artificially increase the size of the cap (we understand now that this is done empirically to enforce the data to agree with the Neumann and Dirichlet conditions on the lateral surface). More intriguing, it was doubted that SCHA could simultaneously solve for the horizontal and radial components with comparable accuracy (Langel and Hinze, 1998, p. 132; Thébault and Gaya-Piqué, 2008). In fact, troubles arise because the incomplete setting of the boundary value problem leads to an incomplete set of basis functions with respect to the relevant function space defined on Ω. Mathematically, it is sufficient to demonstrate the lack of completeness of SCHA by finding one single counter-example. The following Dirichlet problem

$$\nabla^2(V) = 0 \text{ on } \Omega \tag{9.98a}$$

$$V = f \text{ on } \partial_{\theta_0}\Omega \tag{9.98b}$$

$$V = 0 \text{ on } \partial_a\Omega \cup \partial_b\Omega, \tag{9.98c}$$

for instance, would have a null SCHA expansion on the spherical cap; the true solution being obviously not the null function.

Figure 9.8 illustrates one peculiarity of SCHA. We apply the original formalism of Haines (1985a) who, once more, introduces both bases derived from the Neumann and the Dirichlet boundary value problem. We use only the core field part of Z_{all} so that the data do not contain crustal field contributions. This helps us to illustrate the major deficiency of SCHA. The data are represented to 380 km wavelength only because reaching the 100 km spatial resolution was not possible (this resolution is reached in Fig. 9.2 and 9.4). The contradiction between the horizontal and vertical component, as well as non-orthogonality between basis functions, grew up so much that SCHA increasingly failed in representing the field and become more and more unstable. Since the data are equally distributed and dense, regularization based on minimum norm solution is helpless. It suggests that SCHA does not converge towards SH as it should do in the case of an infinite expansion (see also the above discussion about the lack of completeness of SCHA). Such problems are less prominent in many situations, when considering residual fields (even though they are proportional to the strength of the magnetic field, the model error may be of the order of the data noise and thus, discarded) or when considering very large caps.This latter case is better understood by noting that $\lim_{\theta_0 \to \pi/2} \nu(m,k) = n$, where n is an integer degree of SH; thus SCHA becomes an even closer relative to SH for large caps.

9.4.2.3 R-SCHA as a Boundary Value Problem

The Revised SCHA (R-SCHA) is a proposal that should remedy the drawback of SCHA (e.g., Thébault et al., 2004). We give here a general form of the complete boundary value problem. A general BVP, adapted to the domain Ω described in Section 9.4.2.1 and to the Laplace equation, would write (Reddy, 1998, section 8.3)

$$\nabla^2 V = 0 \text{ on } \Omega \tag{9.99a}$$

$$\alpha V + \beta \frac{\partial V}{\partial n} = G \text{ on } \partial\Omega = \partial_{\theta_0}\Omega \cup \partial_a\Omega \cup \partial_c\Omega, \tag{9.99b}$$

where α, β, G are given functions on $\partial\Omega$ (see Fig. 9.7). A first limitation arises if the BVP is to be resolved with the method of variable separation that requires α and β being constant on each piece $\partial_a\Omega$, $\partial_c\Omega$, $\partial_{\theta_0}\Omega$ of the closed surface $\partial\Omega$ but allows however these constants to be different on each piece. The method of variable separation requires in addition to split up the initial BVP into two simpler, partially homogeneous, independent BVP problems

$$\nabla^2 V_1 = 0 \text{ on } \Omega \tag{9.100a}$$

$$\frac{\alpha_{\theta_0}}{r} V_1 + \beta_{\theta_0} \frac{\partial V_1}{\partial n} = 0 \text{ on } \partial_{\theta_0}\Omega \tag{9.100b}$$

$$\alpha_a V_1 + \beta_a \frac{\partial V_1}{\partial n} = G_a \text{ on } \partial_a\Omega \tag{9.100c}$$

$$\alpha_c V_1 + \beta_c \frac{\partial V_1}{\partial n} = G_c \text{ on } \partial_c\Omega, \tag{9.100d}$$

$$\nabla^2 V_2 = 0 \text{ on } \Omega \tag{9.101a}$$

$$\frac{\alpha_{\theta_0}}{r} V_2 + \beta_{\theta_0} \frac{\partial V_2}{\partial n} = G_{\theta_0} \text{ on } \partial_{\theta_0}\Omega \tag{9.101b}$$

$$\alpha_a V_2 + \beta_a \frac{\partial V_2}{\partial n} = 0 \text{ on } \partial_a\Omega \tag{9.101c}$$

$$\alpha_c V_2 + \beta_c \frac{\partial V_2}{\partial n} = 0 \text{ on } \partial_c\Omega, \tag{9.101d}$$

where, for the sake of clarity, the function G has been subscripted according to the piece of boundary involved. Clearly, due to the linearity of the problem, the sum $V = V_1 + V_2$ is a solution of the initial BVP (Eq. 9.99). The BVP defined by the set of Eq. (9.100), (respectively Eq. 9.101), will be termed BVP1 (respectively BVP2) hereafter.

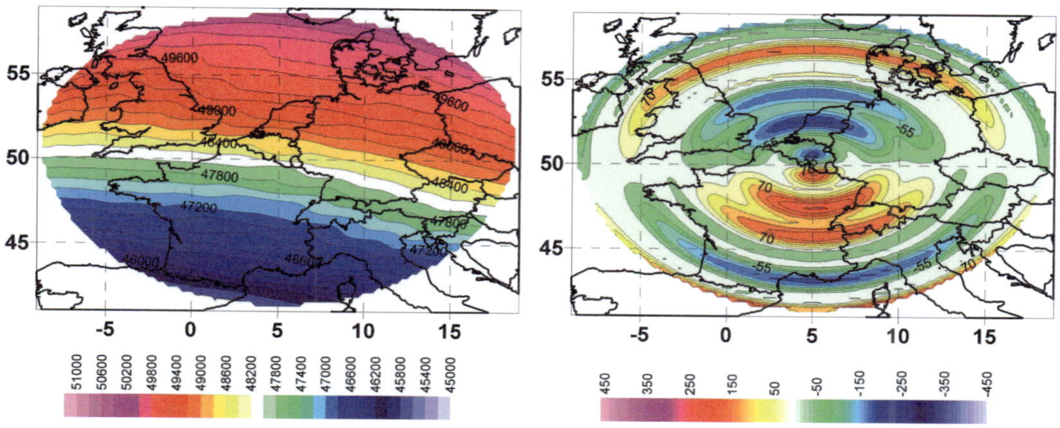

Fig. 9.8 Example of Spherical Cap Harmonic Analysis by least-squares inversion of the Z vector core component. The obtained SCHA model (*top-left*) and the residuals between the model and data (*top-right*) are discussed in the text

BVP1 was solved in Section 9.4.2.2 and will not be discussed any further. BVP2 was extensively discussed elsewhere in two particular cases: ($\alpha_{\theta_0} = \alpha_a = \alpha_c = 0$; Thébault et al., 2004) and ($\beta_{\theta_0} = \alpha_a = \alpha_c = 0$; Thébault et al., 2006a; 2006b). We thus limit ourselves to the changes inferred by the more general boundary conditions (101c and 101d). The most striking difference between BVP2 and the SCHA formulation (typically BVP1) is the Sturm-Liouville problem arising for the radial function $R(r)$ that writes

$$- d_r \left(r^2 d_r R(r) \right) = \lambda R(r) \text{ on }]a, c[\qquad (9.102a)$$

$$\alpha_a R(a) - \beta_a R'(a) = 0 \qquad (9.102b)$$

$$\alpha_c R(c) + \beta_c R'(c) = 0, \qquad (9.102c)$$

Define $L^2\left(]a,c[\right)$ the Hilbert space on the interval $]a,c[$ endowed with the inner product $\langle f, g \rangle = \int_a^c f(r)g(r)dr$. The operator $D = -d_r\left(r^2 d_r\right)$ appearing on the left-hand side of (102a) is a particular case of a regular Sturm-Liouville operator that is self-adjoint on the space of the functions of $C^2\left([a,c]\right) \cap L^2\left(]a,c[\right)$ fulfilling conditions (Eq. 9.102b) and (Eq. 9.102c). Therefore, the eigenvalues λ are real. If, in addition, the pairs (α_a, β_a), (α_c, β_c) have the same sign, the operator is positive (i.e., $\langle D(f), f \rangle \geq 0$) and λ is real positive or null. This positivity property, which is not really important for our purpose, derives from the expression of D, which includes the minus sign, in accordance with the general form of a Sturm-Liouville operator (see Eq. 9.83). Note that the sign change in Eq. (9.102a) does not follow the convention adopted in Thébault et al., (2004, 2006a). The general solution of (102a) may be still formally written

$$R(r) = A_1 \left(\frac{r}{a}\right)^{\nu} + A_2 \left(\frac{a}{r}\right)^{\nu+1}, \qquad (9.103)$$

with $\lambda = -\nu\,(\nu + 1)$ when $\lambda \neq 1/4$, and

$$R(r) = \left(A_1 \ln\left(\frac{r}{a}\right) + A_2\right)\left(\frac{a}{r}\right)^{1/2}, \qquad (9.104)$$

when $\lambda = 1/4$, that is when $\nu = -1/2$. As usual, the values of ν are such that the BVP2 (Eq. 102) has a non trivial null solution. They are the roots of an equation, which resolution relies upon approximate numerical methods in the general case. However, analytical solutions can be straightforwardly derived if we adopt more restrictive boundary conditions

$$a\frac{\alpha_a}{\beta_a} = -c\frac{\alpha_c}{\beta_c} = \alpha, \qquad (9.105)$$

which assumes non zero values for β_a and β_c but includes the Neumann homogeneous boundary conditions ($\alpha_a = \alpha_c = 0$), when $\alpha = 0$, and

$$\frac{\beta_a}{a\alpha_a} = -\frac{\beta_b}{b\alpha_c} = \alpha, \qquad (9.106)$$

which assumes non zero values for α_a and α_c but includes the homogeneous Dirichlet boundary conditions ($\beta_a = \beta_c = 0$) when $\alpha = 0$. Hereafter, the conditions defined in Eq. (9.105) (respectively Eq. 9.106) are referred to as case 1 (respectively case 2). Case 1 leads to eigenfunctions R_p, up to a multiplying constant given by

$$R_p(r) = \sqrt{\frac{a}{r}}\left[\frac{\pi p}{S(\alpha + 1/2)}\cos\left(\frac{\pi p}{S}\ln\left(\frac{r}{a}\right)\right) + \sin\left(\frac{\pi p}{S}\ln\left(\frac{r}{a}\right)\right)\right], \qquad (9.107)$$

that are associated to the eigenvalues

$$\lambda = \frac{1}{4} + \left(\frac{p\pi}{S}\right)^2 \quad p \in \mathbb{N}^*, \qquad (9.108)$$

where $S = \ln(c/a)$, or eigenfunctions

$$R_\alpha(r) = \left(\frac{r}{a}\right)^\alpha, \qquad (9.109)$$

associated to the eigenvalue $\lambda = -\alpha(\alpha + 1)$, which is null if $\alpha = 0$ or $\alpha = -1$ and negative if $\alpha \notin]-1, 0[$. We note that there is only one possibly negative eigenvalue. Interestingly, the eigenfunction associated to the negative eigenvalue (i.e., for $\alpha \notin]-1, 0[$) have the same shape as the basis functions of BVP1 but verify nevertheless the boundary conditions of BVP2. Case 2 leads to eigenfunctions

$$R_p(r) = \sqrt{\frac{a}{r}}\left[\frac{\pi \alpha p}{S(1 + \alpha/2)}\cos\left(\frac{\pi p}{S}\ln\left(\frac{r}{a}\right)\right) + \sin\left(\frac{\pi p}{S}\ln\left(\frac{r}{a}\right)\right)\right], \qquad (9.110)$$

that are associated to the eigenvalues given by Eq. (9.108) or to

$$R(r) = \left(\frac{r}{a}\right)^{1/\alpha}, \qquad (9.111)$$

associated to the possibly negative eigenvalue $\lambda = -\frac{1}{\alpha}\left(\frac{1}{\alpha} + 1\right)$. In this last case, the complete solutions of BVP2 take again the form Eq. (9.96b) with $\nu(m, k)$ replaced by $1/\alpha$. Of course, this basis functions exist only in the case $\alpha \neq 0$. Let us summarize the shape of the solutions for the problem BVP2. In every case the basis functions may be written in the complex form

$$\psi_p^m(r, \theta, \varphi) = \gamma_p^m(r, \varphi) K_p^m(\cos\theta), \qquad (9.112a)$$

$$\psi_\alpha^m(r, \theta, \varphi) = \gamma_\alpha^m(r, \varphi) P_\alpha^m(\cos\theta). \qquad (9.112b)$$

The $\gamma_p^m(r, \varphi)$ and $\gamma_\alpha^{c,m}(r, \varphi)$ functions are defined by

$$\gamma_p^m(r, \varphi) = R_p(r) e^{im\varphi}; \gamma_\alpha^m(r, \varphi) = R_\alpha(r) e^{im\varphi} \quad \text{with } m \in \mathbb{N} \qquad (9.113)$$

$K_p^m(\cos\theta)$ are the Mehler or conical functions described in Thébault et al. (2004, 2006a) (see also Gil et al., 2009 for a recent numerical discussion) and $P_\alpha^m(\cos\theta)$ generalized Legendre functions of real degree $P_\alpha^m(\cos\theta)$. As usual, the complex notation is kept for simplicity but we consider only real functions as solutions. The expressions of $R_p(r)$, $R_\alpha(r)$, eigenvalues λ and, hence, of $K_p^m(\cos\theta)$, $P_\alpha^m(\cos\theta)$, depend on the boundary conditions.

9.4.2.4 Orthogonality Properties, Uniqueness and Completeness

We now examine to which extent the orthogonality properties valid in SHA with respect to the inner product (Eq. 9.19) are valid for SCHA and R-SCHA functions. The orthogonality properties of $\psi_{i,k}^m$ and $\psi_{e,k}^m$ rely on those of $\beta_k^m(\theta, \varphi)$ (Eq. 9.95) on the spherical cap $S_{\theta_0}(1)$. The proofs were given by Lowes (1999). For R-SCHA, there is an extra complication with respect to SH due to the orthogonality properties of, and between, the basis functions $\psi_p^m(r, \theta, \varphi)$, $\psi_\alpha^m(r, \theta, \varphi)$ (see Eq. 9.112) as well as between these letters and family of functions ψ_k^m. Functions $\psi_p^m(r, \theta, \varphi)$ (or $\psi_\alpha^m(r, \theta, \varphi)$) are not orthogonal to each other with respect to the inner product defining $L^2(\Omega)$ because $R_p(r)$ and $R_{p'}(r)$ (Eq. 9.107 or Eq. 9.110) are orthogonal with respect to the inner product $\int_a^c R_p(r) R_{p'}(r) dr$ not with respect to $\int_a^c r^2 R_p(r) R_{p'}(r) dr$ as derived from Eq. (9.19). Orthogonality properties are restored if we resort to a weighted inner product and to a weighted Sobolev space $W^1(\Omega)$ defined by the functions f on Ω possessing the properties: $\frac{f(\mathbf{r})}{(a^2+r^2)^{1/2}}$ and every partial derivative $\partial_{x_i}(f)$ belong to $L^2(\Omega)$ (Dautray and Lions, 1988, Chapter XI, p. 649). Knowing that on the domain Ω under consideration, $a \le r \le c$, the denominator $\left(a^2+r^2\right)^{1/2}$ may be equivalently replaced by r, $W^1(\Omega)$ is a Hilbert space for the following (real) inner product

$$\langle f, g\rangle_{W^1} = \int_{\Omega} \left[\frac{f(\mathbf{r})g(\mathbf{r})}{r^2} + \vec{\nabla} f \cdot \vec{\nabla} g\right] d\tau. \quad (9.114)$$

For the same reasons as for the SH analysis the subspace $W^1(\Omega)$ is not relevant for harmonic functions since the potential is not the measured quantity. It is thus judicious to define an inner product based only on the gradients. Thus, let us denote again $U(\Omega)$ the subspace of the harmonic functions of $W^1(\Omega)$ and provide $U(\Omega)$ with the inner product defined by Eq. (9.19). As it defines only a semi-norm, it is possible to put further constraints on $U(\Omega)$ in order to derive a true norm. We did not explore this possibility but leave it for future investigations.

The basis functions ψ_p^m, ψ_α^m (Eq. 9.112) are orthogonal with respect to the inner product (Eq. 9.114), both terms of the integrand being null. The same property holds true for each family of the basis functions $\psi_{i,k}^m(r,\theta,\varphi)$ and $\psi_{e,k}^m(r,\theta,\varphi)$, (Eq. 9.96) but not necessarily between the families. The orthogonality between the families of functions arising from BVP1 or BVP2, namely pairs like $(\psi_k^m, \psi_{p'}^{m'})$ are obviously holds true for $m \neq m'$ but we need to compute

$$I = \left\langle \psi_k^m, \psi_p^m \right\rangle_U = \int_{\Omega} \vec{\nabla}\psi_k^m \cdot \vec{\nabla}\psi_p^m d\tau. \quad (9.115)$$

Writing $\psi_k^m = \{\psi_{i,k}^m(r,\theta,\varphi),\ \psi_{e,k}^m(r,\theta,\varphi)\}$ and using the Green identity Eq. (9.24), Eq. (9.115) may be transformed into

$$I = \int_{\partial\Omega} \psi_k^m \cdot \frac{\partial \psi_p^m}{\partial n} d\sigma = \int_{\partial_{\theta_0}\Omega} \frac{1}{r}\left(\partial_\theta \psi_p^m\right)_{\theta_0} \psi_k^m d\sigma - \int_{\partial_a\Omega} \left(\partial_r \psi_p^m\right)_a \psi_k^m d\sigma + \int_{\partial_c\Omega} \left(\partial_r \psi_p^m\right)_c \psi_k^m d\sigma. \quad (9.116)$$

Taking the general form of the boundary conditions (9.100b), (9.101c), (9.101d) but restricting them to the particular forms $a\frac{\alpha_a}{\beta_a} = -c\frac{\alpha_c}{\beta_c} = \alpha_1$ or $\frac{\beta_a}{a\alpha_a} = -\frac{\beta_c}{b\alpha_c} = \alpha_2$, it turns out that I vanishes only in the cases $(\alpha_1 = \beta_{\theta_0} = 0)$ or $(\alpha_2 = \alpha_{\theta_0} = 0)$. These conditions are respectively equivalent to $\left(\alpha_a = \alpha_c = \beta_{\theta_0} = 0\right)$, a Neumann condition on $\partial_a\Omega \cup \partial_b\Omega$ and a Dirichlet condition on $\partial_{\theta_0}\Omega$, and to $\left(\beta_a = \beta_c = \alpha_{\theta_0} = 0\right)$, a Dirichlet condition on $\partial_a\Omega \cup \partial_b\Omega$ and a Neumann condition on $\partial_{\theta_0}\Omega$. Conditions $\left(\beta_a = \beta_c = \alpha_{\theta_0} = 0\right)$ are hereafter denoted model M_1 and conditions $\left(\alpha_a = \alpha_c = \beta_{\theta_0} = 0\right)$ model M_2.

Considering a function V belonging to $U(\Omega)$, its expansion on the bases $\left\{\psi_{i,k}^m, \psi_{e,k}^m,\right\}$ and $\{\psi_p^m, \psi_\alpha^m\}$ is the sum of the following double series (Thébault et al., 2006a)

$$S_V = a\sum_{m=0}^{\infty}\sum_{k=1}^{\infty}\left(G_k^{i,m}\psi_{i,k}^m + G_k^{e,m}\psi_{e,k}^m\right) + a\sum_{m=0}^{\infty}\sum_{p=1}^{\infty}\left(G_p^m\psi_p^m\right) + a\sum_{m=0}^{\infty}\left(G_\alpha^m\psi_\alpha^m\right). \quad (9.117)$$

The gradients of V are orthogonal in Ω only in the cases described by M_1 and M_2. This provides a mean to estimate the Gauss coefficients separately by

$$\begin{aligned} G_k^{i,m}\left\|\psi_{i,k}^m\right\|_U^2 &= \left\langle V, \psi_{i,k}^m\right\rangle_U ; G_k^{e,m}\left\|\psi_{e,k}^m\right\|_U^2 = \left\langle V, \psi_{e,k}^m\right\rangle_U \\ G_p^m\left\|\psi_p^m\right\|_U^2 &= \left\langle V, \psi_p^m\right\rangle_U ; G_\alpha^m\left\|\psi_\alpha^m\right\|_U^2 = \langle V, \psi_\alpha^m\rangle_U, \end{aligned} \quad (9.118)$$

Equation (9.118) provides the essential argument against the ability of regional modelling technique to discriminate between internal and external magnetic fields with respect to the Earth's surface. Considering the expansion of V in SH (Eq. 9.28) is may be readily shown that setting $q_n^m = 0$ does not impose $G_k^{e,m} = 0$. This demonstrates that the "external" coefficients do not have the same meaning in SH and in R-SCHA formalisms. Regarding the completeness of R-SCHA expansion, the demonstration relies on the completeness of the bases β_k^m and γ_p^m on their respective spaces. The completeness is derived from the spectral properties of operators like ∇_S^2. Good accounts of the properties of this kind of operators may be found in Dautray and Lions (1988, Chapter VIII). R-SCHA is not designed to deal with single surface measurements. A good account of the ability of R-SCHA to process multi-level data is given in Thébault et al., (2006b).

9.4.3 Boundary Effects

Boundary effects, as already stated in Section (9.2.3), are closely related to uniform convergence (see Haines, 1990, for examples in one-dimensional spaces). Within the frame of generalized Fourier series, the boundary effects are nothing else than the expression of the Gibbs phenomenon, well-known and investigated at length in the case of the Fourier expansion of periodic functions. In this latter case, various summing methods may be used in order to accelerate the convergence rate and reduce the Gibbs effect (see for instance Robin, 1958, vol. II, Chapter VI, Hobson, 1965, Chapter VII, Jerri, 1998, section 3.5) which could probably be adapted in some cases in two dimensions (e.g., Thébault, 2006 who applied the Fejér partial sum theorem). Things are however a great deal more complicated with multi-dimensional series, more specifically with two-dimensional infinite series in the present case. Gonzalez-Velasco (1995, section 9.2) explored in details the case of harmonic expansion on a rectangular domain which involves periodic functions and showed, with a simple manageable example, how the complexity increases from the one-dimensional to two-dimensional situations. In particular, he stressed that uniform convergence depends on continuity property of the second mixed derivative ∂^2_{xy}. The difficulties are still enhanced in the case of SCHA and R-SCHA expansions due to the transcendental nature of the basis functions. Haines (1985a) claimed uniform convergence depending on consistency between the boundary conditions fulfilled by the basis functions and those verified by the potential to be approximated, referring to Sturm-Liouville theorem (see Section 9.4.1.1). This theorem is valid in the context of one-dimensional Sturm-Liouville problems only. To our knowledge, there is no extension to multi-dimensional problems, as illustrated by RHA and the two-dimensional ordinary Fourier series involved. Therefore, including both Neumann and Dirichlet conditions in SCHA does not even ensure a uniform convergence of the solution (but we admit that in practice they do converge faster).

Uniform convergence conditions have been set up for SH expansions. In that case, one may involve the first Harnack theorem mentioned in Section (9.2.3) (Kellog, 1929, p.248) which connects uniform convergence inside the domain to uniform convergence on its boundary. When the domain is a sphere or a shell, uniform convergence on the boundary relies on properties of Laplace series. The addition theorem of spherical harmonics allows transforming the two-dimensional series in degree l and order m into a one-dimensional series in l involving a Legendre expansion (see Kellog, 1929, chap. X and Hobson, 1965, chapter VII, for details). This is a mathematically well-founded simplification not possible in the case of SCHA or R-SCHA, although addition theorems exist for generalized Legendre functions (Hobson, 1965, chap. VIII, Robin, 1958, vol. III, chap. VII). According to Jerri (1998, Section 3.5), further investigations illustrating the link between rate of convergence and boundary conditions fulfilled by the potential to be approximated, could be carried out for instance with models M_1 and M_2 defined in the previous section. This investigation has not yet been performed.

9.4.4 Infinite Conical Domain

We define the infinite conical domain Ω_∞ as the domain described in Section (9.4.2.1) bounded by a sphere of infinite radius c. In order to investigate the changes brought to the expression of the basis functions for the bounded domain, we solve the following boundary value problem which is similar to problem M_2 defined in Section (9.4.2.4)

$$\nabla^2 (V) = 0 \text{ on } \Omega_\infty \tag{9.119a}$$

$$V = G_{\theta_0}(r, \varphi) \text{ on } \partial_{\theta_0}\Omega_\infty \tag{9.119b}$$

$$\frac{\partial V}{\partial n} = G_a(\theta, \varphi) \text{ on } \partial_a\Omega_\infty \tag{9.119c}$$

$$V \text{ and } \vec{\nabla} V \longrightarrow 0 \text{ when } r \to \infty. \tag{9.119d}$$

The problem is again split up into two subproblems with partially homogeneous boundary conditions (compare to Eqs. 9.100 and 9.101)

$$\nabla^2 V_1 = 0 \text{ on } \Omega_\infty \tag{9.120a}$$

$$V_1 = 0 \text{ on } \partial_{\theta_0}\Omega \tag{9.120b}$$

$$\frac{\partial V_1}{\partial n} = G_a \text{ on } \partial_a\Omega \tag{9.120c}$$

$$V_1 \text{ and } \vec{\nabla} V_1 \longrightarrow 0 \text{ when } r \to \infty, \tag{9.120d}$$

$$\nabla^2 V_2 = 0 \text{ on } \Omega_\infty \tag{9.121a}$$

$$V_2 = G_{\theta_0} \text{ on } \partial_{\theta_0}\Omega_\infty \tag{9.121b}$$

$$\frac{\partial V_2}{\partial n} = 0 \text{ on } \partial_a\Omega_\infty \tag{9.121c}$$

$$V_2 \text{ and } \vec{\nabla} V_2 \longrightarrow 0 \text{ when } r \to \infty. \tag{9.121d}$$

Basis functions derived from BVP (Eqs. 9.120) are the same as those of the BVP (Eqs. 9.100) except for the functions $\psi^m_{e,k}$ (Eq. 9.96b) which do not vanish at infinity and have therefore to be discarded. The main difference with the case of the bounded domain comes from the solutions of the second BVP (Eq. 9.121) and more specifically from the changes in Eq. (9.102) which now writes

$$-d_r\left(r^2 d_r R(r)\right) = \lambda R(r) \text{ on }]a, \infty[\tag{9.122}$$

$$R'(a) = 0 \tag{9.123}$$

$$R(r) \text{ and } R'(r) \to 0 \text{ when } r \to \infty. \tag{9.124}$$

It turns out that the eigenvalues are no longer a discrete set of complex numbers. They build up a continuum of the form

$$\lambda = \frac{1}{4} + y^2 = \nu(\nu + 1), \tag{9.125}$$

where y is a real number, positive or null and the roots ν write

$$\nu = -\frac{1}{2} + y \text{ with } y \geqslant 0. \tag{9.126}$$

The radial functions, denoted $R_y(r)$, take the form

$$R_y(r) = \sqrt{\frac{a}{r}}\left[2y\cos\left(y\ln\frac{r}{a}\right) + \sin\left(y\ln\frac{r}{a}\right)\right] \quad \text{when } y > 0, \tag{9.127a}$$

$$\text{and } R_y(r) = \sqrt{\frac{a}{r}}\left[\ln\frac{r}{a} + 2\right] \quad \text{when } y = 0, \tag{9.127b}$$

The basis functions, equivalent to those given by Eq. (9.112a), write now

$$\psi_y^m(r, \theta, \varphi) = \gamma_y^m(r, \varphi) K_y^m(\cos\theta), \tag{9.128}$$

with

$$\gamma_y^m(r, \varphi) = R_y(r) e^{im\varphi} \quad m \in \mathbb{N}, \tag{9.129}$$

where the complex form, as before, is kept for simplicity. In the particular case $y = 0$, the Mehler function K_y^m may be more clearly written $P^m_{-1/2}$ which is a particular generalized Legendre function with real degree.

Splitting the exponential form of the φ-function into real-valued trigonometric function, the Fourier-like expansion of a potential V on the basis functions ψ_y^m more explicitly writes

$$S_V = a\sum_{m=0}^{\infty}\left[\cos m\varphi \int_{y=0}^{\infty} G^m(y) R_y(r) K_y^m(\cos\theta) dy + \sin m\varphi \int_{y=0}^{\infty} H^m(y) R_y(r) K_y^m(\cos\theta) dy\right]. \tag{9.130}$$

The integral factors are the equivalent of inverse Fourier transforms, the coefficients $G^m(y)$ and $H^m(y)$ being now functions of the real variable y instead of being indexed terms of a series. Thus, the formalism for the infinite cone is derived from that of the bounded cone in very much the same way as the Fourier transform may be derived from the ordinary Fourier series when the periodic interval is stretched out to infinity. The functions $G^m(y)$ and $H^m(y)$ might thus be interpreted as generalized Fourier transforms of the potential V. The space $U(\Omega)$ mentioned in Section (9.4.2.4) is still the functional frame. However, the domain being now unbounded, some care must be taken regarding the existence of the inner products $\langle f, g\rangle$ defined by Eq. (9.114) and $\langle f, g\rangle_U = \int_{\Omega_\infty}\left(\vec{\nabla} f \cdot \vec{\nabla} g\right) d\tau$. Likewise, care must be exercised in the use of Green's identity (Eq. 9.24). The computation of $G^m(y)$ and $H^m(y)$ is alike the bounded case if the basis functions are still orthogonal with respect to the inner product $\langle \cdot, \cdot \rangle_U$. Assuming that Green' identity holds true, we have

$$\left\langle \psi_y^m, \psi_{y'}^{m'} \right\rangle_U = \int\limits_{\partial\theta_0\Omega_\infty} \psi_y^m \left(\frac{\partial \psi_{y'}^{m'}}{\partial n} \right)_{\partial\theta_0\Omega_\infty} \sin\theta_0 r dr d\varphi \tag{9.131a}$$

$$= \delta_{m,m'} \left(1 + \delta_{m,0}\right) \pi \sin\theta_0 K_y^m(\theta_0) \partial_{\theta_0} \left(K_y^m\right) \int\limits_a^\infty R_y(r) R_{y'}(r) dr, \tag{9.131b}$$

where $\delta_{m,m'}$, $\delta_{m,0}$ are the Kronecker symbols. It may be shown, using the Fourier transform of the Heaviside function that

$$\int\limits_a^\infty R_y(r) R_{y'}(r)\, dr = 2\pi a (yy' + \frac{1}{4}) \delta\,(y - y'), \tag{9.132}$$

where $\delta\,(y)$ is the Dirac distribution. Thus, ψ_y^m, $\psi_{y'}^m$ are orthogonal in the generalized sense defined by Eq. (9.132). On the other hand, ψ_y^m and $\psi_{i,k}^m$ are still orthogonal due to the boundary conditions (Eqs. 9.120 and 9.121) they respectively fulfill. Taking into account the orthogonality property expressed by Eq. (9.132), it is now straightforward to compute $G^m\,(y)$ and $H^m\,(y)$. For instance

$$G^m\,(y) = \frac{\sin\theta_0 d_{\theta_0} \left(K_y^m\right)}{a \left\| R_y K_y^m \cos m\varphi \right\|_U^2} \int\limits_a^\infty R_y(r) dr \int\limits_0^{2\pi} V(r, \theta_0, \varphi) \cos m\varphi d\varphi. \tag{9.133}$$

Thébault (2008) used an hybrid variant of this method to construct a time-varying magnetic field model over France for the epochs between 1965 and 2007.5, restricting the expansion on the ψ_y^m to the only term $y = 0$ as this term at least was necessary in order to comply with basic properties of the magnetic field, and keeping the so-called external basis function in order to balance the incompleteness induced by this restriction. Therefore, the basis function corresponding to this latter approximation is, strictly speaking, not complete.

Figure 9.9 displays an application of the infinite cone restricting the expansion (Eq. 9.130) to $y = 0$ that is referred to as R-SCHA2D. The maximum series expansion in Eqs. (9.96a) and (9.96b) are defined to resolve the Z_{all} data to 100 km wavelengths. The model fits Z_{all} correctly both for the main and crustal fields. Part of the residuals are due to wavelengths smaller than 100 km but one can see the presence of circular edge effects near the Southern boundary. This is mostly caused by the choice of the Dirichlet boundary condition set in Eq. (9.120b) that makes the Z component converge slower than the horizontal component but the restriction to $y = 0$ may likely be responsible for some part of the residuals. Since the basis functions are orthogonal we could, in principle, compute a power spectrum (which does not make sense in case of aliasing). The total field can be fairly well represented but we cannot ascertain that the upward/downward continuation will be stable, unless we deal with magnetic fields with very specific properties, because the restriction to $y = 0$ may very well hold at the data surface but not anymore at another radius.

9.4.5 Slepian Functions

We now finish our overview of regional modelling with the Slepian functions. These functions originate from a problem in information theory, dealing with the optimal concentration of a signal in both the time and frequency domains (see Simons et al., 2006, for references). They may be introduced in two ways. First, by adopting the viewpoint of strictly band-limited functions (up to degree L in terms of spherical harmonics) which is an approach comparable to the SH expansion. Second, by making use of the concept of strictly spatially localized functions, which is closer to regional modelling like SCHA and R-SCHA. We restrict ourselves to the first approach, which takes a simple algebraic form, part of which has already been seen in the above paragraphs.

Let be H_L the space defined in Section (9.3.2) and $K_L\,(\rho)$ the subspace of $L^2\left(S_\rho\right)$ of the band-limited spherical harmonic functions defined on the sphere $S\,(\rho)$ ($\rho = R_E$ or simply r). $L^2\left(S_\rho\right)$ is endowed with the inner product defined in Eq. (9.16). H_L and $K_L\,(\rho)$ have the same dimension, namely $L\,(L+2)$. The functions $\psi_l^m\,(R\widehat{\mathbf{r}})$ are identical to $\beta_l^m\,(\widehat{\mathbf{r}})$ and are therefore orthonormal on $K_L\,(R)$, whereas $\psi_l^m\,(r\widehat{\mathbf{r}})$ are orthogonal on $K_L\,(r)$ (see Eq. 9.18a for the definition of ψ_l^m -the subscript i having been dropped).

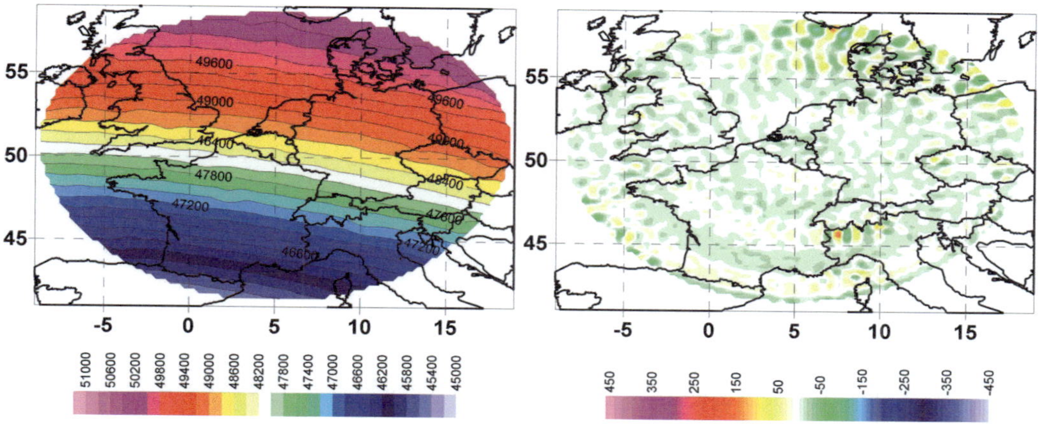

Fig. 9.9 Modelling of the magnetic field and the crustal field using an approximate expansion resembling to the solution of the infinite cone. This approach is called R-SCHA2D. On the left are shown the residuals. Units are in nT

As in Section (9.3.1.2), we illustrate the Slepian technique with the expression of the radial component rB_r. According to Eq. (9.52) and (9.53), where the maximum degree is L, the components $rB_{L,r}(r\widehat{\mathbf{r}})$ and $RB_{L,R}(R\widehat{\mathbf{r}})$ write

$$rB_{L,r}(r\widehat{\mathbf{r}}) = R\sum_{lm}^{L} p_l^m \psi_l^m (r\widehat{\mathbf{r}}) = R\sum_{lm}^{L} p_l^m(r)\, Y_{R,l}^m (R\widehat{\mathbf{r}}), \tag{9.134a}$$

$$RB_{L,R}(R\widehat{\mathbf{r}}) = R\sum_{lm}^{L} p_l^m \psi_l^m (R\widehat{\mathbf{r}}) = R\sum_{lm}^{L} p_l^m Y_{R,l}^m (R\widehat{\mathbf{r}}), \tag{9.134b}$$

with

$$p_l^m(r) = \left(\frac{R}{r}\right)^{l+1} (l+1)\, g_l^m = \left(\frac{R}{r}\right)^{l+1} p_l^m. \tag{9.135}$$

We define $Y_{R,l}^m(R\widehat{\mathbf{r}})$ as in Eq. (9.57) in order to better stress that they are functions defined on S_R. The expression $\sum\limits_{lm}^{L}$ hereafter stands for $\sum\limits_{l=1}^{L}\sum\limits_{m=-l}^{l}$ according to the convention adopted by Simons and Dahlen (2006). The relation between $rB_{L,r}(r\widehat{\mathbf{r}})$ and $RB_{L,R}(R\widehat{\mathbf{r}})$ expanded in terms of the SH $Y_{R,l}^m(R\widehat{\mathbf{r}})$, hence the upward and downward continuation, has been discussed in (section 9.3.1.2).

9.4.5.1 Slepian Functions in $K_L(R)$

We are now looking for a set of basis functions of $K_L(R)$ localized in a region $\Sigma_R \subset S_R$. These functions, defined as $g(\widehat{\mathbf{r}})$, maximize the space energy ratio (Simons et al., 2006; Simons and Dahlen, 2006)

$$\lambda = \frac{\int\limits_{\Sigma_R} \left[g(R\widehat{\mathbf{r}})\right]^2 d\sigma}{\int\limits_{S_R} \left[g(R\widehat{\mathbf{r}})\right]^2 d\sigma}. \tag{9.136}$$

As g belong to $K_L(R)$, there are at most $L(L+2)$ linearly independent functions. Their expansion on the basis $\left\{Y_{R,l}^m\right\}$ writes

$$g_k(R\widehat{\mathbf{r}}) = \sum_{lm}^{L} \gamma_{l,k}^m Y_{R,l}^m (R\widehat{\mathbf{r}}). \tag{9.137}$$

For simplicity, the double indices (l, m) are mapped to a single index j according to the rule

$$j(l,m) = l^2 + l + m, \tag{9.138}$$

and the coefficients $\gamma_{l,k}^m$ can be written C_{jk}, $j = 1, \ldots, L(L+2)$. The column vector $\Gamma_k = C_{\cdot,k}$ belonging to $\mathbb{R}^{L(L+2)}$ contains the components of the vector $g_k(R\widehat{\mathbf{r}})$ on the basis $\left\{Y_{R,l}^m(R\widehat{\mathbf{r}}) \text{ or } Y_{R,j}(R\widehat{\mathbf{r}})\right\}$. Using the mapping from $K_L(R)$ onto $\mathbb{R}^{L(L+2)}$ just described, Simons et al. (2006) showed that the vectors Γ_k are the eigenvectors of the algebraic eigenvalue problem

$$\mathbf{D}(\Gamma) = \lambda\Gamma, \tag{9.139}$$

where $\mathbf{D}$ is the $L(L+2) \times L(L+2)$ — dimensional matrix whose elements are given by

$$D_{lm,l'm'} = \frac{1}{4\pi} \int_{\Sigma_1} Y_{R,l}^{m}(R\hat{\mathbf{r}})\, Y_{R,l'}^{m'}(R\hat{\mathbf{r}})\, d\sigma. \qquad (9.140)$$

Σ_1 is the radial projection of Σ_R onto S_1. According to the mapping defined by Eq. (9.138), the elements of $\mathbf{D}$ may be indexed D_{ij}, $i = 1, \ldots, L(L+2)$, $j = 1, \ldots, L(L+2)$. $\mathbf{D}$ is the matrix of a symmetric (i.e., self-adjoint), positive operator on $\mathbb{R}^{L(L+2)}$, which range is $\mathbb{R}^{L(L+2)}$. Hence, there are $L(L+2)$ positive eigenvalues λ_k, associated to $L(L+2)$ orthogonal eigenvectors Γ_k whose components are $(C_{i,k})_{i=1,\ldots,L(L+2)} = \left(\gamma_{l,k}^{m}\right)$. These eigenvectors may be normalized. Hence, the columns of the matrix $\mathbf{C}$ verify the property

$$\sum_{\alpha=1}^{L(L+2)} C_{\alpha j} C_{\alpha k} = \sum_{lm}^{L} \gamma_{l,j}^{m} \gamma_{l,k}^{m} = \delta_{jk}, \qquad (9.141)$$

and the matrix $\mathbf{C}$ maps the basis $\left\{Y_{R,l}^{m}\right\}$ onto the basis $\{g_k\}$

$$g_k(R\hat{\mathbf{r}}) = \sum_{j=1}^{L(L+2)} C_{kj}^{T} Y_{R,j}(R\hat{\mathbf{r}}), \qquad (9.142)$$

where C is an unitary matrix and therefore $\mathbf{C}^{-1} = \mathbf{C}^{T}$. Conversely

$$Y_{R,j}(R\hat{\mathbf{r}}) = \sum_{k=1}^{L(L+2)} C_{jk} g_k(R\hat{\mathbf{r}}) \quad \text{or} \quad Y_{R,l}^{m}(R\hat{\mathbf{r}}) = \sum_{k=1}^{L(L+2)} \gamma_{l,k}^{m} g_k(R\hat{\mathbf{r}}). \qquad (9.143)$$

Thus, the Slepian functions are constructed such as to verify the property

$$\frac{1}{4\pi} \int_{S_1} g_j(R\hat{\mathbf{r}})\, g_k(R\hat{\mathbf{r}})\, d\sigma = \delta_{jk}. \qquad (9.144)$$

In addition, they have the nice property of being likewise orthogonal on the region Σ_R (Simons and Dahlen, 2006)

$$\frac{1}{4\pi} \int_{\Sigma_1} g_j(R\hat{\mathbf{r}})\, g_k(R\hat{\mathbf{r}})\, d\sigma = \lambda_j \delta_{jk}. \qquad (9.145)$$

As one may conjecture, this property plays a central role in the inverse problem. As expected, when Σ_R tends to cover the whole sphere, every eigenvalue tends towards 1 and $g_k(R\hat{\mathbf{r}})$ tends towards $Y_{R,l}^{m}(R\hat{\mathbf{r}})$. Technical details about the actual calculation of the Slepian functions are to be found in Simons et al. (2006). They show in particular that the mathematics are definitely simpler if Σ_1 is a circular cap. The component $RB_{L,R}(R\hat{\mathbf{r}})$ which expansion on the basis $\left\{Y_{R,l}^{m}(R\hat{\mathbf{r}})\right\}$ is given by Eq. (9.134b) may be likewise expanded on the basis $\{g_k(R\hat{\mathbf{r}})\}$

$$RB_{L,R}(R\hat{\mathbf{r}}) = R \sum_{k=1}^{L(L+2)} s_k g_k(R\hat{\mathbf{r}}). \qquad (9.146)$$

According to the well-known algebraic rules for basis change, s_k and p_l^m are linked by

$$s_k = \sum_{j=1}^{L(L+2)} C_{jk} p_j = \sum_{lm}^{L} \gamma_{l,k}^{m} p_l^{m}, \qquad (9.147a)$$

$$p_j = \sum_{k=1}^{L(L+2)} C_{jk} s_k \ \text{or}\ p_l^{m} = \sum_{k=1}^{L(L+2)} \gamma_{l,k}^{m} s_k. \qquad (9.147b)$$

9.4.5.2 Slepian Functions in $K_L(r)$

In order to calculate $rB_{L,r}(r\hat{\mathbf{r}})$, which expression in terms of solid spherical harmonics is given by Eq (9.134a), we may search likewise an expansion using the functions $g_k(R\hat{\mathbf{r}})$. Therefore $rB_{L,r}(r\hat{\mathbf{r}})$ writes

$$rB_{L,r}(r\hat{\mathbf{r}}) = R \sum_{k=1}^{L(L+2)} s_k(r)\, g_k(R\hat{\mathbf{r}}). \qquad (9.148)$$

The functions $s_k(r)\, g_k(R\hat{\mathbf{r}})$ are orthogonal with respect to the inner product defined in Eq. (9.16) on the space $L^2(S_r)$. This property guarantees the uniqueness of the functions $s_k(r)$. Let us calculate them in terms of the constant p_l^m. We obtain

$$s_k(r) = \sum_{lm}^{L} \gamma_{l,k}^m p_l^m \left(\frac{R}{r}\right)^{l+1}, \qquad (9.149)$$

which shows that the radial functions $s_k(r)$ are significantly more complicated than the $p_l^m(r)$ defined in Eq. (9.135). This will require that multi-level data have to be modelled with different functions $s_k(r)$. In order to write $s_k(r)$ in terms of $s_k(R)$, we replace p_l^m by its expression given by Eq. (9.147b)

$$s_k(r) = \sum_{j=1}^{L(L+2)} \left[\sum_{lm}^{L} \gamma_{l,k}^m \gamma_{l,j}^m \left(\frac{R}{r}\right)^{l+1}\right] s_j. \qquad (9.150)$$

9.4.5.3 Potential Field Estimation On Σ_r

Let us turn back to the eigenvalues λ_k of Eq. (9.136). They are clearly in the interval [0, 1]. The largest values correspond to the Slepian functions most concentrated in the area Σ_R or equivalently Σ_r. The so-called "spherical Shannon number" defined by (Simons and Dahlen, 2006)

$$N = \sum_{k=1}^{L(L+1)} \lambda_k = (L+1)^2 \frac{A}{4\pi}, \qquad (9.151)$$

where A is the area of Σ_R divided by R^2, provides an estimate of the number of Slepian functions to be retained in the expansion. The reduction of basis functions according to the eigenvalue magnitude, together with the property expressed by Eq. (9.145) makes the Slepian basis attractive for the inverse problem. We illustrate this point by considering the inverse problem consisting in estimating the coefficients $s_k(r)$ knowing rB_r $(= rB_{r,\text{mes}})$ on Σ_r. As usual, in the least-squares approach, we minimize a functional with respect to the coefficients $s_k(r)$ of the expansion given by Eq. (9.148). In the present case, the functional writes

$$d^2 = \frac{1}{4\pi r^2} \int_{\Sigma_r} \left[rB_{L,r}(r\hat{\mathbf{r}}) - rB_{r,\text{mes}}(r\hat{\mathbf{r}})\right]^2 r^2 d\sigma. \qquad (9.152)$$

In practice, it is advisable to use a truncated sum. The Shannon number gives an indication of how to select the minimal eigenvalue but other choices may be made (see Simons and Dahlen, 2006 for a discussion) and therefore, we write J the maximal index (therefore, λ_J is the smallest eigenvalue). Replacing $rB_{L,r}(r\hat{\mathbf{r}})$ by the truncated expansion, we obtain

$$\begin{aligned} d^2 = & \tfrac{R^2}{4\pi} \sum_{j=1}^{J} \sum_{k=1}^{J} s_j(r)\, s_k(r) \int_{\Sigma_1} g_j(R\hat{\mathbf{r}})\, g_k(R\hat{\mathbf{r}})\, d\sigma \\ & - \tfrac{2Rr}{4\pi} \sum_{j=1}^{J} s_j(r) \int_{\Sigma_1} g_j(R\hat{\mathbf{r}})\, B_{r,\text{mes}}(r\hat{\mathbf{r}})\, d\sigma \\ & + \tfrac{r^2}{4\pi} \int_{\Sigma_1} \left(B_{r,\text{mes}}(r\hat{\mathbf{r}})\right)^2 d\sigma. \end{aligned} \qquad (9.153)$$

Taking into account Eq. (9.145), d^2 becomes

$$\begin{aligned} d^2 = & R^2 \sum_{j=1}^{J} \lambda_j \left(s_j(r)\right)^2 - \frac{2Rr}{4\pi} \sum_{j=1}^{J} s_j(r) \int_{\Sigma_1} g_j(R\hat{\mathbf{r}}) \\ & B_{r,\text{mes}}(r\hat{\mathbf{r}})\, d\sigma + \frac{r^2}{4\pi} \int_{\Sigma_1} \left(B_{r,\text{mes}}(r\hat{\mathbf{r}})\right)^2 d\sigma. \end{aligned} \qquad (9.154)$$

The normal equations write

$$\lambda_j s_j(r) = \frac{1}{4\pi} \left(\frac{r}{R}\right) \int_{\Sigma_1} g_j(R\hat{\mathbf{r}})\, B_{r,\text{mes}}(r\hat{\mathbf{r}})\, d\sigma. \qquad (9.155)$$

Equation (9.155) shows clearly the importance of the property expressed by Eq. (9.145) and of a good selection of the eigenvalue and Slepian eigenfunction set. Of course, this presentation is a rather elementary approach to the inverse problem. The reader is referred to Simons and Dahlen (2006) for further developments and to Simons et al. (2009) for an application to the modelling of the Bangui anomaly.

Figure (9.10) shows the residuals between the SH synthetic data Z_{all} and the Slepian reconstruction to $L = 200$. Note its similarity with Figures (9.9-right) and (9.2). The Slepians do rather well in reconstructing the total field in this case. Some subtleties are worth being mentioned. First, the Slepian reconstruction was performed here within a spherical cap but this is not required. Among the methods presented so far, this flexibility is rather unique and allows adjusting a model to very specific geometry (that is in general imposed by the available data distribution often correlated with the boundaries of countries). Second, the inverse problem is numerically well conditioned

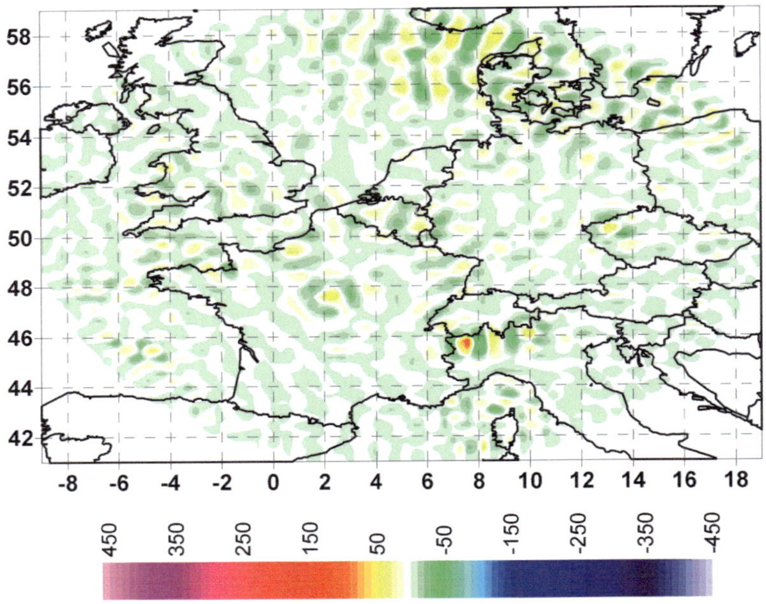

Fig. 9.10 Residuals in nT between the Z_{all} synthetic data and the slepian reconstruction using L = 200

thanks to the orthogonality of the Slepian functions and this allows estimating power spectra (Simons, 2010). Note, however, that a good *a priori* knowledge on the data error is required to avoid modelling the noise and select the optimal number of Slepian functions. At last, by virtue of Eq. (9.145) and its associated comments, Slepian functions are currently powerful for spectral analysis of surface collocated data but efforts are being made towards implementing the technique in order to process simultaneously the three components of the magnetic field vector data measured at different altitudes (Beggan and Simons, 2009).

9.5 Conclusions

We presented in a formal manner different techniques under development for modelling the Earth's magnetic field with, in its wide acceptation, local functions. We showed that the methods based on functions with global support are in fact different realizations of a unique expression given by Eqs. (9.37) and (9.38); they will then differ according to the chosen kernel and regularization. We then illustrated that most approaches provide similar result when a sufficiently large set of perfect magnetic field data at a single surface is available. The techniques are, however, not equivalent from a practical point of view when moving away from this ideal situation. The differences come up because the approaches discussed rely on sometimes fundamentally different concepts. Some do not necessarily solve Laplace equation (techniques inherited from signal processing, for instance), others do not converge uniformly (regional modelling without the appropriate boundary conditions) and, in general, none of them allow internal/external field separation (when applied over a portion of the sphere only). Methods with global support are arguably ill suited for dealing with magnetic field signals with local characteristics but are likely to allow internal/external field separation when they are applied at the global scale since they encompass the internal magnetic field sources of the Earth. Conversely, we do not see how regional modelling could be superior to SH for representing the large-scale fields (unless there is a data distribution issue, of course) because the lateral boundaries may introduce convergence difficulties such as Gibbs phenomenon. Roughly speaking, regional or global approaches thus require signals with wavelengths consistent with the dimension of the studied region. It does not mean that they will fail but that they will necessitate incorporating *a priori* information and regularization.

As one can realize by the preceding pages none of the technique is user-friendly. They require time of adaptation and sometimes new coding from scratch. Understanding the groundwork of each philosophy first (global or regional support), then of each approach, to be able to pick up the technique the most relevant with respect to a particular magnetic dataset is a tedious work that explains well why local functions have not been more widely adopted so far even though they are, in principle, dedicated to detecting small-scale features detection that would be otherwise smoothed out in SH. Other difficulties, not detailed here, arise due to the real data accuracy, distribution of noise, artefacts or biases that require a specific expertise.

Yet, we argue that developing these techniques in the framework of geomagnetism is worth the effort, at least for two simple reasons. One is practical as in the forthcoming years a significantly large amount of high quality satellite data will complement the already large available dataset (Friis-Christensen et al., 2006). However, the amount of near-surface data will not grow as rapidly and the issue of near-surface data spatial distribution will remain critical. In addition, we should not forget that potential fields, in particular magnetic fields, are ones of the few remotely accessible internal properties in planetary explorations. Among some planetary magnetic fields, at least for the Moon and Mars, the contribution from the crust and thus small scales dominate other internal field contributions (e.g., Langlais et al., 2009). Local analysis should be there particularly effective. The second reason is to our point of view too often overlooked. The temptation is big to evaluate, or validate, the robustness of a regional model by comparing results with those provided by SH models. It is customary to prejudge that SH are more or less robust because models indeed showed remarkable fidelity over the last years, especially for the lithospheric field. One should keep in mind, however, that the similarities obtained between SH models may also reflect the self-consistency of the SH procedure (including the data processing) rather than the physics of the magnetic field. Early models of lithospheric fields in SH, based on different procedures, are in fact different (Thébault et al., 2010). In that respect, regional schemes may also be used to challenge the robustness of the standard SH models, to assess regionally their compatibility with dense near-surface measurements, and to verify that magnetic field features are indeed not bound to one specific way of representing the data.

Regional modelling is in its infancy and we do not have the necessary hindsight to state the context in which it is unquestionably superior to SH. Until now regional models have been generally presented as prototypes or used simply for mapping the magnetic field at national scales. Little serious work has been carried out regarding the possible significance of the residuals obtained between their results and equivalent SH models. Investigating if the mismatches show persistent features, if they are independent from the local method used and if they contain time-variability or even periodicity, etc. is ultimately a scope of regional modelling that is likely to offer geophysical novelties. This compels more development, one of the most urgent being probability the ability to define a geomagnetic field spectrum and to separate the sources at a regional scale as this would open new ways to characterize the magnetic field sources in the crust. This, we believe, is not necessarily a long call as the forthcoming abundance of magnetic field measurements and always denser compilations will prompt new interests and thus new practitioners in regional magnetic field modelling.

Acknowledgement We kindly thanks C. Beggan, V. Lesur and F. Simons for providing the data of Figs. 9.2 and 9.9 and helpful discussions and G. Plank for his helpful comments. For IPGP, this is contribution 2638.

References

Abramowitz M, Stegun A (1965) Handbook of mathematical functions. Dover, New York NY

Achache J, Abtout A, Le Mouël JL (1987) The downward continuation of Magsat crustal anomaly field over Southeast Asia. J Geophys Res 92(B11):11, 584–11, 596

Alldredge LR (1981) Rectangular harmonic analysis applied to the geomagnetic field. J Geophys Res 86(B4):3021–3026

Alldredge LR (1982) Geomagnetic local and regional harmonic analyses. J Geophys Res 87(B3):921–1926

Alldredge LR (1983) Varying geomagnetic anomalies and secular variation. J Geophys Res 88(B11):9443–9451

Backus G (1986) Poloidal and toroidal fields in geomagnetic field modelling. Rev Geophys 24(1):75–109

Backus G, Parker R, Constable C (1996) Foundations of geomagnetism. Cambridge University Press, Cambridge

Beggan C, Simons FJ (2009) Reconstruction of bandwidth-limited data on a sphere using Slepian functions: applications to crustal modelling. 505-TUE-1700-0728, IAGA Div.V, August, Sopron, Hungary

Chambodut A, Panet I, Mandea M, Diament M, Holschneider M, Jamet O (2005) Wavelet frames: an alternative to spherical harmonic representation of potential fields. Geophys J Int 163:875–899. doi:10.1111/j.1365-246X.2005.02754.x

Chapman S, Bartels J (1940) Geomagnetism. Oxford University Press, Oxford
Coddington EA (1955) Theory of ordinary differential equations. Mc Graw-Hill, New York, NY
Constable CG, Parker RL, Stark PB (1993). Geomagnetic field models incorporating frozen-flux constraints. Geophys J Int 113:419–433
Dautray R, Lions JL (1987, 1988) Analyse mathématique et calcul numérique pour les sciences et les techniques. Masson
De Santis A. (1991) Translated origin spherical cap harmonic analysis. Geophys J Int 106:253–263
De Santis A, Falcone C (1995) Spherical cap models of Laplacian potentials and general fields. In: Sanso F (ed) Geodetic theory today. Springer, New York, NY, pp 141–150
Freeden W, Glockner O, Thalhammer M (1999) Multiscale gravitational field recovery from GPS satellite-to-satellite tracking. Studia Geoph et Geod 43:229–264
Freeden W, Michel V (2000) Least-squares geopotential approximation by windowed Fourier transform and wavelet transform. In: Klees R, Haagmans R (eds) Wavelet geosciences. Springer, Berlin, pp 189–241
Friis-Christensen E, Lühr H, Hulot G (2006) SWARM: A constellation to study the Earth's magnetic field. Earth Planets Space 58:351–358
Gil A, Segura J, Temme NM (2009) Computing the conical function Pm−1/2+it(x). SIAM J Sci Comp 31(3): 1716–1741
Gillet N, Lesur V, Olsen N (2009) Geomagnetic core field secular variation models. Space Sci Rev pp 1–17, doi:10.1007/s11214-009-9586-6
Gonzalez-Velasco EA (1995) Fourier analysis and boundary value problems. Academic, Pacific Grove, CA
Haines GV (1985a) Spherical cap harmonic analysis. J Geophys Res 90(B3):2583–2591
Haines GV (1985b) Spherical cap harmonic analysis of geomagnetic secular variation over Canada 1960–1983. J Geophys Res 90(B14):12563–12574
Haines GV (1990) Modelling by series expansions: a discussion. J Geomagn Geoelectr 42:1037–1049
Hamoudi M, Thébault E, Lesur V, Mandea M (2007) GeoForschungsZentrum Anomaly Magnetic Map (GAMMA): A candidate model for the world digital magnetic anomaly map. Geochem Geophys Geosyst 8:Q06023. doi:10.1029/2007GC001638
Hobson EW (1965) The theory of spherical and ellipsoidal harmonics. Chelsea, New York, NY, second reprint edition
Holschneider M (1995) Wavelets: an analysis tool. Oxford mathematical monographs, Clarendon Press, Oxford
Holschneider M, Chambodut A, Mandea M (2003) From global to regional analysis of the magnetic field on the sphere using wavelets. Phys Earth Planet Inter 135. doi:10.1016/S0031-9201(02)00210-8
Howarth RJ (2001) A History of regression and related model-fitting in the earth sciences (1636–2000). Natl Resour Res 10(4):241–286
Jerri AJ (1998) The Gibbs phenomenon in Fourier analysis, splines and wavelet approximations. Kluwer, Dordrecht
Kellogg OD (1929) Foundations of potential theory. Dover, New York NY
Korhonen J, Fairhead D, Hamoudi M, Hemant K, Lesur V, Mandea M, Maus S, Purucker M, Ravat D, Sazonova T, Thébault E (2007) Magnetic anomalie map of the world/Carte des anomalies magnétiques du monde, 1st edn, 1:50,000,000, CCGM/CCGMW, ISBN 978-952-217-000-2
Langel RA, Estes RH, Mead GD, Fabiano EB, Lancaster ER (1980) Initial geomagnetic field model from MAGSAT vector data. Geophys Res Lett 7(10):793–796
Langel RA (1987) Main field. In: Jacobs JA (ed) Geomagnetism. pp 249–512. Academic, San Diego, CA
Langel RA, Sabaka TJ, Baldwin RT, Conrad JA (1996) The near-Earth magnetic field from magnetospheric and quiet-day ionospheric sources and how it is modelled. Phys Earth Planet Inter 98:235–267
Langel RA, Hinze WJ (1998) The magnetic field of the earth's lithosphere: the satellite perspective. Cambridge University Press, New York NY
Langlais B, Lesur V, Purucker ME, Connerney JEP, Mandea M (2009) Crustal magnetic field of terrestrial planets. Space Sci Rev. doi:10.1007/s11214-009-9557-y
Lesur V, Gubbins D (1999) Evaluation of fast spherical transforms for geophysical applications. Geophys J Int 139: 547–555
Lesur V (2006) Introducing localized constraints in global geomagnetic field modelling. Earth Planets Space 58:477–483
Lesur V, Maus S (2006) A global lithospheric magnetic field model with reduced noise level in the polar regions. Geophys Res Lett 33:L13304. doi:10.1029/2006GL025826
Lesur V, Wardinski I, Rother M, Mandea M (2008) GRIMM: the GFZ reference internal magnetic model based on vector satellite and observatory data. Geophys J Int 173:382–394. doi:10.1111/j.1365-246X.2008.03724.x
Lowes FJ (1999) Orthogonality and mean squares of vector fields given by spherical harmonic potentials. Geophys J Int 136:781–783
Maier T (2003) Multiscale geomagnetic field modelling from satellite data: theoretical aspects and numerical application. Unpublished PhD thesis, University of Kaiserslautern, Germany
Maier T, Mayer C (2003) Multiscale downward continuation of CHAMP FGM-data for crustal field modelling. In: Reigber C, Lühr H, Schwintzer P (eds) First CHAMP mission results for gravity, magnetic and atmospheric studies. Springer, Berlin, pp 288–295
Malin SRC, Düzgit Z, Baydemir N (1996) Rectangular harmonic analysis revisited. J Geophys Res 101(B12):28,205–28,209
Mandea M, Purucker ME (2005) Observing, modeling, and interpreting magnetic fields of the solid earth. Surv Geophys vol 26(4): pp. 415–459, doi:10.1007/s10712-005-3857-x
Maus S, Rother M, Hemant K, Stolle C, Lühr H, Kuvshinov A, Olsen N (2006) Earth's lithospheric magnetic field determined to spherical harmonic degree 90 from CHAMP satellite measurements. Geophys J Int 164:319–330. doi:10.1111/j.1365-246X.2005.02833.x
Maus S, Lühr H, Rother M, Hemant K, Balasis G, Ritter P, Stolle C (2007a) Fifth generation lithospheric magnetic field model from CHAMP satellite measurements. Geochem Geophys Geosyst 8:Q05013. doi:10.1029/2006GC001521
Maus S, Sazonova T, Hemant K, Fairhead JD, Ravat D (2007b) National geophysical data center candidate for the world digital magnetic anomaly map. Geochem Geophys Geosyst 8:Q06017. doi:10.1029/2007GC001643

Maus S, Yin F, Lühr H, Manoj C, Rother M, Rauberg J, Michaelis I, Stolle C, Müller RD (2008) Resolution of direction of oceanic magnetic lineations by the sixth-generation lithospheric magnetic field model from CHAMP satellite magnetic measurements. Geochem Geophys Geosyst 9:Q0702. doi:10.1029/2008GC001949

Maus S, Barckhausen U, Berkenbosch H, Bournas N, Brozena J, Childers V, Dostaler F, Fairhead JD, Finn C, von Frese RRB, Gaina C, Golynsky S, Kucks R, Lühr H, Milligan P, Mogren S, Müller RD, Olesen O, Pilkington M, Saltus R, Schreckenberger B, Thébault E, Caratori Tontini F (2009) EMAG2: A 2-arc min resolution earth magnetic anomaly grid compiled from satellite, airborne, and marine magnetic measurements. Geochem Geophys Geosyst 10:Q08005. doi:10.1029/2009GC002471

Maus S (2010) An ellipsoidal harmonic representation of Earth's lithospheric magnetic field to degree and order 720, Geochem. Geophys. Geosyst. 11:Q06015. doi:10.1029/2010GC003026

Mayer C, Maier T (2006) Separating inner and outer Earth's magnetic field from CHAMP satellite measurements by means of vector scaling functions and wavelets. Geophys J Int 167:1188–1203. doi:10.1111/j.1365-246X.2006.03199.x

Morse PM, Feshbach H (1953) Methods of theoretical physics. Mc Graw-Hill Company

Nakagawa I., Yukutake T (1985) Rectangular harmonic analyses of geomagnetic anomalies derived from MAGSAT data over the area of the Japanese Islands. J Geomagnetics. Geoelectric. 37(10):957–977

Nakagawa I, Yukutake T, Fukushima N (1985) Extraction of magnetic anomalies of crustal origin from Magsat over the area of the Japanese islands. J Geophys Res 90: 2609–2616

Olsen N, Holme R, Hulot G, Sabaka T, Neubert T, Toffner-Clausen L, Primdahl F, Jorgensen J, Leger J-M, Barraclough D, Bloxham J, Cain J, Constable C, Golovkov V, Jackson A, Kotze P, Langlais B, Macmillan S, Mandea M, Merayo J, Newitt L, Purucker M, Risbo T, Stampe M, Thomson A, Voorhies C (2000) ØRSTED initial field model. Geophys Res Lett 27(22):3607–3610

Olsen N, Lühr H, Sabaka TJ, Mandea M, Rother M, Tøffner-Clausen L, Choi S (2006) CHAOS—A Model of Earthś magnetic field derived from CHAMP, Ørsted, and SAC-C magnetic satellite data. Geophys J Int 166:67–75. doi:10.1111/j.1365-246X.2006.02959.x

Olsen N, Mandea M, Sabaka TJ, Tøffner-Clausen L (2009) CHAOS-2 A geomagnetic field model derived from one decade of continuous satellite data. Geophys J Int 179: 1477–1487. doi:10.1111/j.1365-246X.2009.04386.x

Olver FWJ. (1997) Asymptotics and special functions. Peters AK, Natick, Massachusetts

Panet I, Chambodut A, Diament M, Holschneider M, Jamet O (2006) New insights on intraplate volcanism in French Polynesia from wavelet analysis of GRACE, CHAMP, and sea surface data. Geophys J Res 111:B09403. doi:10.1029/2005JB004141

Purucker ME, Whaler W (2007) Crustal magnetism. : In: Kono M (ed) Geomagnetism, Elsevier, Amsterdam, treatise on geophysics, Chapter 6, vol 5. pp 195–237

Reddy, BD (1998) Introductory functional analysis, vol 27. Springer, New York, NY, Texts in applied mathematics

Reigber C, Lühr H, Schwintzer P (2002) CHAMP Mission status. Adv Space Res 30(2), 129–134. doi:10.1016/S0273-1177(02)00276-4

Robin L (1958) Fonctions sphériques de Legendre et fonctions sphéroidales, vol II and III. Gauthier-Villars, Paris

Sabaka TJ, Baldwin RT (1993) Modeling the Sq magnetic field from POGO and MAGSAT satellite and contemporaneous hourly observatory data: Phase I. Contract Report HSTX/G&G9302

Sabaka TJ, Olsen N, Purucker ME (2004) Extending comprehensive models of the Earth's magnetic field with Ørsted and CHAMP data. Geophys J Int 159:521–547. doi:10.1111/j.1365–246X.2004.02421.x

Sabaka TJ, Olsen N (2006) Enhancing comprehensive inversions using the SWARM constellation. Earth Planet Space 58: 371–395

Sabaka TJ, Hulot G, Olsen N (2009) Mathematical properties relevant to geomagnetic field modelling. In: Freeden W, Nashed Z, Sonar T (eds) Handbook of Geomathematics. Springer, Heidelberg, (in press), ISBN 978-3-642-01547-2

Shure L, Parker RL, Backus GE (1982) Harmonic splines for geomagnetic modelling. Phys Earth Planet Inter 28: 215–229

Simons FJ, Dahlen FA (2006) Spherical slepian functions and the polar gap in geodesy. Geophys J Int 166:1039–1061. doi: 10.1111/j.1365-246X.2006.03065.x

Simons FJ, Dahlen FA, Wieczorek MA (2006) Spatiospectral concentration on a sphere. SIAM Rev 48(3):504–536. doi: 10.1137/S0036144504445765

Simons FJ, Hawthorne JC, Beggan CD (2009) Efficient analysis and representation of geophysical processes using localized spherical basis functions. In: Goyal VK, Papadakis M, Van de Ville D (eds) Wavelets XIII. 7446:74460G1-15. doi:10.1117/12.825730

Simons FJ (2010) Slepian functions and their use in signal estimation and spectral analysis. In: Freeden W, Nashed Z, Sonar T (eds) Handbook of geomathematics. Springer, Heidelberg, (in press), ISBN: 978-3-642-01547-2

Stockman R, Finlay CC, Jackson A (2009) Imaging Earth's crustal magnetic field with satellite data: a regularized spherical triangle tessellation approach. Geophys J Int 179: 929–944. doi:10.1111/j.1365-246X.2009.04345.x

Sneeuw N (1994) Global spherical harmonic analysis by least-squares and numerical quadrature methods in historical perspective. Geophys J Int 118:707–716

Thébault E, Schott JJ, Mandea M, Hoffbeck JP (2004) A new proposal for spherical cap harmonic analysis. Geophys J Int 159:83–105

Thébault E (2006) Global lithospheric magnetic field modeling by successive regional analysis. Earth Planets Space 58: 485–495

Thébault E, Schott JJ, Mandea M, (2006a) Revised spherical cap harmonic analysis (RSCHA): validation and properties. J Geophys Res 111:B01102. doi:10.1029/2005JB003836

Thébault E, Mandea M, Schott JJ (2006b) Modelling the lithospheric magnetic field over France by means of revised spherical cap harmonic analysis (R-SCHA). J Geophys Res 111:B05102. doi:10.1029/2005JB004110

Thébault E (2008) A proposal for regional modelling at the Earth's surface, R-SCHA2D. Geophys J Int. doi:10.1111/j.1365-246X.2008.03823.x

Thébault E, Gaya-Piqué L (2008) Applied comparisons between SCHA and R-SCHA regional modelling techniques. Geochem Geophys Geosyst 9:Q07005. doi:10.1029/2008GC001953

Thébault E, Purucker ME, Whaler K, Langlais B, Sabaka TJ (2010) The Magnetic field of the Earth's lithosphere. Space Sci Rev, doi:10.1007/S11214-010-9667-6

Torta JM, Gaya-Piqué LR, De Santis A (2006) Spherical cap harmonic analysis of the geomagnetic field with application for aeronautical mapping. In: Rasson JL, Delipetrov T (eds) Geomagnetics for aeronautical safety: a case study in and around the balkans. Springer, Dordrecht, pp 291–307

Chapter 10

The International Geomagnetic Reference Field

Susan Macmillan and Christopher Finlay

Abstract The International Geomagnetic Reference Field (IGRF) is an internationally agreed and widely used mathematical model of the Earth's magnetic field of internal origin. It is produced and agreed under the auspices of IAGA. We describe its inception in the 1960s and how it has developed since. We also describe the current generation of the IGRF and potential future developments. Maps of the geomagnetic field derived from the IGRF and valid for 2010–2015 are also included.

10.1 Introduction

The International Geomagnetic Reference Field (IGRF) is an internationally agreed and widely used mathematical model of the Earth's magnetic field of internal origin. We describe its inception in the 1960s and how it has developed since. We also describe the current generation of the IGRF (the 11th) and potential future developments.

10.2 Scope of the IGRF

The IGRF is designed to provide an easily accessible approximation, near and above the Earth's surface, to the large-scale part of the Earth's magnetic field which has its origin inside the surface. This field is predominantly that due to electric currents in the Earth's liquid metal core.

Rapid field fluctuations due to variations of electric current systems in the magnetosphere and ionosphere, as well as the weak, smaller scale field due to magnetized crustal rocks are not included in the IGRF.

10.3 Inception and Development

The concept of an IGRF grew out of discussions concerning the presentation of the results of the World Magnetic Survey (WMS) (Barraclough, 1993). The WMS was a deferred element in the programme of the 1957–1958 International Geophysical Year which, during the next 12 years, encouraged magnetic surveys on land, at sea, in the air and from satellites and organised the collection and analysis of the results. At a meeting in 1960, the Committee on World Magnetic Survey and Magnetic Charts of IAGA recommended that, as part of the WMS programme, a global spherical harmonic model of the field be derived using the results of the WMS. This proposal was accepted but another 8 years of argument and discussion followed (see Zmuda (1971) for a summary of this, together with a detailed description of the WMS programme) before the first IGRF was ratified by IAGA in 1969.

The IGRF has now been revised and updated ten times since 1969 and a summary of the revision history is given in Table 10.1 (see also Barraclough (1993) and Barton (1997) and references therein). More details concerning the latest revision—IGRF 11th generation (Finlay et al. 2010)—are given below.

Each generation of the IGRF comprises several constituent models at five-year intervals, each one of

S. Macmillan (✉)
British Geological Survey, Edinburgh EH9 3LA, UK
e-mail: smac@bgs.ac.uk

M. Mandea, M. Korte (eds.), *Geomagnetic Observations and Models*, IAGA Special Sopron Book Series 5,
DOI 10.1007/978-90-481-9858-0_10,

Table 10.1 Summary of IGRF history

Full name	Short name	Valid for	Definitive for	References
IGRF 11th generation (revised 2009)	IGRF-11	1900.0–2015.0	1945.0–2005.0	Finlay et al. (2010)
IGRF 10th generation (revised 2004)	IGRF-10	1900.0–2010.0	1945.0–2000.0	Macmillan and Maus (2005)
IGRF 9th generation (revised 2003)	IGRF-9	1900.0–2005.0	1945.0–2000.0	Macmillan et al. (2003)
IGRF 8th generation (revised 1999)	IGRF-8	1900.0–2005.0	1945.0–1990.0	Mandea and Macmillan (2000)
IGRF 7th generation (revised 1995)	IGRF-7	1900.0–2000.0	1945.0–1990.0	Barton (1997)
IGRF 6th generation (revised 1991)	IGRF-6	1945.0–1995.0	1945.0–1985.0	Langel (1992)
IGRF 5th generation (revised 1987)	IGRF-5	1945.0–1990.0	1945.0–1980.0	Langel et al. (1988)
IGRF 4th generation (revised 1985)	IGRF-4	1945.0–1990.0	1965.0–1980.0	Barraclough (1987)
IGRF 3rd generation (revised 1981)	IGRF-3	1965.0–1985.0	1965.0–1975.0	Peddie (1982)
IGRF 2nd generation (revised 1975)	IGRF-2	1955.0–1980.0	–	IAGA (1975)
IGRF 1st generation (1969)	IGRF-1	1955.0–1975.0	–	Zmuda (1971)

which is designated definitive or non-definitive. Once a constituent model is designated definitive it is called a Definitive Geomagnetic Reference Field (DGRF) and it is not revised in subsequent generations of the IGRF. The non-definitive constituent models are referred to, rather confusingly, as IGRFs.

10.4 Applications and Availability

The original idea of an IGRF had come from global modellers, including those who produced such models in association with the production of navigational charts. However, the IGRF as it was first formulated was not considered to be accurate or detailed enough for navigational purposes.

The majority of users of the IGRF at the time of its inception consisted of geophysicists interested in the geological interpretation of regional magnetic surveys. An initial stage in such work is the removal of a background field, that approximates the field whose sources are in the Earth's core, from the observations. With different background fields being used for different surveys, difficulties arose when adjacent surveys had to be combined. An internationally agreed global model, accurately representing the field from the core, eased this problem considerably.

Another group of researchers who were becoming increasingly interested in descriptions of the geomagnetic field at this time were those studying the ionosphere and magnetosphere and behaviour of cosmic rays in the vicinity of the Earth. This remains an important user community today, with the IGRF being the internal field model of many ionospheric and magnetospheric models (for example, Tsygenenko, 2002).

Geomagnetic coordinate systems are almost exclusively based on the IGRF (Russell, 1971; Hapgood, 1992, 1997). A commonly used axis in these coordinate systems is that of the centred dipole and for 2010.0 from IGRF-11, this is tilted at an angle of 9.99° to the Earth's axis of rotation and has longitude 72.22°W in the northern hemisphere.

Today, there are many on-line calculators available for the IGRF, screenshots of two examples are shown in Figs. 10.1 and 10.2. Most on-line calculators demand position input relative to the surface of the WGS84 reference ellipsoid model of the Earth, and convert the position from a geodetic to a geocentric coordinate system for use in the spherical harmonic expansion. A few (e.g., Fig. 10.1) also permit the input position to be in the geocentric coordinate system.

10.5 Geomagnetic Field Components

The geomagnetic field vector **B** is fully described by an appropriate set of three elements selected from the seven possible elements (Fig. 10.3). The orthogonal set is the northerly intensity X, the easterly intensity Y and the vertical intensity Z (positive downwards). The other elements are the horizontal intensity H, the total intensity F, the inclination angle I, (also called the dip angle and measured from the horizontal plane to the field vector, positive downwards), and the declination angle D (also called the magnetic variation and

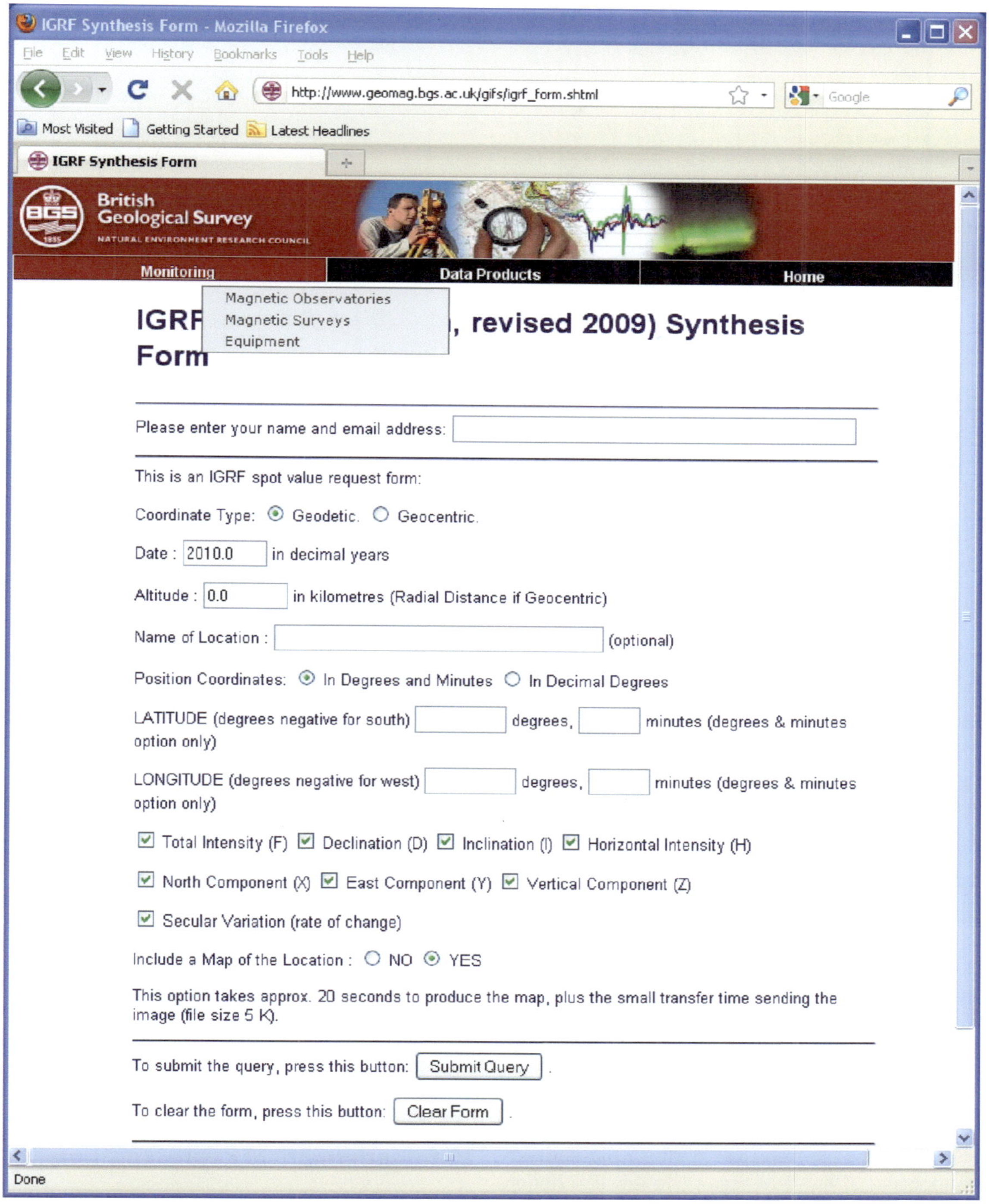

Fig. 10.1 On-line IGRF calculator maintained by the British Geological Survey

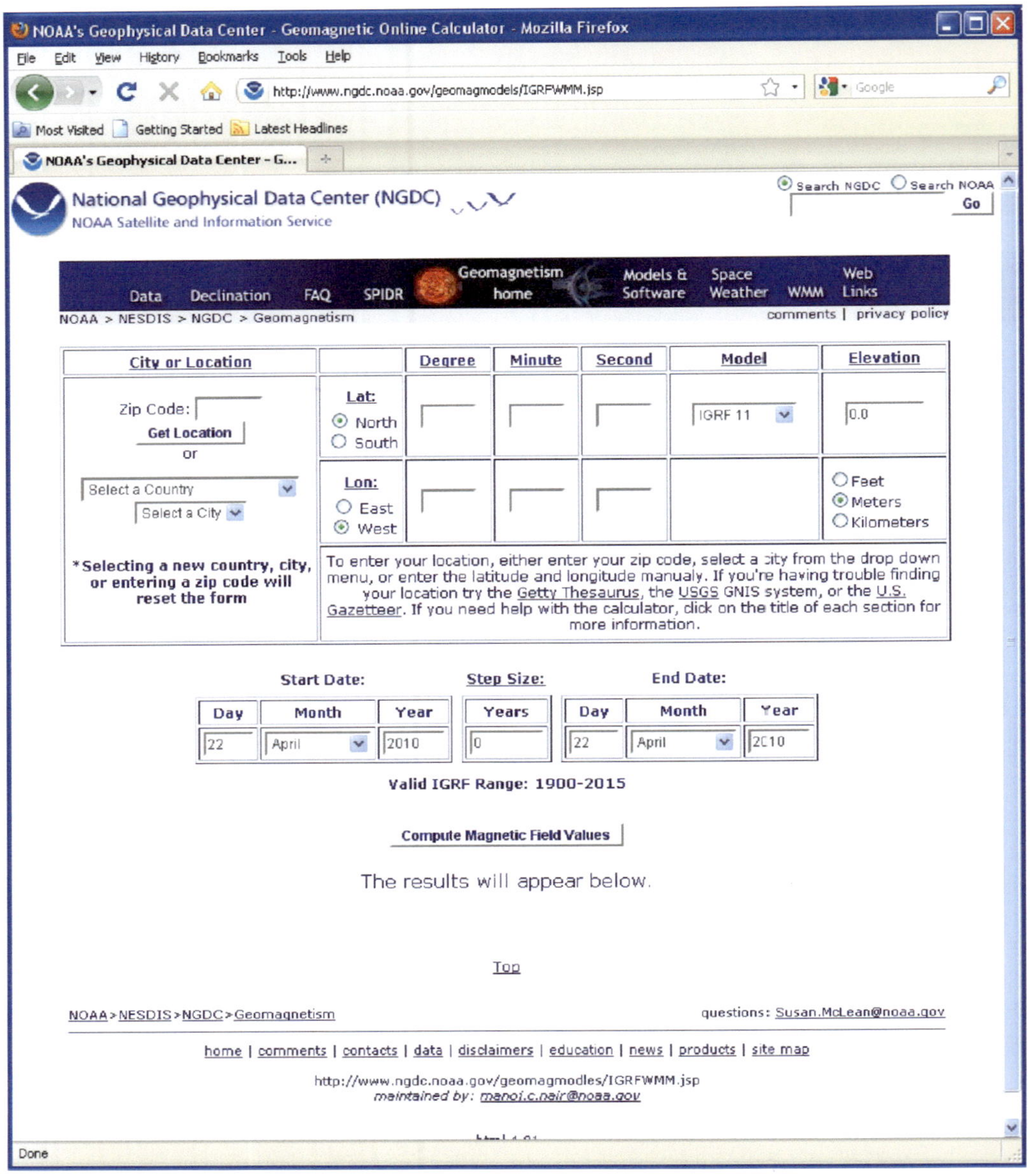

Fig. 10.2 On-line calculator maintained by the US National Oceanographic and Atmospheric Administration

measured clockwise from true north to the horizontal component of the field vector). In this description of X, Y, Z, H, F, I and D, the vertical direction is assumed perpendicular to the WGS84 reference ellipsoid model of the Earth's surface and the clockwise rotational direction is determined by a view from above the Earth. Conventionally the intensities are given in units of nanoTeslas (nT).

10.6 Mathematical Representation

In a source-free region the Earth's magnetic field **B** is the negative gradient of a magnetic potential V that satisfies Laplace's equation:

$$\mathbf{B} = -\nabla V \quad \text{where} \quad \nabla^2 V = 0$$

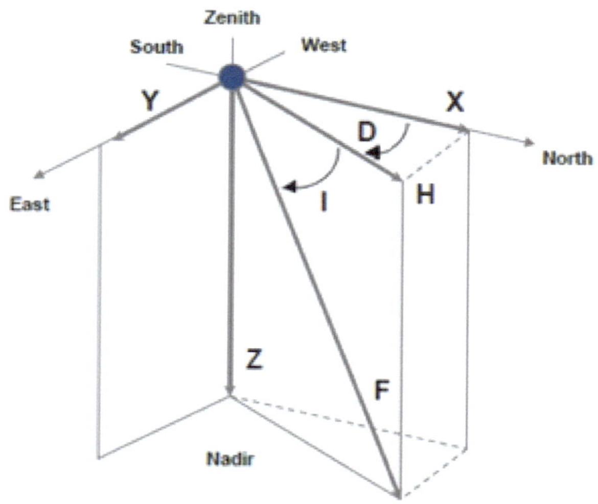

Fig. 10.3 The 7 elements of the magnetic field

Each constituent model of the IGRF is a set of spherical harmonics of degree n and order m, representing a solution to Laplace's equation for the magnetic potential arising from sources inside the Earth at a given epoch; the harmonics are associated with the Gauss coefficients g_n^m and h_n^m:

$$V(r,\theta,\lambda) = a\sum_{n=1}^{n_{\max}}\left(\frac{a}{r}\right)^{n+1}\sum_{m=0}^{n}\left(g_n^m\cos m\lambda + h_n^m\sin m\lambda\right)P_n^m(\theta)$$

In this equation r, θ, λ are geocentric coordinates (r is the distance from the centre of the Earth, θ is the geocentric colatitude, i.e., 90°—latitude, and λ is the longitude), a is a reference radius for the Earth (6371.2 km), $P_n^m(\theta)$ are the Schmidt semi-normalised associated Legendre polynomials and $n_{\max}$ is the maximum degree of the spherical harmonic expansion. Conventionally the units of the Gauss coefficients are nT.

In addition to the DGRFs and IGRFs which represent the main field at 5-year intervals there is always a predictive secular-variation model to allow computation of the magnetic field for some time after the epoch of the last main-field model, generally for 5 years after but sometimes longer. Recent generations of the IGRF have been produced in a timely manner, i.e., before the previous generation was no longer strictly valid, but this was not always the case for the early generations of the IGRF. The predictive secular-variation model comprises Gauss coefficients in units of nT/year. Between the DGRF and IGRF main-field models the magnetic potential, and therefore the magnetic field, is assumed to vary linearly with time.

New constituent models of the IGRF are carefully produced and well documented. The IAGA Working Group charged with the production of the IGRF invites submissions of candidate models several months in advance of decision dates. Detailed evaluations are then made of all submitted models and the final decision is made by the IAGA Working Group. The evaluations are also widely documented. For most generations of the IGRF there is a special issue of a journal containing papers describing the candidate models and the evaluations. For IGRF-10 see volume 57, number 12 of the journal *Earth, Planets and Space*, a similar special issue of *Earth, Planets and Space* is in preparation for IGRF-11 at time of writing.

The coefficients of the new constituent models are derived by taking means (sometimes weighted) of the coefficients of selected candidate models. This method of combining several candidate models has been used in almost all generations as, not only are different selections of available data made by the teams submitting models, there are many different methods for dealing with the fields which are not modelled by the IGRF, for example the ionospheric, magnetospheric and crustal fields.

The constituent main-field models of the most recent generation of the IGRF (Finlay et al. 2010) extend to spherical harmonic degree 10 up to and including epoch 1995.0; thereafter they extend to degree 13 to take advantage of the excellent coverage and quality of satellite data provided by Ørsted and CHAMP. The predictive secular-variation model has extended to degree 8 for all generations of the IGRF to date.

10.7 IGRF 11th Generation (Revised 2009)

For IGRF-11 a Working Group was set up at the 2007 IAGA meeting in Toulouse. During 2009 eight modelling teams around the world worked on production of candidate sets of coefficients for IGRF-11. These were a DGRF set of coefficients for 2005.0, an IGRF set of coefficients for 2010.0 and a predictive secular-variation set of coefficients for 2010.0–2015.0 (Finlay et al. 2010). At the 2009 Sopron meeting of IAGA a report on progress towards IGRF-11 was given, i.e., the teams participating and summaries of their data selections and modelling techniques. The final sets of coefficients were determined later in the year by vote.

The most common approach taken by each modelling team to produce their candidate models was to select data from the satellite and ground-based datasets available, to decide on an appropriate parameterisation of parent models, invert the selected datasets for the parent models, iterate this process several times, and extract or extrapolate the final sets of candidate coefficients. The parent models generally included time-varying signals of external origin, i.e., from outside the Earth in the ionosphere and magnetosphere, and signals from the Earth's crust represented by spherical harmonic degrees greater than 13. This is because any observation of the magnetic field includes signals from the core, the crust and the coupled ionosphere-magnetosphere system. When trying to model the field of internal origin, the biggest challenge at present is probably dealing with the simultaneous presence of time-varying fields produced by the ionosphere and magnetosphere. Most teams select periods of undisturbed data (using a variety of indices and solar wind data) on the night-side of the Earth in order to reduce the ionospheric and magnetospheric contamination but there is no general agreement as to how best to choose these periods. (There is a trade-off between good spatial and temporal coverage and not using contaminated data.) Then modellers generally attempt an estimation of any remaining signal in the data of external origin, often relying on other data (indices) to do so. The results are variable.

One predictive secular variation candidate model, however, was derived from the assimilation of a parent model into a numerical geodynamo model, showing how this area of ongoing research is now starting to find application.

The evaluation process involved several independent assessments followed by a vote. The final IGRF-11 coefficients are available from the IAGA web page at http://www.ngdc.noaa.gov/IAGA/vmod/.

The coefficients which are revised for IGRF-11 are listed in Table 10.2.

Table 10.2 The new coefficients (in nT and nT/year) in IGRF-11

g/h	n	m	2005.0	2010.0	SV
g	1	0	−29554.63	−29496.5	11.4
g	1	1	−1669.05	−1585.9	16.7
h	1	1	5077.99	4945.1	−28.8
g	2	0	−2337.24	−2396.6	−11.3
g	2	1	3047.69	3026.0	−3.9
h	2	1	−2594.50	−2707.7	−23.0
g	2	2	1657.76	1668.6	2.7
h	2	2	−515.43	−575.4	−12.9
g	3	0	1336.30	1339.7	1.3
g	3	1	−2305.83	−2326.3	−3.9
h	3	1	−198.86	−160.5	8.6
g	3	2	1246.39	1231.7	−2.9
h	3	2	269.72	251.7	−2.9
g	3	3	672.51	634.2	−8.1
h	3	3	−524.72	−536.8	−2.1
g	4	0	920.55	912.6	−1.4
g	4	1	797.96	809.0	2.0
h	4	1	282.07	286.4	0.4
g	4	2	210.65	166.6	−8.9
h	4	2	−225.23	−211.2	3.2
g	4	3	−379.86	−357.1	4.4
h	4	3	145.15	164.4	3.6
g	4	4	100.00	89.7	−2.3
h	4	4	−305.36	−309.2	−0.8
g	5	0	−227.00	−231.1	−0.5
g	5	1	354.41	357.2	0.5
h	5	1	42.72	44.7	0.5
g	5	2	208.95	200.3	−1.5
h	5	2	180.25	188.9	1.5
g	5	3	−136.54	−141.2	−0.7
h	5	3	−123.45	−118.1	0.9
g	5	4	−168.05	−163.1	1.3
h	5	4	−19.57	0.1	3.7
g	5	5	−13.55	−7.7	1.4
h	5	5	103.85	100.9	−0.6
g	6	0	73.60	72.8	−0.3
g	6	1	69.56	68.6	−0.3
h	6	1	−20.33	−20.8	−0.1
g	6	2	76.74	76.0	−0.3
h	6	2	54.75	44.2	−2.1
g	6	3	−151.34	−141.4	1.9
h	6	3	63.63	61.5	−0.4
g	6	4	−14.58	−22.9	−1.6
h	6	4	−63.53	−66.3	−0.5
g	6	5	14.58	13.1	−0.2
h	6	5	0.24	3.1	0.8
g	6	6	−86.36	−77.9	1.8
h	6	6	50.94	54.9	0.5
g	7	0	79.88	80.4	0.2
g	7	1	−74.46	−75.0	−0.1
h	7	1	−61.14	−57.8	0.6
g	7	2	−1.65	−4.7	−0.6

Table 10.2 (continued)

g/h	*n*	m	2005.0	2010.0	SV
h	7	2	−22.57	−21.2	0.3
g	7	3	38.73	45.3	1.4
h	7	3	6.82	6.6	−0.2
g	7	4	12.30	14.0	0.3
h	7	4	25.35	24.9	−0.1
g	7	5	9.37	10.4	0.1
h	7	5	10.93	7.0	−0.8
g	7	6	5.42	1.6	−0.8
h	7	6	−26.32	−27.7	−0.3
g	7	7	1.94	4.9	0.4
h	7	7	−4.64	−3.4	0.2
g	8	0	24.80	24.3	−0.1
g	8	1	7.62	8.2	0.1
h	8	1	11.20	10.9	0.0
g	8	2	−11.73	−14.5	−0.5
h	8	2	−20.88	−20.0	0.2
g	8	3	−6.88	−5.7	0.3
h	8	3	9.83	11.9	0.5
g	8	4	−18.11	−19.3	−0.3
h	8	4	−19.71	−17.4	0.4
g	8	5	10.17	11.6	0.3
h	8	5	16.22	16.7	0.1
g	8	6	9.36	10.9	0.2
h	8	6	7.61	7.1	−0.1
g	8	7	−11.25	−14.1	−0.5
h	8	7	−12.76	−10.8	0.4
g	8	8	−4.87	−3.7	0.2
h	8	8	−0.06	1.7	0.4
g	9	0	5.58	5.4	0.0
g	9	1	9.76	9.4	0.0
h	9	1	−20.11	−20.5	0.0
g	9	2	3.58	3.4	0.0
h	9	2	12.69	11.6	0.0
g	9	3	−6.94	−5.3	0.0
h	9	3	12.67	12.8	0.0
g	9	4	5.01	3.1	0.0
h	9	4	−6.72	−7.2	0.0
g	9	5	−10.76	−12.4	0.0
h	9	5	−8.16	−7.4	0.0
g	9	6	−1.25	−0.8	0.0
h	9	6	8.10	8.0	0.0
g	9	7	8.76	8.4	0.0
h	9	7	2.92	2.2	0.0
g	9	8	−6.66	−8.4	0.0
h	9	8	−7.73	−6.1	0.0
g	9	9	−9.22	−10.1	0.0
h	9	9	6.01	7.0	0.0
g	10	0	−2.17	−2.0	0.0
g	10	1	−6.12	−6.3	0.0
h	10	1	2.19	2.8	0.0
g	10	2	1.42	0.9	0.0
h	10	2	0.10	−0.1	0.0
g	10	3	−2.35	−1.1	0.0

Table 10.2 (continued)

g/h	n	*m*	2005.0	2010.0	SV
h	10	3	4.46	4.7	0.0
g	10	4	−0.15	−0.2	0.0
h	10	4	4.76	4.4	0.0
g	10	5	3.06	2.5	0.0
h	10	5	−6.58	−7.2	0.0
g	10	6	0.29	−0.3	0.0
h	10	6	−1.01	−1.0	0.0
g	10	7	2.06	2.2	0.0
h	10	7	−3.47	−4.0	0.0
g	10	8	3.77	3.1	0.0
h	10	8	−0.86	−2.0	0.0
g	10	9	−0.21	−1.0	0.0
h	10	9	−2.31	−2.0	0.0
g	10	10	−2.09	−2.8	0.0
h	10	10	−7.93	−8.3	0.0
g	11	0	2.95	3.0	0.0
g	11	1	−1.60	−1.5	0.0
h	11	1	0.26	0.1	0.0
g	11	2	−1.88	−2.1	0.0
h	11	2	1.44	1.7	0.0
g	11	3	1.44	1.6	0.0
h	11	3	−0.77	−0.6	0.0
g	11	4	−0.31	−0.5	0.0
h	11	4	−2.27	−1.8	0.0
g	11	5	0.29	0.5	0.0
h	11	5	0.90	0.9	0.0
g	11	6	−0.79	−0.8	0.0
h	11	6	−0.58	−0.4	0.0
g	11	7	0.53	0.4	0.0
h	11	7	−2.69	−2.5	0.0
g	11	8	1.80	1.8	0.0
h	11	8	−1.08	−1.3	0.0
g	11	9	0.16	0.2	0.0
h	11	9	−1.58	−2.1	0.0
g	11	10	0.96	0.8	0.0
h	11	10	−1.90	−1.9	0.0
g	11	11	3.99	3.8	0.0
h	11	11	−1.39	−1.8	0.0
g	12	0	−2.15	−2.1	0.0
g	12	1	−0.29	−0.2	0.0
h	12	1	−0.55	−0.8	0.0
g	12	2	0.21	0.3	0.0
h	12	2	0.23	0.3	0.0
g	12	3	0.89	1.0	0.0
h	12	3	2.38	2.2	0.0
g	12	4	−0.38	−0.7	0.0
h	12	4	−2.63	−2.5	0.0
g	12	5	0.96	0.9	0.0
h	12	5	0.61	0.5	0.0
g	12	6	−0.30	−0.1	0.0
h	12	6	0.40	0.6	0.0
g	12	7	0.46	0.5	0.0
h	12	7	0.01	0.0	0.0

Table 10.2 (continued)

g/h	n	*m*	2005.0	2010.0	SV
g	12	8	−0.35	−0.4	0.0
h	12	8	0.02	0.1	0.0
g	12	9	−0.36	−0.4	0.0
h	12	9	0.28	0.3	0.0
g	12	10	0.08	0.2	0.0
h	12	10	−0.87	−0.9	0.0
g	12	11	−0.49	−0.8	0.0
h	12	11	−0.34	−0.2	0.0
g	12	12	−0.08	0.0	0.0
h	12	12	0.88	0.8	0.0
g	13	0	−0.16	−0.2	0.0
g	13	1	−0.88	−0.9	0.0
h	13	1	−0.76	−0.8	0.0
g	13	2	0.30	0.3	0.0
h	13	2	0.33	0.3	0.0
g	13	3	0.28	0.4	0.0
h	13	3	1.72	1.7	0.0
g	13	4	−0.43	−0.4	0.0
h	13	4	−0.54	−0.6	0.0
g	13	5	1.18	1.1	0.0
h	13	5	−1.07	−1.2	0.0
g	13	6	−0.37	−0.3	0.0
h	13	6	−0.04	−0.1	0.0
g	13	7	0.75	0.8	0.0
h	13	7	0.63	0.5	0.0
g	13	8	−0.26	−0.2	0.0
h	13	8	0.21	0.1	0.0
g	13	9	0.35	0.4	0.0
h	13	9	0.53	0.5	0.0
g	13	10	−0.05	0.0	0.0
h	13	10	0.38	0.4	0.0
g	13	11	0.41	0.4	0.0
h	13	11	−0.22	−0.2	0.0
g	13	12	−0.10	−0.3	0.0
h	13	12	−0.57	−0.5	0.0
g	13	13	−0.18	−0.3	0.0
h	13	13	−0.82	−0.8	0.0

10.8 Global Magnetic Field Patterns

Global maps of the magnetic elements, based on IGRF-11 and valid for the period 2010.0 to 2015.0, are shown in Figs. 10.4, 10.5, 10.6, 10.7, 10.8, 10.9, and 10.10.

Using IGRF-11 to compute the root mean square magnetic field vector at the Earth's surface through time arising from all spherical harmonic terms ($n \leq 10$), the centred dipole terms ($n = 1$) and the non-dipole terms ($1 < n \leq 10$), gives Fig. 10.11. It can be seen that since 1900 the Earth's magnetic field is weakening overall, by reduction of the dipole field. However the non-dipolar part is strengthening, though to a lesser extent. This may have consequences for the trajectories of energetic charged particles that enter the Earth's magnetosphere. One manifestation of this increase in the non-dipole field is the deepening, and westwards movement, of the South Atlantic Anomaly (Macmillan et al. 2009), a region where the Earth's magnetic field is weaker than elsewhere (see Fig. 10.8). In this region energetic charged particles are able to penetrate closer to the Earth, and cause a radiation hazard for satellites passing through the region.

10.9 Limitations

The limitations of the IGRF are discussed in a "health warning" available from the IAGA web page at http://www.ngdc.noaa.gov/IAGA/vmod/. The accuracy of the IGRF is considered to be limited by a combination of two types of error, namely error of commission where there is a difference between the IGRF and the part of the field that it is attempting to model, and error of omission where the error is the part of the field that the IGRF is not attempting to model. The difficulty is that it is not easy to directly separate those parts of the observed magnetic field due to the different sources since they each produce signals spanning a range of wavelengths and frequencies. In fact the separation can only properly be done through co-estimation of all sources.

The errors of commission are estimated mainly by comparing different generations of the IGRF at dates common to both. They vary considerably with time. Recent constituent models of the IGRF for epochs when satellite data were available are thought to be within 5-10 nT root mean square (rms) of the true value, the true value in this case being the internal field up to spherical harmonic degree n_{max} at the Earth's surface. Other constituent models are thought to be within 50–300 nT of the true value.

The error of omission is dominated by the crustal field and the rms value is estimated to be 200–300 nT. At high latitudes and on the day-side of the Earth the ionospheric and magnetospheric fields will become more significant.

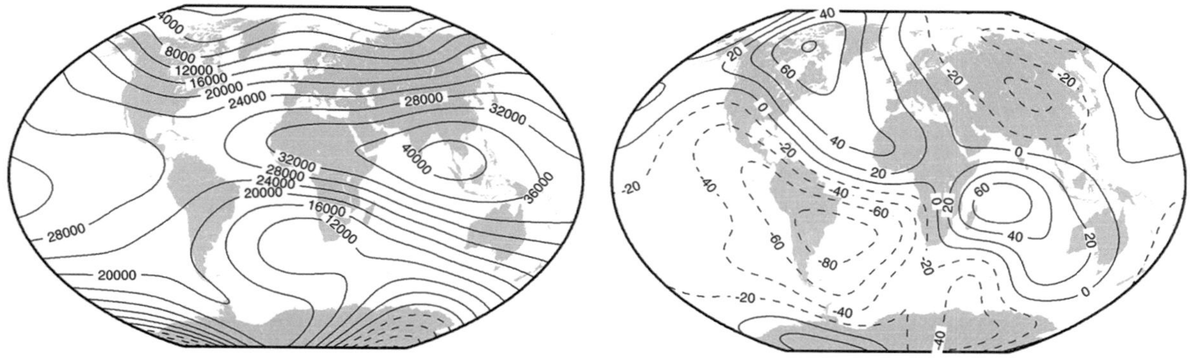

Fig. 10.4 Northerly intensity *X* (nT) at 2010.0 and its rate of change (nT/year) for 2010.0–2015.0 computed from IGRF-11. Map projection is Winkel Tripel

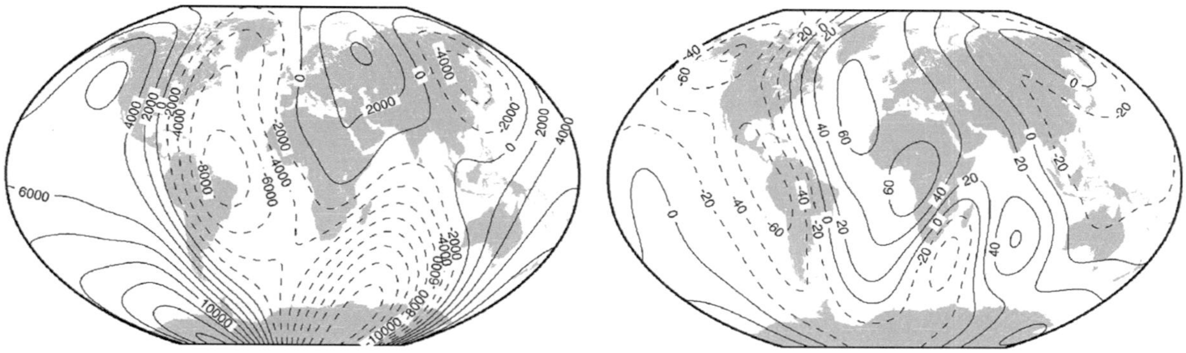

Fig. 10.5 Easterly intensity *Y* (nT) at 2010.0 and its rate of change (nT/year) for 2010.0–2015.0 computed from IGRF-11

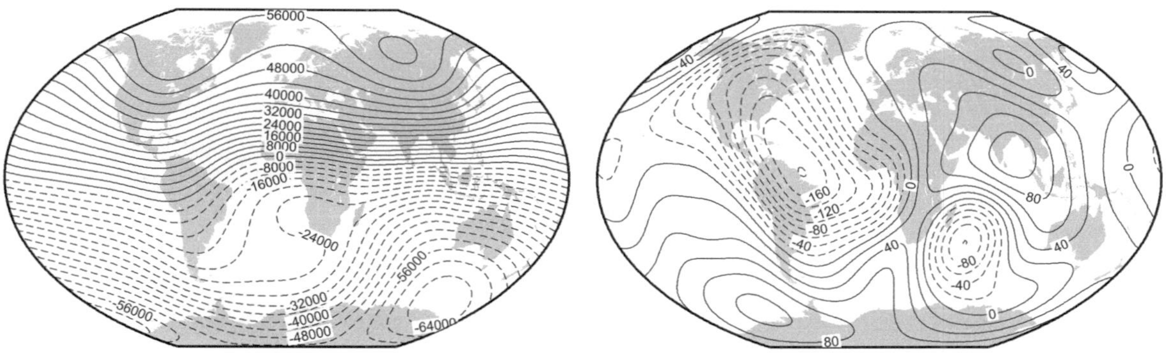

Fig. 10.6 Vertical intensity *Z* (nT) at 2010.0 and its rate of change (nT/year) for 2010.0–2015.0 computed from IGRF-11

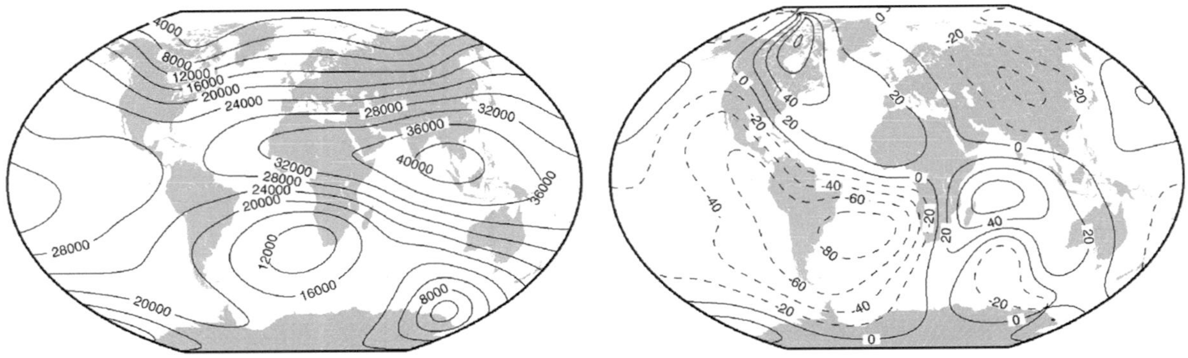

Fig. 10.7 Horizontal intensity *H* (nT) at 2010.0 and its rate of change (nT/year) for 2010.0–2015.0 computed from IGRF-11

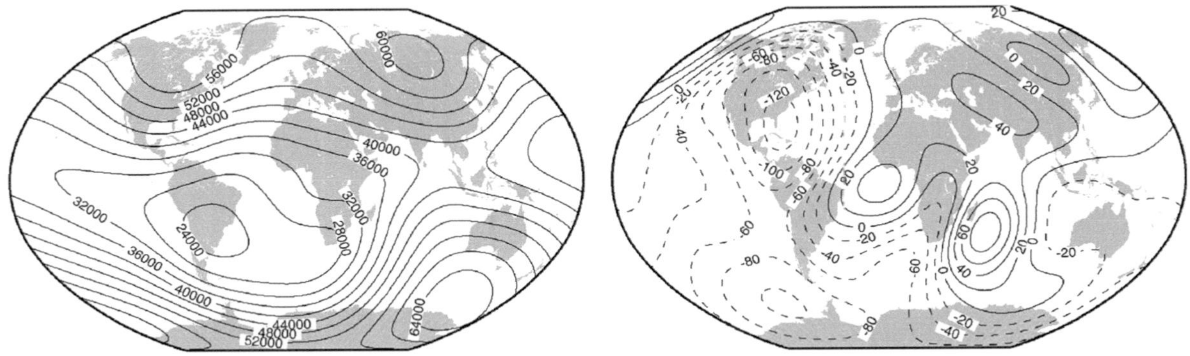

Fig. 10.8 Total intensity *F* (nT) at 2010.0 and its rate of change (nT/year) for 2010.0–2015.0 computed from IGRF-11

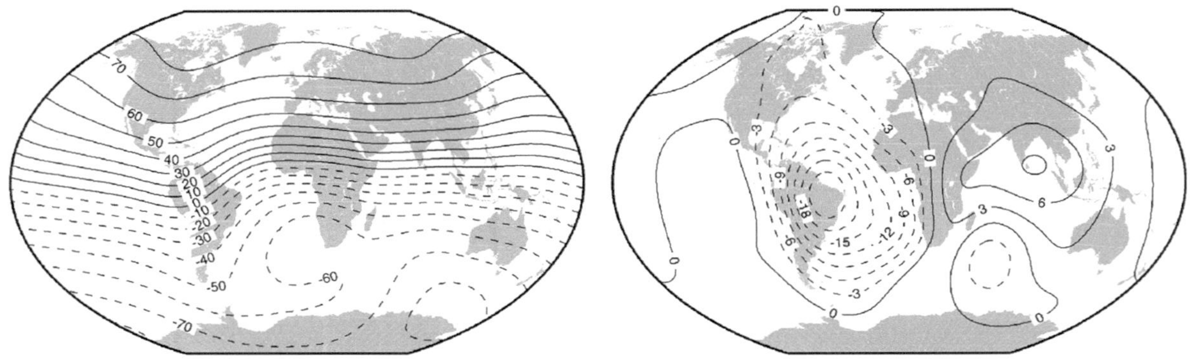

Fig. 10.9 Inclination *I* (degrees) at 2010.0 and its rate of change (arc-minutes/year) for 2010.0–2015.0 computed from IGRF-11

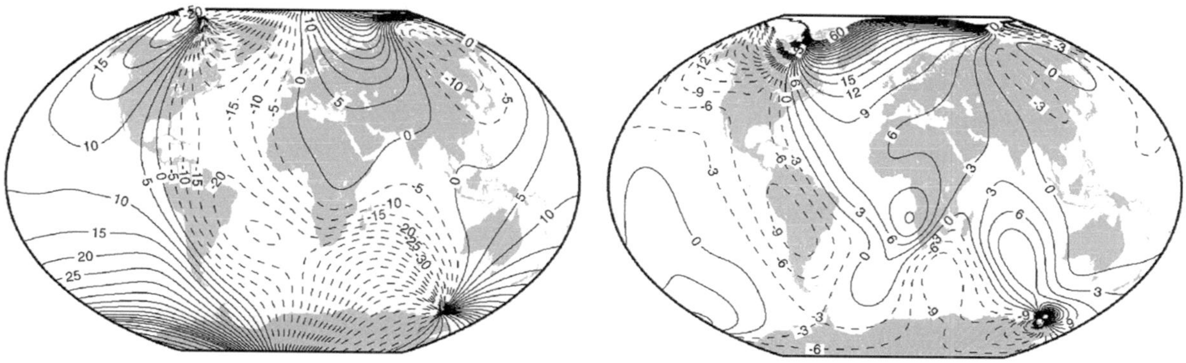

Fig. 10.10 Declination *D* (degrees) at 2010.0 and its rate of change (arc-minutes/year) for 2010.0–2015.0 computed from IGRF-11. (Declination is not defined at the geographic poles or magnetic dip poles)

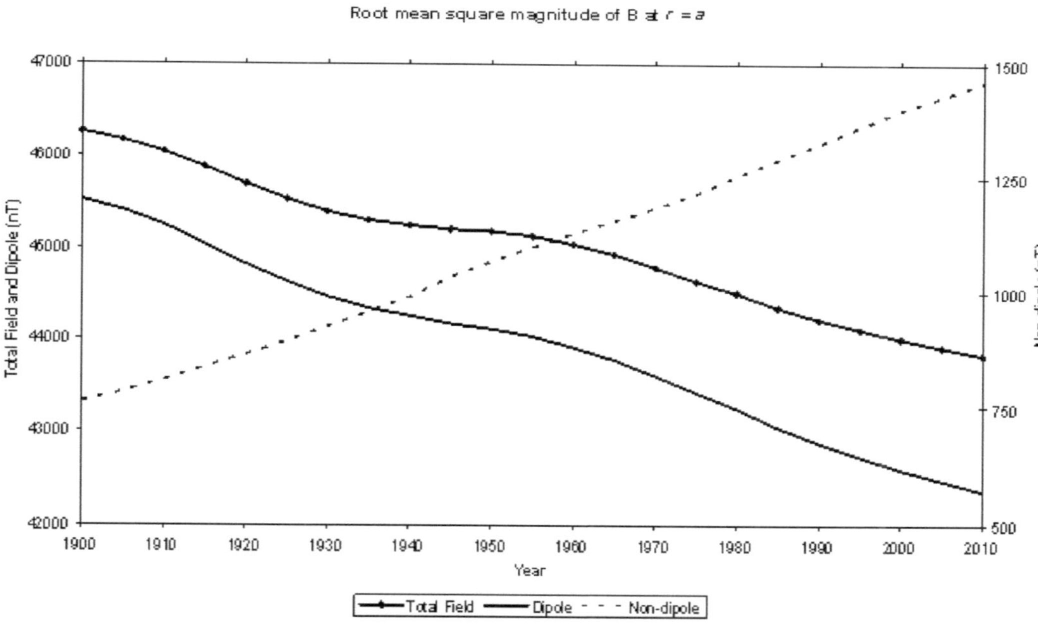

Fig. 10.11 The decline of the whole, and dipolar part of the magnetic field at the Earth's surface and the growth of the non-dipolar part since 1900, computed from IGRF-11

10.10 Future

Firstly, no model of the geomagnetic field can be better than the data on which it is based. An assured supply of high-quality data distributed evenly over the Earth's surface is therefore a fundamental prerequisite for a continuing and acceptably accurate IGRF. Data from magnetic observatories continue to be the most important source of information about time-varying fields. However their spatial distribution is poor and although data from other sources such as repeat stations, the high-level vector aeromagnetic survey Project MAGNET programme lasting from 1953 to 1994, and marine magnetic surveys have all helped to fill in the gaps, the best spatial coverage is provided by near-polar satellites. Measurements made by the

POGO satellites (1965–1971), Magsat (1979–1980), POGS (1990–1993), Ørsted (1999–), SAC-C (2001–2004) and CHAMP (2000-) have all been utilised in the production of the IGRF. They have ultimately been responsible for the improved quality of recent IGRF revisions.

Secondly, the future of the IGRF depends on the continuing ability of the groups who have contributed candidate models to the IGRF revision process to produce global magnetic field models. This ability is dependent on the willingness of the relevant funding authorities to continue to support this type of work.

Thirdly, the continued interest of IAGA is a necessary requirement for the future of the IGRF. This is assured as long as there is, as at present, a large and diverse group of IGRF-users around the world. One reason why the IGRF has gained the reputation it has is because it is endorsed and recommended by IAGA, the recognised international organisation for geomagnetism.

Finally the extension of the predictive secular-variation model to spherical harmonic degree 13 (same as the main field) will be considered for the IGRF-12. The decision will be based on whether the European Space Agency Swarm mission is delivering good-quality data by the time of the model revision and the success of some trial predictive secular-variation models that were submitted at the time of production of the IGRF-11.

References

Barraclough DR (1993) History and Development of the IGRF and DGRF series of geomagnetic reference field models. BGS Technical Report WM/93/26

Barraclough, DR (1987) International geomagnetic reference field: the fourth generation. Phys Earth Planet In 48:279–292

Barton CE (1997) International geomagnetic reference field: the seventh generation. J Geomagn Geoelectr 49: 123–148

Finlay CC, Maus S, Bondar T, Chambodut A, Chernova TA, Chulliat A, Golovkov VP, Hamilton B, Hamoudi M, Hulot G, Langlais B, Lesur V, Lowes FJ, Lühr H, Kuang W, Macmillan S, McLean S, Mandea M, Manoj C, Menvielle M, Michelis I, Olsen N, Rauberg J, Rother M, Sabaka TJ, Thebault E, Thomson AWP, Toffner-Clausen L, Wardinski I, Zvereva TI (2010) The 11th-generation international geomagnetic reference field. Geophys J Int (in press)

Hapgood MA, (1992) Space physics coordinate transformations: a user guide. Planet Space Sci 40:711–717

Hapgood MA, (1997) Corrigendum to space physics coordinate transformations: a user guide. Planet Space Sci 45:1047

IAGA Division I Study Group on geomagnetic reference fields (1975) international geomagnetic reference field (1975). J Geomagnetics Geoelectrics 27:437–439

Langel RA, (1992) International geomagnetic reference field: the sixth generation. J Geomagn Geoelectr 44: 679–707

Langel RA, Barraclough DR, Kerridge DJ, Golovkov VP, Sabaka TJ, Estes, RH (1988) Definitive IGRF models for 1945, 1950, 1955, and 1960. J Geomagn Geoelectr 40: 645–702

Macmillan S, Maus S, Bondar T, Chambodut A, Golovkov V, Holme R, Langlais B, Lesur V, Lowes F, Lühr H, Mai W, Mandea M, Olsen N, Rother M, Sabaka T, Thomson A, Wardinski I (2003) The 9th-generation international geomagnetic reference field. Geophys J Int 155:1051–1056

Macmillan S, Maus S (2005) International geomagnetic reference field—the tenth generation. Earth Planets Space 57:1135–1140

Macmillan S, Chris T, Alan T (2009) Ascension and Port Stanley geomagnetic observatories and monitoring the South Atlantic Anomaly. Ann Geophys 52:83–96

Mandea M, Macmillan S (2000) International geomagnetic reference field—the Eighth Generation, 2000. Earth Planets Space 52:1119–1124

Peddie NW (1982) International geomagnetic reference field: the third generation. J Geomagn Geoelectr 34:309–326

Russell CT (1971) Geophysical coordinate transformations. Cosmic Electrodyn 2:184–196

Tsyganenko NA (2002) A model of the near magnetosphere with a dawn-dusk asymmetry—1. Mathematical Structure. J Geophys Res 107:A8. doi:10.1029/2001JA000219

Zmuda AJ (1971) The international geomagnetic reference field 1965. In: Zmuda AJ (ed) World magnetic survey 1957–1969, IAGA Bulletin, No. 28, IUGG, Paris, pp 147–204

Chapter 11

Geomagnetic Core Field Models in the Satellite Era

Vincent Lesur, Nils Olsen, and Alan W.P. Thomson

Abstract After a brief review of the theoretical basis and difficulties that modelers are facing, we present three recent models of the geomagnetic field originating in the Earth's core. All three modeling approaches are using recent observatory and near-Earth orbiting survey satellite data. In each case the specific aims and techniques used by the modelers are described together with a presentation of the main results achieved. The three different modeling approaches are giving similar results. For a snap shot of the core magnetic field at a given epoch and observed at the Earth's surface, the differences between models are generally small. They do not exceed 16 nT which gives an idea of the accuracy of the models. Secular variation models are robustly resolved up to spherical harmonic degree 13, but only on time scale as large as 10 years. On time scale of a year, secular variation models are resolved only up to degree 8 or 9. For higher time derivatives of core field models, only the very first degrees are robustly derived.

11.1 Introduction

Although the compass needle had been in use for a very long time for navigation purposes, it was only in the first half of the 19th century that it became possible to measure the absolute strength of the magnetic field. In the same era the modern way of describing the main magnetic field of the Earth was established by Gauss (in 1839). Gauss represented the magnetic field by the gradient of a potential, expanded in a series of spherical harmonic functions. He applied this method to derive spherical harmonic coefficients (nowadays called the Gauss coefficients) up to spherical harmonic degree and order four. In order to limit the number of necessary calculations, Gauss required evenly spaced data and therefore his model was constructed by calculating values from contour charts of declination (D), inclination (I) and total intensity (F). Although other methods were proposed, this intermediate step of drawing contour charts was used until the wide-spread introduction of computers in the middle of the 20th century (Langel 1987). Nowadays, geomagnetic field models are constructed by directly fitting a massive amount of data, usually by the method of least squares. However, the basic mathematical representation technique is still based on Gauss' approach.

The main part of the geomagnetic field (up to spherical harmonic degree 13) is dominated by contribution from the Earth's core. The terms "main field" and "core field" are therefore often used as synonyms. Although the crust also contributes to terms of degree lower than 13, this part is negligible compared to the core contribution. In following we will therefore ignore the difference between main and core field and denote the large-scale part of the observed field (up to degree 13) as "core field".

The main sources of data for building models of the geomagnetic main field have been, until very recently, from magnetic observatories and repeat stations. These data are still extremely useful even over time periods, such as the present era, where satellite data are also available, although the role of the ground-based data has changed. In terms of satellite data, the first satellite to obtain near-earth magnetic field measurements was

V. Lesur (✉)
Helmholtz Center, GFZ German Research Centre for Geosciences, Telegrafenberg, F 453, 14473 Potsdam, Germany
e-mail: lesur@gfz-potsdam.de

M. Mandea, M. Korte (eds.), *Geomagnetic Observations and Models*, IAGA Special Sopron Book Series 5,
DOI 10.1007/978-90-481-9858-0_11,

Sputnik 3 in 1958, although the measured data were of very poor accuracy. The Cosmos 49 satellite that flew in 1964 is often cited as the first survey satellite but it only obtained field measurements in a limited region. The first global mapping of Earth's magnetic field was carried out by a series of OGO satellites between 1965 and 1971 (Cain et al. 1966); however only the field intensity was measured by these satellites. The first vector mapping survey was done by the Magsat satellite, launched in 1979 (Langel et al. 1982). This satellite carried both vector and scalar magnetometers and flew between November 1979 and May 1980 at a mean altitude that decayed from 465 km to 330 km. The analysis of this first set of global high quality vector magnetic data led to significant progress in our description and understanding of the magnetic field. Despite these successes, the next set of high quality satellite vector data came much later with the Ørsted satellite launched in 1999 (Olsen 2007a). In between, only magnetic data of lower quality were available, through the Polar Orbiting Geophysical Satellite (POGS) mission that flew between 1990 and 1993. Ørsted was designed for a nominal lifetime of only 14 months, but after 11 years in orbit the satellite still provides data — although since 2005 only of the field intensity. Two other satellites were launched in 2000. The first was the CHAMP satellite (Reigber et al. 2002). This continues to provide high quality vector and scalar data. The second satellite was SAC-C. However following a technical problem during launch, only scalar data have ever been retrieved from SAC-C, and only for the years 2001 to 2004.

Numerous models of the geomagnetic fields of internal origin have been derived since Gauss' epoch. A review of older models up to the mid 80's can be found in Barraclough (1976), Langel (1987). For more recent reviews, we refer to Hulot et al. (2007), Jackson and Finlay (2007) or to Gillet et al. (2009) where the authors discuss the information provided by core field models that cover the historical era. In the present manuscript, we therefore do not want to repeat these reviews but simply present, with some new details, three recently published models. This serves to illustrate the practical problems faced, and solutions offered, by modelers particularly concerned with deriving up-to-the-minute models of the Earth's field based on the latest available data. We describe what were the aims of these modelers, the techniques that were used and, of course, we discuss their resulting models. In the following section we present an overview of the geomagnetic field modeling problem; the subsequent section gives a short survey of core field models followed by a more detailed description of each of the three models. Some conclusions and a discussion on the outlook for future modeling efforts is given in the final section.

11.2 Overview and Theory

In this section a short description of the different contributions to the magnetic field is given. This serves to highlight the difficulties that face the modeler when trying to derive a model from a given set of data. A rough description of the 'typical' modeling approach is also given.

The main contributions to each measurement of the geomagnetic field, either at the Earth's surface or few hundred kilometers above it, are magnetic fields generated by sources in the core, in the lithosphere, in the ionosphere and the magnetosphere. The ionosphere is connected to the magnetosphere, especially at polar latitudes, through field-aligned currents (FACs) that also contribute to the magnetic field. The currents induced in the conducting Earth by the time variations of the external field (i.e., field generated in the ionosphere and the magnetosphere) also generate a magnetic signal. Finally, there is the magnetic field contribution generated by the movement of conducting material — e.g., seawater — through the Earth's main magnetic field. In any case, close to the surface the core field signal is strongly dominant. Its strength varies from slightly more that 20000 nT in the south Atlantic to more than 60000 nT near the poles. The field of the "magnetic lithosphere" is generally only a fraction of the core field but may reach comparable amplitudes in some locations, and it decreases strongly with increasing distance from lithospheric sources. The field generated in the ionosphere at mid and low latitudes is strongly dependent on the local time and becomes weak during local night. At high latitudes, over the polar caps, the magnetic field is also dependent on the local time but is mainly controlled by solar and geomagnetic activity. The fields generated in the magnetosphere are essential large scale, close to Earth (due to the large distance to the source), but can involve rapid temporal variations.

Measurements of the magnetic field used for core field modeling are made either from a satellite platform or at ground-based observatories and repeat survey points. Near surface measurements (e.g., by observatories) can be considered to be taken in a source-free region and therefore the magnetic field vector **B** can be described as the negative gradient of potentials of internal and external origin. We therefore write:

$$\begin{aligned} \mathbf{B} &= -\nabla(V_i + V_e) \\ V_i &= a\sum_{l=1}^{L_i}\sum_{m=-1}^{l} g_l^m(t)\left(\tfrac{a}{r}\right)^{l+1} Y_l^m(\theta,\phi) \\ V_e &= a\sum_{l=1}^{L_e}\sum_{m=-1}^{l} q_l^m(t)\left(\tfrac{r}{a}\right)^{l} Y_l^m(\theta,\phi) \end{aligned} \tag{11.1}$$

where θ, ϕ, r, t are the geocentric co-latitude, longitude, radius and time. V_i and V_e are the parts of the potential produced by sources internal, resp. external, to a sphere of radius a. The reference radius is taken to be $a = 6371.2$km, as recommended by IAGA for field models (but note that some modelers use a different reference radius). $Y_l^m(\theta,\phi)$ are the Schmidt quasi-normalized Spherical Harmonics (SH). We use the convention that negative orders, $m < 0$, are associated with $\sin(|m|\phi)$ terms whereas zero or positive orders, $m \geq 0$, are associated with $\cos(m\phi)$ terms.

As the satellites are flying through a conducting plasma that itself generates a magnetic field, the magnetic field cannot be described by means of a Laplacian potential and a representation in terms of poloidal and toroidal contributions is required. Because the parametrization in Eq. (11.1) is valid only for potential fields, the non-potential field contributions from in-situ currents in the plasma are typically ignored and are regarded as noise.

The internal Gauss coefficients $g_l^m(t)$ in Eq. (11.1) are time dependent for the lowest SH degrees because they mainly describe the core field. However the way this time dependence is parameterized varies significantly from one model to another. We note that all contributions to a Laplacian potential magnetic field can be parameterized according to Eq. (11.1), but this may require a particularly complicated time dependence of the Gauss coefficients and high maximum spherical harmonic degrees.

The essential difficulty in core field modeling is the separation of the different contributions to the magnetic field. The non-potential fields have, outside the polar cap, relatively small amplitudes. Over the polar cap itself most, but not all, modelers use field intensity data to minimize the field contributions from field-aligned currents because these do not contribute to the magnetic field in the direction of the main field. Therefore, these non-potential field contributions are generally not too difficult to handle. The first main difficulty facing the modeler then comes with the separation of contributions from internal and external sources. For this, the modeler has to assume some kind of regularity (or smoothness) property of the field of internal origin. The extent to which the field model has to be smooth to achieve the separation, depends on the data distribution. Even with modern satellite data this is a challenge especially for the long-wavelength field as the data coverage is rather coarse in local time, although usually excellent in longitude and latitude (cf. Olsen et al. 2010a). In particular, if one tries to extract a signal over a short period of time, the very high temporal variability of the field of external origin makes the separation particularly difficult.

The second difficulty facing the modeler is to distinguish the individual signals of internal origin. Currents in the ionospheric E-layer (90–150 km altitude) are seen at satellite altitude as an internal source of the magnetic field. However, their magnetic field contribution is very small during night times (induced contributions are however still present, even when the ionospheric, inducing, currents vanish during night). Then, unless the ionospheric field is co-estimated with other contributions, the usual approach is to select data for local-times around midnight. This selection process is remarkably efficient but, as the satellites are slowly drifting with local-times, results in periods without satellite data. A separation of the field of lithospheric origin from the (static part of the) core field is not possible from magnetic data alone. For spherical harmonic degree smaller than 13, the field is dominated by processes in the core, whereas above degree 16, its static part is assumed to be mainly generated in the lithosphere. The observed field for degrees around 14 contains about similar contributions from core and lithosphere. Of course, the lithospheric field is nearly independent of time on decadal time scales, and it is, in principal, possible to extract Secular Variation (SV) estimates at SH degree higher than 13. Finally, the remaining significant internal contributions to a magnetic field measurement are the fields induced in the electrically conducting Earth by the time-varying external fields. Again, these contributions cannot be easily separated from other contributions. The usual approach consists of giving a well defined time dependence to these induced fields. This time dependence is based on an *a priori* large scale external field model and a predefined model of the mantle conductivity. Both the external field source structure and mantle conductivity are not yet well known and the separation of the contributions from the induced fields remains one of the main challenges in core magnetic field modeling.

11.3 Detailed Description

The very large amount of data provided by recent satellite missions, as well as the quality of the data set and its coverage, has led to a renewal of techniques applied by modelers for identifying and describing the magnetic field generated in the Earth's core. In deriving long term models, such as CALS7K (Korte and Constable 2004) or GUFM (Jackson et al. 2000), which are based on paleomagnetic and/or historical data, the small amount of data, its relatively poor and uneven coverage and the scarcity of vector data all prevent the use of data selection techniques such as are applied to satellite data. Of course, such long term models are smooth in space and time and the separation of the different sources to the magnetic field is particularly difficult to achieve. Magnetic field models constructed in the last few decades have suffered from similar difficulties, outside of the six month period during which Magsat flew. The wide spread installation of magnetic observatories from the turn of the 20th century, with good baseline control and permanent monitoring of the magnetic field, provides an invaluable initial data set to modelers. Today however there remains too few observatories and their distribution on Earth is too uneven to obtain core magnetic field models as detailed as those that can be derived from satellite data. Therefore, by taking data from Magsat and incorporating most of the available magnetic data sets, the Comprehensive Model (CM) was developed at the beginning of the 1990s (Sabaka and Baldwin 1993; Langel et al. 1996). Its latest version, CM4, covers 43 years (Sabaka et al. 2002, 2004) — i.e., 1960 to 2002. During the development of the different versions of the Comprehensive Model, numerous new ideas and modeling techniques have been introduced. The CM4 model (Sabaka et al 2004) may not use the most recent years of CHAMP and Ørsted data, but it stands as a reference model and has been widely used in studies of the core field.

Over the last ten years, since the launch of Ørsted, many different models developed by different groups have been proposed. From the initial core field models derived from Ørsted data (Olsen et al. 2000; Olsen 2002) up to the candidate models for the IGRF-10 (Olsen et al. 2005; Maus et al. 2005; Lesur et al. 2005), the temporal parameterization of the core field models was typically based on a polynomial of maximum degree 3. The CHAOS model (Olsen et al. 2006) was the first core field model to be based only on satellite data that used B-splines for describing the time dependence. In later versions, CHAOS has been extended to encompass all the available satellite data of these last 10 years. This series of models, known to be an accurate representation of the core magnetic field, is presented in the next subsection below. The GRIMM series of models (Lesur et al. 2008, 2010b) also uses a B-spline temporal representation and, although their derivation is based on different techniques, the GRIMM models are very similar to the CHAOS models. As an example, Figure 11.1 displays an estimate of the radial component of the core field and its secular variation at the Core Mantle Boundary (CMB). We discuss the GRIMM series of models in the second sub-section. Finally, a particular effort has been made at the British Geological Survey to test and develop new data selection techniques. Their latest model, MEME08 (Thomson et al. 2010), is described in the last sub-section. The MEME08 temporal representation technique — piecewise linear representation — is also different and does not impose a continuity condition on core field temporal variations. This results in a series of models which are less correlated in time.

11.3.1 CHAOS Model Series

11.3.1.1 Aims

The goal of the CHAOS model series is to provide a good representation of the recent geomagnetic field by making use of multi-year continuous time series of high-precision satellite observations. In particular, the models aim at describing core field changes with high

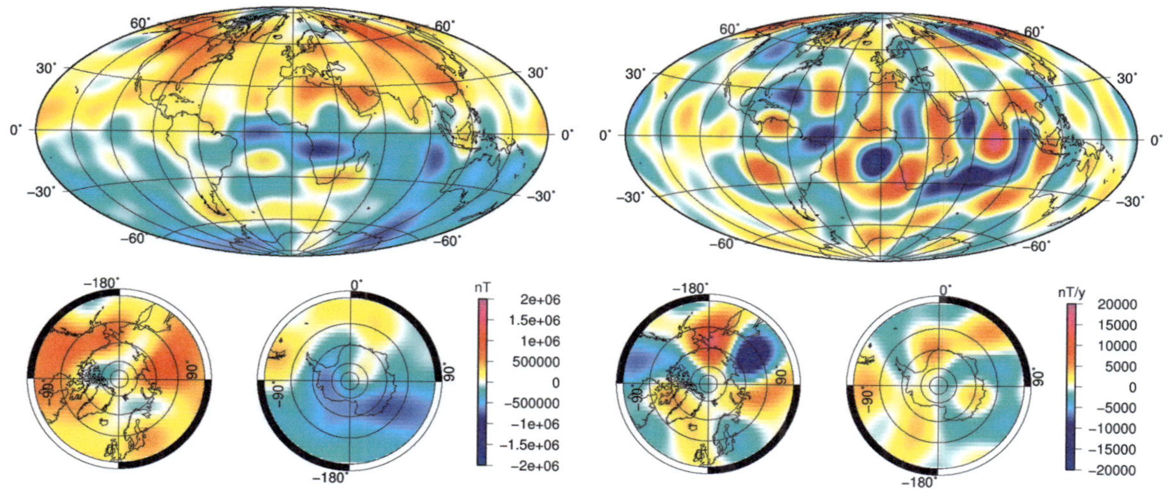

Fig. 11.1 Vertical down component of the core field and its SV for year 2005, at the CMB. The core field model is truncated to SH degree 13, whereas the SV model is truncated to SH degree 12. We point out that the SV model has a diverging spectrum at the CMB and therefore including higher SH degrees may lead to a significantly different map. The model presented here is the GRIMM model Lesur et al. (2010b), however maps of other models would not be very different

spatial resolution of the first time derivative (linear secular variation), and high temporal resolution (rapid field changes). Recognizing the unpredictability and chaotic nature of the Earth's magnetic field, the model series is called CHAOS, which stands for CHAMP, *Ørsted and* SAC-C *model of Earth's magnetic field*. The first version in the series, called CHAOS (Olsen et al. 2006), was based on 6.5 years (March 1999 to December 2005) of Ørsted, CHAMP and SAC-C satellite data; CHAOS-2 (Olsen et al. (2009) – see Olsen and Mandea (2008) for a description of a predecessor called xCHAOS) is derived from 10 years of satellite data (until March 2009) augmented by observatory monthly means for 1997 to 2006. The latest version, CHAOS-3 (Olsen et al. 2010b), is based on almost 11 years of satellite data (up to December 2009) and 13 years of ground observatory data.

11.3.1.2 Technique

Data selection

Ørsted, CHAMP and SAC-C satellite data are selected for quiet geomagnetic conditions as defined by the following criteria: First, for data at all latitudes the D_{st}-index should not change by more than 2 nT h^{-1}. At non-polar latitudes (equator-ward of 60° dipole latitude) $Kp \leq 2o$ has to be fulfilled. For regions poleward of 60° the merging electric field, E_m, at the magnetopause should be <0.8 mV m^{-1}. Only data from dark regions (sun 10° below horizon) are used, to reduce contributions from ionospheric currents. Vector data are taken for dipole latitudes equator-ward of $\pm 60°$, to avoid the disturbing effect of field-aligned currents, which only influence the vector components but not field intensity. Scalar data are used for regions poleward of $\pm 60°$ or if attitude data are not available. Non-polar CHAMP data are only taken from local time past midnight, to avoid the influence of the diamagnetic effect of dense plasmas. Due to their higher altitudes, a corresponding rejection of pre-midnight data is not necessary for Ørsted and SAC-C. Contrary to most other satellite-based field models, which use the vector components in an Earth- or Sun-fixed coordinate system, the CHAOS models are based on vector and attitude data in the instrument frame. Alignment of vector data requires a precise determination of the rotation (Euler angles) between the star imager (providing attitude information) and the vector magnetometer, which are then applied to all satellite vector data. This process requires a model of the ambient magnetic field, the accuracy of which *to be known at the time and position of each data point* is the limiting factor for the alignment. To avoid the inconsistency of deriving a field model from vector data that have been aligned using a different (pre-existing) magnetic field model, vector data alignment is done as part of the CHAOS modeling effort.

In addition to satellite data, annual differences of observatory monthly means of the North, East and downward components (X, Y, Z) have been used for CHAOS-2 and CHAOS-3 (annual difference means that the value at time t is obtained by taking the difference between those at $t+6$ months and $t-6$ months, thereby eliminating an annual variation in the data). For CHAOS-2 traditional monthly means of the years 1997 to 2006 were used, while for CHAOS-3 a new scheme for removal of external field from the observations has been applied in order to derive revised monthly means for the years 1997 to 2009. To account for correlated errors due to external field contributions, the vector components of each observatory are weighted according to their 3×3 data covariance matrix (including non-diagonal elements, i.e., correlation between the different components, cf. Wardinski and Holme (2006)).

Model parameterization and regularization

The mathematical model that was fitted to the satellite data consists of two parts: spherical harmonic expansion coefficients (cf. Eq. (11.1)) describing the magnetic field vector in a geophysical coordinate system (for instance the *Earth-Centered Earth-Fixed (ECEF)* frame) and sets of Euler angles needed to rotate the vector readings from the magnetometer frame to the star imager frame. The magnetic field vector in the geophysical frame, $\mathbf{B} = -\nabla V$, is derived from a magnetic scalar potential $V = V_i + V_e$ consisting of a part, V_i, describing internal (core and crustal) sources, and a part, V_e, describing external (mainly magnetospheric) sources (including their Earth-induced counterparts). Both are expanded in terms of spherical harmonics, cf. Eq. 11.1.

CHAOS is the first satellite-only model that describes the time changes of the core field by splines. However, only coefficients up to degree $l = 14$ were parameterized by splines (of order 4), and the knot spacing of 12 months was rather coarse. The later versions use a spline representation of degrees up to $l = 20$, a knot spacing of 6 months, and splines of order 5 (for CHAOS-2), resp. order 6 (for CHAOS-3). Higher spherical harmonic degrees (up to $L_i = 50$ for CHAOS and up to $L_i = 60$ for CHAOS-2 and -3) are assumed to be static.

The external potential, V_e, describes large-scale magnetospheric sources and is parameterized according to

$$\begin{aligned} V_e = {} & a \sum_{l=1}^{2} \sum_{m=0}^{l} \left(q_l^m \cos mT_d + s_l^m \sin mT_d \right) \\ & \times \left(\tfrac{r}{a}\right)^l P_l^m(\cos\theta_d) \\ & + a \sum_{m=0}^{1} \left(\hat{q}_1^m \cos T_d + \hat{s}_1^m \sin T_d \right) \\ & \times \left\{ E_{\rm st}(t) \left(\tfrac{r}{a}\right) + I_{\rm st}(t) \left(\tfrac{a}{r}\right)^2 \right\} P_1^m(\cos\theta_d) \\ & + a \sum_{l=1}^{2} q_l^{0,\rm GSM} R_l^0(r,\theta,\phi) \end{aligned} \tag{11.2}$$

where θ_d and T_d are dipole co-latitude and dipole local time, respectively, and $E_{\rm st}, I_{\rm st}$ are time series of the decomposition of the $D_{\rm st}$-index, $D_{\rm st}(t) = E_{\rm st}(t) + I_{\rm st}(t)$, into external and induced parts, respectively (Maus and Weidelt 2004; Olsen et al. 2005).

The first two lines of this equation represent an expansion in the *Solar Magnetic (SM)* coordinate system and describe mainly contributions from the magnetospheric ring current. The expansion in *Geocentric Solar Magnetospheric (GSM)* coordinates used in the last term describes contributions from magnetotail and magnetopause currents. The functions R_l^0 are modifications of spherical harmonics to account explicitly for induced field contributions due to the wobble of the GSM z-axis with respect to the Earth's rotation axis (see Olsen et al. (2006) for details).

Large-scale magnetospheric fields that are not described by $E_{\rm st}(t), I_{\rm st}(t)$ are accounted for by solving for time-varying degree-1 coefficients in bins of 12 h length (for q_1^0), resp. 5 days length (for q_1^1 and s_1^1). For the latest model version, CHAOS-3, this results in 6,411 coefficients describing the external field.

Finally, an in-flight instrument calibration is performed by co-estimating the Euler angles of the rotation between the coordinate systems of the vector magnetometer and of the star sensor that provide attitude information. To account for the thermo-mechanical instabilities of the magnetometer/star-sensor system of the CHAMP satellite, Euler angles are solved in bins of 10 days (for Ørsted two sets of Euler angles are determined).

For the latest model version, CHAOS-3, the total number of model parameters is 16,920 (internal field) + 6,411 (external field) + 639 (Euler angles) = 23,970. These model parameters are estimated by means of a regularized *Iteratively Reweighted Least-Squares* approach using Huber weights.

No spatial regularization is applied, but the time-dependence of the core field is damped in the following way: For CHAOS, $\left\langle |\ddot{\mathbf{B}}|^2 \right\rangle$, the mean square magnitude of the second time derivative of **B** integrated over the Earth's surface and averaged over time, is minimized. For CHAOS-2, this quantity is minimized at the core-mantle boundary. For CHAOS-3, $\left\langle |\partial^3 \mathbf{B}/\partial t^3{}^2| \right\rangle$, the third time derivative of the squared magnetic field intensity (and the second time derivative at the model endpoints) is regularized at the core-mantle boundary. This is similar to the regularization used for GRIMM (cf. Eqs. 11.3 and 11.4) apart from the fact that field intensity $|\mathbf{B}|$ rather than B_r is minimized.

11.3.1.3 Results and Discussion

As mentioned above, one of the goals of the CHAOS model series is to provide a good estimate of the first time derivative of small-scale core field structures. Due to the dominance of the crustal field at smaller scales, it is not possible to determine the static part of the core field for spherical harmonic degrees above $l = 14$. However, since the lithospheric field is time independent (at least on the time scales considered here), the *time changes* of the core field are, in principle, observable at all spatial wavelengths. As shown in Fig. 11.2, the various versions of the CHAOS model series resolve the first time derivative coefficients beyond $l = 13$, with lowest noise level for the more recent model versions. This demonstrates the possibility to infer the time change (secular variation) of the core field down to smaller scales (smaller than 1600 km at core surface) than the (static) core field itself.

An assessment of the ability of CHAOS-3 to model rapid core field changes is possible by comparing with an independent month-by-month spherical harmonic model. Fig. 11.3 shows time series of the first time derivative, $\dot{g}_l^m, \dot{h}_l^m$, of some internal Gauss coefficients for $l = 3$ (top) and $l = 6$ (bottom). The black symbols present annual differences of the coefficient determined from a monthly model obtained from CHAMP "virtual observatory" monthly means between January 2001 and December 2009, determined using the approach described in Mandea and Olsen (2006) and Olsen and Mandea (2007b). The curves show model values from CHAOS-3. The scatter of the individual monthly solutions (black dots) is largest for the zonal coefficients (left panel), which is probably due to contamination by the magnetic field contributions from polar ionospheric currents. (Note that only zonal coefficients contribute to the field at the geographic poles.) Sectorial terms g_l^l, h_l^l describing low-latitude processes show much lesser scatter. The temporal regularization chosen for CHAOS-3 (which increases with degree l but is independent on order m) results in reasonable time changes of the zonal coefficients but is obviously not able to fully describe the rapid changes of higher degree sectorial coefficients like h_6^6.

11.3.2 GRIMM Models

11.3.2.1 Aims

GRIMM is an acronym for the GFZ Reference Internal Magnetic Model and it aims at describing two of the main internal sources of the geomagnetic field: the

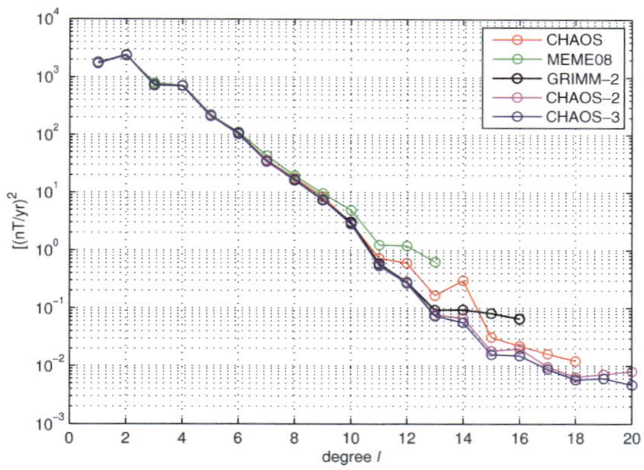

Fig. 11.2 Lowes-Mauersberger spectra of the first time derivative (secular variation) at Earth's surface and epoch $t = 2005.0$, for various magnetic field models

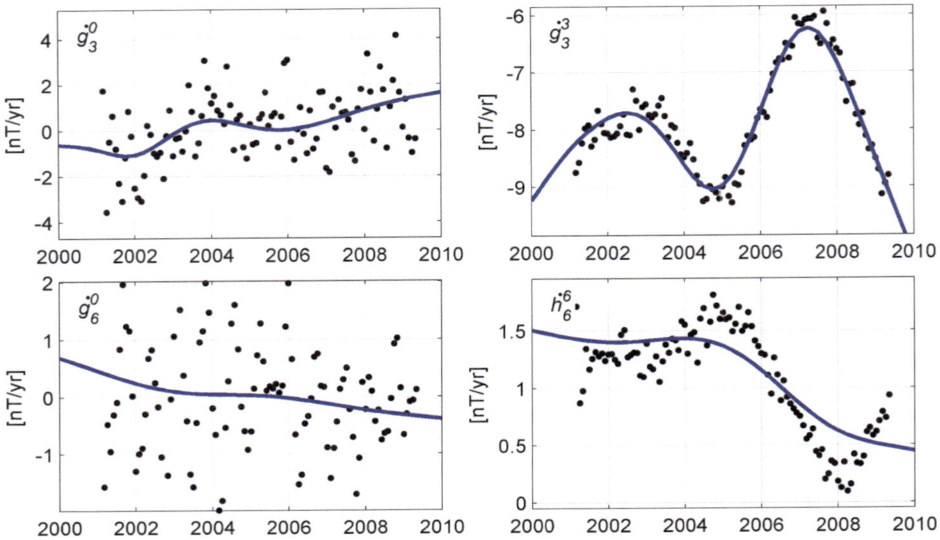

Fig. 11.3 First time derivatives of some internal Gauss coefficients for $l = 3$ (*top*), resp. 6 (*bottom*), in nT yr^{-1}. Symbols represent annual differences of time series of Gauss coefficients obtained from CHAMP "virtual observatory" monthly means while the blue curves show predictions of the *CHAOS-3* model

core and the lithosphere. If, for its first version (Lesur et al. 2008), a co-estimation of the models describing these two fields was a natural objective, in the second version (Lesur et al. 2010b), they have been modeled sequentially. The reason for this evolution lies in the characteristics of the data set. With a time span of almost 10 years, the CHAMP satellite magnetic data set covers a substantial part of the past solar cycle. However the magnetically quiet recent years — i.e., 2007, 2008, 2009 — combined with the low altitude of the satellite provides a remarkable data set for lithospheric field modeling. Therefore, if including the years 2001 to 2006, proves to be beneficial for building models of the core, it however generates difficulties when dealing with the lithospheric model and hence, the core and lithosphere fields have been modeled sequentially. We note also that modeling part of the ionospheric field and part of the FACs signal was attempted in the first version. However, here we only present results regarding the core field.

The core field modeling approach has also slightly shifted between the two GRIMM versions. In the first version the emphasis was in mapping the core field at the Earth surface, whereas the second version the emphasis is more oriented towards core field studies. The main difference between these approaches is in the way the core field inversion process is regularized, either at the Earth's surface or at the core surface. However, both versions of the model were built with two, often conflicting, objectives. Firstly the observatory and satellite data are selected to minimize the noise due to the poorly modeled large-scale external field. Secondly, the selection is done such that the data distribution is good enough for an accurate description of the core field temporal evolution. Of course these objectives are contradictory because a very strict data selection will reduce significantly the noise in the data but then the temporal evolution of the core field may be difficult to model because of the lack of data. Therefore the characteristic of the GRIMM model series is its specific data selection built such that the temporal evolution of the core field remains accurately described. Otherwise, as for the CHAOS series of models, the temporal evolution is described by B-splines, a model of the external magnetic field is co-estimated — but we will see below that this external field model is not robust — and a quadratic smoothing semi-norm is applied as a regularization process. A more detailed technical description is given next, and then the main results obtained with these models are presented and discussed.

11.3.2.2 Techniques

The data selection process applied to the available observatory and satellite data differs depending on the magnetic latitude of the observation point. In any case, data are selected for magnetically quiet days and for positive values of the Z component of the Interplanetary Magnetic Field (IMF-Bz). For GRIMM, the active days are simply identified using the VMD (Vector Magnetic Disturbances index; Thomson and Lesur (2007)) even if other indexes are often used by other modelers– see particularly the description in Section 11.3.3.

At mid- and low magnetic latitudes (i.e., in between magnetic latitudes ±55°) only, night time, X and Y components of the vector magnetic data in Solar Magnetic (SM) system of coordinates are used. Using only night-time data leads to a discontinuous temporal distribution of satellite data where, in the case of CHAMP, a 60 to 70 day period without data follows 50 to 60 days with high data density. This is an effect of the drift in local time of the satellite. Observatory data are not affected. The unused Z, SM, component is aligned with the Earth main field dipole direction and hence is the direction most disturbed by large-scale external magnetic field. It is shown in Lesur et al. (2008) that using two of the three SM components of each magnetic vector data is enough to derive a robust model of the magnetic field generated inside the Earth. Therefore, by using X and Y component only, there is no loss of temporal resolution for the core field model but the level of noise in the data is significantly reduced. The price to pay for this is an incomplete description of the large scale external field. It can be shown (see Lesur et al. (2008)) that a part of the external field signal cannot be extracted from the selected data, furthermore, the proper separation of internal and external contributions is possible only for external field modeled up to SH degree 2. Therefore, the model of the external field co-estimated together with the model of internal origin cannot be easily re-used and is not distributed.

At high latitudes – i.e., outside the interval ±55° magnetic latitudes – the approach is different. At high latitudes the main source of noise is a combination of signals generated in the ionosphere and by the FACs. The usual approach to minimize this noise is to use night-time total intensity data. However, in that case, the night time selection leads to a data gap of several months in the data set and a loss of temporal resolution in the core field model – we point out here that all modelers do not agree over this point (see particularly the CHAOS-2 model Olsen et al. (2009)). In order to avoid this gap in the GRIMM data selection process, the CHAMP satellite data are selected at all local times, independently of the sun orientation. Hoping for a better separation of the different contributions to the magnetic field, only vector data are used. This has a cost. The level of noise in high latitude selected data is high, particularly in the X and Y NEC – i.e., North, East – directions. This high level of noise could limit the accuracy of the SV model estimate for the shortest wavelengths, but the GRIMM model seems not to be less accurate at these wavelengths than any other model. Similarly, using day-side data apparently does not affect significantly the separation of ionospheric and core field signals at high latitudes.

The GRIMM models parameterization (Lesur et al. 2008, 2010b) is similar to that used by other modelers. The time dependent Gauss coefficients for the field of internal origin in Eq. (11.1) are parameterized from SH degree 1 to 16 using B-splines of order 6 (order 5 for the first version of the model). Splines nodes are one year apart. The external field coefficients are estimated up to SH degree 2 (SH degree 1 in the first version). Their time parameterization is defined by a piece-wise linear polynomial with nodes three months apart and a dependence to the VMD. Gauss coefficients are estimated using a reweighted least square algorithm and an L_1 measure of the misfit to the data. The inversion process of the most recent version of the GRIMM model, is regularized by minimizing a measure of the third time derivative of the core model:

$$\Phi_{t3} = \lambda_{t3} \int_{\mathcal{T}} \int_{\Omega_c} |\partial_t^3 B_r|^2 \mathrm{d}\omega \; \mathrm{d}t \tag{11.3}$$

where $\mathcal{T}$ is the model time span 2000–2011, Ω_c is the spherical surface with radius $c = 3485$ km (i.e., the estimated Earth's core radius) and B_r is the radial component of the magnetic field model. In order to avoid spurious effects near the end point of the model, a measure of the second time derivative of the core model is minimized for epochs 2000.0 and 2011.0:

$$\Phi_{t2}(t) = \lambda_{t2} \int_{\Omega_c} |\partial_t^2 B_r|^2 \mathrm{d}\omega. \tag{11.4}$$

We point out that both integrals are calculated at the Earth's core surface that is one of the differences with regards to the first GRIMM version and that minimizing the integral 3 is consistent with the order 6 B-splines used for the temporal representation of the Gauss coefficients. The integral (11.4) is introduced just to minimize edge effects at the end points of the model time-span.

11.3.2.3 Results and Discussion

The quality and coverage of the satellite data leads to robust estimation of core field models with associated SV. The GRIMM model is therefore not much different of other derived models and the radial component of the core field model is mapped at the CMB for epoch 2005.0 in Fig. 11.1. Some differences are visible for SV models particularly at high latitudes but, roughly, most of the models agree on the general aspect of the SV at the CMB when truncated to SH degree 12. Fig. 11.1 presents a map of the SV for year 2005.0 at the CMB.

GRIMM, and particularly its second version, differs from other models in its description of the second time derivative of the core field i.e., Secular Acceleration (SA) and its temporal variability. The power spectra of GRIMM core field, SV, SA and third time derivative are shown in Fig. 11.4. The core field power spectrum increases from SH degree 13 due to the contribution from the lithospheric field. The SV spectrum increases unrealistically from SH degree 14. This is likely to be an instability of the model. Both SA and third time derivative have a converging spectrum due to the constraints applied on the model. Recent studies (Lesur et al. 2010a) have shown that such a decrease is not compatible with the frozen-flux hypothesis (Roberts and Scott 1965). However, results show that:

- The SA has a high temporal variability
- Both the SV and SA derived from the GRIMM model are averages over time, and the amount of averaging is dependent on the wavelength.

The first of these points is relatively new. Before the modern satellite era, Barraclough and Malin (1979) recognized that their SA model derived from observatory data has a limited temporal validity, however, they estimated that it could be used to extrapolate the SV over a decade. For the 10th version of the IGRF (Lesur et al. 2005; Olsen et al. 2005; Maus et al. 2005), SA models were derived from several years of satellite data and used for extrapolation purpose. Nowadays, the SA is seen as so variable, that its usage for extrapolation purpose can be questioned.

Despite its large energy in its third time derivative, The GRIMM model is temporally a smooth model; nonetheless it shows some remarkable features of the SA evolution. As an example, Fig. 11.5 presents the growth and decay of a significant deceleration anomaly in the Atlantic between years 2004 and 2007. So far, flow models at the CMB do not satisfactorily explain such variations of the SA. The CHAOS-2 model (Olsen

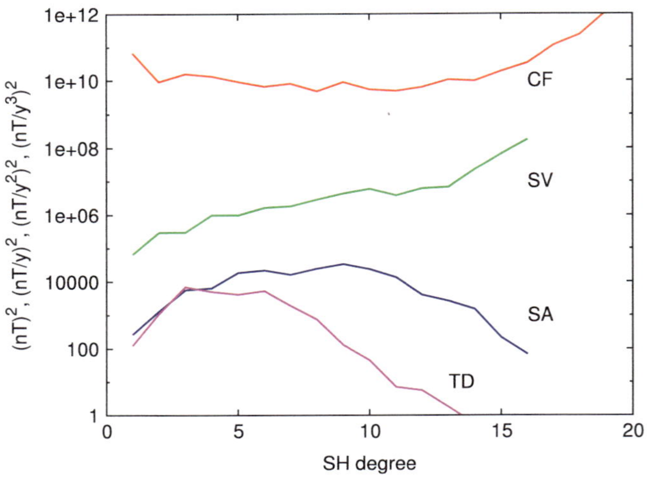

Fig. 11.4 Power spectra of the GRIMM model for year 2005 at the CMB. Are plotted the core field (CF), SV, SA and third derivative (TD) spectra

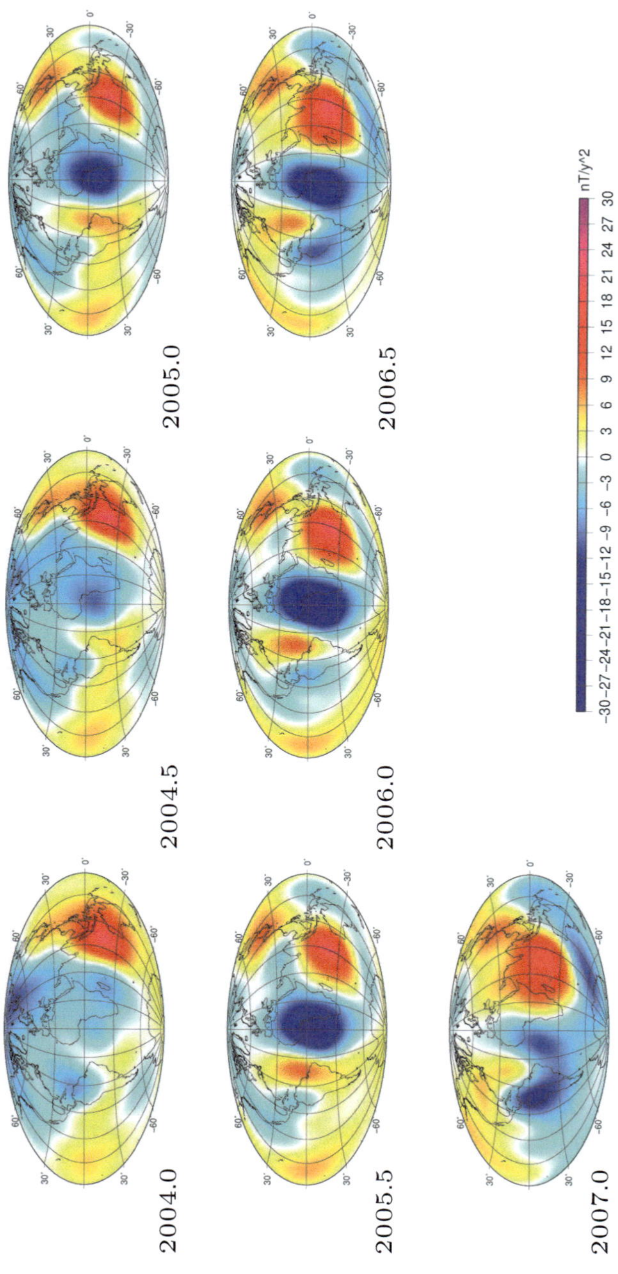

Fig. 11.5 Maps of the vertical down component of the SA at the Earth's surface ($a = 6371.2$ km) from year 2004 to 2007

et al. 2009) presents a much higher variability than GRIMM. This difference comes mainly from the two first SH degrees, and particularly from the $g_1^0(t)$ Gauss coefficient. The CHAOS-2 model allows for a better fit to the observatory data, but that carries the risk of a contamination of the core field model with signals of external origin or their internally induced counterpart. This interesting issue is currently studied and we expect significant progress in the coming years.

11.3.3 BGS Models

11.3.3.1 Aims

BGS produces and updates a number of global magnetic field models. These are: an International Geomagnetic Reference Field candidate model (here referred to as IGRF-BGS: Hamilton et al. (2010), Lesur et al. (2005)); the co-authored World Magnetic Model (Maus et al. 2009, WMM); and an annual BGS Global Magnetic Model (BGGM), intended for oil industry applications. Recently MEME08 (The Model of Earth's Magnetic Environment, 2008) has been developed, intended for study of Earth properties (Thomson et al. 2010).

MEME08 is the most detailed of the recent BGS models, in terms of parameterization and likely accuracy and it is described here. In some respects MEME08 can be viewed as a "parent model" to the other BGS models, as recent versions of these other models have followed broadly similar data selection and model parameterization strategies. We therefore discuss MEME08 with respect to known Earth structure and to other recent models. We also outline areas of research intended to improve the accuracy of future BGS magnetic models. MEME is intended to be an evolving model, in the sense that changes in model parameterization and data handling will occur as global magnetic modeling techniques and scientific understanding improves.

11.3.3.2 Techniques

Data Sources, Selection and Weighting

The MEME08 data selection identifies satellite samples (Champ and Oersted 20-s data, 1999–2007) and observatory samples (hourly mean data, 1999–2007) with minimal contribution from field-aligned, Sq and electrojet current systems. Ideally, the data selection produces samples for modeling with contributions only from core, lithosphere and large-scale quiet magnetospheric sources, as reflected in the model parameterization scheme. The most recent revised models - IGRF-BGS, WMM and BGGM — have extended the data selections of MEME08 to include satellite and observatory data from 2008 and 2009. Satellite data are calibrated vector and scalar components in the North-East-Down geographic coordinate system.

Magnetic indices Kp, Dst, VMD, IE, PC, Sector-A are used in the MEME08 data selection. Kp and Dst are well known, long-established mid and low-latitude indices, widely used in global field modeling. VMD is the vector magnetic disturbance index (Thomson and Lesur 2007), IE is the Image magnetometer chain version of the AE index (Viljanen and Hakkinen 1997), PC is the Polar Cap index (North and South, Troshichev et al. (1988)), and Sector-A are A-indices in specific geographical longitude sectors (Menvielle and Berthelier 1991). Solar wind speed and interplanetary field orientation, as measured by the NASA ACE spacecraft, are further data filters and data are taken only on the night-side of the Earth, identified by local time and solar zenith angle. For some applications (e.g., IGRF-BGS) satellite data are sampled at 60 seconds. For MEME08, a data residual filter is also applied to satellite samples, where the filter passes only data within ± 2 standard deviations from an earlier BGS model. Finally, gaps in the resulting satellite spatial distribution are filled with slightly more active data (e.g., Sector-A $< 2+$, compared with the quietest data selection data-set where Sector-A $< 2-$) in what is described as a "two-pass" data selection strategy (Thomson and Lesur 2007).

A novel weighting scheme has been developed for MEME08. Vector and scalar component data inverse weights are assigned for each satellite sample. These are the root-sum-of-squares of the along-track sample standard deviation and the Local Area Vector Activity — "LAVA" — index at each sample point. For a given satellite sample, its LAVA index is derived from nearest-neighbor ground-level observatory data, interpolated to the satellite ground position, inverse-weighted by distance. LAVA measures the large-scale external field variation at the satellite ground position on a scale of 0–10. At the same time the sample standard deviation measures local field activity within ±75 kilometers of the satellite position and at satellite altitude. This novel weighting scheme is applied to satellite data only and, because of the weight properties (e.g., Fig. 11.4 of Thomson et al. (2010)), MEME08 is derived from vector data at all latitudes.

As an example of this down-weighting scheme, Fig. 11.6 shows standard deviation (SD) weight components, averaged in 1 degree bins (i.e., without the LAVA weight). We show here both quiet-time Champ and Ørsted data used in the IGRF-BGS (Hamilton et al. 2010), with data selected between 1999 and 2009. One sees how the difference in altitude between the satellites affects the SD geographical weight distribution, particularly in the auroral and polar zones. It is also clear how very short wavelength lithospheric field signals may be interpreted as "noise" by this process, e.g., over the Bangui (Central Africa) anomaly. The SD weights, per satellite sample, are calculated over approximately 150 km of satellite ground track, centered on the sample point. This is a much shorter wavelength than is represented by IGRF-BGS (or even by MEME08), so has a smoothing effect. However, for future higher degree spherical harmonic models further thought will be necessary to define an appropriate scale over which to define the SD weights.

Model Parameterization and Solution

The spherical harmonic model for MEME08 contains a degree 13 internal field, with piecewise-linear secular variation between seven nodes at approximately 1.0 year, or 1.5 year intervals, between 2000.0 and 2007.5. The node positions are determined by the temporal density of data. MEME08 also contains a degree 14–60 static lithospheric field and a degree 1 external field, with VMD dependence and a piecewise-linear time variation, again with seven nodes and with further individual external harmonic terms representing 24-h, semi-annual and annual periodicities.

There are 5205 model parameters in MEME08 and the model inversion requires seven iterations with an L1 norm, from an initial model vector derived from an L2 solution. No damping or regularization is involved, though the data are (tesseral) weighted by latitude according to the method of Lesur et al.

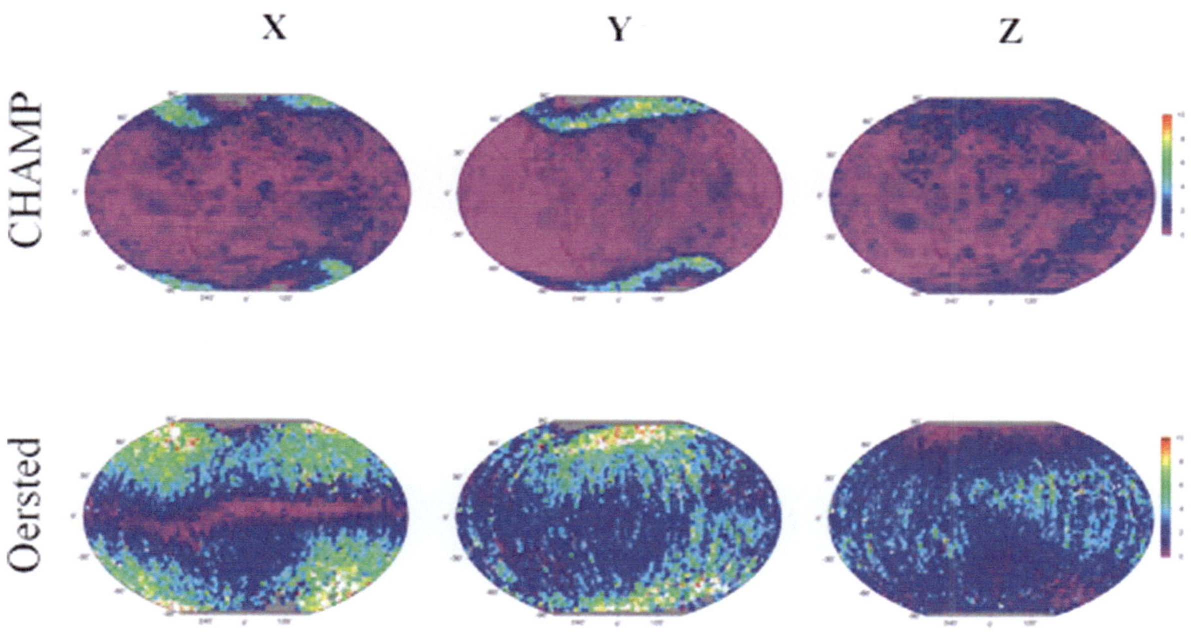

Fig. 11.6 Standard deviation (SD) down-weights (nT) used in IGRF-BGS Hamilton et al. (2010). Component X,Y and Z are in the geocentric North, East and vertically down directions. Data are average in one-degree bins, for quiet-time data selection from the CHAMP and Ørsted satellites between 1999 and 2009

(2005). MEME08 is truncated from an initial degree 100 model.

11.3.3.3 Results and Discussion

Results

The MEME08 core field model power spectrum is shown in Fig. 11.5 of Thomson et al. (2010). In respect of the core field component of MEME08 (i.e., spherical harmonic degree < 13), we find that

1. The MEME08 model is generally consistent with other recent core field models (xCHAOS: Olsen and Mandea (2008), GRIMM: Lesur et al. (2008)) that have different approaches to external field rejection.
2. The coherency of MEME08 with other recent models is above 0.99 below spherical harmonic degree 13 (Fig. 11.7 of Thomson et al. (2010)).
3. Power spectral differences with respect to xCHAOS and GRIMM are greatest (but still less than about 1 nT^2) at harmonic degrees 1, 5, 7 and 11. In comparison, core field differences between xCHAOS and GRIMM only exceed about 1 nT^2 at degrees 7 and 11.
4. The MEME08 secular variation spectrum is more clearly "noisy" above about degree 11 (i.e., around 1 $(nT\ year^{-1})^2$), compared to other models that contain smoother models of temporal variation in the core field.

Here, Fig. 11.7 and 11.8 show, respectively, the vertical core field at the Earth's surface from the MEME08

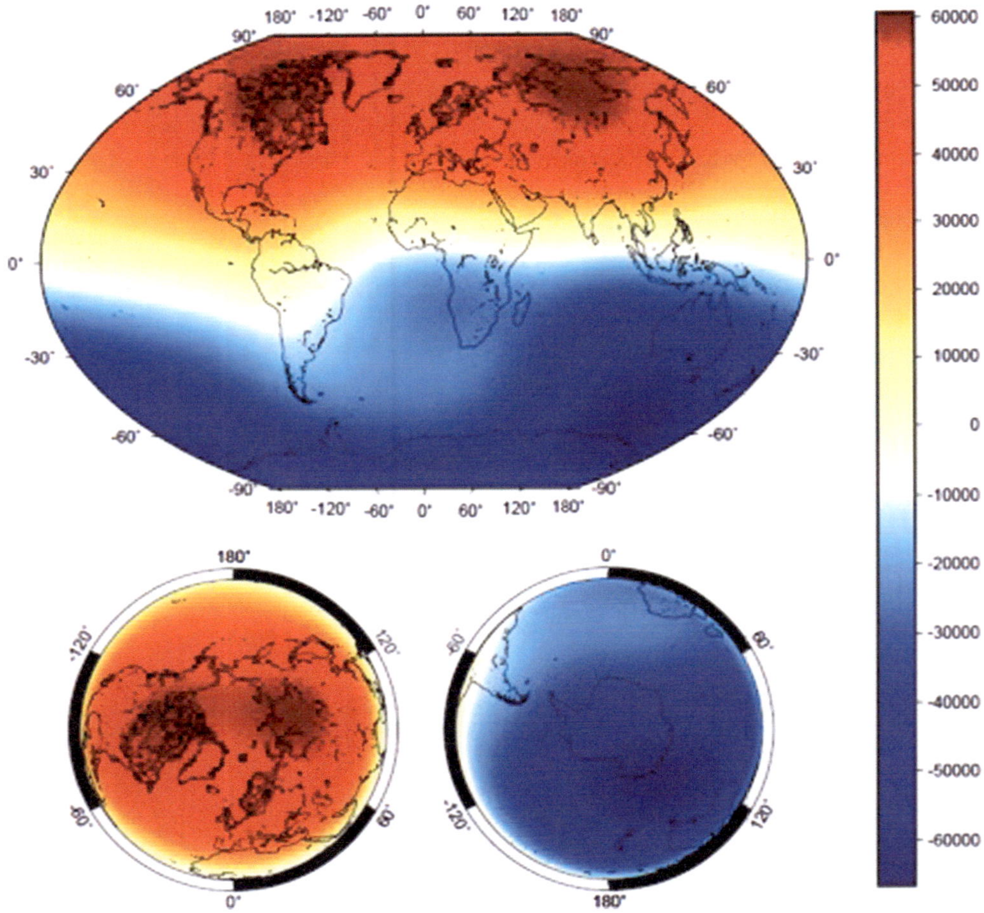

Fig. 11.7 The vertical down component of the core field of MEME08 at the Earth's surface at 2004.5 (in nT)

model, and the vertical component of the difference field, GRIMM-MEME08, again at the Earth's surface. In Fig. 11.8 polar and auroral zone differences are clear, presumably a consequence of the weighting used in MEME08. Interestingly we also find core field differences organized around the Pacific Rim and in the Indian Ocean, which may have consequences for interpreting field changes at the core-mantle boundary.

Discussion

There are a number of improvements planned for BGS models in the next few years, including

1. A higher order spline for the temporal variations arising from the core (to aid studies of core flow change).
2. A higher degree magnetospheric model, derived in an appropriate Sun-fixed coordinate system (to aid magnetospheric studies and to further minimize aliasing between internal and external field sources).
3. Study of improved data selection and weighting, particularly in the polar regions (to aid recovery of higher degree lithospheric structure, particularly at high latitudes). We note that MEME08 was notably different from other recent models in the Polar Regions: this will therefore be investigated further.

Further ahead we envisage the use of selected day-side magnetic data, implying detailed modeling of the Sq ionospheric current systems and a

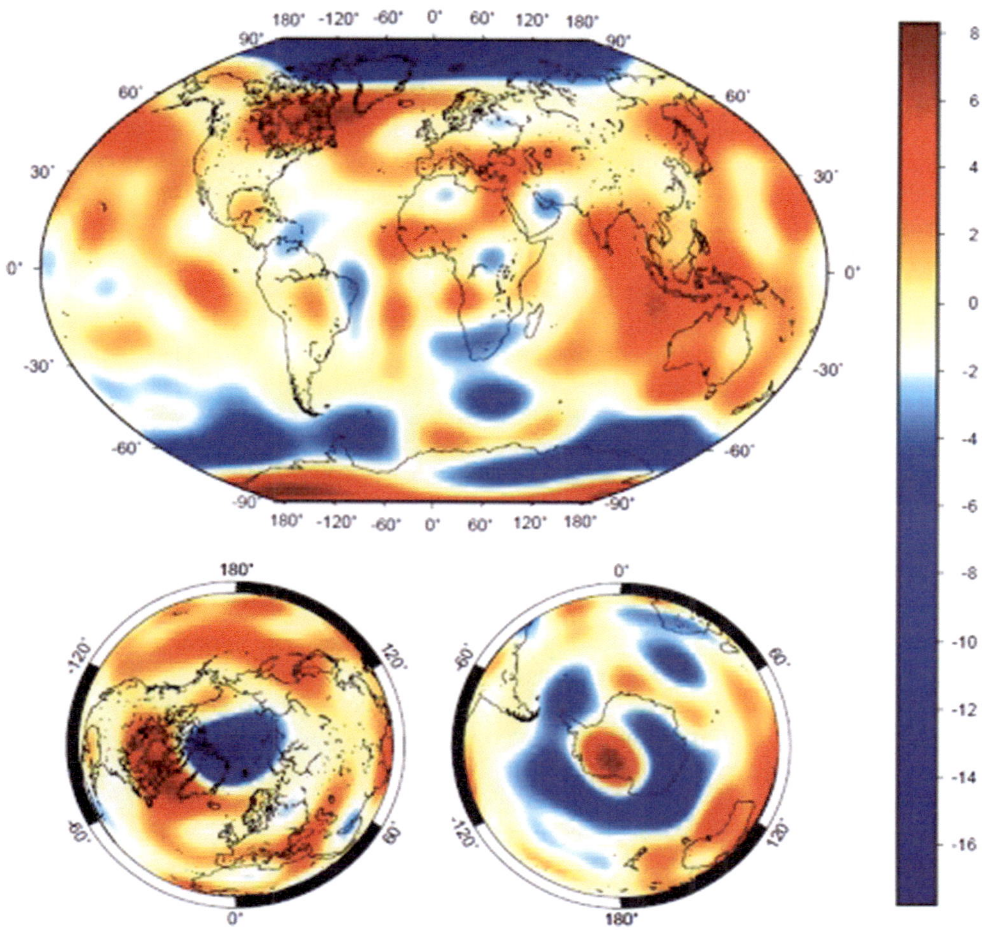

Fig. 11.8 Vertical down component of the difference in the core field between GRIMM and MEME08, at the Earth's surface at 2004.5 (in nT)

re-appraisal of the magnetospheric model methodology. As our understanding of localized ionospheric and field-aligned currents improves, for example during the ESA Swarm satellite mission era, we also envisage developing appropriate models for such sources. Detailed ocean tide and 3D mantle conductivity models remain further areas of modeling interest, as does predictive secular variation modeling from core-flows.

11.4 Conclusion

We have presented, firstly, an overview of the difficulties found in the derivation of core field models from geomagnetic data and, secondly, we have discussed three recently published models that are derived mainly from satellite data. Technically, the three different modeling approaches are broadly similar. The main difference in terms of core field parameterization is the time dependency assumed for the Gauss coefficients. This has a strong effect on the resulting core field model, with each parameterization hopefully revealing different aspect of the core magnetic field time behavior. For the external fields and their associated induced counterparts all modelers use a rather large-scale parameterization and, again, differences are mainly associated with the time dependencies. It is not clear how the different ways these fields are parameterized in MEME08, CHAOS or GRIMM affect the resulting core field model. Overall, the main differences between the three approaches remain the data selection techniques and these differences are particularly apparent at high latitudes. However, the fact that data sets covering different magnetic activity ranges, different local times and having such different noise levels, all lead to rather similar core field models give us confidence in the accuracy of the derived models.

So, where do we stand in term of accuracy of the static and different time derivative of the core field models?

For the static part of the field, differences between the large-scale part (up to degree $l = 13$) of good quality models generally do not exceed 13 to 16 nT at any place on the Earth's surface; the largest differences being observed at high latitudes. In term of Gauss' coefficients this corresponds to differences generally less than 0.7 nT. Such bounds hold for the years 2001.5 to 2009.0 (i.e., nearly the full era of available modern satellite data) and have been observed between different candidates to the DGRF-2005 model. Regarding SV models, accuracy estimates are more difficult to derive. For models like GRIMM (resp. CHAOS), SV estimates are given up to SH degree 16 (resp. 20) but from degree above 8, they are strongly averaged over time. Typically, at degree 12 the SV estimates are average over the full range of available data (i.e., 10 years). Furthermore, it is observed that the SV models derived so far need spatial regularization before being downward continued to the CMB for degrees higher than 13 and therefore their small wavelength components are probably not valid everywhere on the Earth surface. At SH degree 8 or 9, the SV models are likely to be valid for time scales around a year and it is only around degree 5 that this time scale is reduced to several months. The accuracy of Gauss' coefficients of degree 1 is lower than one could expect because of the difficulty in separating the large scale magnetospheric contributions using data spanning only few months. Finally, it should be noticed that the accuracy of derived SV models significantly decreases towards the ends of the data time span. For higher time derivatives, such models are even less accurate. Thus SA models are likely to be poor for SH degrees higher than 6, even if a good coherency is found between, for example, GRIMM-2 and CHAOS-3 up to degree 9. SA models are also strongly averaged in time for degrees as low as 5 or 6. There is no agreement so far on Gauss coefficient estimates for the degree 1 of the SA models.

As we have seen, the SA models of the core magnetic field remain of limited accuracy, nonetheless they are a important source of information for core studies as they allow one to investigate temporal variations of the liquid outer core. Novel approaches are therefore being developed in order to increase the robustness and accuracy of these models. For example, deriving models by imposing constraints associated with the physics of core dynamics is one possibility that has been proposed by Lesur et al. (2010a). Assimilation methods (Canet et al. 2009; Fournier et al. 2007; Liu et al. 2007) are another very promising approach. We hope that significant progress will continue to be made in the coming years through the combination of new techniques and the availability of new high quality satellite data sets.

Acknowledgement AWPT would like to acknowledge the contributions of his BGS colleagues in global field modeling activities. The authors collectively would like to acknowledge the provision of geomagnetic satellite and magnetic observatory data from national institutes.

References

Barraclough DR (1976) Spherical harmonic analysis of the geomagnetic secular variation—a review of methods. Phys Earth Planet Int 12:365–382

Barraclough DR, Malin SRC (1979) Geomagnetic secular acceleration. Geophys J R Astr Soc 58:785–793

Cain JC, Langel RA, Hendricks SJ (1966) First magnetic field results from the OGO-2 satellite. Goddard Space Flight Center, NASA

Canet E, Fournier A, Jault D (2009) Forward and adjoint quasi-geostrophic models of the geomagnetic secular variation. J Geophys Res 114, B11101. doi:10.1029/2008JB006189

Fournier A, Eymin C, Alboussière T (2007) A case for variational geomagnetic data assimilation: insights from a one-dimensional, non-linear, and sparsely observed MHD system. Nonlinear Process Geophys 14(3):163–180

Gillet N, Lesur V, Olsen N (2009) Geomagnetic core field secular variation models. Space Sci Rev. doi:10.1007/s11214-009-9586-6 (In press)

Hamilton B, Macmillan S, Thomson A (2010) The BGS magnetic field candidate models for the 11th generation IGRF. Earth Planet Space (In press)

Hulot G, Sabaka TJ, Olsen N (2007) The present field. In: Kono M (ed) Treatise on geophysics, vol 5:33–75. Elsevier, Amsterdam

Jackson A, ART. Jonkers, Walker MR (2000), Four centuries of geomagnetic secular variation from historical records. Phil Trans R Soc Lond A 358:957–990

Jackson A, Finlay C (2007) Geomagnetic secular variation and applications to the core. In: Kono M (ed) Treatise on geophysics, vol 5:147–193.05. Elsevier, Amsterdam

Korte M, Constable CG (2004) Continuous geomagnetic field models for the past 7 millennia II: CALS7K. Geochem Geophys Geosyst. doi:10.1029/2004GC000801

Langel RA, Ousley G, Berbert J, Murphy J, Settle M (1982) The Magsat mission. Geophys Res Lett 9:243–245

Langel R (1987) The main field. In: Jacobs JA (ed) Geomagnetism, vol 1. Academic, San Diego, CA, Orlando, pp 249–512

Langel RA, Sabaka TJ, Baldwin RT, Conrad JA (1996) The near-earth magnetic field from magnetospheric and quiet-day ionospheric sources and how it is modeled. Phys Earth Planet Int 98:235–267

Lesur V, Macmillan S, Thomson A (2005) The BGS magnetic field candidate models for the 10th generation IGRF. Earth Planets Space 57:1157–1163

Lesur V, Wardinski I, Rother M, Mandea M (2008) GRIMM—The GFZ reference internal magnetic model based on vector satellite and observatory data. Geophys J Int. doi:10.1111/j.1365-246X.2008.03724.x

Lesur V, Wardinski I, Asari S, Minchev B, Mandea M (2010a) Modelling the Earth's core magnetic field under flow constraints. Earth Planets Space 62:503–516. doi: 10.547/eps.2010.02.010

Lesur V, Wardinski I, Hamoudi M, Rother M (2010b) The second generation of the GFZ reference internal magnetic field model: GRIMM-2. Earth Planet Space (In print). doi: 10.547/eps.2010.07.007

Liu D, Tangborn A, Kuang W (2007) Observing system simulation experiments in geomagnetic data assimilation. J Geophys Res 112(B08103). doi: 10–10292006004691

Mandea M, Olsen N (2006) A new approach to directly determine the secular variation from magnetic satellite observations. Geophys Res Lett 33:15306. doi:10.1029/2006GL026616

Maus S, Weidelt P (2004) Separating the magnetospheric disturbance magnetic field into external and transient internal contributions using a 1D conductivity model of the Earth. Geophys Res Lett 31:12614. doi:10.1029/2004GL020232

Maus S, McLean S, Dater D, Lühr H, Rother M, Mai W, Choi S (2005) NGDC/GFZ candidate models for the 10th generation international geomagnetic reference field. Earth Planets Space 57:1151–1156

Maus S, Macmillan S, McLean S, Hamilton B, Thomson A, Nair M (2009) The US/UK world magnetic model for 2010–2015, Technical report, NOAA, 2009. Technical Report NESDIS/NGDC

Menvielle M, Berthelier A (1991) The derived planetary indices: description and availability. Rev Geophys 29:415–432

Olsen N, Holme R, Hulot G, Sabaka T, Neubert T, Tøffner-Clausen L, Primdahl F, Jørgensen J, Léger JM, Barraclough D, Bloxham J, Cain J, Constable C, Golovkov V, Jackson A, Kotzé P, Langlais B, Macmillan S, Mandea M, Merayo J, Newitt L, Purucker M, Risbo T, Stampe M, Thomson A, Voorhies C (2000) Ørsted initial field model. Geophys Res Lett 27(22):3607–3610

Olsen N (2002) A model of the geomagnetic field and its secular variation for epoch 2000 estimated from Ørsted data. Geophys J Int 149:454–462

Olsen N, Sabaka T, Lowes F (2005) New parameterisation of external and induced fields in geomagnetic field modelling, and a candidate model for IGRF 2005. Earth Planets Space 57:1141–1149

Olsen N, Lühr H, Sabaka TJ, Mandea M, Rother M, Tøffner-Clausen L, Choi S (2006) CHAOS—a model of Earth's magnetic field derived from CHAMP, Ørsted, and SAC-C magnetic satellite data. Geophys J Int 166:67–75. doi:10.1111/j.1365-246X.2006.02959.x

Olsen N (2007a) Encyclopedia of geomagnetism and paleomagnetism. In: Gubbins D, Herrero-Bervera E (eds). Springer, Heidelberg, pp 743–746, Chap Ørsted

Olsen N, Mandea M (2007b) Investigation of a secular variation impulse using satellite data: the 2003 geomagnetic jerk. Earth Planet Sci Lett 255:94–105. doi:10.1016/j.epsl.2006.12.008

Olsen N, Mandea M (2008) Rapidly changing flows in the Earth's core. Nat Geosci 1(6):390–394

Olsen N, Mandea M, Sabaka TJ, Tøffner-Clausen L (2009) CHAOS-2—a geomagnetic field model derived from one decade of continuous satellite data. Geophys J Int. doi:10.1111/j.1365-246X.2009.04386.x

Olsen N, Hulot G, Sabaka TJ (2010a) The geomagnetic field—its various field contributions and the data that enables their investigation. In: Freeden W, Nashed Z, Sonar T (eds) Handbook of geomathematics. Springer, Heidelberg

Olsen N, Mandea M, Sabaka TJ, Tøffner-Clausen L (2010b) The CHAOS-3 geomagnetic field model and candidates for the 11th Generation of IGRF. Earth Planet Space. (In print)

Reigber C, Lühr H, Schwintzer P (2002) CHAMP mission status. Adv Space Res 30(2):129–134

Roberts PH, Scott S (1965) On the analysis of secular variation, 1, A hydromagnetic constraint: theory. J Geomag Geoelectr 17:137–151

Sabaka T, Baldwin R (1993) Modeling the Sq magnetic field from POGO and Magsat satellite and contemporaneous hourly observatory data: phase I, Contract report HSTX 9302, Hughes STX Corp. for NASA/GSFC Contract NAS5-31 760

Sabaka TJ, Olsen N, Langel RA (2002) A comprehensive model of the quiet-time, near-Earth magnetic field: phase 3. Geophys J Int 151:32–68. doi:10.1046/j.1365-246X.2002.01774.x

Sabaka TJ, Olsen N, Purucker ME (2004) Extending comprehensive models of the Earth's magnetic field with Ørsted and CHAMP data. Geophys J Int 159:521–547. doi:10.1111/j.1365-246X.2004.02421.x

Thomson A, Lesur V (2007) An improved geomagnetic data selection algorithm for global geomagnetic field modelling. Geophys J Int 169:951–963. doi:10.1111/j.1365-246X.2007.03354.x

Thomson A, Hamilton B, Macmillan S, Reay S (2010) A novel weighting method for satellite magnetic data and a new global magnetic field model. Geophys J Int 181: issue 1 250–260. doi:10.1111/j.1365-246X.2010.04510.x

Troshichev OA, Andresen VG, Vennerstrøm S, Friis-Christensen E (1988) Magnetic activity in the polar cap—a new index. Planet Space Sci 36:1095–1102

Viljanen A, Hakkinen L (1997) Image magnetometer network, in satellite-ground based coordination. Technical report, ESA, Noordwijk, SP-1198

Wardinski I, Holme R (2006) A time-dependent model of the Earth's magnetic field and its secular variation for the period 1980–2000. J Geophys Res 111(B10):12101. doi:10.1029/2006JB004401

Chapter 12

Interpretation of Core Field Models

Weijia Kuang and Andrew Tangborn

Abstract In this chapter we review several recent research results on the observed geomagnetic secular variation and secular acceleration, the core flow models inferred from these observations, and their implications, in particular those of the torsional oscillations, on short period secular variation and on the dynamical properties inside the core. We also provide a comprehensive review on the recent development in geomagnetic data assimilation, and its applications to predict future secular variation. Most of the reviewed research results are either reported in IAGA General Assembly in Soporan in 2009, or in the period between this and the previous IAGA conference.

12.1 Introduction

Since POGO satellites in mid 1960s, satellites have been used to measure global geomagnetic field for nearly a half century. In particular, the launch of Ørsted in 1999 started a decade-long continuous monitoring of the geomagnetic environment from space. The influx of magnetic data from Ørsted, CHAMP and SAC-C have enabled geomagnetic communities to model the global geomagnetic field accurately at high spatial and temporal resolutions (Sabaka et al. 2004; Olsen et al. 2006, 2009; Maus et al. 2006; Lesur et al. 2008), in particular its secular variation (SV) and secular acceleration (SA).

W. Kuang (✉)
Planetary Geodynamics Laboratory, NASA Goddard Space Flight Center, Greenbelt, MD, USA
e-mail: Weijia.Kuang-1@nasa.gov

These field models are critical for understanding the dynamical processes in the Earth's core, since much of the observed geomagnetic field is of internal origin (the core field), which is generated and maintained by convection in the Earth's fluid outer core (geodynamo). Therefore, the measured SV and SA are manifestations of various core dynamical processes. Utilizing geomagnetic observations to gain insight on the core dynamics, and on the mechanisms responsible for the observed SV and SA, has been one of the main efforts in geomagnetic studies.

Among the goals is to infer core flow right beneath the core-mantle boundary (CMB) using the geomagnetic field model output. In addition to fit the observed SV and SA, the inferred core flow is also used to interpret other geodynamic observables, such as the length of day (LOD) (Jault et al. 1988; Holme and Whaler, 2001). The high resolution global field models from the satellite era have lead to new core flow models with more details and complexities (Hulot et al. 2002; Eymin and Hulot, 2005; Holme and Olsen, 2006; Olsen and Mandea, 2008; Beggan et al. 2009).

Satellite data significantly improves the observation of geomagnetic jerks, or SV impulses that are, simply speaking, the directional changes of SA on sub-annual time scales. Although the jerk events were detected early on (Courtillot and Le Mouël, 1984), their global distributions are difficult to be identified with the ground observatory data. The continuous satellite measurements can now be used to determine accurately their morphologies and occurrences (Ballani et al. 2005; Olsen and Mandea, 2007). These have lead to intensive studies on determining their origins in the deep interior (Bloxham et al. 2002; Holme and Olsen, 2006; Olsen and Mandea, 2008; Pinheiro and Jackson, 2008; Wardisnki et al. 2008).

M. Mandea, M. Korte (eds.), *Geomagnetic Observations and Models*, IAGA Special Sopron Book Series 5,
DOI 10.1007/978-90-481-9858-0_12,

Independent from the geomagnetic observations, numerical geodynamo simulation is another approach to understand the core dynamics. Though dynamo theory was first proposed nearly a century ago (Larmor, 1919), fully nonlinear dynamo simulation has only become available over the past 15 years (Glatzmaier and Roberts, 1995; Kageyama and Sato, 1997; Kuang and Bloxham, 1997). But this effort has been by far disconnected from the geomagnetic field studies.

A new development has been shaping up in the past 3 years to combine geomagnetic observations and dynamic models to better understand core dynamics and to predict future geomagnetic field variations (Sun et al. 2007; Liu et al. 2007; Fournier et al. 2007; Kuang et al. 2008, 2009, 2010; Canet et al. 2009). Similar to this development, attempts are also made to use core flow models to predict SV (Maus et al. 2008; Beggan and Whaler, 2009). These can be generally called geomagnetic data assimilation. The potential of these efforts can not be over-estimated: dynamic models will be assessed and improved based on observational constraints; and the improved models can then be used to forecast more accurately future SV and to improve the qualities of past geomagnetic records. In particular, forecast accuracy can provide independent, yet very important assessments on hypotheses and approximations used in core dynamics models.

In this chapter, we will review the recent research results on high resolution core flow models inferred from the observed SV and SA with an emphasis on interpretation of geomagnetic jerks, on geomagnetic data assimilation and SV forecasts. Many of the results were presented in the 11th IAGA Scientific Assembly in Sopron. But we add also some background theories and mathematics for better comprehension.

This chapter is organized as follows: a brief description of core dynamics and geodynamo is given first; followed by discussions on the core flow models inferred from surface observations, on torsional oscillations and on interpretation of geomagnetic jerks. After that, we review the results on geomagnetic data assimilation, and prediction of future SV. Discussion is given at the end of the chapter.

12.2 Core Dynamics and Geodynamo

Much of the geomagnetic field observed at the Earth's surface is of internal origin. This part of the field, the core field, is generated and maintained by convection in the Earth's iron-rich liquid outer core (geodynamo). The convection in the core is powered by gravitational energy released from differentiation and secular cooling of the planet (and possibly from other sources). Therefore, proper interpretation of the core field models (from observations) depends on and helps understanding of the core dynamical states.

To help understand the results in the rest of this chapter, we provide first theoretical descriptions and mathematical equations of the core dynamics. Variations of these equations have been used for core flow models, numerical simulations and data assimilation.

In the leading order approximation, the core fluid is Boussinesq, i.e., the density variation is negligible except in the gravitational effect; its material properties are also uniform and constant (on the geodynamo time scales). With these approximations, the core dynamical processes can be described by the following set of nonlinear partial differential equations (defined in the reference frame co-rating with the solid mantle):

$$\frac{\partial \mathbf{v}}{\partial t} + \mathbf{v}\cdot\nabla\mathbf{v} + 2\mathbf{\Omega}\times\mathbf{v} = -\nabla\frac{p}{\rho_0} + \frac{1}{\mu\rho_0}(\mathbf{B}\cdot\nabla)\mathbf{B} + \frac{\rho}{\rho_0}\mathbf{g} + \nu\nabla^2\mathbf{v}, \tag{12.1}$$

$$\frac{\partial \mathbf{B}}{\partial t} = \nabla\times(\mathbf{v}\times\mathbf{B}) + \eta\nabla^2\mathbf{B}, \tag{12.2}$$

$$\frac{\partial \rho}{\partial t} = -\mathbf{v}\cdot\nabla\rho + \kappa\nabla^2\rho, \tag{12.3}$$

where $\mathbf{B}$ is the magnetic field, $\mathbf{v}$ is the velocity field, ρ_0 is the mean core fluid density, ρ is the spatially-temporally varying density distribution,$\mathbf{\Omega}$ is the mean angular velocity of the Earth, μ is the core fluid magnetic permeability, ν is the kinematic viscosity, η is the magnetic diffusivity, and κ is the diffusivity for the density anomaly. Certainly, these equations can be more complicated if additional geophysical processes are considered, e.g., in a non-uniform rotating reference frame (i.e., a time varying $\mathbf{\Omega}$). The variation of the density ρ can arise from temperature differences inside the core or at the boundaries, and or from chemical processes, e.g., release of lighter constituents at the ICB and at the CMB (e.g., Amit et al. 2008; Aubert et al. 2009; Buffett and Seagle, 2010)

Eqs. (12.1)–(12.3) of the (leading order approximation) core states are still mathematically too complicated: obtaining analytical solutions is nearly impossible, unless substantial simplifications are made to the equations. Numerical simulations with specified boundary conditions and initial states ("forward modeling") are currently the only way to obtain the fully nonlinear solutions of the system.

In numerical modeling, the magnetic field **B** and the velocity field **v** are often decomposed into the poloidal and toroidal components:

$$\mathbf{B} = \nabla \times (T_B \mathbf{1}_r) + \nabla \times \nabla \times (P_B \mathbf{1}_r), \qquad (12.4)$$

$$\mathbf{v} = \nabla \times (T_v \mathbf{1}_r) + \nabla \times \nabla \times (P_v \mathbf{1}_r), \qquad (12.5)$$

where $\mathbf{1}_r$ is the unit vector in the radial direction, T and P are the toroidal and poloidal scalars, respectively. If the Earth's outer core is approximated as a perfectly spherical shell, then these scalars can be conveniently described by spherical harmonic expansions

$$\begin{bmatrix} P_v \\ T_v \\ P_B \\ T_B \end{bmatrix} = \sum_{0 \le m \le l} \begin{bmatrix} v_l^m(r,t) \\ \omega_l^m(r,t) \\ b_l^m(r,t) \\ j_l^m(r,t) \end{bmatrix} Y_l^m(\theta,\phi) + C.C., \qquad (12.6)$$

in the spherical coordinate of the radius r, the co-latitude θ and the longitude ϕ. In (12.6), Y_l^m are the complex spherical harmonic functions of degree l and order m, and $C.C.$ implies the complex conjugate part. The varying density ρ can be expanded similarly. In this description, the core state is then specified by the spherical harmonic coefficients in (12.6).

Connection between the core state and surface geomagnetic observations can be made through the properties of the magnetic field. In an electrically insulating domain, **B** is a potential field, much simpler than (12.4):

$$\mathbf{B} = -\nabla V, \qquad \text{and} \qquad \nabla \cdot \mathbf{B} = \nabla^2 V = 0. \qquad (12.7)$$

The Earth's solid mantle can be approximated to leading order an electrical insulator. In this region,

$$V(r,\theta,\phi) = \sum_{l,m} \left(\frac{r_b}{r}\right)^{l+1} c_l^m \, Y_l^m(\theta,\phi) + C.C. \qquad (12.8)$$

$$= \sum_{l,m} \left(\tfrac{r_e}{r}\right)^{l+1} d_l^m \, Y_l^m(\theta,\phi) + C.C. \qquad (12.9)$$

for $r_e \ge r \ge r_b$. In the above equation, r_b is the mean radius of the bottom of the insulating mantle, e.g., the CMB, r_e is the mean radius of the Earth's surface, c_l^m and d_l^m are the spectral coefficients defined at r_b and at r_e, respectively. In particular,

$$c_l^m = \left(\frac{r_e}{r_b}\right)^{l+1} d_l^m . \qquad (12.10)$$

For the Earth, $r_e/r_b \approx 2$. By (12.4), (12.6), (12.7) and (12.8),

$$b_l^m(r_b) = \frac{r_b}{l} c_l^m . \qquad (12.11)$$

The relations (12.10) and (12.11) provide some important information: first, given the surface geomagnetic measurements $\{d_l^m\}$ at r_e, one could find the corresponding coefficients $\{c_l^m\}$ at the bottom of the mantle r_b with the simple downward continuation (12.10), and thus part of the core state $\{b_l^m\}$ via (12.11). However, the toroidal field in the core, i.e., the spectral coefficients $\{j_l^m\}$ in the expansion (12.6), cannot be obtained directly from surface observations.

The other is the strong spatial damping of the core signals through the solid mantle: the higher degree of the coefficient c_l^m (thus the finer core field structures) at the bottom of the mantle r_b, the faster it decays at the surface (d_l^m). In fact, d_l^m for the degrees $l > 14$ are buried into the crustal magnetic signals (Langel and Estes, 1982), giving an upper limit on the direct information for the core state from observations.

However, as described in the next section, more information about the core state can be obtained if both the field and its SV are better utilized.

12.3 Core Flow and High Frequency SV

Temporal variation of the core field arises from two distinct processes, advection (via emf in induction) and dissipation (due to finite electrical conductivity), as shown in (12.2). One would expect that information about the core flow could be extracted from the observed core field and its SV at the surface. The first effort was made by Roberts and Scotts (1965). Their work has laid the foundation for core flow modeling studies since then.

In their study, Roberts and Scotts (1965) proposed the "frozen-flux" approximation in which contributions of the magnetic dissipation to SV is assumed negligible based on the considerations that the spatial scales of the field are sufficiently large, and that the time scales in considerations are sufficiently short. But, with this approximation, the core flow beneath the CMB cannot be uniquely determined (Roberts and Scott, 1965; Backus, 1968). To eliminate such non-uniqueness, various constraints are added to core flow. Reader can find more details from recent reviews, e.g., Holme (2007).

The decade-long continuous satellite geomagnetic measurements have provided accurate and high resolution (in both space and time) mapping of the magnetic field. These have lead to more complex core flow models that could interpret the observed SV and SA, in particular the short-period SV such as geomagnetic jerks (impulses).

In this section, we review some of recent research results on core flow models and their interpretation of geomagnetic jerks. Formulations in this sections are made consistent with the geodynamo equations (12.1), (12.2) and (12.3).

12.3.1 Core Flow Inferred from Satellite Magnetic Data

Time variation of the magnetic field **B** in an electrically conducting fluid is governed by the induction equation (12.2). Its radial component at the CMB is used in the core flow studies,

$$\frac{\partial B_r}{\partial t} = -\nabla \cdot (\mathbf{v}_H B_r) + \frac{1}{r^2}\eta\left(\frac{\partial^2}{\partial r^2} + \frac{\hat{L}}{r^2}\right)\left(r^2 B_r\right), \tag{12.12}$$

where $\mathbf{v}_H$ represents the horizontal components of the core flow velocity field $\mathbf{v}$, B_r is the radial component of **B**, and

$$\hat{L} \equiv \frac{1}{\sin\theta}\frac{\partial}{\partial\theta}\sin\theta\frac{\partial}{\partial\theta} + \frac{1}{\sin^2\theta}\frac{\partial^2}{\partial\phi^2} \tag{12.13}$$

is the angular momentum operator in the spherical coordinate (r, θ, ϕ). Continuity of **B** across the CMB (i.e., no surface current) implies that B_r and $\partial B_r/\partial t$ beneath the CMB can be obtained from their values measured at the surface. However, $\partial^2(r^2 B_r)/\partial r^2$ in (12.12) is unknown.

With the "frozen-flux" approximation (Roberts and Scott, 1965), (12.12) is simplified to

$$\frac{\partial B_r}{\partial t} = -\nabla \cdot (\mathbf{v}_H B_r). \tag{12.14}$$

with the only unknown $\mathbf{v}_H$ (the quantity to be evaluated). But, (12.14) alone can not determine $\mathbf{v}_H$ uniquely (Roberts and Scott, 1965; Backus, 1968). Therefore, additional approximations (or constraints) on $\mathbf{v}_H$ are necessary to eliminate the non-uniqueness, e.g., steady flow (Gubbins, 1982), tangentially geostrophic flow (Le Mouël, 1984), etc. More complex constraints are discussed recently (Holme and Whaler, 2001; Amit and Olson, 2004; Pais et al. 2004). We refer the reader to a recent review by Holme (2007) for detailed descriptions on various core flow models.

Continuous satellite measurements from 1999 have provided more accurate and higher resolution core field and SV, i.e., B_r and $\partial B_r/\partial t$ in (12.14). For example, the maximum degree l of the SV Gauss coefficients is less than 10 with only the observatory data (Langel et al. 1986). However, $l = 16$ in the most recent field models from satellite data (Olsen et al. 2009). Similar high resolution field models can also be obtained from satellite data via local algorithms, e.g., the "virtual observatories" proposed by Mandea and Olsen (2006). In this approach, the difference between the satellite measured magnetic field $\mathbf{B}^m$ and the background static field $\mathbf{B}^\mathbf{M}$ is approximated by a potential

$$\delta\mathbf{B} \equiv \mathbf{B}^m - \mathbf{B}^M = -\nabla V(x, y, z), \tag{12.15}$$

$$\begin{aligned} V = {} & V_x(x - x_i) + V_y(y - y_i) + V_z(z - z_i) \\ & + V_{xx}(x - x_i)^2 - V_{yy}(y - y_i)^2 \\ & - \left(V_{xx} + V_{yy}\right)(z - z_i)^2 + V_{xy}(x - x_i)(y - y_i) \\ & + V_{xz}(x - x_i)(z - z_i) \\ & + V_{yz}(y - y_i)(z - z_i), \end{aligned} \tag{12.16}$$

in a spatial domain centered at (x_i, y_i, z_i). Good agreement between the satellite measurements and ground observatory data are found with the domain dimension of 400 km and monthly means (Mandea and Olsen, 2006). Olsen and Mandea (2007) are able to use this algorithm to determine the global distribution of the 2003 geomagnetic jerk. However, Beggan et al. (2009)

cautioned that the accuracies of such approach depends on appropriate data selection.

More accurate SA can be obtained from higher time resolution SV and can therefore be used for core flow modeling, as shown in the equation by taking the time derivative of (12.14):

$$\frac{\partial^2 B_r}{\partial t^2} = -\nabla \cdot \left(\mathbf{v}_H \frac{\partial B_r}{\partial t}\right) - \nabla \cdot \left(\frac{\partial \mathbf{v}_H}{\partial t} B_r\right). \quad (12.17)$$

One such application is by Olsen and Mandea (2008). They found from their studies that of various core flow models, the relaxed tangentially geostrophic flow and the relaxed helical flow are better in explaining the short-period SV (e.g., 2003 geomagnetic jerk), and the SA, as shown in Fig. 12.1.

12.3.2 Torsional Oscillations in the Core

From various core flow models inferred from magnetic measurements, torsional oscillations are often extracted for the studies of short period SV, LOD variations and core-mantle interactions. In core dynamics, the torsional oscillations are perhaps the fastest waves driven by the Lorentz force. Because of their detectability beneath the CMB and their relations to the core field in the deeper interior, they have been intensively studied for decades.

The origin of the torsional oscillations could be attributed to Taylor's (1963) analysis on the geodynamo. If the fluid inertia and the viscous effect are negligible in the momentum balance (12.1) in the core, then we have

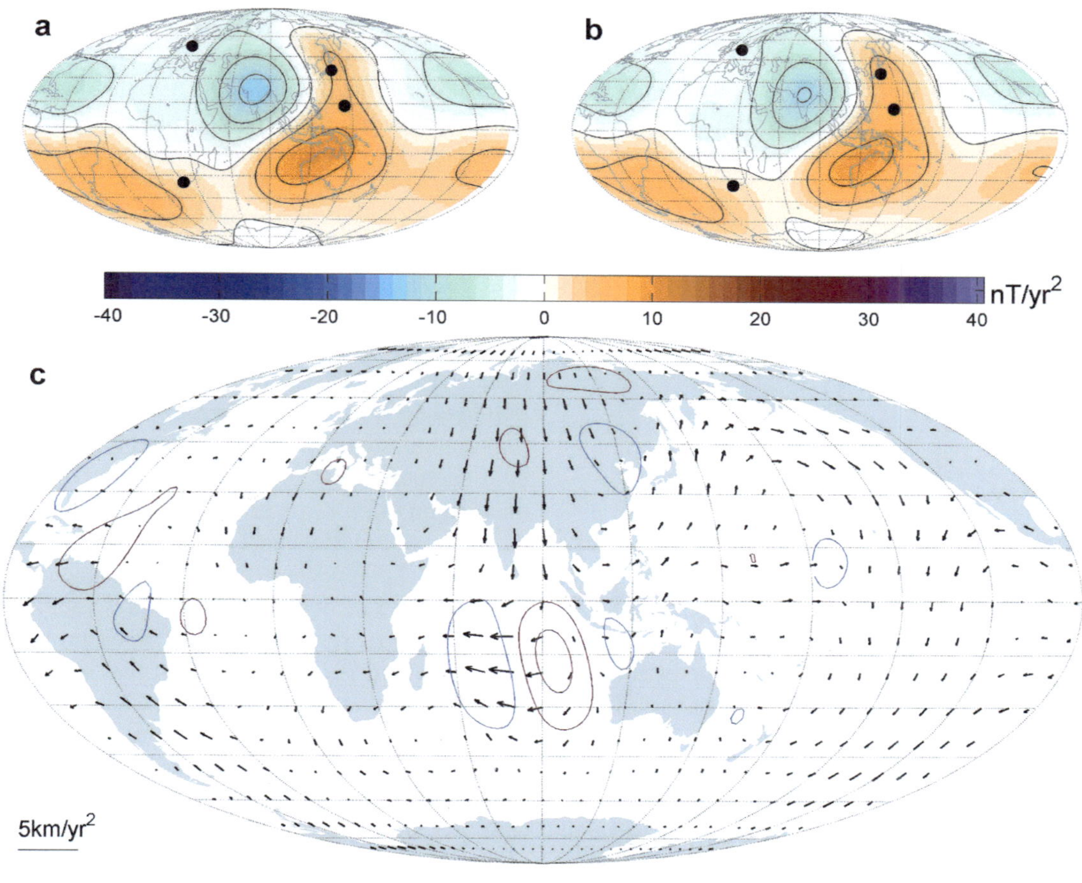

Fig. 12.1 The observed SA (**a**) and (**b**) that are recovered by the relaxed tangential geostrophic core flow (**c**) from Olsen and Mandea (2008). Reprint is permitted by Nature Publishing Group

$$2\mathbf{\Omega} \times \mathbf{v} = -\nabla p + \frac{1}{\mu\rho_0} (\mathbf{B} \cdot \nabla) \mathbf{B} + \frac{\rho}{\rho_0} \mathbf{g}. \quad (12.18)$$

This is the so-called the magnetostrophic balance. The Taylor's constraint can then be derived by taking the ϕ-component of (12.18) and integrating over the cylindrical surfaces A co-axial with the rotation axis of the Earth,

$$T_B \equiv \frac{1}{\mu\rho_0} \iint_A [(\mathbf{B} \cdot \nabla) \mathbf{B}]_\phi \, dS = 0. \quad (12.19)$$

This is equivalent to the vanishing of the vertical Lorentz torque on A. Taylor (1963) argued that in the Earth's core, the constraint (12.19) is not exactly satisfied, and suggested that any departure from (12.19) could be balanced by the fluid inertia reintroduced back to (12.18),

$$\frac{\partial}{\partial t} \iint_A v_\phi dS = T_B, \quad (12.20)$$

i.e., finite Lorentz torques on the cylindrical surfaces lead to rapidly varying, cylindrical zonal flow in the core, the so-called torsional oscillations. Later, Braginsky (1967, 1976) formulated in more detail the magnetic waves in the Earth's core, including the torsional oscillations, by examining small perturbations to an axisymmetric geodynamo state.

It is very interesting to notice that in the cylindrical coordinate system (s, ϕ, z), T_B can be written as

$$\begin{aligned} T_B = & \frac{1}{\mu\rho} \left(\frac{\partial}{\partial s} + \frac{2}{s} \right) \iint_A B_\phi B_s dS \\ & + \frac{1}{\mu\rho} \left[\left(B_z B_\phi \right)^T - \left(B_z B_\phi \right)^B \right], \end{aligned} \quad (12.21)$$

where the superscripts T and B imply the zonal mean values at the top and bottom intersections of A and the CMB; the subscripts s and z imply the radial and vertical components in cylindrical coordinates, respectively. In deriving (12.19), the fluid incompressibility is used. Obviously, if the full momentum balance (12.1) is considered, the viscous effect will also be added to (12.20). In this case, the viscous drag could also balance finite T_B, i.e., Model-Z states (Braginsky and Roberts, 1987; Braginsky, 1989). With certain assumptions on the core state and lower mantle properties, (12.21) can be simplified as a function of the "measured" magnetic field at the CMB and the rms B_s on the cylindrical surfaces A (e.g., Zatman and Bloxham, 1997; Jault 2003).

In addition to early theoretical studies (Taylor 1963; Braginsky, 1967, 1976), torsional oscillations are also examined via numerical simulations (Kuang, 1999; Dumberry and Bloxham, 2003; Jault, 2008; Wicht and Christensen, 2010) and observations (Jault et al. 1988; Zatman and Bloxham, 1997; Holme and Whaler, 2001; Buffett et al. 2009).

Jault et al. (1988) took an elegant yet simple approach to extract torsional oscillations from the core flow inferred from SV data. In their approach, they first identify the axisymmetric and equatorially symmetric time-varying zonal flow. They then assume this part of the flow invariant vertically in the outer core. Therefore, they were able to calculate the total axial angular momentum of the outer core. This approach is very successful in matching the core angular momentum variation with that of the solid mantle from the observed LOD variation on decadal time scales. A variation of this approach by Holme and Whaler (2001) can extend this angular momentum conservation further back in time. These results demonstrate kinematically that the observed decadal LOD variation is due to the angular momentum exchange between the liquid outer core and the solid mantle. Similar approaches of determining torsional oscillations can be found also in later studies (Zatman and Bloxham, 1997; Bloxham et al. 2002; Buffett et al. 2009).

Inferred torsional oscillations are also used to examine the magnetic field properties deep inside the outer core. For example, by utilizing the torsional oscillation frequencies and several assumptions, e.g., a steady background state (that the torsional oscillations are built upon) that does not interact with the oscillations, the drag on the CMB is proportional to the zonal flow, etc., Zatman and Bloxham (1997) obtained the strength of the radial component $B_s(s)$ of the magnetic field in the core as a function of the distance s from the rotation axis. Their results show that B_s is on the order of 10^{-4} T, and is the strongest on the tangent cylinder (the cylindrical surface co-axial with the rotation axis and tangent to the inner core at the equator). More recent studies of Gillet et al. (2010) suggest a much stronger field strength of 4×10^{-3} T deep inside the core.

Recently, torsional oscillations are also considered to explain the observed geomagnetic jerks. Studies by Jackson (1997) and by Davis and Whaler (1997) showed that a time dependent core flow is necessary to explain the observed geomagnetic jerks. Bloxham et al. (2002) explicitly used in their studies the following core flow

$$\mathbf{v}_H = \mathbf{v}_0 + \mathbf{a}_0 t + \sum_{i=1}^{3} \mathbf{v}_T(\omega_i) \tag{12.22}$$

that comprises of a steady flow $\mathbf{v}_0$, a steady acceleration $\mathbf{a}_0$ and three torsional oscillations $\mathbf{v}_\mathrm{T}$ with different frequencies ω_i. They are able to successfully fit the three geomagnetic jerks occurred in 1969, 1978 and 1991 with this core flow model. But it seems that the torsional oscillations may not be sufficient to explain the 2003 geomagnetic jerk. Olsen and Mandea (2008) found in their study that there was a strong cross-equatorial acceleration of the core flow beneath the Indian ocean accompanying the jerk (see Fig. 12.1). If this feature is not an artifact, then it is certainly not originated from the torsional oscillations. Regardless, these studies strongly suggest that the geomagnetic jerks are due to interactions of the core field and the rapidly time-varying core flow.

It should be pointed out that, not all axisymmetric and equatorially symmetric zonal flow inferred from magnetic data are torsional oscillations. For example, let us revisit again the magnetostrophic balance (12.18). The ϕ-component of $\nabla\times$(12.18) is

$$\begin{aligned}\frac{\partial v_\phi}{\partial z} =& \frac{1}{2\mu\rho_0}\left\{\frac{\partial}{\partial s}\left[(\mathbf{B}\cdot\nabla)B_z\right]\right.\\ &\left.-\frac{\partial}{\partial z}\left[(\mathbf{B}\cdot\nabla)B_s - \frac{B_\phi^2}{s}\right]\right\} \equiv F_B.\end{aligned} \tag{12.23}$$

Obviously the axisymmetric but equatorially antisymmetric part of F_B generates an axisymmetric and equatorially symmetric zonal flow. In particular, the contribution from the zonal field B_ϕ^2 in F_B could be dominant in the core. There is no observational information of the field inside the core, however, as shown in Fig. 12.2, this part of F_B appears in numerical geodynamo simulation solutions. Thus, this non-torsional zonal flow could also be important in explaining geomagnetic jerks and other short period SV.

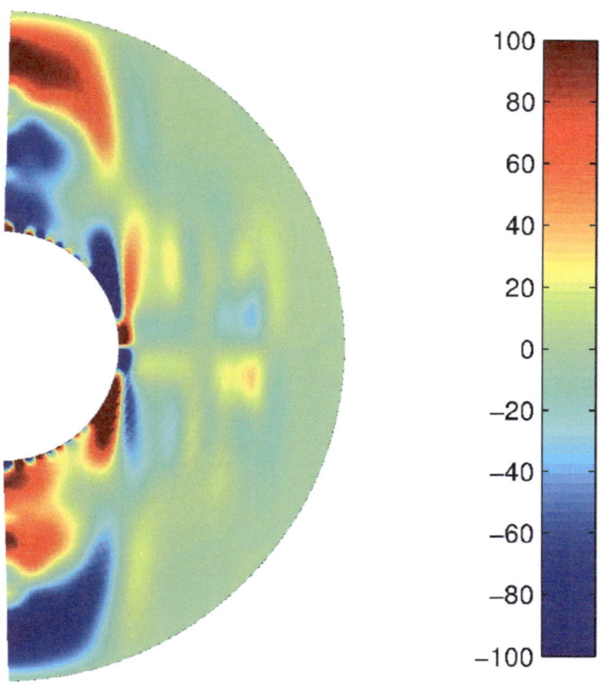

Fig. 12.2 A snapshot of the axi-symmetric F_B that is anti-symmetric about the equator. This is from a numerical dynamo solution from the MoSST core dynamics model (Kuang and Chao, 2002; Jiang and Kuang, 2008)

12.4 Prediction of Geomagnetic Secular Variation

In the past few years, a new research direction has been developing: combining theoretical models and surface observations to predict future SV. The models used in these studies include the core flow models inferred from the measured SV in the past (Maus et al. 2008; Beggan and Whaler, 2009), simplified core dynamics models (Canet et al. 2009), and numerical geodynamo models (Kuang et al. 2008, 2009, 2010).

This new development deserves special attention for several reasons: geomagnetic data can provide constraints on the approximations and assumptions used in the theoretical models; and the improved models (i.e., the models adjusted to the observational constraints) can be used to better predict future changes of the geomagnetic field.

Assimilation of data and models is not new. It has been developed and applied in meteorology and oceanography over many decades. Therefore, methodologies developed in those fields can be transplanted directly or with certain modifications to geomagnetic data assimilation. The transplant and modifications need, however, better understanding of unique properties of the geomagnetic field and and core dynamics models.

In this section, we shall provide a brief, but very comprehensible description of the mathematics of data assimilation. Then we describe the research results reported in the recent past.

12.4.1 *Mathematical Fundamentals of Data Assimilation*

Any geophysical system can be defined by a set of variables $\mathbf{x}$ which is called the "state vector" or simply the "state" in data assimilation. Variation of the state vector $\mathbf{x}$ can be described by a set of the differential equations which, in the simplest description, are of the form

$$\frac{\partial \mathbf{x}}{\partial t} = \mathbf{M} \cdot \mathbf{x}, \tag{12.24}$$

where the matrix $\mathbf{M}$ describes various forces in the system and depends in general on the state vector $\mathbf{x}$ as well.

Observations of geophysical systems are generally only part of the state vector, so that the observed data, described by a vector $\mathbf{y}$, is a subset of $\mathbf{x}$, in addition with observational errors ϵ^o:

$$\mathbf{y} = \mathbf{H} \cdot \mathbf{x} + \epsilon^o, \tag{12.25}$$

where $\mathbf{H}$ is the projection operator that maps the full state vector into the observation subspace.

Assimilation methods can generally be divided into sequential and variational approaches. In sequential assimilation, observations are combined with the current state estimate, or forecast ($\mathbf{x}^f$), to produce a new state estimate, called the analysis ($\mathbf{x}^a$). The analysis is then used as the initial state for the next model run that will produce new forecast. Sequential techniques include Optimal Interpolation (OI) which uses prescribed error covariances; the Kalman Filter (or extended Kalman Filter), which evolves the error covariances using linearized model equations; and the ensemble Kalman Filter, which estimates the time evolving forecast error covariances from a perturbed ensemble of model runs. The steps involved in sequential methods at time t_i (the analysis time) are:

$$\mathbf{x}_i^a = \mathbf{x}_i^f + \mathbf{K}_i \left(\mathbf{y}_i - \mathbf{H}\mathbf{x}_i^f \right) \tag{12.26}$$

where the gain matrix $\mathbf{K}$ is given by

$$\mathbf{K}_i = \mathbf{P}_i^f \mathbf{H}^T \left[\mathbf{H}\mathbf{P}_i^f \mathbf{H}^T + \mathbf{R} \right]^{-1}, \tag{12.27}$$

where $\mathbf{P}_i^f$ is the forecast error covariance and $\mathbf{R}$ is the observation error covariance.

Variational data assimilation differs from sequential assimilation in that the goal is a global adjustment of the model trajectory through the simultaneous assimilation of observations over an assimilation window. Thus, in the variational approach, it is possible that observations at a later time could influence the model solution at earlier time. The fit between model trajectory and observations is done through the minimization of a cost function:

$$J(\mathbf{x}) = \tfrac{1}{2} \left(\mathbf{x} - \mathbf{x}^{\mathbf{b}} \right) \mathbf{B}^{-1} \left(\mathbf{x} - \mathbf{x}^b \right) + \tfrac{1}{2} \sum_{i=0}^{n} \left[\mathbf{H}\mathbf{x}(t_i) - \mathbf{y}_i^o \right]^T \mathbf{R}^{-1} \left[\mathbf{H}\mathbf{x}(t_i) - \mathbf{y}_i^o \right] \tag{12.28}$$

where $\mathbf{x}^b$ is the firsts guess or background state and $\mathbf{B}$ is the background error covariance. In this approach the observations are incorporated into the assimilation at the measurement time (t_i). The cost function $J(\mathbf{x})$ in (12.28) is minimized relative to the estimated errors **B** and **R** so as to give the best fit to the data and model. Further details on both sequential and variational methods can be found in Kalnay (2003).

12.4.2 Application of Core Flow Models in SV Forecast

Aimed at providing a more accurate SV forecast for the international geomagnetic reference field (IGRF), Maus et al. (2008) attempted to use core flow inferred from satellite data to hindcast SV and examined the hindcast and those from direct temporal extrapolation of the data. Their motivation is very clear: with high resolution, accurate satellite geomagnetic measurements, one could better determine the core surface flow $\mathbf{v}_H$ and its acceleration $\partial \mathbf{v}_H/\partial t$ by (12.14) and (12.17) for given SV and SA. The inferred core flow could then be used to determine the SV either before or after the observation epoch.

In their studies, they considered the simplified version of (12.22)

$$\mathbf{v}_H = \mathbf{v}_0 + \mathbf{a}_0\, t, \tag{12.29}$$

i.e., only a steady flow $\mathbf{v}_0$ and a steady acceleration $\mathbf{a}_0$. The acceleration $\mathbf{a}_0$ is either zonal, purely toroidal, or zero (i.e., a steady core flow). They are inferred with the SV and SA Gauss coefficients from the field model Pomme-3.0 at the epoch 2003.0. The hindcast procedure is straight forward: given the initial state $B_r(t_0)$ and $\mathbf{v}_H(t_0)$ ($t_0 = 2003.0$), the SV $\dot{B}_r(t_0)$ is then determined via (12.14). The field at the earlier time $t_0 - \delta t$ is then updated via Taylor expansion

$$B_r(t_0 - \delta t) = B_r(t_0) - \delta t \dot{B}_r(t_0);$$

and the flow $\mathbf{v}_H(t_0 - \delta t)$ is evaluated via (12.29). Repeat this procedure and one can then hindcast SV in the past (or forecast in future with a positive δt).

Their results are very interesting. First, they showed that the time varying component in (12.29) provides the best fit to the observed SV and SA. However, it is the steady core flow model that provides the most accurate hindcast in their studies. In the first look, these seem contradictory, since a better core flow model is expected to provide more accurate predictions. But the reality is opposite. Maus et al. (2008) credited such poor performance of the time varying core flow models to the 2003 geomagnetic jerk that reversed the trend of SV.

But the deeper reason may be the omission of model response to observations. As described in the Section 12.4.1, theoretical (numerical) models will respond to assimilation of data. A forecast will be more accurate if such response is taken into consideration in obtaining the gain matrix **K** or the cost function $J(\mathbf{x})$ in data assimilation. Since the acceleration $\mathbf{a}_0$ in (12.29) models the observed SV and SA much better, it is more sensitive to the observations, and thus needs to be updated more frequently (to be closer to the true state). Such updating is in particular necessary when SV changes rapidly, e.g., the geomagnetic jerk in 2003. Without such updates, $\mathbf{a}_0$ moves further away from truth with t, and the prediction accuracies would worsen.

Lesur and Wardinski (2009) noticed the importance of an accurate time-dependent core flow model for the forecast approach of Maus et al. (2008). One improvement can be made via co-estimation of core flow models and geomagnetic field models (e.g., Gillet et al. 2010). Beggan and Whaler (2009) went a step further from Maus et al. (2008) by introducing an ensemble approach. In their study, they used only the steady core flow model, i.e., $\mathbf{a}_0 = 0$ in (12.29). The advantage of using the Ensemble approach is to take account of model errors and observational errors in their forecasting system, thus improving forecast accuracies, as shown in their work. However, the model response to observations is still not included in their analysis.

Though limited in scope, these two studies do provide a new method for understanding the core flow inferred from surface magnetic data. It could potentially lead to improved core flow models by minimizing forecast errors.

In addition to model responses, interactions between the magnetic field and the core flow are also not considered. The latter would eventually limit our understanding of the core dynamics, and the mechanisms responsible for the geomagnetic secular variation.

12.4.3 Assimilation with Simple Dynamical Models

A different approach from the application of the core flow models to forecast future SV is to establish a data assimilation system with a core dynamics model (e.g., a geodynamo model). Response of this model to surface observations could then provide important insight on how close the model output is to the true core states, and what improvement can be made to the model to reduce the differences between model solutions and the observed field.

But the geodynamo in the Earth's outer core is a very complex, strongly nonlinear process, and is described by a set of nonlinear partial differential equations (12.1)–(12.3). Numerical dynamo models are computationally expensive (and the expense can increase easily by an order of magnitude in data assimilation runs). Instead of rushing into the full dynamo models, it is therefore very instructive and pragmatic to consider first simplified dynamical models to obtain some basic knowledge on model responses to observational constraints, and their implications for geomagnetc forecasting.

Sun et al. (2007) took such approach by developing a simplified one-dimensional MHD system that includes some important features of the full dynamo system,

$$\frac{\partial b}{\partial t} = -v\frac{\partial b}{\partial x} - v\left(\frac{\partial B_0}{\partial x}\right) + q\frac{\partial^2 b}{\partial x^2}, \tag{12.30}$$

$$\frac{\partial v}{\partial t} = -v\frac{\partial v}{\partial x} + Rb + E\frac{\partial^2 v}{\partial x^2}, \tag{12.31}$$

in a finite spatial domain $[0, 2\pi]$. In this system b is the magnetic field, v is the velocity field, B_0 is a prescribed background field, E is the fluid viscosity and q is the magnetic diffusivity. They used this system to understand the response of the dynamical system to sparse observations, and changes of the model solutions due to assimilation. Both are of the fundamental importance to geomagnetic data assimilation. For example, in a numerical dynamo model with a modest spatial resolution $50 \times 50 \times 50$, the core state vector $\mathbf{x}$ includes 6.25×10^5 variables (components). But geomagnetic observations could only provide the Gauss coefficients of the poloidal field up to degree $L = 13$, or 104 variables.

Sun et al. (2007) employed a sequential assimilation algorithm (as described in Section 12.4.1) with this system. The gain matrix $\mathbf{K}$ in their analysis is calculated with an ensemble of 100 model runs. They carried out several experiments, called the observing system simulation experiments (OSSEs) in data assimilation, that can be basically grouped into three scenarios: (1) observation of (b, v) on sparsely spatial grid points; (2) observations of v only; and (3) observations of b only.

Their results can be summarized as follows. Any observation of b (Scenarios 1 and 3) can help improve the model: the difference between the forecast and the truth is smaller than that between the free model runs (without assimilation) and the truth. This difference decreases with the assimilation time, i.e., when more data are assimilated. The improvement is significant even in the extremely sparse observation case: b is only observed at the boundary points. They also found that the cross-correlation between the magnetic field b and v helps improving the model. However, observation of v alone (Scenario 2) does not help improve the model (no reduction in the solution differences), even if v is fully observed at all grid points. The model improvement becomes significant after a short assimilation period. But this period depends on observation quality, for fewer observations, the system responds more slowly.

Following the same approach, Fournier et al. (2007) later studied the same simplified system (12.30, 12.31), but with the variational assimilation algorithm (as described in Section 12.4.1). They carried out additional OSSEs with uneven temporal distributions of the observations to better emulate geomagnetic observations. Similar to Sun et al. (2007), they found that observations of the magnetic field always improve the model estimate of the velocity field (assumed unobservable). In addition, they also found that more accurate observations at a later time improve model forecast before that. The latter could not be demonstrated with the sequential data assimilation algorithms. But it is very interesting to geomagnetic community since, if proved to be valid with more realistic core dynamics models, satellite magnetic measurements can be used to improve the past geomagnetic field models.

More recently, Canet et al. (2009) made a substantial advance on the work of Fournier et al. (2007) by developing a variational assimilation system with a quasi-geostrophic core dynamics model. In this model, the "frozen flux" approximation is still applied to the magnetic induction (12.2),

$$\frac{\partial \mathbf{B}}{\partial t} = \nabla \times (\mathbf{v} \times \mathbf{B}). \tag{12.32}$$

But the core flow is assumed quasi-geostrophic, and is solved iteratively with the following simplified momentum Eq. (12.1)

$$\mathbf{v} = \mathbf{u}_0 + \mathbf{u}_1, \tag{12.33}$$

$$2\Omega \times \mathbf{u}_0 = -\nabla p_0, \tag{12.34}$$

$$\left(\frac{\partial}{\partial t} + \mathbf{u}_0 \cdot \nabla\right)\mathbf{u}_0 + 2\Omega \times \mathbf{u}_1 = -\nabla p_1 + \frac{1}{\mu\rho_0}(\mathbf{B}\cdot\nabla)\mathbf{B}. \tag{12.35}$$

The system is tested with several OSSEs: one with the synthetic data (i.e., data from their quasi-geostrophic flow forward modeling solutions), one with a simple nonzonal and equatorially symmetric steady core flow inferred from geomagnetic data, and one with torsional oscillations. In all cases, they demonstrated that the observations have improved model forecast accuracies. In particular, the accuracy increases with the assimilation time period T (i.e., with more observations).

12.4.4 Data Assimilation with Full Geodynamo Models

The results of Sun et al. (2007) and Fournier et al. (2007) are very encouraging. However, they did not demonstrate if similar results could be obtained with a full dynamo model. For this purpose, Liu et al. (2007) extended the work of Sun et al. (2007) by considering the MoSST core dynamics model (Kuang and Bloxham, 1999; Kuang and Chao, 2003; Jiang and Kuang, 2008). In this model, the Eqs. (12.1)–(12.3) are all nondimensionalized. For example, the momentum equation is of the form

$$R_o\left(\frac{\partial}{\partial t} + \mathbf{v}\cdot\nabla\right)\mathbf{v} + \mathbf{1}_z \times \mathbf{v} = -\nabla p + (\mathbf{B}\cdot\nabla)\mathbf{B} + R_{th}\Theta\mathbf{r} + E\nabla^2\mathbf{v}, \tag{12.36}$$

where Θ is the nondimensional temperature perturbation, R_o is the magnetic Rossby number that describes the fluid inertia, E is the Ekman number for the viscous effect and R_{th} is the Rayleigh number measuring the buoyancy effect. The spheric harmonic coefficients in (12.6) are all defined at the discrete radial grid points $\{r_i\}$.

Liu et al. (2007) performed a series of OSSEs in which synthetic data are made from the dynamo model runs with a larger Rayleigh number $R_{\text{th}} = 15,000$. Simulation without assimilation (free running model, or nature run) is carried out at a smaller Rayleigh number $R_{\text{th}} = 1450$ for comparisons. To mimic geomagnetic data assimilation, the synthetic data at the top of the electrically conducting D"-layer are assimilated into the dynamo solutions via a sequential assimilation algorithm. In particular, the maximum spherical harmonic degree of the data is less than that of the truncation order of the solutions. In their experiments, the assimilation periods are also different so that they can examine the convergence rate of the assimilated solutions to the true states.

They found that the assimilated model output is very different from those of the nature run. The *rms* errors between the assimilated solutions and the true states are much smaller than those between the free model run solutions and the true states. In particular, the errors decrease as the assimilation time increases, i.e., the model solutions are gradually pulled towards the true states by assimilation. The assimilation solutions also changed deep in the outer core (and far away from the boundary where the synthetic data are provided).

The studies of Sun et al. (2007) and Liu et al. (2007) are part of the collaborative effort to develop the first geomagnetic data assimilation system, the MoSST_DAS (Kuang et al. 2008). This system includes three components: the MoSST core dynamics model (dynamo model); an assimilation component utilizing an optimal interpolation algorithm; and a geomagnetic observation component based on three field models, CALS7K (Korte and Constable, 2005), gufm1 (Jackson et al. 2000) and CM4 (Sabaka et al. 2004), which combined provide over 7000 years of the Gauss coefficients.

Utilization of the field model product, not the original data, implies that MoSST_DAS will incorporate all features of the field models. In addition, since most field models do not provide error bounds on the coefficients, this implies that in MoSST_DAS these coefficients are assumed perfect (i.e., perfect observation). This, as argued by Kuang et al. (2008), should be a good first order approximation for geomagnetic data assimilation because the the dynamo model errors are very likely much larger than those of the field models.

Kuang et al. (2008) also assimilated only the geomagnetic field directions in MoSST_DAS, based on the consideration that the non-dimensional parameters in (12.36) appropriate for the Earth's core are either too small compared to those in simulation (e.g., E and R_o), or unknown (e.g., R_{th}) , rendering any simple and accurate relationship between the field magnitudes of the numerical solutions and of the observations. They also argue that utilization of the scaling rules from dynamo simulation (Olson and Christensen, 2006) may introduce additional errors in identifying model responses to observations (Kuang et al. 2009).

They first tested MoSST_DAS with the geomagnetic field coefficients from 1900 to 2000 (Kuang et al. 2009), and found that the poloidal field from the assimilation is very different from that from the unconstrained model (without assimilation). In addition, the difference between the model forecast and the observations decreases rapidly over the first 40 years as more data are assimilated. Further examination shows that the toroidal magnetic field and the velocity field (both non-observable) are also changed inside the core. Though the true core state is mostly unknown, the improved forecast accuracy suggests that the model solutions are drawn closer to the truth by assimilation.

Most recently, Kuang et al. (2010) used MoST_DAS to forecast 5-year SV from 2010 to 2005, as part of the community effort for IGRF-11. In this effort, they extended the observation component of MoSST_DAS with the addition of CHAOS_2s field model (Olsen et al. 2009). Their assimilation results are benchmarked with other extrapolation based forecasts in earlier periods. In Fig. 12.3 are their forecast of 5-year mean SV from 2010 to 2015. Their results will be cross-examined in several years.

12.5 Discussion

In this chapter we have reviewed several recent research results on understanding the dynamical processes in the Earth's outer core with the core flow models inferred from observed SV and SA, and on geomagnetic data assimilation and its application to future SV forecast.

In addition to the comprehensive field modeling approach in which magnetic signals from different sources are co-estimated (Sabaka et al. 2004), Mandea and Olsen (2006) proposed a different approach, the "virtual observatories" to optimize spatial/temporal resolutions of satellite measurements. From (12.15) to (12.16), it seems like this approach is a sequential iterative approach in which signals are separated first based on prior knowledge, and thus very simple and effective to obtain high resolution SV from satellite data (Olsen and Mandea, 2007, 2008). But data sampling and

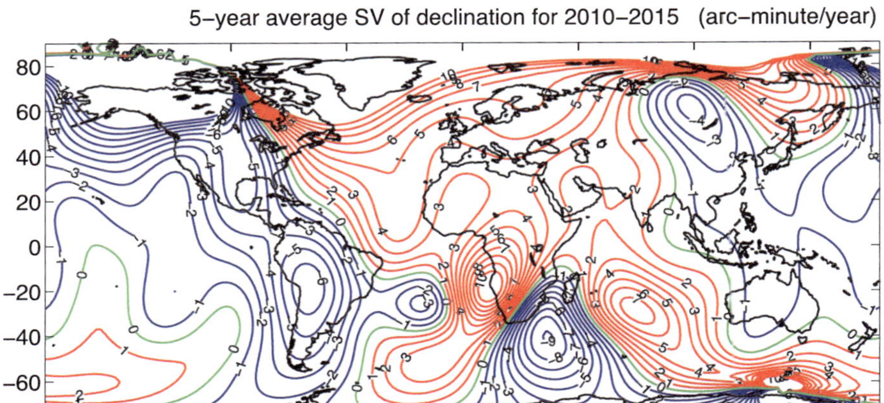

Fig. 12.3 Predicted 5-year average SV of the declination from 2010 to 2015 by MoSST_DAS (provided by Kuang and Wei)

separation of external and internal SV signals are still to be improved for this approach (Beggan et al. 2009).

Among various core flow models, torsional oscillations have been extensively studied for their unique pro-perties, as shown in (12.20), and their significances in understanding the core-mantle angular momentum exchanges (Jault et al. 1988), the properties inside the core (Zatman and Bloxham, 1997; Buffett et al. 2009; Gillet et al. 2010), and more recently, the geomagnetic jerks (Bloxham et al. 2002). Torsional oscillations have also been considered in geomagnetic data assimilation studies (Canet et al. 2009).

Geomagnetic data assimilation has been growing very fast in the past three years: the research has advanced from very simple systems for concept developments (Sun et al. 2007; Liu et al. 2007; Fournier et al. 2007; Maus et al. 2008)), to more comprehensive studies with complex models (Kuang et al. 2008, 2009; Beggan and Whaler, 2009; Canet et al. 2009), and to forecast of future SV for geomagnetic communities (Kuang et al. 2010). But, it is also clear that geomagnetic data assimilation is still in its early stage, and more effort is needed in this area. Regardless, geomagnetic data assimilation is an important approach to introduce dynamics into geomagnetic field modeling, and to include geomagnetic observations in numerical geodynamo modeling for better understanding of the dynamical states in the core.

Though not described in this chapter, there are also many other important results on core flow modeling and on their geodynamic applications to other surface geodynamic observables. The pioneering results of Jault et al. (1988) lead this effort in explaining the decadal LOD variation. Fang et al. (1996) expanded this effort by considering deformation and gravity changes due to the core pressure loading on the CMB. Dumberry and Bloxham (2004), and Dumberry (2010) attempted to estimate time-variable gravity from various core flow models. These efforts have enriched investigations of core dynamics from multi-disciplinary surface observations. Geomagnetic data assimilation could potentially help these efforts, e.g., providing assimilated dynamo solutions to estimate core contributions to time variable gravity (e.g., Jiang et al. 2007).

Acknowledgements We thank Terence Sabaka, Nils Olsen and Mioara Mandea for their help on this manuscript. This work is supported by the NSF Collaborative Mathematical Geophysics (CMG) program under the grant EAR-0327875, by the NSF Collaborative Research CSEDI under the grant EAR-0757880, and by the NASA Earths Surface and Interior Program.

References

Amit H, Olson P (2004) Helical core flow from geomagnetic secular variation. Phys Earth Planet Inter 147:1–25

Amit H, Aubert J, Hulot H, Olson P (2008) A simple model for mantle-driven flow at the top of the Earth's core. Earth Planet Space 60:845–854

Aubert J, Amit H, Hulot H, Olson P (2009) Thermochemical flows couple the Earth's inner core growth to mantle heterogeneity. Nature 454:758–761

Backus GE (1968) Kinematics of secular variation in a perfectly conducting core. Phil Trans R Soc Lond A263:239–266

Ballani L, Wardinski I, Stromeyer D, Greiner-Mai H (2005) Time structure of the 1991 magnetic jerk in the core-mantle boundary zone by inverting global magnetic data supported by satellite measurements. In: Earth Observation with CHAMP Results from three years in orbit. Springer, New York, NY

Beggan CD, Whaler KA (2009) Forecasting change of the magnetic field using core surface flows and ensemble Kalman Filtering. Geophys Res Lett 36. doi:10.1029/2009GL0399

Beggan CD, Whaler KA, Macmillan S (2009) Biased residuals of core flow models from satellite-derived `virtual observatories'. Geophys J Int 177:463–475

Bloxham J, Zatman S, Dumberry M (2002) The origin of geomagnetic jerks. Nature 420:65–68

Braginsky SI (1967) Magnetic waves in the Earth's core. Geomag Aeron 7:851–859

Braginsky SI (1976) Torsional magnetohydrodynamic vibrations in the Earth's core and variations in day length. Geomag Aeron 10:1–8

Braginsky SI (1989) The Z model of the geodynamo with an inner core and the oscillations of the geomagnetic dipole. Geomag Aeron 29:98–103

Braginsky SI and Roberts PH (1987) A model-Z geodynamo. Geophys Astrophys Fluid Dyn 38:327–349

Buffett BA and Seagle CT (2010) Stratification of the top of the core due to chemical interactions with the mantle. J Geophys Res 115. doi:10.1029/2009JB006751

Buffett BA, Jackson A (2009) Inversion of torsional oscillations for the structure and dynamics of Earths core. Geophys J Int doi:10.11111/j.1365-246X.2009.04129.x

Canet E, Fournier A, Jault D (2009) Forward amd adjoint quasi-geostrophic models of the geomagnetic secular variation. J Geophy Res 114. doi:10.1029/2008JB006189

Courtillot V, Le Mouël JL (1984) Geomagnetic secular variation impulses. Nature 311:709–716

Davis RG and Whaler KA (1997) The 1969 geomagnetic impulse and spin-up of the Earth's liquid core. Phys Earth Planet Inter 103:181–194

Dumberry M (2010) Gravity variations induced by core flows. Geophy J Int 180:635–650

Dumberry M and Bloxham J (2003) Torque balance, Taylors constraint and torsional oscillations in a numerical model of the geodynamo. Phys Earth Planet Inter 140:29–51

Dumberry M and Bloxham J (2004) Variations in the Earth's gravity field caused by torsional oscillations in the core. Geophy J Int 159:417–434

Eymin C, Hulot G (2005) On core surface flows inferred from satellite magnetic data. Phys Earth Planet Inter 152:200–220

Fournier A, Eymin C, Alboussiere T (2007) A case for variational geomagnetic data assimilation: insights from a one-dimensional, nonlinear, and sparsely observed MHD system. Nonlinear Proces Geophys 14:163–180

Gillet N, Jault D, Canet E, Fournier A (2010) Fast torsional waves and strong magnetic field within the Earth's core. Nature doi:10.1038/nature09010

Gillet N, Lesur V, Olsen N (2010) Geomagnetic core field secular variation models. Space Sci Rev doi:10.1007/s11214-009-9586-6

Glatzmaier GA, Roberts PH (1995) A three-dimensional self-consistent computer simulation of a geomagnetic field reversal. Nature 377:203–209

Gubbins D (1982) Finding core motions from magnetic observations. Phil Trans R Soc Lond A306:247–254

Holme R (2007) Large-scale flow in the core. In: Olson P (ed) Core dynamics, treatise on geophysics, vol 8. Elsevier, Amsterdam: 107–130

Holme R, Whaler KA (2001) Steady core flow in an azimuthally drifting frame. Geophys J Int 145:560–569

Holme R, Olsen N (2006) Core surface flow modeling from high-resolution secular variation. Geophys J Int 166: 518–528

Hulot G, Eymin C, Langlais B, Mandea M, Olsen N (2002) Small-scale structure of the geodynamo inferred from Oersted and Magsat satellite data. Nature 416:620–623

Jackson A (1997) Time-dependency of tangentially geostrophic core surface motions. Phys Earth Planet Int 103: 293–311

Jackson A, Jonkers ART, Walker MR (2000) Four centuries of geomagnetic secular variation from historical records. Phil Trans R Soc Lond A358:957–990

Jault D (2003) Electromagnetic and topographic coupling, and LOD variations. In: Jones CA, Soward AM, Zhang K (eds) Earth's core and lower mantle. Taylor and Francis, London, pp 56–76

Jault D (2008) Axial invariance of rapidly varying diffusionless motions in the Earths core interior. Phys Earth Planet Int 166:67–76

Jault D, Gire C and LeMouël JL (1988) Westward drift, core motions and exchanges of angular momentum between core and mantle. Nature 333:353–356

Jiang W, Kuang W (2008) An MPI-based MoSST core dynamics model. Phy Earth Planet Inter 170:46–51

Jiang W, Kuang W, Chao BF, Cox C (2007) Understanding time-variable gravity due to core dynamical processes with numerical geodynamo model. In: Dynamic planet 2005. IAG Proc 130:473–479

Kageyama A, Sato T (1997) Generation mechanism of a dipole field by a magnetohydrodynamical dynamo. Phys Rev E 55:4617–4626

Kalnay E (2003) Atmospheric modelingm, data assimilation and predictability. Cambridge University Press, Cambridge, UK

Korte M, Constable CG (2005) The geomagnetic dipole moment over the last 7000 years—new results from a global model. Earth Planet Sci Lett 236:348–358

Kuang W (1999) Force balances and convective state in the Earth's core. Phys Earth Planet Inter 116:65–79

Kuang W, Bloxham J (1997) An Earth like numerical dynamo model. Nature 389:371–374

Kuang W, Bloxham J (1999) Numerical modeling of magnetohydrodyanmic convection in a rapidly rotating spherical shell: weak and strong field dynamo actions. J Comp Phys 153:51–81

Kuang W, Chao BF (2003) Geodynamo modeling and core-mantle interaction. In: Dehandt V et al. (eds) The core-mantle boundary region. Geodyn Series 9 31:193–212

Kuang W, Tangborn A, Jiang W, Liu D, Sun Z, Bloxham J, Wei Z (2008) MoSST_DAS: the first generation geomagnetic data assimilation framework. Commun Comput Phys 3: 85–108

Kuang W, Tangborn A, Wei Z, Sabaka T (2009) Constraining a numerical geodynamo model with 100 years of surface observations. Geophys J Int 179:1458–1468 doi:10.1111/j.1365-246X.2009.04376.x

Kuang W, Wei Z, Holme R, Tangborn A (2010) Prediction of geomagnetic field with data assimilation: a candidate secular variation model for IGRF-11, Earth Planets Space (2010)

Langel RA, Estes RH (1982) A geomagnetic field spectrum. Geophys Res Lett 9:250–253

Langel RA, Kerridge DJ, Barraclough DR, Malin SRC (1986) Geomagnetic temporal change: 1903–1982, A spline representation. J Geomag Geoelectr 38:573–579

Larmor J (1919) How could a rotating body such as the Sun become a magnet. Rep Br Assn Advan Sci 159–160

Le Mouël JL (1984) Outer core geostrophic flow and secular variation of Earths magnetic field. Nature 311:734–735

Lesur V, Wardinski I, Rother M, Mandea M (2008) GRIMM: the GFZ reference internalmagnetic model based on vector satellite and observatory data. Geophys J Int 173:382–394

Lesur V, Wardinski I (2009) Comment on "Can core-surface flow models be used to improve the forecast of the Earth's main magnetic field?" In: Stefan Maus, Luis Silva, Gauthier Hulot (eds). J Geophys Res 114 doi:10.1029/2008JB006188

Liu D, Tangborn A, Kuang W (2007) Observing system simulation experiments in geomagnetic data assimilation. J Geophys Res 112. doi:10.1029/2006JB004691

Mandea M, Olsen N (2006) A new approach to directly determine the secular variation from magnetic satellite observations. Geophys Res Lett doi:10.1029/2006GL026616

Maus S, Rother M, Stolle C, Mai W, Choi S, Lühr H (2006) Third generation of the Potsdam magnetic model of the Earth. Geochem Geophy Geosys doi:10.1029/2006GC001269

Maus S, Silva L, Hulot G (2008) Can core-surface flow models be used to improve the forecast of the Earth's main magnetic field? J Geophy Res 113. doi:10.1029/2007JB005199

Olsen N, Mandea M (2007) Investigation of a secular variation impulse using satellite data: the 2003 geomagnetic jerk. Earth Planet Sci Lett 255:94–105

Olsen N, Mandea M (2008) Rapidly changing flows in the Earth's core. Nature Geosci 1:390–394

Olsen N, Lür H, Sabaka TJ, Mandea M, Rother M, Tøffner-Clausen L, Choi S (2006) CHAOS–a model of the Earth's magnetic field derived from CHAMP, Ørsted and SAC-C magnetic satellite data. Geophys J Int 166: 67–75

Olsen N, Mandea M, Sabaka TJ, Tøffner-Clausen L (2009) CHAOS-2: a geomagnetic field model derived from one decade of continuous satellite data. Geophys J Int. 179:1477–1487 doi:10.1111/j.1365-246X.2009.04386.x

Olson P, Christensen UR (2006) Dipole moment scaling for convection-driven planetary dynamos. Earth Planet Sci Lett 250:561–571

Pais MA, Oliveria O, Nogueira F (2004) Nonuniqueness of inverted coremantle boundary flows and deviations from tangential geostrophy. J Geophys Res. doi:10.1029/2004JB003012

Pinheiro K, Jackson A (2008) Can a 1-D mantle electrical conductivity model generate magnetic jerk differential time delays? Geophys J Int 173:781–792

Roberts PH and Scott S (1965) On analysis of the secular variation, 1: a hydromagnetic constraint: theory. J Geomagnetic Geoelectric 17:137–151

Sabaka TJ, Olsen N, Purucker ME (2004) Extending comprehensive models of the Earth's magnetic field with Ørsted and CHAMP data. Geophys J Int 159:521–547

Sun Z, Tangborn A, Kuang, W (2007) Data assimilation in a sparsely observed one-dimensional modeled MHD system. Nonlin Process Geophys 14:181–192

Taylor JB (1963) The magnetohydrodynamics of a rotating uid and the Earth's dynamo problem. Proc R Soc Lond A274:274–283

Wardinski I, Holme R, Asari S, Mandea M (2008) The 2003 geomagnetic jerk and its relation to the core surface flows. Earth Planet Sci Lett 267:468–481

Wicht J, Christensen UR (2010) Torsional oscillations in dynamo simulations. Geophy J Int 181:1367–1380

Zatman SA, Bloxham J (1997) Torsional oscillations and the magnetic field within the Earth's core. Nature 388: 760–763

Chapter 13

Mapping and Interpretation of the Lithospheric Magnetic Field

Michael E. Purucker and David A. Clark

Abstract We review some of the controversial and exciting interpretations of the magnetic field of the earth's lithosphere occurring in the four year period ending with the IAGA meeting in Sopron in 2009. This period corresponds to the end of the Decade of Geopotential Research, an international effort to promote and coordinate a continuous monitoring of geopotential field variability in the near-Earth environment. One of the products of this effort has been the World Digital Magnetic Anomaly Map, the first edition of which was released in 2007. A second, improved, edition is planned for 2011. Interpretations of the lithospheric magnetic field that bear on impacts, tectonics, resource exploration, and lower crustal processes are reviewed. Future interpretations of the lithospheric field will be enhanced through a better understanding of the processes that create, destroy, and alter magnetic minerals, and via routine measurements of the magnetic field gradient.

13.1 Introduction

The magnetic field originating in the earth's lithosphere is part of the earth's magnetic field complex, a dynamic system (*Friis-Christensen et al.* 2009) dominated by the interaction of the earth's magnetic field dynamo with that of the sun's. The lithospheric field is dominated by static (on a human time scale) contributions that typically represents less than 1% of the overall magnitude of the magnetic field complex, and originate from rocks in the crust and locally, the uppermost mantle. Interpretation of the lithospheric magnetic field is used in (1) structural geology and geologic mapping, and extrapolation of surface observations of composition and structure, (2) resource exploration and 3) plate tectonic reconstructions and geodynamics.

This article is designed as a review describing recent progress in mapping and interpreting the lithospheric magnetic field, and also includes some highlights from the 2009 IAGA meeting in Sopron, Hungary. Since IAGA meets every four years, we have designed this review to highlight progress in the four year period from 2005 through 2009, although references to earlier important works are not neglected, especially in the area of resource exploration. Several reviews bearing on the mapping and interpretation of the lithospheric magnetic field have appeared between 2005 and 2009. Review articles within books and encyclopedias have included those within the Encyclopedia of Geomagnetism and Paleomagnetism (Gubbins and Herrero-Bervera 2007) and the Treatise of Geophysics (*Schubert* 2007). The Encyclopedia included articles on the Crustal Magnetic Field (D. Ravat, pp. 140–144), Depth to Curie temperature (M. Rajaram, pp. 157–159), Magnetic anomalies for Geology and Resources (C. Reeves and J. Korhonen, pp. 477–481), Magnetic Anomalies, Long Wavelength (M. Purucker, pp. 481–483), Magnetic Anomalies, Marine (J. Heirtzler, pp. 483–485), and Magnetic Anomalies, modeling (J. Arkani-Hamed, pp. 485–490). The Treatise of Geophysics included articles on 'Crustal

M.E. Purucker (✉)
Raytheon at Planetary Geodynamics Lab, Goddard Space Flight Center, Code 698, Greenbelt, MD 20771, USA
e-mail: michael.e.purucker@nasa.gov

M. Mandea, M. Korte (eds.), *Geomagnetic Observations and Models*, IAGA Special Sopron Book Series 5,
DOI 10.1007/978-90-481-9858-0_13,

Magnetism' (Purucker and Whaler 2007), on the 'Source of Oceanic Magnetic anomalies and the geomagnetic polarity timescale' (Gee and Kent 2007) and on 'Plate Tectonics' (Wessel and Müller 2007). A series of workshops at the International Space Science Institute (Bern, Switzerland) in 2008 and 2009 on Planetary Magnetism (2008) and Terrestrial Magnetism (2009) has resulted in a review article on the earth's magnetic lithosphere (Langlais et al. 2010). Reviews in journals in this time frame include those of Nabighian et al. (2005), Mandea and Purucker (2005), and Robinson et al. (2008).

This review will highlight some of the controversial and exciting areas relating to the interpretation of the lithospheric magnetic field. We begin with the World Digital Magnetic Anomaly Mapping project (Korhonen et al. 2007), the first truly global compilation of lithospheric magnetic field observations. This sets the stage for the discussion of impact processes, and the magnetization and demagnetization processes involved. After briefly reviewing the magnetic record of terrestrial impact craters, we discuss recent quantitative and theoretical work in the area, both terrestrial and extra-terrestrial. We then go on to review some of the new interpretations at the Vredefort, Lonar, and Sudbury structures, with possible implications for the extraterrestrial record, especially at Mars. We next review some of the interpretations of magnetic data for tectonics, and structural geology and geologic mapping. Included within this section is recent work suggesting that parts of the uppermost mantle, especially in the vicinity of subduction zones, may be magnetic. If true, this may have important implications as a predictive tool for the spatial localization of large megathrust earthquakes and associated tsunamis. Following this is a review of interpretations for resource exploration, especially minerals, geothermal resources, and water. We also highlight some of the new developments in predictive mineral exploration models. This is followed by a review of the interpretation of lower crustal processes, motivated by exciting new work on ilmenite-hematite intergrowths by S. McEnroe and colleagues, and on the effects of pressure on magnetization by S. Gilder and colleagues.

13.2 World Digital Magnetic Anomaly Map

The first version of the World Digital Magnetic Anomaly Map (WDMAM), published by the Commission for the Geologic Map of the World (CGMW), summarizes our publicly available mapping knowledge of the lithospheric magnetic field of the Earth (Fig. 13.1) as of 2007. Prior to that compilation, there had been publicly available regional and continental scale digital compilations, and several global analog compilations. As befitting a subject with significant economic importance, commercial groups have also produced compilations for the exploration community. The digital data and metadata of the WDMAM are at 3 min of arc spacing, and 5 km above the WGS84 ellipsoid. They are available in grid and map form at http://ftp.gtk.fi/WDMAM2007. The map grew out of the peer review of several candidate models (Maus et al. 2007b; Hamoudi et al. 2007; Hemant et al. 2007). The NOAA model (Maus et al. 2007b) was selected as the base model, and subsequent changes were made to this base map prior to its publication as the WDMAM (Korhonen et al. 2007). Two versions of the WDMAM are available, A and B. The A version fills areas without near-surface data with a downward-continued CHAMP model (Maus et al. 2007a) whereas the B version uses model data derived from marine ages to fill in marine areas without near-surface data (Purucker et al. 2007). The B version is shown in the printed map available from the CGMW. The major data sets utilized for the WDMAM, their spatial resolution, and online links are available at http://www.agu.org/pubs/eos-news/supplements/2007/25-263.shtml and on the printed map. There are also a series of products derived from the WDMAM (equivalent source, Reduced to Pole, and analytic signal) that are available at http://dapple.geosoft.com. While the lithospheric field represented by the WDMAM may be quasi-static, the maps of that field continue to improve. Examples include a new full spectrum magnetic anomaly grid of the United States (Ravat et al. 2009), and a new global marine magnetic anomaly data set (Quesnel et al. 2009). We thus expect that there will be updates

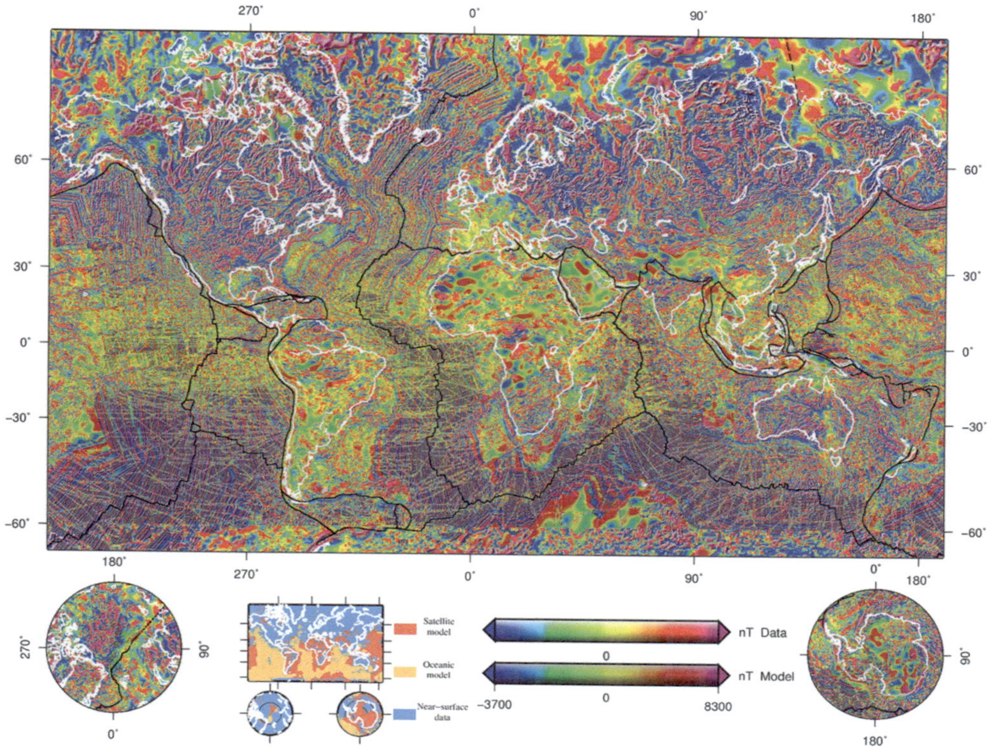

Fig. 13.1 Version B of the World Digital Magnetic Anomaly Map, with Mercator and polar stereographic projections, from Korhonen et al. (2007). The distribution of data sources is shown in the index map. Ridges, fracture zones, and trenches are shown in black

to the WDMAM, with both formally released products from the WDMAM organization, and informal releases of interim products from individual research groups, such as the recent releases from NOAA at http://geomag.org. The next update is scheduled to be released in 2011.

13.3 Impacts

Impact cratering produces two classes of craters, the smaller 'simple' and the larger 'complex' (Melosh 1989). The two types can be distinguished based on their morphologies, and the transition between the two occurs at diameters of between 2 and 4 km on the Earth. Complex craters have central peaks, wall terraces, and a much smaller depth/diameter ratio than simple craters. They have undergone more collapse than the simple craters, and the transition diameter is inversely proportional to the local (planetary) gravitational acceleration. At even larger diameters the central peak evolves into a central peak ring. Multi-ring craters are a type of complex crater characterized by multiple, large inward-facing scarps, and are most clearly developed on the Moon. Unlike the transition from simple to complex, or from complex to peak ring, multi-ring craters do not seem to scale with the local gravitational acceleration. The other crater type worthy of mention on the Earth is the 'inverted sombrero' often seen in km-size terrestrial craters and characterized by a disturbed central zone surrounded by a shallow moat. Atmospheric interactions may contribute to this distinctive shape, as discussed by *Melosh* (1989).

The magnetic signature of impact craters can be complex, but in general two types of features are often apparent (Pilkington and Hildebrand 2003).

Short-wavelength, relatively intense magnetic anomalies that occur near the center of the structure are the first of these types of features. Impact craters also disrupt the pre-existing magnetic signature, and that disruption is the second feature that can sometimes be recognized (Spray et al. 2004). The relatively intense magnetic anomalies occurring within the crater can be attributed to (1) uplifted magnetic lithologies, often basement, (2) magnetized impact melt rocks or breccia, (3) hydrothermal activity, (4) shock remanent magnetization or demagnetization, or (5) some combination of the above. Although variable, it is often the case that terrestrial impact structures are characterized by broad magnetic lows (Grieve and Pilkington 1996). Two useful guides to the variability of the magnetic signature are provided by the works of Ugalde et al. (2005), and Cowan and Cooper (2005). Numerical modeling using 2-D hydrocodes predict the distribution of pressure and temperature from which inferences can be made about the final magnetization distribution (Ugalde et al. 2005). While very useful, this model does not take into account later hydrothermal processes, which can significantly alter the magnetization distribution, and are often the source of significant ore deposits (Grant 1984; Clark 1997; Clark 1999).

Recent work on the utility of the magnetic method over terrestrial impacts includes the work of Pilkington and Hildebrand on estimating the size of the transient and disruption cavity. These sizes can be directly related to the energy release associated with impact. Weak lower and upper bounds are placed on these quantities by establishing the sizes of two parameters: (1) the size of the relatively intense features in the interior of the crater, and (2) the size of the region where magnetic features have been disrupted. The authors suggest, based on 19 complex terrestrial structures, that the collapsed disruption cavity is about half the size of the crater diameter.

Of critical importance to the interpretation of the magnetic signature is the coherence scale, or size of a region of coherent magnetization (Lillis et al. 2010; Carporzen et al. 2005). The high-frequency and relatively intense magnetic features seen in the interior of impact basins, when upward-continued, often result in broad magnetic lows because adjacent coherently magnetized regions effectively cancel out. To complicate matters further, the coherence scale is often asymmetric. A simple example comes from the terrestrial oceans, where strongly magnetized sea-floor 'stripes' are often very narrow (kms) in a direction perpendicular to the spreading axis, but very wide (thousands of kms) in the direction parallel to the spreading axis. When marine magnetic surveys of the oceans are upward-continued to satellite altitude they 'reveal' that oceanic magnetic fields are much weaker than continental magnetic fields (Hinze et al. 1991). The reality is more complex. Typical oceanic basalts are much more magnetic than typical continental granitic rocks. Another example, discussed in depth below, comes from the Vredefort impact crater (Carporzen et al. 2005) where aerial measurements of the magnetic field are lower than over surrounding regions, but surface magnetizations from within the crater are large and variable on the cm scale. Finally, it should be noted that there may not be a single coherence scale for a particular region. The coherence scale is dictated by the physical process or processes at work, and multiple processes may result in multiple coherence scales. In certain idealized cases, it is often useful to employ the concept of a matched filter (Syberg 1972; Phillips 1997) to estimate the depths of the principal magnetic sources, and to estimate crudely what a map of the magnetic fields from those sources would resemble. Certain parameters are independent of coherence scale. Ideal body theory helps to establish bounds on quantities such as the magnetization strength required to explain a magnetic field distribution (Parker 1991; 2003, Purucker et al. 2009b).

The Vredefort impact in South Africa, Earth's oldest and largest impact crater, has been the subject of several recent studies (Carporzen et al. 2005; Muundjua et al. 2007) and commentary (Dunlop 2005; Reimold et al. 2008; Muundjua et al. 2008). Carporzen et al. (2005) explain the elevated NRM intensities and Q-ratios typical of many of the exposed rocks at Vredefort as a consequence of short-lived plasmas produced during the impact. They find that paleomagnetic directions from the shocked but unmelted bedrock exposed to these hypothetical plasmas have directions which vary on scales of 10 cm or less. They explain the broad aeromagnetic low over the central portion of the impact (Fig. 13.2) as a consequence of viewing this spatially incoherent magnetic signal from an altitude of 150 m. Carporzen et al. (2005) also find magnetic evidence for lightning in the surface rocks at Vredefort, another example of a plasma phenomenon. According to the authors, lightning can reproduce many, but not all, of the magnetic features of the surface rocks. As

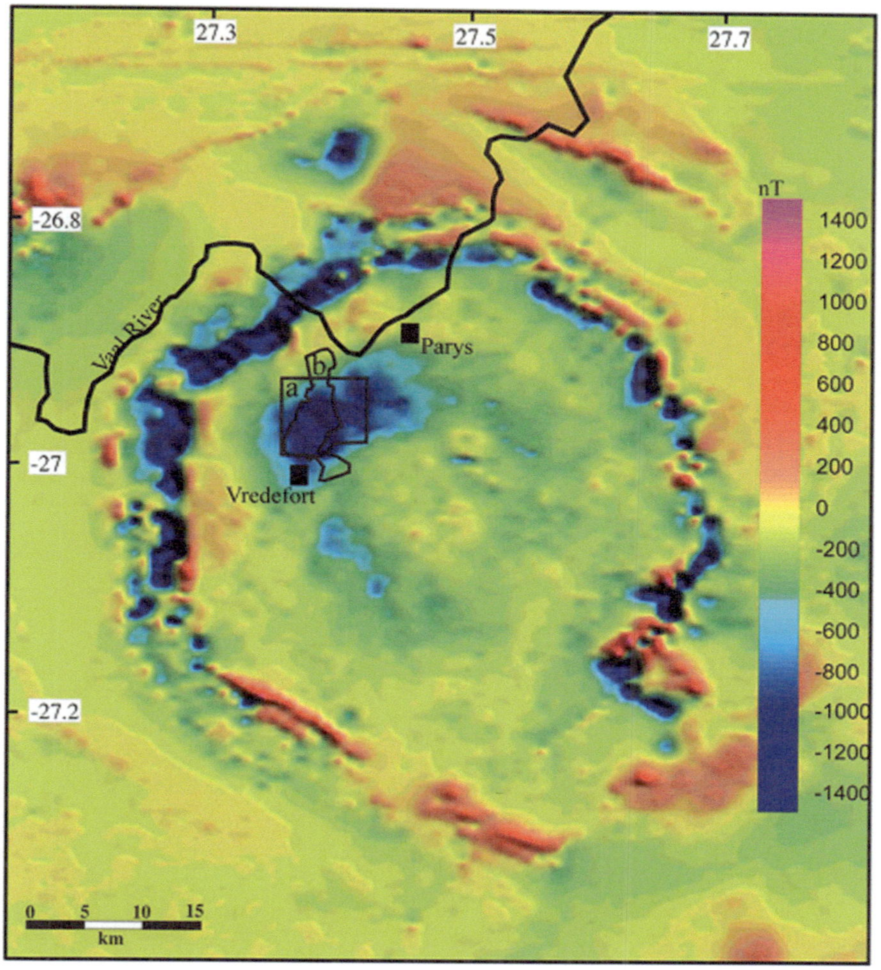

Fig. 13.2 Aeromagnetic anomaly map of the Vredefort impact structure, from Muundjua et al. (2007)

many as a quarter of their samples have been affected by lightning. Graham (1961) was the first to document the pervasive magnetic effects of lightning on surface rocks in South Africa. Carporzen et al. (2005) extrapolate their Vredefort results to the five youngest large impact basins on Mars (Lillis et al. 2010) where very weak magnetic fields have been measured. They suggest that a much smaller coherence wavelength characterized these basins, and the measured magnetic fields do not require the absence of a planetary dynamo when they were created. The Martian observations had previously been taken as evidence (Acuña et al. 1999) that these basins had been demagnetized by the impact, and that the magnetic dynamo had ceased by this time. In addition, it has been observed that the 14 oldest large impact basins on Mars have significant magnetic fields associated with them (Lillis et al. 2008), suggestive of the presence of a magnetic dynamo at this time (Fig. 13.3). To explain the difference in terms of coherence wavelength, and not in terms of the presence or absence of a magnetic dynamo, suggests that another process is at work, perhaps changes in the aqueous alteration environment (Lillis et al. 2010).

The 1.85 Ga impact that produced the Sudbury structure struck a region of the southern Canadian shield characterized by late Archean and early Proterozoic faulting, and dike emplacement. Spray et al. (2004) document the termination of the magnetic signature of the 2.47 Ga Matachewan dike swarm as it reaches Ring 2 of the impact structure, some 65 km from the center of the impact (Fig. 13.4). Post impact magnetic dikes at 1.24 Ga are not terminated.

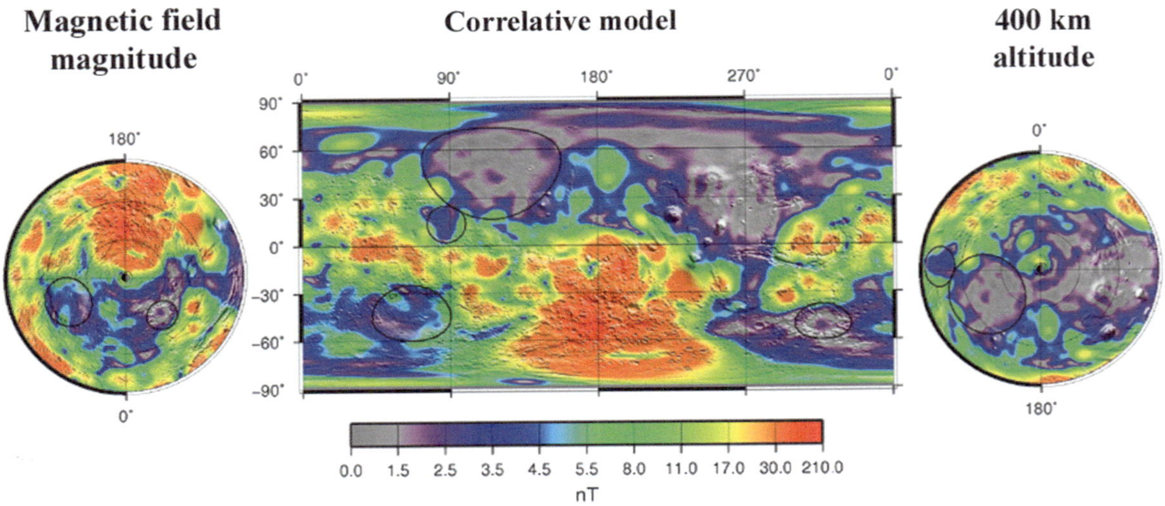

Fig. 13.3 Magnetic anomaly map of Mars, adapted from Lillis et al. (in press). The circles represent the visually determined locations of the youngest large impact basins on Mars

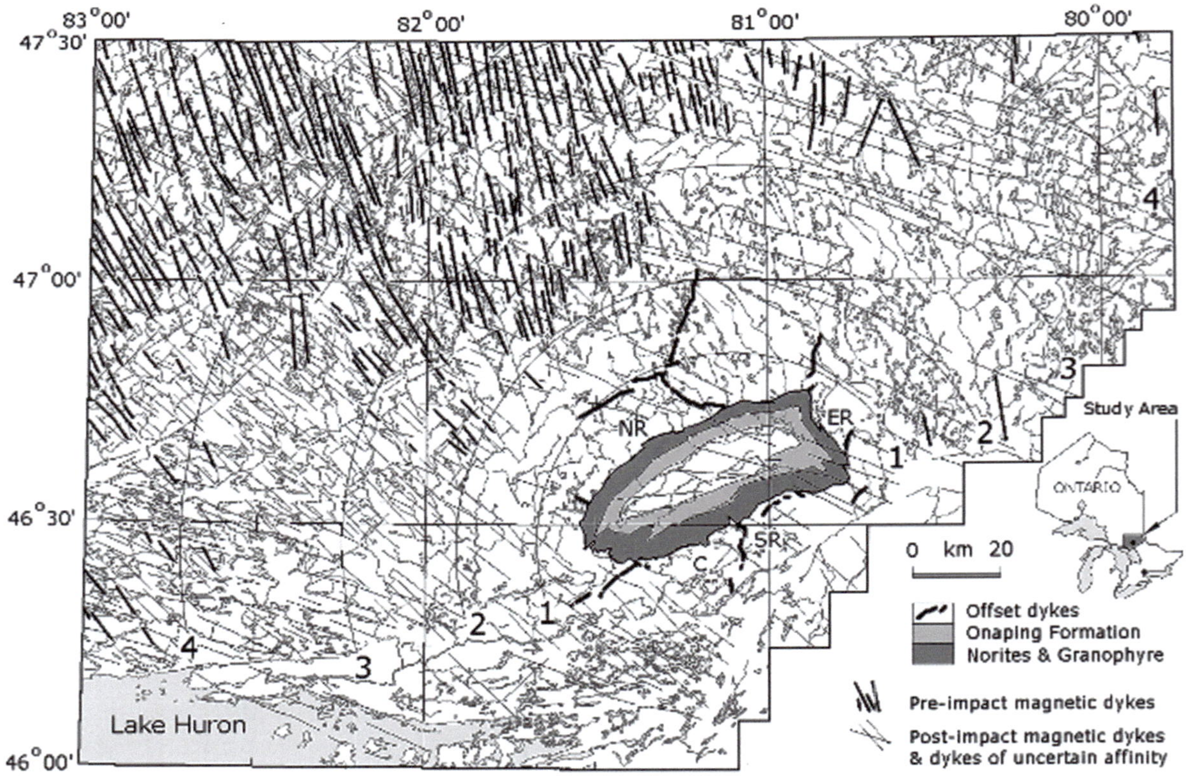

Fig. 13.4 Distribution of magnetic dikes, and ring structures, from Spray et al. (2004)

The authors interpret these observation in terms of shock demagnetization, and further interpret Ring 2 to correspond to a shock isobar pressure of between 1 and 10 GPa, depending on whether the magnetism of the dikes is dominated by induced or remanent magnetization.

A recent magnetic study of the Lonar impact structure (Louzada et al. 2008) document the magnetic processes active at this simple, young (< 50 ka) crater formed in the Deccan basalts. In this 1.88 km diameter crater, shocked ejecta blocks exhibit a slightly elevated coercivity. No evidence of shock remanent magnetization (Gattacceca et al. 2008), shock demagnetization, or transient, plasma-related processes, such as have been suggested around larger impact structure, was identified.

13.4 Tectonics

Interpretations of magnetic field observations for tectonics, structural geology, and geologic mapping have a long history (Reeves 2007), and a recent special issue of Tectonophysics (Singh and Okuma 2009) highlights the utility of magnetic field observations in the understanding of complex crustal structure. The US Geological Survey has long been active in this area, and their current efforts include an ongoing program to evaluate seismic hazards in the Seattle (USA) region. The shallow earthquakes in this active forearc basin can be devastating, and paleoseismology studies indicate the presence of a M7+ earthquake some 1100 years ago on the Seattle fault, accompanied by a tsunami. Integrated magnetic studies (Blakely et al. 2002) have focused on recognizing these shallow faults, and tracing them in areas of poor exposure. Recent work in the Puget lowland (Sherrod et al. 2008) and to the west in the Olympic peninsula (Blakely et al. 2009) continues to unravel the complexities, and highlights the advances that can be made by an integrated geological and geophysical approach, which includes LIDAR, magnetics, gravity, and paleoseismological studies. For example, the Saddle Mountain deformation zone (Blakely et al. 2009) in the Olympic peninsula has been shown to have been active at approximately the same time as the Seattle fault, some 1100 years ago, suggesting a kinematic linkage between the two fault zones. The interpretation favored by the authors suggests that the Seattle and Saddle Mountain zones form the boundaries of the northward advancing Seattle uplift.

Recent work (Blakely et al. 2005) suggests that parts of the uppermost mantle, especially in the vicinity of subduction zones, may be magnetic. At critical depths of 40 to 50 km, subducting ocean crust goes through important metamorphic changes that release large amounts of water into overriding mantle rocks. Introduction of water into the mantle produces serpentinite (Peacock et al. 2002), a highly magnetic, low-density rock (Fig. 13.5).

Thermal models (Oleskevich et al. 1999) indicate that, in many of the subduction zones of the world, this part of the mantle is cooler than the Curie temperature of magnetite, the most important magnetic mineral in serpentinite, and thus large volumes of mantle in subduction-margin settings should be magnetic. The World Digital Magnetic Anomaly Map (Fig. 13.6) does indeed show large-amplitude magnetic signatures over many of the world's subduction forearcs, including the Aleutian Islands, southern Alaska, Cascadia, Central America, and the Kurile Islands. Certainly these near-surface magnetic anomalies are caused in large part by upper crustal lithologies, and they have been recognized since the time of the U.S satellite MAGSAT (Frey 1982). However, detailed analysis of a number of these subduction zones (Cascadia, Nankai, southern Alaska, Aleutians, and Central America) indicates that the magnetic anomalies also include long-wavelength components originating from mantle depths. These mantle-depth anomalies are thought to be caused by highly magnetic serpentinite in the mantle above the subducting slab (Blakely et al. 2005; Manea and Manea 2008).

Not all subduction zones exhibit high-amplitude magnetic anomalies, reflecting geothermal and geochemical complexities. Part of this may be the result of inadequacies in the WDMAM maps (Thébault et al. in press), which form the background of several of the illustrations here. In the WDMAM, the oceanic component of the B map has been supplemented by models derived from the Digital Age map of the oceans, and the polarity reversal timescale. The details of both A and B maps, even in places where marine magnetic surveys have been conducted, are compromised by the inability to separate spatial from temporal variations, a consequence of the absence of base stations in marine magnetic surveys. Future generations of this map will result in a more objective and useful product.

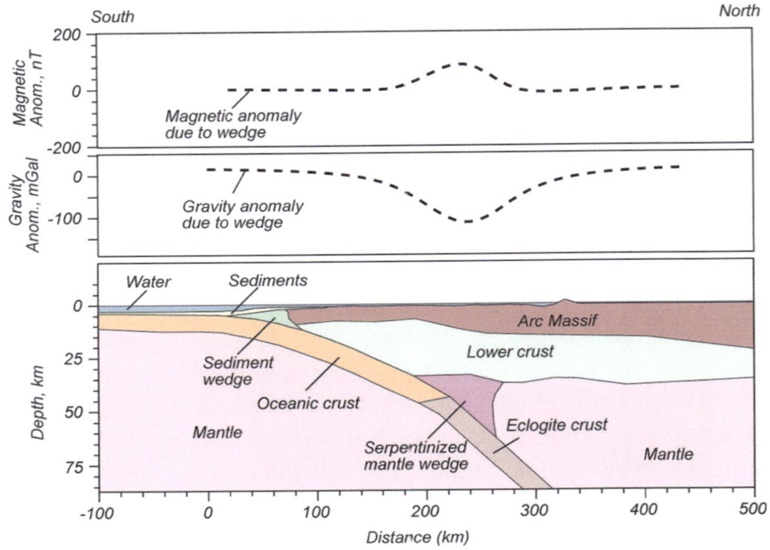

Fig. 13.5 Highly simplified crust and upper mantle model of the Aleutian subduction zone and related serpentinite mantle wedge, showing predicted near-surface contributions of the wedge to magnetic and gravity anomalies

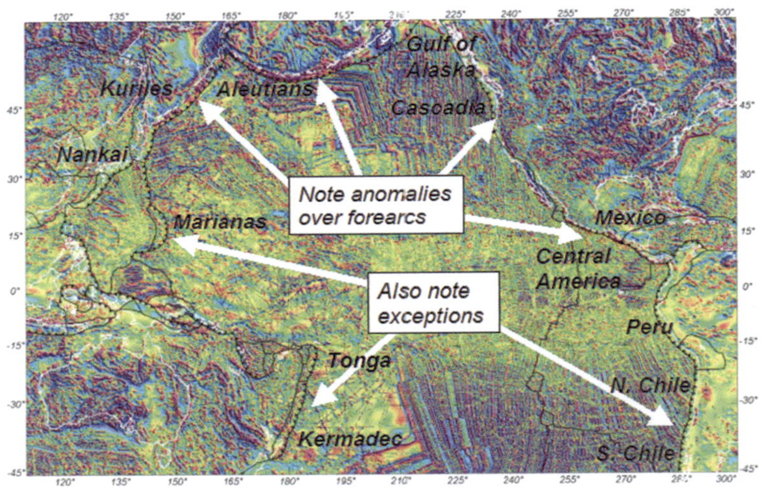

Fig. 13.6 Magnetic anomalies of the Circum Pacific, showing the location of subduction zone magnetic anomalies. Source: World Digital Magnetic Anomaly Map (Korhonen et al. 2007)

The presence of serpentinite in subduction margins has two important links to large and giant earthquakes, and associated tsunamis. First, dewatering the subducting slab is thought to embrittle the slab, reactivate pre-existing faults and other structures, and produce within-slab earthquakes (Kirby et al. 1996; Peacock et al. 2002).Thus, we expect to see a spatial association between this type of earthquake and mantle magnetic anomalies (Hyndman and Peacock 2003; Blakely et al. 2005). Second, in cool subduction margins, the down-dip limit of megathrust earthquakes (M 8.0–9.6) is controlled by the slab's first encounter with serpentinized mantle (Oleskevich et al. 1999). Again, we expect to see a spatial association between these devastating earthquakes and mantle magnetic anomalies. For example, the devastating 2004 and 2009 Sumatra-Andaman earthquakes are spatially associated with long-wavelength magnetic anomalies and thus consistent with the predicted pattern. Long recurrence intervals on megathrust earthquakes make current seismic compilations an unreliable guide to the location of past earthquakes, although non-volcanic tremors can be used, at least in part.

The existence of serpentinized mantle is well demonstrated in a few subduction margins. At Cascadia, for example, anomalously low mantle velocities have been interpreted as evidence for serpentinization of the mantle wedge (Bostock et al. 2002; Brocher et al. 2003), and these low-velocity zones are located directly beneath static long-wavelength magnetic anomalies (Blakely et al. 2005). However, in many of the subduction zones of the world, including the Aleutian Islands (Fig. 13.7), where a proposed magnetic survey (Serpent) would be conducted, seismic data appropriate for these studies are unavailable. If it can be demonstrated that long-wavelength magnetic anomalies are a reliable predictor of the presence of serpentinized mantle, then high-altitude magnetic surveys, such as the Serpent survey proposed to NASA by Purucker et al. (2009a) provide the promise of mapping hydrated mantle at subduction zones worldwide, thereby illuminating zones spatially and causally associated with both megathrust and within-slab earthquakes.

In the Antarctic, aeromagnetic surveys play a much larger role than elsewhere in deciphering tectonics because exposures of basement rocks are rare. The interpretation of a new survey over the Admiralty Block of the Transantarctic Mountains by Ferraccioli et al. (2009b) adds to our understanding of the relationships there between Cenozoic magmatism, faulting, and rifting. Fault zones here are defined by magnetic lineaments, and these help to define transtensional fault systems which may have served to localize the McMurdo volcanics. Farther inland, interpretations of high-frequency aeromagnetic anomalies within the Wilkes subglacial basin (Ferraccioli et al. 2009a) suggest the presence of large volumes of Jurassic tholeiites which may be related to rifting. By analogy with the Cordillera of North America, the authors infer that the Wilkes basin contains fold and thrust belts and a former backarc basin. These features may represent the transition between the Precambrian East Antarctic craton and the Ross orogenic belt. On the other side of the Antarctic continent, Shepherd et al. (2006) delineated subglacial geology via a combined aeromagnetic and radio echo sounding survey over three tributaries of Slessor Glacier in the East Antarctic. They tentatively identified Jurassic dikes and sills intruding the Precambrian block here, and a post-Jurassic(?) sedimentary basin with a significant accumulation of sediment. Ice motion above the inferred sedimentary basin is seen to be different in character, comprising basal sliding and/or a deforming layer of sediment, than that above the remainder of the survey area.

In the Sinai peninsula, Rabeh and Miranda (2008) interpret a new high-resolution aeromagnetic survey, in conjunction with GPS and seismic data. They find systematic trends in the depth to the magnetic basement, and in the magnetically defined structural trends.

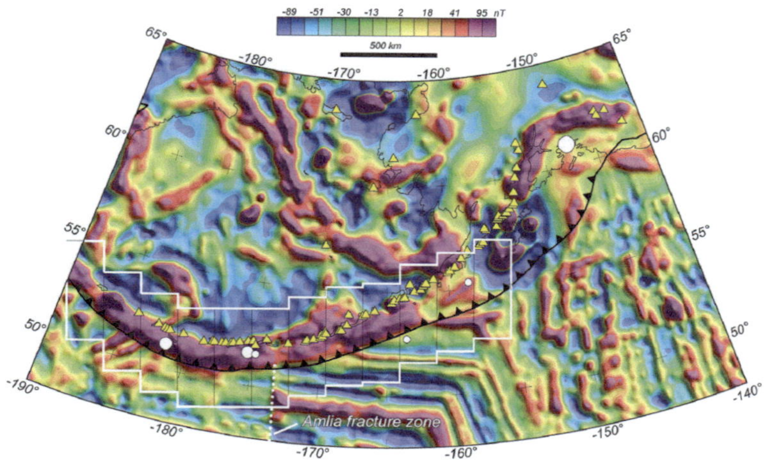

Fig. 13.7 Proposed high altitude Global Hawk magnetic surveys (Purucker and SerpentTeam 2010) over the Aleutian Islands and surrounding regions, outlined by the white polygon. Yellow triangles locate arc volcanoes, and white circles are historic megathrust earthquakes with magnitude greater than 8.0. The Amlia fracture zone is indicated by the dotted white line. The color base map shows an approximation of magnetic anomalies observed at 20 km altitude. Anomalies were calculated by analytically continuing the World Digital Magnetic Anomaly Map to 20 km altitude, from its nominal 5 km altitude

The depth to basement increases to the west and north, reaching some 4 km deep at the north end of the study area.

Aeromagnetic, gravity, geologic, and remote sensing data were combined in the Eljufra region of Libya by Saadi et al. (2008) to define geologic structures and outline hydrothermally altered basalt. Analytic signal determinations of the magnetic field were used to estimate the location and depths of magnetic contacts.

Aeromagnetic surveys often provide unparalleled views of faults in sedimentary basins. For example, Grauch and Hudson (2007) find that prominent low-amplitude (5–15 nT) linear anomalies are often associated with surficially hidden faults that offset basin-fill sediments in the central Rio Grande rift of north-central New Mexico (USA). They also find that the linear anomalies are not the consequence of chemical processes acting within the fault zone, but rather due to the tectonic juxtaposition of magnetically different strata across the fault. They develop a set of simple graphical, mathematical, and conceptual models to help them determine parameters of direct interest to structural geology.

Drenth and Finn (2007) have also recognized hidden faulting in the Pine Canyon caldera of Big Bend National Park, along the US-Mexico border. The caldera-filling Pine Canyon rhyolite can be used as a magnetic marker because it is reversely magnetized. The authors use this marker to assess the thickness of the caldera fill, and suggest that it is controlled by buried faults evident in the magnetic survey.

Magnetic surveys, interpreted in conjunction with gravity and radiometric data, can also delineate basin architecture and tectonic evolution, as illustrated by the study of the Neocomian Rio do Peixe basin of NE Brazil (de Castro et al. 2007). The Rio do Peixe is a tripartite basin developed during the opening of the South Atlantic Ocean. Many pre-existing faults within the basement complex were reactivated during basin development, and the magnetics also serves to delineate the thickness of the sedimentary packages in these asymmetrical half-graben basins.

The utility of high-resolution airborne magnetic data in the interpretation of tectonic processes is borne out by the analysis of such a survey along a 120-km-long section of the Dead Sea Fault in Jordan and Israel (ten Brink et al. 2007). This fault is poorly delineated on the basis of surface morphology, or micro-seismic activity, although damaging earthquakes have struck along this fault as recently as AD 1458. The fault is clearly seen on maps of the first vertical derivative, indicating a shallow source for the anomalies. The authors interpret these 5–20 nT anomalies as originating from the alteration of magnetic minerals due to groundwater within the fault zone. Based on modeling of the magnetic observations, the width of the shallow fault zone is several hundred meters wide. On a regional scale, the authors observe no igneous intrusions related to the fault zone, and confirm previous interpretations of 107–111 km of left-lateral offset across the fault.

Magnetic techniques continue to play a major role in delineating plate tectonic processes in the marine realm. Maia et al. (2005) document the interaction between the Foundation hotspot and the Pacific-Antarctic ridge within the South Pacific. Analysis of the magnetic anomaly data document a difference between the age of hotspot-related seamounts, and the underlying oceanic crust. This difference suggests that the ridge has approached the hotspot at a rate of 40 km Ma^{-1}. This is in good agreement with published radiometric dates.

13.5 Resource Exploration

Over the last three decades there have been dramatic improvements in the quality of magnetic surveys, accompanied by lowering of acquisition costs. Rapid developments in image processing, 3D visualization, computer-intensive enhancements of magnetic data and semi-automated interpretation methods have ensured that magnetic data is routinely acquired and used by geologists as well as geophysicists, at every stage of hard rock mineral exploration programs, from regional area selection to prospect scale exploration. Recognition of the information about the sedimentary section, as well as the crystalline basement, that is obtainable from modern high resolution magnetics has also led to greater use of magnetics in hydrocarbon exploration programs, although potential field methods will always remain subsidiary to seismic methods in that arena. Nabighian et al. (2005) have provided a comprehensive overview of the historical development of magnetics in exploration and the recent state of the art. Accordingly, we will concentrate on some new developments in magnetic exploration and some hitherto unpublished work on magnetic signatures of mineralization.

Robinson et al. (2008) have reviewed magnetic and other geophysical methods for hydrogeological research, with suggestions for future research directions. Tectonic and structural interpretations derived from aeromagnetic and Landsat thematic mapper (TM) data sets form the basis for an ambitious program of groundwater exploration (Ranganai and Ebinger 2008) in the arid southern Zimbabwe craton (Africa). The lack of primary permeability and porosity in this crystalline basement terrain results in poor overall groundwater potential. However, available groundwater is localized by the presence of faults, fractures, dikes, and deeply weathered regions. These features are often recognizable through enhanced aeromagnetic and/or thematic mapper observations, and the authors utilize these to identify lineaments, and place them in the context of the regional structural geology. They develop a model in which the aeromagnetic data is used to map faults and fractures of considerable depth extent which may be open to groundwater (under tension) while the TM lineaments are typically closed to groundwater (under compression) and define recharge areas. The authors predict that coincident magnetic and TM lineaments, and continuous structures associated with large catchment basins, will be most favorable for groundwater. The sparse record of existing borehole data, some of which is of questionable quality, suggest a relationship between productivity and spatial proximity to faults and dikes, but proximity does not guarantee productivity. The trends of the NNE and NW sinistral faults in the Chilimanzi plutons can be traced from higher elevation areas in the north that represent the watershed, to lower areas in the arid south. Since regional groundwater flows mostly follow the dominant topographic gradient, these structures were identified as the most promising in terms of sustainable ground water resources.

Aeromagnetic data can also be utilized to infer heat flow within the crust, via determination of the depth to the Curie isotherm, the depth at which rocks lose their permanent and induced magnetism. When these determinations are from active geothermal areas, they provide important constraints on the depth to the heat source, and its extent. Espinosa-Cardena and Campos-Enriquez (2008) make such a determination from the Cerro Prieto geothermal area of NW Mexico. They find that the Curie point ranges from 14 to 17 km depth, slightly deeper than previous studies, but supported by seismic, gravity, and heat flow measurements.

Province- and continental-scale compilations of magnetic data sets provide a useful framework for identifying regional crustal structures that control distribution of mineralization and favorable geological environments (e.g., Hildenbrand et al. 2000; Chernicoff and Nash 2002; Chernicoff et al. 2002; Betts et al. 2004; Sandrin and Elming 2006; Airo and Mertanen 2008; Austin and Mertanen 2008, 2009; Anand and Rajaram 2006; Anand et al. 2009), particularly if these data sets are integrated with other geophysical data.

By utilizing magnetic and gravity data in an integrated geological and geophysical study, Blakely et al. (2007) establish that the White River area of Washington exhibits many similarities to the Goldfield mining district of Nevada, home to one of the largest epithermal gold deposits in North America. To date, White River has produced only silica commercially, but deep weathering, young surficial deposits, and dense vegetation have hindered the evaluation of its economic potential for base and precious metals in the near surface. The magnetic data was invaluable in defining structural controls on hydrothermal alteration in both areas, but especially at White River because of poor exposures. The deposits are penecontemporaneous products of the Cascade Arc some 20 Ma ago. Gravity and magnetic data were instrumental in locating the intrusive body beneath both regions that presumably was the source of fluids and heat to the overlaying calc-alkaline volcanic rocks. Magnetic susceptibility measurements at White River demonstrate the destruction of magnetic minerals in the altered rocks, and provide a way of estimating the depth extent of alteration (230–390 m). The White River altered area is located between two magnetically identified faults, in a temporary extensional stress regime.

Magnetic petrological studies of magnetic stratigraphy within layered intrusions (e.g., Ferré et al. 2009; McEnroe et al. 2009a) should improve detailed mapping within such intrusions, particularly beneath cover or between widely spaced drill holes, with evident applications to exploration for such commodities as Cr, PGEs, Ni, Cu , V and Ti.

The magnetic properties of igneous rocks that are genetically related to metalliferous mineralization vary systematically with the ore metals and deposit style (Clark 1999). Table 13.1 summarizes generalized

Table 13.1 Magnetic properties of Unaltered Weakly Altered Intrusive Rocks related to Mineralization

Lithology	Strongly Oxidized (NNO-HM)			Strongly Reduced (≤QFM)		
	k (10^3 SI)	NRM	Associated mineralization	k (10^{-3} SI)	NRM	Associated mineralization
Syenogranite, Alkali granite	1–30	Weak, VRM, $Q < 1$	Mo, Mo-W, (Au)	0.1–0.3	V. weak, VRM, $Q << 1$	Sn, Sn-W
Monzogranite (Adamellite), Qtz monzonite	3–40	VRM, $Q < 1$	Cu, Cu-Mo, Au	0.1–0.4	V. weak, VRM, $Q << 1$	
Granodiorite, Monzonite, Tonalite	20–70	VRM, $Q < 1$	Cu, Cu-Mo, Cu-Mo-Au, Au	0.2–0.5	Weak; $Q << 1$	
Qtz diorite, Qtz monzodiorite	25–90	VRM + TRM; $Q < 1$	Cu-Au, Cu-Au-Mo, Au	0.4–0.6	Weak; $Q << 1$	
Monzodiorite, diorite	30–100	TRM+(VRM); $Q \sim 1$ ($Q < 1$)	Cu-Au, Au	0.5–0.8	Weak; $Q << 1$	
Gabbro, Norite, Alkali gabbro	40–160	TRM + (VRM); $0.5 < Q < 10$	Fe, Ti, V	0.6–1.3	Weak; $Q << 1$	Cr, PGEs

Rocks that have undergone deuteric alteration and/or minor rock-buffered hydrothermal alteration of normal type and intensity, as well as essentially unaltered rocks, are included here. Susceptibilities of unaltered and unmetamorphosed volcanic rocks are similar to those of their corresponding intrusive rocks, e.g., for a given igneous suite k(andesite) ≈ k(diorite). NNO—HM indicates crystallization at oxygen fugacities between the Ni—NiO and hematite-magnetite buffers; ≤QFM indicates crystallization at oxygen fugacities at or below the quartz-fayalite-magnetite buffer. NRM = natural remanent magnetization, TRM = thermoremanent magnetization, VRM = viscous remanent magnetization, Q = Koenigsberger ratio.

results from a magnetic petrophysical database (Clark et al. 2004). Although for a given rock type the total range of susceptibilities given in Table 13.1 can be quite large, the general trends are clear. Within individual provinces, and in particular within specific igneous suites, the variability is much less. Understanding the magnetic signatures of magmatic-hydrothermal systems associated with mineralization requires detailed consideration of the effects of different alteration types on a range of protoliths. Some examples are given below.

Clark et al. (2004) produced a major study of the magnetic signatures of porphyry copper deposits, volcanic-hosted epithermal gold deposits, and iron-oxide copper-gold (IOCG) deposits. Although magnetic surveys are an integral part of exploration programmes for porphyry, epithermal and IOCG deposits, the magnetic signatures of these deposits and mineralized systems are extremely variable and exploration that is based simply on searching for signatures that resemble those of known deposits is rarely successful. However, the reasons for this variability are reasonably well understood and are summarized below.

A number of well-known geological models of porphyry and epithermal deposits are routinely used in exploration, even though most deposits fail to match the idealized models closely, due to post-emplacement tectonic disruption and rotations, asymmetric alteration zoning due to emplacement along a contact between contrasting country rock types, and so on. These complications are taken into account by exploration geologists as geological information about a prospect accumulates. The variability of magnetic signatures of these deposits reflects strong dependence of magnetic signatures on local geological setting, departures of real mineralized systems from idealized geological models, the direction and intensity of the geomagnetic field, which varies over the Earth, and differing magnetic environments (host rock magnetization, regional gradients, interference from neighbouring anomalies etc.). To tackle this problem Clark et al. (2004) developed the concept of predictive magnetic exploration models that are specific to the local geological environment and history, and are based on magnetic petrological principles (Clark 1997, 1999) applied to standard geological models, and on magnetic petrophysical data and detailed modeling of selected deposits for which detailed magnetic and geological data are available.

Significant geological factors that affect magnetic signatures include tectonic setting and its influence on magma composition and mode of emplacement; influence of pre-existing structures on the geometry and depth of emplacement; and the crucial influence of host rock composition on alteration assemblages, including secondary magnetic minerals, and on the stability of primary magnetic minerals.

Magnetic signatures reflect not only the local geological setting at the time of emplacement, but also post-emplacement modification of deposits. Post-emplacement tilting of porphyry and epithermal systems and dismemberment by faulting are very common and drastically modify the geophysical signatures. Burial of a deposit by younger sedimentary or volcanic rocks also modifies the anomaly pattern. Conversely, exhumation and partial erosion of the system produces a very different magnetic signature. In older deposits, metamorphism can substantially modify the magnetic mineralogy of the deposits and host rocks, with concomitant changes in the magnetic anomaly pattern. Although the majority of porphyry and epithermal deposits are relatively young and, at most, weakly metamorphosed, some relatively ancient deposits in metamorphosed terrains are known. There is a strong possibility that some older porphyry and epithermal deposits occur that have not been recognized, because effects of metamorphism and deformation have obscured their true nature.

Given a comprehensive magnetic petrophysical database and the understanding of the geological factors that create and destroy magnetic minerals in porphyry systems, however, the magnetic effects of the above-mentioned geological complications are quite predictable. Variations in signatures due to varying geomagnetic inclination across the globe are best handled by calculating reduced-to-pole (RTP) signatures that can be compared with RTP processed survey data from high and moderate latitudes or in low latitudes, where RTP processing is unstable, by calculating reduced-to-equator (RTE) that can be compared with observed signatures, particularly if they are also reduced to the equator.

The porphyry copper model of Lowell and Guilbert (1970) has been highly influential in exploration programs and has been successfully applied in many different areas. It should be remembered, however, that the model is based on a reconstruction of the San Manuel and Kalamazoo porphyry deposits in Arizona, which originally formed a single intrusion-centred orebody with concentric zoning, before being tilted and disrupted by faulting (Lowell 1968; Force et al. 1995). The present disposition of intrusive rocks and alteration zones in and around these orebodies differs greatly from the idealized model, but when the displacement along the San Manuel fault is removed and the intact porphyry system restored to the vertical, it is apparent that the system originally conformed closely to the model. Figure 13.8(a) shows a model of a Laramide type deposit, associated with a high sulfur quartz monzonite magma that intruded weakly magnetic felsic rocks and subsequently was tilted and dismembered by faulting in a similar fashion to the San Manuel-Kalamazoo system. The predicted magnetic signature is shown in Fig. 13.8(b)–(d). This model, and those in subsequent figures, were created using the *Noddy*™ structural history modeling program (Jessel 2001).

Such post-emplacement disruption of porphyry systems is common. Wilkins and Heidrick (1995) report that approximately 45% of the deposits of the southwestern North American porphyry copper province have been significantly faulted, extended and rotated during Oligocene and Miocene time. Tilting through more than 0° is common. Geissman et al. (1982) have quantified rotations in the Yerington district using paleomagnetism. A paleomagnetic study of the Porgera Intrusive Complex by Schmidt et al. (1997) showed that the upper levels of this complex have been disrupted by thin-skinned tectonics. The exposed intrusions have undergone substantial, but varying, degrees of tilting and rotation about vertical axes. Lum et al. (1991) point out the prevalence of local block rotations that distort outcrop patterns of high level intrusions and porphyry and epithermal alteration systems in the tectonically very active SW Pacific. Rotation rates of 20°–30° in 100,000 years are unexceptional.

A more straightforward modification of a zoned alteration system is afforded by the giant Chuquicamata porphyry copper deposit in Chile (Lindsay et al. 1995), which has been bisected by a major fault, leaving a mineralized system with zoned alteration juxtaposed against unaltered intrusive rocks. Mineralized systems are susceptible to dismemberment, because major faults that controlled emplacement of intrusions and flow of hydrothermal

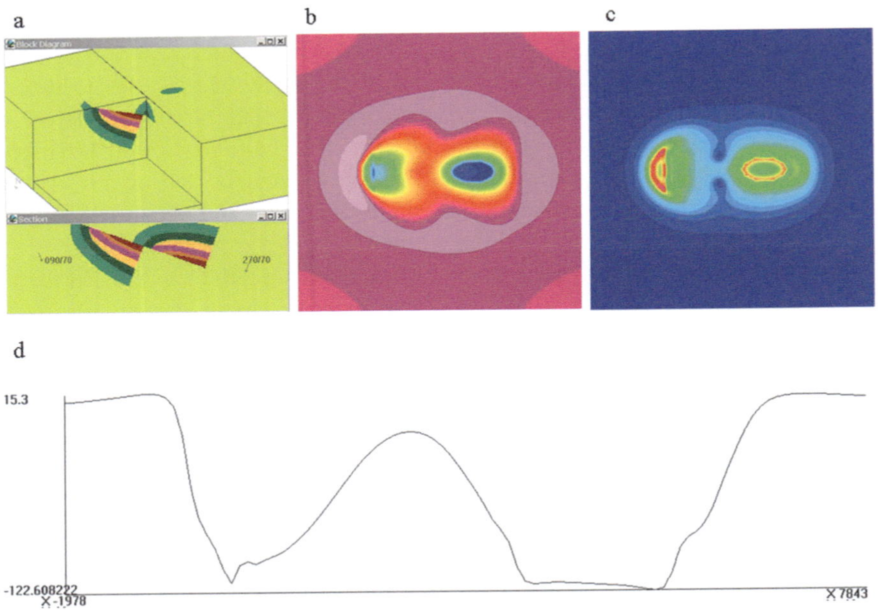

Fig. 13.8 (**a**) Laramide type quartz monzonite porphyry model with concentric zoning, emplaced vertically into felsic rocks and subsequently tilted and faulted. Alteration zones are inner potassic core (*red*), outer potassic (*orange*), ore shell (*pink*), phyllic/argillic (*yellow*), strong propylitic (*dark green*), and weak propylitic (*light green*), (**b**) calculated RTP magnetic signature of model, (**c**) calculated analytic signal amplitude of RTP magnetic signature, (**d**) profile of calculated RTP magnetic signature across model

fluids, as at Chuquicamata, are often reactivated during or after deposition of mineralization.

Clark et al. (2004) summarized data from 50 deposits for which some information on magnetic signatures was available. The quality of the data is highly variable, making definitive statistical conclusions problematic. Taking the data at face value, however, they found that approximately 50% of deposits exhibited local RTP magnetic highs associated with mineralizing intrusions, approximately 70% had highs associated with alteration (usually potassic alteration with magnetite), approximately 70% also exhibited local magnetic lows over magnetite-destructive alteration zones, approximately 20% had highs associated with skarns, and 40–60% of deposits were associated with magnetic lineaments that appear to represent structural controls on mineralization. Recognition of structural controls, in particular, is very sensitive to the quality of the data. Many deposits show more than one of these features, accounting for the fact that the proportions add up to much more than 100%.

Of the deposits for which high resolution magnetic data are available, 12 show well developed concentric zoning patterns. Ten of these, the majority from the Goonumbla cluster, New South Wales, have "doughnut" patterns, with central alteration low surrounded by an annular alteration high (Clark and Schmidt 2001). Two, the Bajo de la Alumbrera Cu-Au deposit and the Anabama Hill Cu prospect, have "archery target" patterns, with central alteration highs surrounded by annular alteration lows.

Although the number of well-characterized empirical examples of these types of zoned signature is low, the likely occurrence of similar signatures can be inferred from other information, e.g., the distribution of magnetite reported for some deposits. Predictive models are designed to bridge the gap between purely geological models (both idealized models and detailed deposit descriptions) and empirical magnetic signatures. This process has suggested the following conclusions for porphyry deposits that have not been significantly modified by post-emplacement tectonism or metamorphism:

(i) The majority of gold-rich porphyry copper deposits (classic morphology, quartz-monzonite zoning pattern) hosted by magnetic mafic-intermediate volcanics are predicted to have large (> 1000 nT) bullseye high RTP anomalies over the potassic core, with incipient to prominent development of the archery target signature, depending on the extent of the phyllic zone, *providing erosion has exposed or nearly exposed the potassic zone*. This signature should be easily detectable beneath 100 m of sedimentary cover, and even beneath a similar thickness of magnetic volcanics.

(ii) For a completely buried, uneroded or slightly eroded, gold-rich porphyry copper system the signature is basically an alteration low due to the large volume of magnetite-destructive alteration surrounding the deeply buried magnetic core. At intermediate levels of exposure a more complex pattern of a central high surrounded by an alteration low occurs, with the relative amplitude of the high and low dependent on the erosion level.

(iii) Similar deposits emplaced into weakly magnetic felsic rocks or unreactive rocks, such as quartzites or shales, are characterized by a strong bullseye high, without a surrounding low.

(iv) If emplaced into limestone the bullseye high associated with the potassically altered intrusion is likely to be supplemented by skarn anomalies (possibly remanently magnetized) associated with proximal magnetite-garnet skarn in favourable horizons, with discrete anomalies associated with distal skarn bodies, developed near the marble interface in structurally controlled zones. The skarn signature should be more strongly developed if the host rocks are dolomitic.

(v) Alkalic porphyry Cu-Au deposits typically exhibit diorite model zonation, with poorly developed phyllic zones, and produce strong bullseye highs over the potassic core.

(vi) In areas of greater crustal influence on magmas (e.g., the Laramide province), those magmas with relatively high sulphur content generate large volumes of magnetite-destructive alteration, in contrast to low sulfur magmas, for which magnetite is associated with potassic alteration. Porphyry Cu and Cu-Au deposits of the former type are associated with alteration lows, if emplaced into magnetic host rocks, or very weak signatures if emplaced into non-magnetic host rocks.

(vii) Giant porphyry copper deposits of the Atacama desert are characterized by large volumes of magnetite-destructive alteration, with locally developed magnetite-bearing potassic alteration, and thick overlying supergene blankets. The signature of such deposits, when hosted by moderately magnetic rocks, is an areally extensive alteration low, with a typical amplitude of approximately 100 nT. Such deposits will be visible to magnetics if they are covered by non-magnetic overburden, but cover by magnetic volcanics renders them difficult to see. When hosted by non-magnetic rocks the magnetic signature is inconspicuous, apart from local highs associated with remnant zones of potassic alteration within the broad zones of phyllic overprinting. These deposits are ringed by chargeable zones due to pyrite-bearing propylitic halos.

(viii) Phyllic alteration produced by magmatic, rather than meteoric, fluids (e.g., the Goonumbla, New South Wales, deposits) tends to be "inside-out" with respect to the potassic zone, producing a doughnut magnetic signature. Another source of this reverse zoning pattern may be structurally controlled access of meteoric fluids to deeper portions of a deposit. This type of signature is to be expected in two main settings: Volcanic morphological models, with small intrusive spines within comagmatic volcanics, tapped off a large mother magma chamber (e.g., Goonumbla), and plutonic/batholithic porphyry deposits.

(ix) Reduced porphyry Au-(Cu) and reduced intrusion-related gold deposits are characterized by incomplete doughnut signatures on a scale of kilometers, due to distal pyrrhotite-bearing mineralization developed in favourable sites, around a weakly magnetic intrusion.

As an illustration of a specific category of predictive magnetic exploration model we will discuss gold-rich porphyry copper deposits, which have been intensively studied and for which genesis, structural controls, overall morphology and alteration zoning patterns are quite well understood (Sillitoe 2000).

Predictive magnetic models for gold-rich porphyry copper deposits illustrated here conform to general geological models of this type of deposit and are closely based on deposits that may be regarded as archetypes for particular settings. In particular, the

Table 13.2 Canonical Magnetic Model of a Gold-Rich Porphyry Copper Deposit with a Magnetite-Rich Potassic Core, Hosted by Mafic-Intermediate Oxidized Igneous Rocks

Zone	Diameter* (m)	Width* (m)	Depth extent (m)	Susceptibility (SI)
Inner potassic	360	360	2400	0.351
Outer potassic	600	120	2500	0.173
Phyllic/argillic	1000	200	3000	0.003
Strong propylitic	1200	100	3000	0.007
Weak propylitic	1500	150	3000	0.027
Andesite/Basalt/Diorite/Gabbro	Very large	Very large	3000	0.043

*Diameters and widths of zones are maxima (at a depth 2000 m below the top of the phyllic zone for the propylitic and phyllic zones, and 1000 m below the top of the phyllic zone for the potassic zones).

model adopted for mafic-intermediate host rocks is based upon the Bajo de la Alumbrera deposit in Argentina (Guilbert 1995), and the model for carbonate host rocks is based upon Grasberg/Ertsberg (Papua New Guinea) (MacDonald and Arnold 1994; McDowell et al. 1996; Potter 1996). The assumed zoning is concentric with a magnetite-rich potassic core surrounded by a shell of phyllic alteration passing outwards into propylitic alteration (in silicate host rocks) or zoned skarn alteration (in a carbonate host).

The geometry of a gold-rich porphyry copper model, hosted by intermediate-mafic oxidized igneous rocks (nominally andesite), with a magnetite-rich potassic core is shown in Fig. 13.9(a). This type of model is mostly applicable to relatively mafic systems in island arc environments, or to those associated with alkaline (e.g., high-K calc-alkaline to shoshonitic) magmatism in continental settings. In Fig. 13.9(a) there has been insufficient erosion to expose the deposit. The top of the mineralization lies 500 m below the surface and the only sign of the mineralized system at the surface is a patch of propylitic alteration that could easily be overlooked or, if observed, assumed to be of little significance. The inner potassic zone is strongly mineralized and magnetite-rich. It is surrounded by an outer potassic zone that contains less abundant, but still significant, magnetite. The inner potassic zone represents relatively intense development of quartz-magnetite-K feldspar veins, whereas the outer potassic zone corresponds to biotite-K feldspar-quartz-magnetite alteration. A shell of magnetite-destructive phyllic alteration with very low susceptibility envelops the potassic zones. At upper levels this alteration may grade into intermediate argillic and shallow advanced argillic alteration, but the magnetic properties are equivalent for these alteration types and a single shell is sufficient to model the effects. The phyllic zone is surrounded by a zone of intense propylitic alteration, which is partially magnetite-destructive, which passes out into weak propylitic alteration and then into unaltered andesite. The dimensions and susceptibilities of the zones are given in Table 13.2. The predicted reduced-to-pole (RTP) magnetic signature of this model is shown in Fig. 13.9b–c.

After 500 m of erosion (Fig. 13.9d) a patch of phyllic, surrounded by propylitic, alteration is exposed, but the mineralization is only subcropping. Removal of a 1 km thickness of rock exposes the mineralized core of the system and its surrounding alteration zones, as at Bajo de la Alumbrera.

An alternative model with lesser secondary magnetite is shown in Fig. 13.9g. Its RTP magnetic signature is shown in Fig. 13.9h–i. This model type is generally applicable to less strongly oxidized or relatively felsic systems, or to low-medium K calc-alkaline associations, typically in areas with thick continental crust.

Other models include deposits hosted by different country rocks, including weakly magnetic felsic igneous (or metaigneous) rocks, as shown in Fig. 13.9j–l, unreactive sedimentary rocks (e.g., quartzites), and carbonates. Quartzites (unaltered and within the propylitic and phyllic zones) and unaltered carbonates have essentially zero susceptibility.

The predicted RTP magnetic signatures for many deposit types depend strongly on the level of exposure. For exposed systems within magnetic intermediate-mafic igneous host rocks, as at Bajo de la Alumbrera, a strong central high is surrounded by a relatively weak annular low over the phyllic zone, gradually returning to background levels over the propylitic zone. For a completely buried system, however, the signature is basically an alteration low due to the large volume of magnetite-destructive alteration surrounding the deeply buried magnetic core. At intermediate

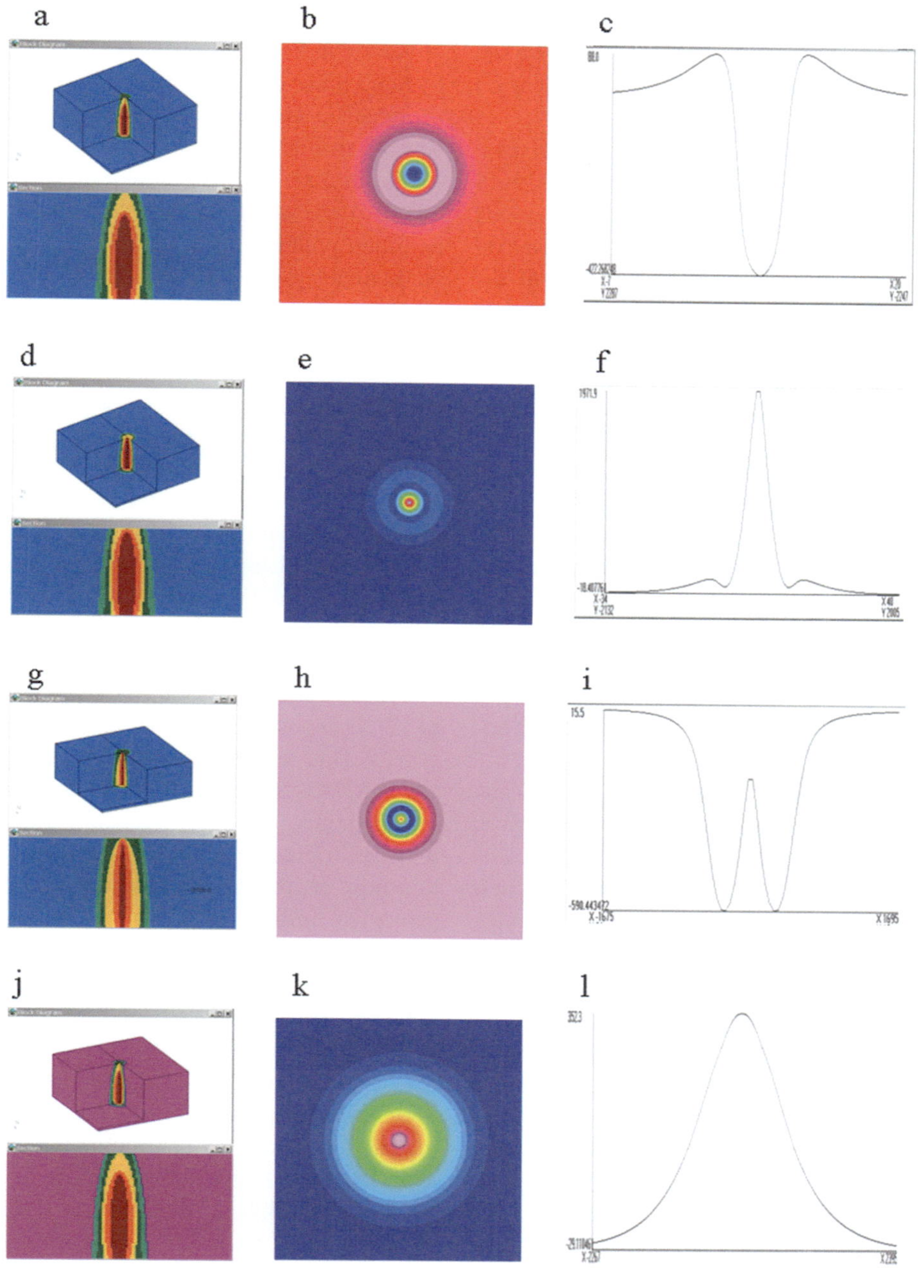

Fig. 13.9 Examples of canonical exploration models and their predicted magnetic signatures. (**a**) Gold-rich porphyry copper with magnetite-rich potassic core, emplaced into mafic-intermediate volcanics, uneroded, (**b**) RTP magnetic signature of model (**a**), (**c**) profile of RTP magnetic signature across model (**a**), (**d**) Same as model (**a**), with 500 m removed by erosion, (**e**) RTP magnetic signature of model (**d**), (**f**) profile of RTP magnetic signature across model (**d**), (**g**) As for model (**a**), with less secondary magnetite in potassic core, (**h**) RTP magnetic signature of model (**g**), (**i**) profile of RTP magnetic signature across model (**g**), (**j**) Gold-rich porphyry copper with magnetite-rich potassic core, emplaced into felsic igneous rocks, uneroded, (**k**) RTP magnetic signature of model (**j**), (**l**) profile of RTP magnetic signature across model (**j**)

levels of exposure a more complex pattern of a central high surrounded by an alteration low occurs, with the relative amplitude of the high and low dependent on the erosion level.

For quartzites and similar host rocks, on the other hand, the signature is a simple high over the mineralized, potassically altered intrusion, for all levels of exposure. The anomaly becomes gradually weaker and broader with increasing depth of burial of the magnetic core. The signature for moderately magnetic felsic host rocks is intermediate between the signatures for mafic hosts and unreactive sedimentary host rocks, but the annular low around the central high is poorly developed, due to the relatively low magnetization contrast between the magnetite-destructive alteration zones (phyllic and strong propylitic) and the unaltered felsic rocks. Carbonate host rocks have the central high, with highs also developed over magnetite-rich proximal Cu-(Au) skarns, in the inner garnet-rich zone, and distal Au skarns, near the marble contact. These skarns tends to occur as discrete magnetite-bearing skarn bodies, developed within favourable horizons and localized by overall zonation patterns and by structural controls.

Figure 13.10a–c shows effects of burial under a significant thickness of magnetic volcanics of the deposit model shown in Fig. 13.9(a), demonstrating that such deposits should be clearly detectable beneath volcanic cover. Fig. 13.10d–f shows distortion of the magnetic signature produced by post emplacement tilting. Figure 13.10g–i illustrates a porphyry system similar to that of Fig. 13.9(a) but emplaced into carbonate rocks, and Fig. 13.10j–l shows the effects of asymmetric alteration zoning produced by emplacement into a contact between magnetic mafic rocks and carbonates.

With regard to epithermal gold deposits, intense epithermal-style alteration, whether low- or high-sulphidation, is invariably magnetite-destructive. The magnetic signature is strongly dependent on the host rocks. Epithermal alteration systems hosted by magnetic volcanic rocks are characterized by smooth, flat magnetic low zones within the overall busy magnetic texture (Irvine and Smith 1990, Feebrey et al. 1998). Similar systems within non-magnetic sedimentary rocks have negligible magnetic expression. High sulphidation systems may have a diffuse intrusion + alteration high due to a deeper porphyry system within a few hundred meters to a few kilometres of the deposit. This may be more prominent if post-formation faulting has brought the intrusion closer to the surface, or the porphyry and epithermal systems are telescoped by rapid uplift during formation. Upward continuation of magnetic maps may help identify deep intrusions, beneath magnetic volcanics, that are possible sources for mineralizing fluids (Gunn et al. 2009). Magnetics can also be useful for detecting structural corridors and fluid pathways that are related to more localized epithermal systems. Radiometrics can detect K-rich alteration (e.g., adularia-sericite) associated with some exposed or subcropping epithermal deposits. Although in favourable circumstances magnetics is a useful tool for defining hydrothermal systems that are prospective for volcanic-hosted epithermal gold deposits, delineation of the ore zones is not possible. Electrical methods are generally more useful for detecting conductive or chargeable mineralization and defining resistive zones of silicification. Gravity methods are sometimes useful for defining lower density alteration, or in some cases enhanced density due to silicification of porous rocks.

Smith (2002) has reviewed geophysical signatures of IOCG deposits, with an emphasis on magnetics and gravity. Due to the high densities of iron oxide minerals, IOCG deposits are invariably associated with gravity highs, up to 20 mgal for large deposits. The magnetic signatures of IOCG deposits, and their relationships to the gravity anomalies, depend on the geological history and are much more variable. The iron oxide zonation pattern of IOCG systems (Hitzman et al. 1992; Wall and Gow 1995) indicates that the magnetic sources overall tend be deeper and more laterally extensive than the gravity sources. For this reason detailed analysis of the anomalies should reveal a somewhat deeper and/or broader magnetic zone than the anomalous density distribution, as at the archetypical Olympic Dam, South Australia, deposit (Esdale et al. 2003). Thus a better term for the relationship between the magnetic and gravity anomalies is "superposed", rather than "coincident", anomalies.

Redox conditions during deposition and alteration overprinting, which controls the abundances and proportions of magnetite and hematite in IOCG deposits and their alteration envelopes are a crucial control on the magnetic signatures. Magnetite, formed under relatively reducing conditions, has very high susceptibility compared to hematite. Hematite has low susceptibility and also has fairly weak remanence, unless it has formed at, or been taken to, very high temperatures

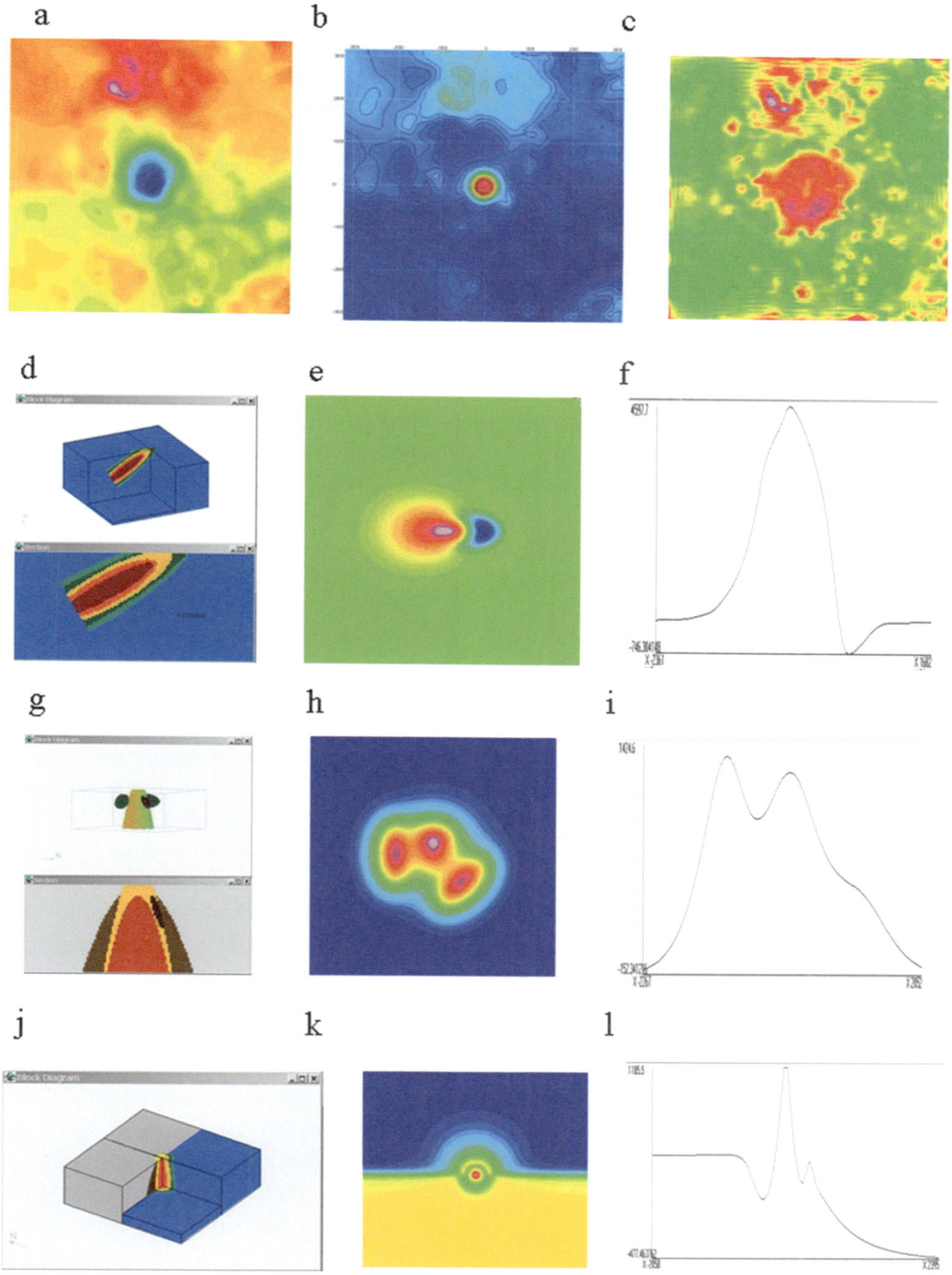

Fig. 13.10 Examples of canonical exploration models and their predicted magnetic signatures. (**a**) RTP magnetic signature of model in Fig. 13.9(**a**), buried beneath 100 m of magnetic volcanics (**b**) As for (**a**), but upper 1000 m of deposit removed by erosion, then covered by 100 m of magnetic volcanics (**c**) Analytic signal amplitude (total gradient) of RTP magnetic signature in (**a**), (**d**) Model of Fig. 13.9(**a**), with post-formation tilt through 60 degrees, (**e**) RTP magnetic signature of model (**d**), (**f**) profile of RTP magnetic signature across model (**d**), (**g**) Gold-rich porphyry copper, emplaced into carbonate rocks with magnetite-rich potassic core plus proximal and distal skarns, eroded 500 m, (**h**) RTP magnetic signature of model (**g**), (**i**) profile of RTP magnetic signature across model (**g**), (**j**) Gold-rich porphyry copper with magnetite-rich potassic core, emplaced into contact between mafic volcanics and carbonate rocks, eroded 500 m, (**k**) RTP magnetic signature of model (**j**), (**l**) profile of RTP magnetic signature across model (**j**)

and thereby acquired a thermoremanent magnetization. Monoclinic pyrrhotite, formed under reducing conditions with moderate sulphur fugacity, has moderate susceptibility, but tends to carry intense remanence. The oxidation state of the host sequence appears to influence the oxidation state of the IOCG deposits, as well as the redox state of the source (e.g., magmatic fluids) and the paleodepth (more oxidized at shallow depths). IOCG deposits tend to occur in relatively oxidized crustal provinces, characterized by strong anomalies and complex magnetic patterns (Haynes 2000). Within the regional magnetic variability, the deposits lie within or near relatively strong magnetic highs, associated with semi-regional magnetite-producing sodic and sodic-calcic alteration systems.

In a typical vertically zoned IOCG system, magnetite-destructive, hematite-rich hematite-sericite-chlorite-carbonate (HSCC) alteration dominates upper levels, whereas magnetite-rich alteration (potassic at intermediate depths, grading into deeper sodic or sodic-calcic alteration) dominates at depth. Thus the current erosion level determines whether the exposed or near-surface portions of the system are hematite-rich or magnetite-rich. If the level of exposure is sufficiently deep, overprinting magnetite-rich potassic alteration tends to be more focussed around the deposits, enhancing the magnetic highs. On the other hand, if upper levels of the system have been retained, magnetite-destructive hematite-rich alteration hosts the mineralization. In this case the signature is a relatively smooth pattern, which can be a local magnetic low (depending on the host rock magnetization) within the overall broad high associated with deeper and/or more laterally extensive magnetite-dominant alteration. Tilting of a vertically zoned system, or upfaulting of the deeper magnetite-rich portion, may juxtapose the magnetite and hematite zones, producing juxtaposed, rather than "coincident" gravity and magnetic anomalies.

High grade regional or contact metamorphism of hematite-rich zones can impart an intense thermoremanence to the hematite, which can cause large magnetic anomalies. This appears to explain the strong magnetic anomalies exhibited by massive hematite bodies of the Mount Woods Inlier that have been contact metamorphosed, such as the Peculiar Knob deposit (Schmidt et al. 2007), whereas similar unmetamorphosed massive hematite at Prominent Hill produces no discernible magnetic anomaly.

At continent to province scale favourable tectonic settings for ancient deposits may be recognisable from regional potential field data sets, supplemented by seismic, magnetotelluric or other deep-penetrating methods. Ancient buried subduction zones are characterized by arc-parallel linear belts of magnetic highs, corresponding to magnetite-series granitoid provinces, and lows, corresponding to ilmenite-series granitoid provinces or sedimentary basins. Subduction-related magnetite-series belts are much more prospective for IOCG and porphyry copper-(gold) deposits. Belts of reduced, ilmenite-series granitoids are prospective for Sn-(W) and also for intrusive-related Au and reduced porphyry Au-(Cu) deposits. Within belts of magnetite-series granitoids, Cu-Au is associated with more magnetic magmatic-hydrothermal systems than Cu-Mo; W-Mo-Bi and Au in tin provinces is much less magnetic. In oxidized Au-bearing systems, Au mineralization is often associated with the felsic end of magmatic evolution and is then associated locally with a weaker magnetic character and higher radioelement contents. Continental rift settings, which are also prospective for IOCG deposits, may be associated with rift-parallel regional gravity and magnetic highs along the ancient continental margin, with a quiet magnetic zone outboard of the regional highs and relatively busy magnetic patterns inboard of the margin (Gunn 1997). Large intrusions associated with bimodal magmatism, which is characteristic of anorogenic environments, including several IOCG provinces, can be seen in magnetic and gravity images, due to the contrasting physical properties of mafic and felsic intrusions.

At a regional scale major structures that control the emplacement of mineralizing or heat-engine magmas, or which channel flow of crustal fluids, are often evident in suitably processed gravity and magnetic data sets (e.g., Sandrin et al. 2007). These features may also be visible in satellite imagery. Intersections of lineaments appear to be particularly favourable for IOCG mineralization. Structural controls at a range of scales, from semiregional to prospect scale, may be evident in detailed magnetic data. Identification of favourable orientations of structures may be possible if senses of movement, block rotations etc. are known. Anomaly offsets and abrupt changes of trend in magnetic images can help to define tectonic movements. Paleomagnetic studies can also be useful for defining rotations and tilting within and around deposits (Geissman et al.

1982; Force et al. 1995; Schmidt et al. 1997) and to define distinct magmatic and hydrothermal alteration events recorded by remanent magnetization (Schmidt et al. 1997; Clark and Lackie 2003; Astudillo et al. 2010).

A large range of magma compositions within a comagmatic suite is indicative of substantial fractional crystallization, which can partition metals into late stage fluids, potentially concentrating them to levels that can produce economic ore deposits. For this reason fractional crystallization is favourable for development of intrusive-related mineralization. Strongly zoned oxidized intrusions produced by fractional crystallization exhibit zoned magnetic signatures and, if exposed or subcropping, zoned radiometric patterns. Similarly, multiple/nested intrusions, with a substantial range of magnetic properties, densities and radioelement contents, particularly when there are geophysical indications of an underlying magma chamber, are also favourable indicators of fractional crystallization.

Well-developed contact aureoles around intrusions are indicative of emplacement of high-temperature, melt-rich magma capable of undergoing substantial fractional crystallization. Strong contact aureole effects produce substantial mineralogical changes in the metamorphosed and metasomatized host rocks, often with pronounced changes in magnetic susceptibility (particularly increased susceptibility due to creation of secondary magnetite and/or pyrrhotite). Strong remanent magnetization of contact aureoles is also suggestive of high temperature emplacement or substantial metasomatism. Prominent contact aureole magnetic signatures, particularly if they are clearly zoned or show strong local overprinting are therefore also favorable indicators of potentially mineralized magmatic-hydrothermal systems.

Future advances in magnetic exploration will rely firstly on better understanding of the processes that create, destroy and alter magnetic minerals in mineralized environments, based on magnetic petrological studies, accompanied by more comprehensive magnetic property databases and detailed case studies of magnetic signatures of mineralized environments. Secondly, improvements in acquisition, processing and interpretation of the crustal magnetic field will enable more geological information to be extractable from magnetic surveys. Although small incremental improvements in conventional surveys will continue, the most dramatic advances are likely to involve gradiometry, particularly measurements of the full gradient tensor by new generation highly sensitive instruments, to obtain more detailed information about structures and subtly varying magnetization patterns in the shallow crust.

In a seminal paper Pedersen and Rasmussen (1990) discuss in some detail the practical problems encountered in the collection and processing of gradient tensor data and the benefits obtainable from acquisition of tensor data. For instance, these workers point out that resolution is enhanced compared to conventional magnetic surveys and that rotational invariants calculated from tensor data have attractive properties for interpretation. Christensen and Rajagopatan (2000) suggested that the next breakthrough in magnetic exploration is likely to be the measurement of the gradient tensor and demonstrated the utility of analytic signal amplitudes (total gradients) of vector components, which can be derived directly from gradient tensor components, for locating boundaries and interpreting source geometries in the presence of remanence. Schmidt and Clark (2006) have summarized the multiple benefits of gradiometry in general and tensor gradiometry in particular.

Recently the first practical low temperature (liquid helium cooled) SQUID-based system for geophysical gradient tensor surveys has been developed by the Jena (Germany) group (Stolz et al. 2006). The intrinsic noise of the LTS planar gradiometers developed by this group is 0.2 $\mathrm{pT\,m^{-1}}$ (integrated between 0.01 Hz and 10 Hz). Noise spectral density of the full tensor gradiometer system in motion is about 1–10 $\mathrm{pT\,m^{-1}}(\sqrt{\mathrm{Hz}})^{-1}$ over a frequency range of 0.1 – 2 Hz in a bird towed beneath a helicopter and approximately ten times higher for installation on a fixed wing aircraft.

The discovery of high temperature (liquid nitrogen temperatures and above) superconducting materials in the late 1980s has created opportunities for cheaper, smaller devices that can be readily transported and refilled, but retain very high sensitivities. Liquid nitrogen cooled SQUIDs and gradiometers are very sensitive, with noise levels that are about an order of magnitude higher than those of low T SQUIDs. Clark et al. (1998) suggested the use of combined vector field and gradient tensor measurements, using high T superconducting devices, for separating contributions of induced and remanent magnetization to magnetic anomalies and for inferring source properties, such as total magnetization direction, remanence direction and Koenigsberger ratio.

A number of tensor gradiometer systems, based on a wide range of different technologies including high temeprature SQUID devices, are under development (e.g., Clem et al. 2001; Humphrey et al. 2005; Leslie et al. 2007; Wiegert et al. 2007; Sunderland et al. 2009; Keenan et al. 2010). We anticipate that over the next decade full gradient tensor systems will be commercialized and be routinely used in next generation magnetic surveys for exploration.

13.6 Interpretation of Lower Crustal Processes

Lower crustal processes are dominated by increasing temperatures, and an important temperature is that associated with the Curie point of magnetite (580°C), above which it loses its permanent and induced magnetism. To the extent that other magnetic minerals dominate in the lower crust, the temperatures of those other magnetic phases will be important for interpretation. Ilmenite-hematite, hematite-magnetite, or titanomagnetite-rich, phases exhibit different Curie, Néel, or unblocking temperatures from pure magnetite (McEnroe et al. 2004), and they extend to 670°C for hematite-rich compositions. Fine-scale exsolution of ilmenite-hematite phases (McEnroe et al. 2009b), and possibly also magnetite-hematite phases (Schmidt et al. 2007), significantly increases the magnetic remanence and coercivity from typical multi-domain values. If these lamellae are not resorbed by temperature and pressure conditions in the lower crust, then a much greater range of magnetic mineral phases may be present. Experiments by McEnroe et al. (2004) suggest that the lamellae may be stable at lower crustal temperatures and pressures.

Increasing pressures also have an effect on the magnetic properties of single and multi-domain magnetite (Gilder et al. 2004) and titanomagnetite (Gilder and Le Goff 2008). Both saturation remanent magnetization and coercivity increase markedly in titanomagnetites at typical lower crustal pressures. The percentage of Ti in the titanomagnetite structure seems to control the increase in magnetization and coercivity, with the highest increases associated with the highest amounts of Ti.

Much work continues to be devoted to the difficult question of determining the depth to the Curie and Néel isotherms, and with comparing results from different approaches. Works utilizing standard approaches (Spector and Grant 1970) include those of Bilim (2007), Bektas et al. (2007), and Maden (2009) in Turkey, Trifonova et al. (2009) in Bulgaria, ChunFeng et al. (2009) and Xu-Zhi et al. (2006) in China, Prutkin and Saleh (2009) in Egypt, and Stampolidis et al. (2005) in Albania. A fractal approach based on the formulation of Maus et al. (1997) was used in the western United States by Bouligand et al. (2009), and a similar approach was used in California by Ross et al. (2006). Ravat et al. (2007) compares several spectral approaches, while Rajaram et al. (2009) compares the spectral approach with an approach that integrates seismic, heat flow, and satellite magnetic data sets (Purucker et al. 2007).

13.7 Summary

The interpretation of terrestrial impact structures continues to garner much attention because of its relevance to the interpretation of extraterrestrial impacts. These impacts are the targets of robotic exploration by NASA, ESA, and the national space agencies of Japan, India, and China.

The importance of the coherence scale, or size of a region of coherent magnetization, can not be overemphasized, both in the terrestrial and extraterrestrial examples. It is often the case that observations of a feature are made from only a single altitude. A change in that altitude can often make a dramatic difference in what features are available for interpretation, and 'color' the interpretation in subtle ways.

Future advances in magnetic exploration are critically dependent on a better understanding of the processes that create, destroy, and alter magnetic minerals. Comprehensive magnetic property databases are also a requirement for future advances in magnetic exploration.

The magnetic technique has often been faulted as having too little resolution. Recent advances in the measurement, processing, and interpretation of gradient data utilizing high temperature SQUID devices on helicopters and fixed-wing aircraft offer the prospect of dramatic improvements in the resolution of our magnetic imagery. Similar improvements in our view of the magnetic lithosphere from near-Earth space will be inaugurated with the Swarm constellation (Friis-Christensen et al. 2009). Swarm will utilize highly sensitive Helium and fluxgate magnetometers flying in constellation to make its gradient field measurements.

Acknowledgments We would like to acknowledge the support of the SERPENT proposal team, especially R. Blakely (USGS) for Fig. 13.5, Fig. 13.6, and Fig. 13.7 and R. Bracken (USGS) for Fig. 13.7.

References

Acuña M et al (1999) Global distribution of crustal magnetization discovered by the Mars global surveyor MAG/ER experiment. Science 284(5415):790–793

Airo M, Mertanen S (2008) Magnetic signatures related to orogenic gold mineralization. J Appl Geophys 64:14–24

Anand S, Rajaram M (2006) Aeromagnetic data analysis for the identification of concealed uranium deposits: a case history from Singhbhum uranium province, India. Earth Planet Space 58:1099–1103

Anand S, Rajaram M, Majumdar T, Bhattacharyya R (2009) Structure and tectonics of 85 E Ridge from analysis of geopotential data. Tectonophysics 478:100–110

Astudillo N, Roperch P, Townley B, Arriagada C, Chauvin A (2010) Magnetic polarity zonation within the El Teniente copper-molybdenum porphyry deposit, central Chile. Mineralium Deposita 45:23–41

Austin J, Blenkinsop T (2008) The Cloncurry Lineament: Geophysical and geological evidence for a deep crustal structure in the eastern succession of the Mount Isa Inlier. Precambrian Res 163:50–68

Austin J, Blenkinsop T (2009) Local to regional scale structural controls on mineralization and the importance of a major lineament in the eastern Mount Isa Inlier. Australia: review and analysis with autocorrelation and weights of evidence. Ore Geol Rev 35:298–316

Bektas O, Ravat D, Bueyueksarac A, Bilim F, Ates A (2007) Regional geothermal characterisation of East Anatolia from aeromagnetic, heat flow and gravity data. Pure Appl Geophys 164(5):975–998. doi:10.1007/s00024-007-0196-5

Betts P, Barraud J, Lumley J, Davies M (2004) Aeromagnetic patterns of half-graben and basin inversion: implications for sediment-hosted massive sulfide Pb-Zn-Ag exploration. J Struct Geol 26:1137–1156

Bilim F (2007) Investigations into the tectonic of Kutahya-Denizli region, lineaments and thermal structure western Anatolia, from using aeromagnetic, gravity and seismological data. Phys Earth Planet Inter 165(3–4):135–146. doi:10.1016/j.pepi.2007.08.007

Blakely R, Wells R, Weaver C, Johnson S (2002) Location, structure, and seismicity of the Seattle fault zone, Washington: evidence from aeromagnetic anomalies, geologic mapping, and seismic-reflection data. Geol Soc Am Bull 114(2): 169–177

Blakely R, Brocher T, Wells R (2005) Subduction-zone magnetic anomalies and implications for hydrated forearc mantle. Geology 33(6):445–448. doi:10.1130/G21447.1

Blakely RJ, John DA, Box SE, Berger BR, Fleck RJ, Ashley RP, Newport GR, Heinemeyer GR (2007) Crustal controls on magmatic-hydrothermal systems: A geophysical comparison of White River, Washington, with Goldfield, Nevada. Geosphere 3(2):91–107. doi:10.1130/GES00071.1

Blakely RJ, Sherrod BL, Hughes JF, Anderson ML, Wells RE, Weaver CS (2009) Saddle mountain fault deformation zone, Olympic Peninsula, Washington: western boundary of the Seattle uplift. Geosphere 5(2):105–125. doi:10.1130/GES00196.1

Bostock M, Hyndman R, Rondenay S, Peacock S (2002) An inverted continental Moho and serpentinization of the forearc mantle. Nature 417(6888):536–538

Bouligand C, Glen JMG, Blakely RJ (2009) Mapping Curie temperature depth in the western United States with a fractal model for crustal magnetization. J Geophys Res Solid Earth 114. B11104, doi:10.1029/2009JB006494

Brocher T, Parsons T, Trehu A, Snelson C, Fisher M (2003) Seismic evidence for widespread serpentinized forearc upper mantle along the Cascadia margin. Geology 31(3):267–270

Carporzen L, Gilder S, Hart R (2005) Palaeomagnetism of the Vredefort meteorite crater and implications for craters on Mars. Nature 435(7039):198–201. doi:10.1038/nature 03560

Chernicoff C, Nash C (2002) Geological interpretation of Landsat TM imagery and aeromagnetic survey data, northern Precordillera region, Argentina. J South Am Earth Sci 14:813–820

Chernicoff C, Richards J, Zappettini E (2002) Crustal lineament control on magmatism and mineralization in northwestern Argentina: geological, geophysical and remote sensing evidence. Ore Geol Rev 21:127–155

Christensen A, Rajagopalan S (2000) The magnetic vector and gradient tensor in mineral and oil exploration. Preview 77

ChunFeng L, Bing C, ZuYi Z (2009) Deep crustal structures of eastern China and adjacent seas revealed by magnetic data. Sci China Series D Earth Sci 52(7):984–993. doi:10.1007/s11430-009-0096-x

Clark D (1997) Magnetic petrophysics and magnetic petrology: aids to geologic interpretation of magnetic surveys. AGSO J Aust Geol Geophys 17:83–103

Clark D (1999) Magnetic petrology of igneous intrusions: implications for exploration and magnetic interpretation. Explor Geophys 30:5–26

Clark D, Lackie M (2003) Palaeomagnetism of the Early Permian Mount Leyshon Intrusive Complex and Tuckers Igneous Complex, North Queensland, Australia. Geophys J Int 153(3):523–547. doi:10.1046/j.1365-246x.2003.01907.x

Clark D, Schmidt P (2001) Petrophysical properties of the Goonumbla volcanic complex, NSW: implications for magnetic and gravity signatures of porphyry Cu-Au mineralization. Explor Geophys 32:171–175

Clark D, Schmidt P, Coward D, Huddleston M (1998) Remote determination of magnetic properties and improved drill targeting of magnetic anomaly sources by Differential Vector Magnetometry (DVM). Explor Geophys 29:312–319

Clark D, Geuna S, Schmidt P (2004) Predictive magnetic exploration models for porphyry, epithermal, and iron oxide copper-gold deposits: implications for exploration, P700 Final Report, AMIRA International Ltd.

Clem T, Overway D, Purpura J, Bono J, Koch R, Rozen J, Keefe G, Willen S, Mohling R (2001) High-Tc SQUID gradiometer for mobile magnetic anomaly detection. IEEE Trans Appl. Supercond 11:871–875

Cowan D, Cooper G (2005) Enhancement of magnetic signatures of impact craters modeling and petrophysics. In: Kenkmann T, Hörz F, Deutsch A (eds) Large meteorite impacts III, Geological Society of America Boulder, Colorado, Special Paper 384, pp 51–65

de Castro DL, de Oliveira DC, Gomes Castelo Branco RM (2007) On the tectonics of the Neocomian Rio do Peixe Rift Basin, NE Brazil: lessons from gravity, magnetics, and radiometric data. J South Am Earth Sci 24(2–4):184–202. doi:10.1016/j.jsames.2007.04.001

Drenth BJ, Finn CA (2007) Aeromagnetic mapping of the structure of Pine Canyon caldera and Chisos mountains intrusion, Big Bend National Park, Texas. Geol Soc Am Bull 119(11–12):1521–1534

Dunlop D (2005) Planetary science—Magnetic impact craters. Nature 435(7039):156–57. doi:10.1038/435156a

Esdale D, Pridmore D, Coggon J, Muir P, William P, Fritz F (2003) Olympic dam copper-uranium-gold-silver-rare earth element deposit, South Australia: a geophysical case history. In: Dentith MC (ed) Geophysical signatures of South Australian mineral deposits, Centre for Global Metallogeny, University of Western Australia, Publication 31, pp 147–168

Espinosa-Cardena JM, Campos-Enriquez JO (2008) Curie point depth from spectral analysis of aeromagnetic data from Cerro Prieto geothermal area, Baja California, Mexico. J Volcanol Geothermal Res 176(4):601–609. doi:10.1016/j.jvolgeores.2008.04.014

Feebrey C, Hishida H, Yoshioka K, Nakayama K (1998) Geophysical expression of low sulphidation epithermal Au-Ag deposits and exploration implications-examples from the Hokusatsu region of SW Kyushu, Japan. Resour Geol 48:75–86

Ferraccioli F, Armadillo E, Jordan T, Bozzo E, Corr H (2009a) Aeromagnetic exploration over the East Antarctic ice sheet: a new view of the Wilkes Subglacial Basin. Tectonophysics 478(1–2, Sp. Iss. SI):62–77. doi:10.1016/j.tecto.2009.03.013, General Assembly of the International-Association-of-Geodesy/24th General Assembly of the International-Union-of-Geodesy-and-Geophysics, Perugia, Italy, Jul 02–13, 2007

Ferraccioli F, Armadillo E, Zunino A, Bozzo E, Rocchi S, Armienti P (2009b) Magmatic and tectonic patterns over the northern Victoria land sector of the transantarctic mountains from new aeromagnetic imaging. Tectonophysics 478(1–2, Sp. Iss. SI):43–61. doi:10.1016/j.tecto.2008.11.028, General Assembly of the International-Association-of-Geodesy/24th General Assembly of the International-Union-of-Geodesy-and-Geophysics, Perugia, Italy, Jul 02–13, 2007

Ferré E, Maes S, Butak K (2009) The magnetic stratification of layered mafic intrusions: Natural examples and numerical models. Lithos 111:83–94

Force E, Dickinson W, Hagstrum J (1995) Tilting history of the San Manuel-Kalamazoo porphyry system, southeastern Arizona. Econ Geol 90:67–80

Frey H (1982) MAGSAT Scalar anomaly distribution-The global perspective. Geophys Res Lett 9(4):277–280

Friis-Christensen E, Lühr H, Hulot G, Haagmans R, Purucker M (2009) Geomagnetic Research from Space. Eos Trans Agu 90(25):213–214

Gattacceca J, Berthe L, Boustie M, Vadeboin F, Rochette P, De Resseguier T (2008) On the efficiency of shock magnetization processes. Phys Earth Planetary Inter 166(1–2):1–10. doi:10.1016/j.pepi.2007.09.005

Gee J, Kent D (2007) Source of Oceanic Magnetic Anomalies and the Geomagnetic Polarity Timescale. In: Kono M (ed) Geomagnetism, treatise of geophysics Amsterdam, vol 5, Elsevier, 5.12, pp 455–508

Geissman J, Van der Voo R, KL Howard (1982) A paleomagnetic study of the structural deformation in the Yerington district, Nevada. Am J Sci 282:1042–1109

Gilder S, LeGoff M, Chervin J, Peyronneau J (2004) Magnetic properties of single and multi-domain magnetite under pressures from 0 to 6 GPa. Geophys Res Lett 31(10). doi:10.L243061029/2004GL019844

Gilder SA, Le Goff M (2008) Systematic pressure enhancement of titanomagnetite magnetization. Geophys Res Lett 35(10). doi:10.243061029/2008GL033325

Graham K (1961) The re-magnetization of a surface outcrop by lightning currents. Geophys J Royal Astronom Soc 6(1): 85–102

Grant F (1984) Aeromagnetics, Geology, and ore environments, I. Magnetite in igneous, sedimentary, and metamorphic rocks: an overview. Geoexploration 23:303–333

Grauch VJS, Hudson MR (2007) Guides to understanding the aeromagnetic expression of faults in sedimentary basins: lessons learned from the central Rio Grande rift, New Mexico. Geosphere 3(6):596–623. doi:10.1130/GES00128.1

Grieve RAF, Pilkington M (1996) The signature of terrestrial impacts. J Aust Geol Geophys 16:399–420

Gubbins D, Herrero-Bervera E (eds) (2007) Encyclopedia of geomagnetism and paleomagnetism, Springer, The Netherlands

Guilbert J (1995) Geology, alteration, mineralization and genesis of the Bajo de la Alumbrera porphyry copper-gold deposit, Catamarca Province, Argentina. In: Pierce FW, Bolm JG (eds) Porphyry copper deposits of the American Cordillera, Arizona Geological Society Digest 20, pp 646–656

Gunn P (1997) Regional magnetic and gravity responses of extensional sedimentary basins. Agso J Aust Geol Geophys 17(2):115–131

Gunn P, Mackey T, Meixner A (2009) Magnetic exploration. Technical report Pacific Islands Applied Geoscience Commission, SOPAC Technical Bulletin 11

Hamoudi M, Thébault E, Lesur V, Mandea M (2007) GeoForschungsZentrum Anomaly Magnetic Map (GAMMA): a candidate model for the world digital magnetic anomaly map. Geochem Geophys Geosyst 10. doi:10.1029/2007GC001638

Haynes D (2000) Iron oxide copper(-gold) deposits: their position in the ore deposit spectrum and modes of origin. In: Porter TM (ed) Hydrothermal iron oxide copper-gold and related deposits: a global perspective, vol 1. PGC Publishing, Adelaide, pp 71–90

Hemant K, Thébault E, Mandea M, Ravat D, Maus S (2007) Magnetic anomaly map of the world: merging satellite, airborne, marine and ground-based data. Earth Planetary Sci Lett 260:56–71

Hildenbrand T, Berger B, Jachens R, Ludington S (2000) Regional crustal structures and their relationship to the distribution of ore deposits in the western United States, based on magnetic and gravity data. Econ Geol 95:1583–1603

Hinze W, Von Frese R, Ravat D (1991) Mean magnetic contrasts between oceans and continents. Tectonophys 192(1–2): 117–127. doi:10.1016/0040-1951(91)90250-V

Hitzman M, Oreskes N, MT Einaudi (1992) Geological chracteristics and tectonic setting of proterozoic iron oxide (cu-U-Au-REE) deposits. Precambrian Res 1:1

Humphrey K, Horton T, Keene M (2005) Detection of mobile targets from a moving platform using an actively shielded, adaptively balanced SQUID gradiometer. IEEE Trans Appl Supercond 15(2):753–756

Hyndman R, Peacock S (2003) Serpentinization of the forearc mantle. Earth Planetary Sci Lett 212(3–4):417–432. doi:10.1016/S0012-821X(03)00263-2

Irvine R, Smith M (1990) Geophysical exploration for epithermal gold systems. J Geochem Explor 36:375–412

Jessel M (2001) Three-dimensional geological modeling of potential-field data. Comput Geosci 27:455–465

Keenan T, Young J, Foley C, Du J (2010) A high-Tc flip-chip SQUID gradiometer for mobile underwater magnetic sensing. Supercond Sci Technol 23(025,029):7pp

Kirby S, Engdahl E, Denlinger R (1996) Intermediate-depth intraslab earthquakes and arc volcanism as physical expressions of crustal and uppermost mantle metamorphism in subducting slabs. In: Bebout GE et al. (eds) Subduction: top to bottom, American Geophysical Union Geophysical Monograph 96, pp 195–214

Korhonen J, et al. (2007) Magnetic anomaly map of the world, and associated DVD, commission for the geological map of the world, UNESCO, Paris, France, Scale:1:50,000,000

Langlais B, Lesur V, Purucker M, Connerney J, Mandea M (2010) Crustal magnetic fields of terrestrial planets. Space Sci Rev. doi:10.1007/s11214-009-9557-y, 152:223–249

Leslie K, Blay K, Clark D, Schmidt P, Tilbrook D, Bick M, Foley C (2007) Helicopter trial of magnetic tensor gradiometer. In: ASEG 19th International Conference, Perth, Australia

Lillis RJ, Frey HV, Manga M, Mitchell DL, Lin RP, MH Acuña, Bougher SW (2008) An improved crustal magnetic field map of Mars from electron reflectometry: highland volcano magmatic history and the end of the martian dynamo. Icarus 194(2):575–596. doi:10.1016/j.icarus.2007.09.032

Lillis RJ, Purucker ME, Louzada HJS, Stewart-Mukhopadhyay K, Manga M, Frey H (2010) Study of impact demagnetization on Mars using Monte Carlo modeling and multiple altitude data. J Geophys Res Planet. 115, E07007, doi:10.1029/2009JE003556

Lindsayz D, Zentilli M, Rojas de la Rivera J (1995) Evolution of an active ductile to brittle shear system controlling mineralization at the Chuquicamata porphyry copper deposit, northern Chile. Int Geol Rev 37:945–958

Louzada KL, Weiss BP, Maloof AC, Stewart ST, Swanson-Hysell NL, Soule SA (2008) Paleomagnetism of Lonar impact crater, India. Earth Planet Sci Lett 275(3–4):308–319. doi:10.1016/j.epsl.2008.08.025

Lowell J (1968) Geology of the Kalamazoo orebody, San manuel district, Arizona. Econ Geol 63:645–654

Lowell J, Guilbert J (1970) Lateral and vertical alteration-mineralization zoning in porphyry copper deposits. Econ Geol 65:373–408

Lum J, Clark A, Coleman P (1991) Gold potential of the southwest Pacific: Papua New Guinea, Solomon Islands, Vanuatu, and Fiji, Technical report, East-West Center, Honolulu

MacDonald G, Arnold L (1994) Geological and geochemical zoning of the Grasberg Igneous complex, Irian Java, Indonesia. J Geochem Explor 50:179–202

Maden N (2009) Crustal Thermal Properties of the Central Pontides (Northern Turkey) deduced from spectral analysis of magnetic data. Turkish J Earth Sci 18(3):383–392. doi:10.3906/yer-0803-7

Maia M, Dyment J, Jouannetaud D (2005) Constraints on age and construction process of the Foundation chain submarine volcanoes from magnetic modeling. Earth Planet Sci Lett 235(1–2):183–199. doi:10.1016/j.epsl.2005.02.044

Mandea M, Purucker M (2005) Observing, modeling, and interpreting magnetic fields of the solid Earth. Surv Geophys 26(4):415–459. doi:10.1007/s10712-005-3857-x

Manea M, Manea VC (2008) On the origin of El Chichon volcano and subduction of Tehuantepec ridge: a geodynamical perspective. J Volcanol Geotherm Res 175(4, Sp. Iss. SI):459–471. doi:10.1016/j.jvolgeores.2008.02.028

Maus S, Gordon D, Fairhead D (1997) Curie-temperature depth estimation using a self-similar magnetization model. Geophys J Int 129(1):163–168

Maus S, Luehr H, Rother M, Hemant K, Balasis G, Ritter P, Stolle C (2007a) Fifth-generation lithospheric magnetic field model from CHAMP satellite measurements. Geochem Geophys Geosyst 8. doi:10.1029/2006GC001521

Maus S, Sazonova T, Hemant K, Fairhead J, Ravat D (2007b) National geophysical data center candidate for the world digital magnetic anomaly map. Geochem Geophys Geosyst 8(6). doi:10.1029/2007GC001643

McDowell F, McMahon T, Warren P, Cloos M (1996) Pliocene Cu-Aubearing igneous intrusions of the Gunung Bijih (Ertsberg) district, Irian Java, Indonesia: K-Ar geochronology. J Geol 104:327–340

McEnroe S, Langenhorst F, Robinson P, Bromiley G, Shaw C (2004) What is magnetic in the lower crust? Earth Planet Sci Lett 226(1–2):175–192. doi:10.1016/j.epsl.2004.07.020

McEnroe S, Brown L, Robinson P (2009a) Remanent and induced magnetic anomalies over a layered intrusion: Effects from crystal fractionation and magma recharge. Tectonophysics 478:119–134

McEnroe S, Fabian K, Robinson P, Gaina C, Brown LL (2009b) Crustal magnetism, lamellar magnetism and rocks that remember. Elements 5(4):241–246. doi:10.2113/gselements.5.4.241

Melosh H (1989) Impact cratering: a geologic process, Oxford University Press, New York, NY

Muundjua M, Hart RJ, Gilder SA, Carporzen L, Galdeano A (2007) Magnetic imaging of the Vredefort impact crater, South Africa. Earth Planet Sci Lett 261(3–4):456–468. doi:10.1016/j.epsl.2007.07.044

Muundjua M, Galdeano A, Carporzen L, Gilder SA, Hart RJ, Andreoli MAG, Tredoux M (2008) Reply to comment by Reimold WU, Gibson RL, Henkel H on Muundjua et al. (2007), Magnetic imaging of the Vredefort impact crater, South Africa, EPSL 261, pp 456-468

Discussion. Earth Planet Sci Lett 273(3–4):397–399. doi:10.1016/j.epsl.2008.06.044

Nabighian M, Grauch V, Hansen R, LaFehr T, Li Y, Peirce J, Phillips J, Ruder M (2005) 75th Anniversary—The historical development of the magnetic method in exploration. Geophys 70(6):33ND–61ND. doi:10.1190/1.2133784

Oleskevich D, Hyndman R, Wang K (1999) The updip and downdip limits to great subduction earthquakes: thermal and structural models of Cascadia, south Alaska, SW Japan, and Chile. J Geophys Res Solid Earth 104(B7):14,965–14,991

Parker R (1991) A theory of Ideal bodies for Seamount magnetism. J Geophys Res Solid Earth 96(B10)

Parker R (2003) Ideal bodies for Mars magnetic. J Geophys Res Planet 108(E1). doi:10.1029/2001JE001760

Peacock S, Wang K, McMahon A (2002) Thermal structure and metamorphism of subducting oceanic crust: insight into Cascadia intraslab earthquakes and processes, and earthquake hazards, Open-file Report 02-328, 17–24, U.S. Geological Survey

Pedersen L, Rasmussen T (1990) The gradient tensor of potential field anomalies: some implications on data collection and data processing of maps. Geophys 55:1558–1566

Phillips J (1997) Potential-field geophysical software for the PC-version 2.2, Open-file Report 97-725, U.S. Geological Survey

Pilkington M, Hildebrand A (2003) Transient and disruption cavity dimensions of complex terrestrial impact structures derived from magnetic data. Geophys Res Lett 30(21). doi:10.1029/2003GL018294

Potter D (1996) What makes Grasberg anomalous, implications for future exploration. In: Porphyry related copper and gold deposits of the Asia Pacific region, Australian Mineral Foundation, pp 10.1–10.13

Prutkin I, Saleh A (2009) Gravity and magnetic data inversion for 3D topography of the Moho discontinuity in the northern Red Sea area, Egypt. J Geodynam 47(5):237–245. doi:10.1016/j.jog.2008.12.001

Purucker M, SerpentTeam M (2010) Magnetic signatures of serpentinized mantle and mesoscale variability along the Alaska/Aleutian subduction zone, in Abstract book of the European Geoscience Union, Vienna, Austria

Purucker M, Whaler K (2007) Crustal Magnetism. In: Kono M (ed) Geomagnetism, Treatise of Geophysics, vol 5, Chapter 6. Elsevier, Amsterdam, pp 195–236

Purucker M, Sabaka T, Le G, Slavin JA, Strangeway RJ, Busby C (2007) Magnetic field gradients from the ST-5 constellation: Improving magnetic and thermal models of the lithosphere. Geophys Res Lett 34(24). L24306 doi:10.1029/2007GL031739

Purucker M, Olsen N, Sabaka T, HR (2009a) Geomagnetism mission concepts after Swarm, in Abstract book of the Int. Assoc. Geomag. Aeronom. 11th Scientific Assembly, Sopron, Hungary, pp 105

Purucker ME, Sabaka TJ, Solomon SC, Anderson BJ, Korth H, Zuber MT, Neumann GA (2009b) Mercury's internal magnetic field: Constraints on large- and small-scale fields of crustal origin. Earth Planet Sci Lett 285(3–4, Sp. Iss. SI):340–346. doi:10.1016/j.epsl.2008.12.017

Quesnel Y, Catalan M, Ishihara T (2009) A new global marine magnetic anomaly data set. J Geophys Res Solid Earth 114(B04106). doi:10.1029/2008JB006144

Rabeh T, Miranda M (2008) A tectonic model of the Sinai Peninsula based on magnetic data. J Geophys Eng 5(4): 469–479. doi:10.1088/1742-2132/5/4/010

Rajaram M, Anand SP, Hemant K, Purucker ME (2009) Curie isotherm map of Indian subcontinent from satellite and aeromagnetic data. Earth Planetary Sci Lett 281(3–4):147–158. doi:10.1016/j.epsl.2009.02.013

Ranganai RT, Ebinger CJ (2008) Aeromagnetic and Landsat TM structural interpretation for identifying regional groundwater exploration targets, south-central Zimbabwe Craton. J Appl Geophys 65(2):73–83. doi:10.1016/j.jappgeo.2008.05.009

Ravat D, et al. (2009) A preliminary, full spectrum magnetic anomaly grid of the United States with improved long wavelengths for studying continental dynamics: a website for distribution of data, Open-file Report 2009-1258, U.S. Geological Survey

Ravat D, Pignatelli A, Nicolosi I, Chiappini M (2007) A study of spectral methods of estimating the depth to the bottom of magnetic sources from near-surface magnetic anomaly data. Geophys J Int 169(2):421–434. doi:10.1111/j.1365-246X.2007.03305.x

Reeves C (2007) The role of airborne geophysical reconnaissance in exploration geosciences. First Break 19(9):501–508

Reimold WU, Gibson RL, Henkel H (2008) Scientific comment on In: Muundjua et al., 2007: magnetic imaging of the Vredefort impact crater, South Africa, EPSL 261, 456–468 Discussion. Earth Planet Sci Lett 273(3–4): 393–396. doi:10.1016/j.epsl.2008.06.046

Robinson D, et al. (2008) Advancing process-based watershed hydrological research using near-surface geophysics: a vision for, and review of, electrical and magnetic geophysical methods. Hydrol Process 22:3604–3635

Ross HE, Blakely RJ, Zoback MD (2006) Testing the use of aeromagnetic data for the determination of Curie depth in California. Geophys 71(5):L51–L59. doi:10.1190/1.2335572

Saadi NM, Watanabe K, Imai A, Saibi H (2008) Integrating potential fields with remote sensing data for geological investigations in the Eljufra area of Libya. Earth Planets Space 60(6):539–547

Sandrin A, Elming S (2006) Geophysical and petrophysical study of an iron oxide copper gold deposit in northern Sweden. Ore Geol Rev 29:1–18

Sandrin A, Berggren R, Elming S (2007) Geophysical targeting of Feoxide Cu-(Au) deposits west of Kiruna, Sweden. J Appl Geophys 61:92–101

Schmidt P, Clark D (2006) The magnetic gradient tensor: its properties and uses in source characterization. Leading Edge 25(1):75–78

Schmidt P, Clark D, Logan K (1997) Paleomagnetism, magnetic petrophysics and magnetic signature of the Porgera Intrusive Complex, Papua New Guinea. Explor Geophys 28:276–280

Schmidt PW, McEnroe SA, Clark DA, Robinson P (2007) Magnetic properties and potential field modeling of the Peculiar Knob metamorphosed iron formation, South Australia: an analog for the source of the intense Martian magnetic anomalies? J Geophys Res Solid Earth 112(B3). doi:10.1029/2006JB004495

Schubert G (ed) (2007) Treatise on geophysics. Elsevier, Amsterdam

Shepherd T, Bamber J, Ferraccioli F (2006) Subglacial geology in coats land, East Antarctica, revealed by airborne magnetics and radar sounding. Earth Planetary Sci Lett 244(1–2): 323–335. doi:10.1016/j.epsl.2006.01.068

Sherrod BL, Blakely RJ, Weaver CS, Kelsey HM, Barnett E, Liberty L, Meagher KL, Pape K (2008) Finding concealed active faults: extending the southern whidbey Island fault across the Puget Lowland, Washington. J Geophys Res Solid Earth 113(B5). doi:10.1029/2007JB005060

Sillitoe R (2000) Gold-rich porphyry deposits: descriptive and genetic models and their role in exploration and discovery. In: Hagemann, SG, Brown PE (eds) Gold in 2000, Society of Economic Geologists, Littleton, Colorado, vol. 13. Reviews of Economic Geology, pp 315–345

Singh K, Okuma S, Special issue—Magnetic anomalies: tectonophysics. Tectonophysics 478:1–142

Smith R (2002) Geophysics of iron oxide copper-gold deposits. In: Porter TM (ed) Hydrothermal Iron Oxide Copper-Gold and related deposits: a global perspective, vol 2. PGC Publishing, Adelaide, pp 123–136

Spector A, Grant F (1970) Statistical models for interpreting aeromagnetic data. Geophysics 35(2):293–302

Spray J, Butler H, Thompson L (2004) Tectonic influences on the morphometry of the Sudbury impact structure: Implications for terrestrial cratering and modeling. Meteoritics Planetary Sci 39(2):287–301

Stampolidis A, Kane I, Tsokas G, Tsourlos P (2005) Curie point depths of Albania inferred from ground total field magnetic data. Surveys Geophys 26(4):461–480. doi:10.1007/s10712-005-7886-2

Stolz R, Chwala A, Zakosarenko V, Schulz M, Fritzsch L, Meyer H (2006) SQUID technology for geophysical exploration. In: SEG Expanded abstracts 25, pp 894–898

Sunderland A, Golden H, McRae W, Veryaskin A, Blair D (2009) Results from a novel direct magnetic gradiometer. Explor Geophys 40:222–226

Syberg F (1972) A Fourier method for the regional-residual of potential fields. Geophys Prospect 20:47–75

ten Brink US, Rybakov M, Al-Zoubi AS, Rotstein Y (2007) Magnetic character of a large continental transform: an aeromagnetic survey of the dead sea fault. Geochem Geophys Geosyst 8. doi:10.1029/2007GC001582

Thébault E, Purucker M, Whaler K, Langlais B, Sabaka T, The magnetic field of the Earth's lithosphere. Space Sci Rev in press

Trifonova P, Zhelev Z, Petrova T, Bojadgieva K (2009) Curie point depths of Bulgarian territory inferred from geomagnetic observations and its correlation with regional thermal structure and seismicity. Tectonophys 473(3–4):362–374. doi:10.1016/j.tecto.2009.03.014

Ugalde H, Artemieva N, Milkereit B (2005) Magnetization on impact structures-Constraints from numerical modeling and petrophysics. In: Kenkmann T, Hörz, F, Deutsch A (eds) Large meteorite impacts III, Geological Society of America Special Paper 384, pp 25–42

Wall V, Gow P (1995) Some copper-gold ore-forming systems: iron(ic) connections. In: Clark AH (ed) Giant Ore Deposits-II, Controls on the Scale or Orogenic Magmatic-Hydrothermal Mineralization. Proceedings of the Second Giant Ore Deposits Workshop, Kingston, Ontario, Canada, pp 557–582

Wessel P, Müller R (2007) Plate tectonics. In: Watts AB (ed) Crust and lithosphere dynamics, Treatise of Geophysics, vol 6. Elsevier, Amsterdam, pp 6.02, 49–98

Wiegert R, Oeschger J, Tuovila E (2007) Demonstration of a novel manportable magnetic STAR technology for real time localization of unexploded ordnance, in Proceedings of MTS/IEEE Oceans 2007

Wilkins J, Heidrick T (1995) Post Laramide extension and rotation of porphyry copper deposits, southwestern United States. In: Pierce FW, Bolm JG (eds) Porphyry copper deposits of the American Cordillera, Arizona Geological Society Digest 20, pp 109–127

Xu-Zhi H, Ming-Jie X, Xiao-An X, Liang-Shu W, Qing-Long Z, Shao-Wen L, Guo-Ai X, Chang-Ge F (2006) A characteristic analysis of aeromagnetic anomalies and Curie point isotherms in Northeast China. Chinese J Geophys Chinese Edn 49(6):1674–1681

Index

A

B

C

M. Mandea, M. Korte (eds.), *Geomagnetic Observations and Models*, IAGA Special Sopron Book Series 5,
DOI 10.1007/978-90-481-9858-0,

D

E

F

G

H

I

J

K

L

M

MIX
Papier aus verantwortungsvollen Quellen
Paper from responsible sources
FSC® C105338

If you have any concerns about our products,
you can contact us on
ProductSafety@springernature.com

In case Publisher is established outside the EU,
the EU authorized representative is:
Springer Nature Customer Service Center GmbH
Europaplatz 3, 69115 Heidelberg, Germany

Printed by Libri Plureos GmbH
in Hamburg, Germany